Mechanically Responsive Materials for Soft Robotics

Mechanically Responsive Materials for Soft Robotics

Edited by
Hideko Koshima

WILEY-VCH

Editor

Prof. Hideko Koshima
Waseda University
Research Organization for Nano & Life Innovation
513 Wasedatsurumaki-cho, Shinjuku
Tokyo 162-0041
Japan

Library of Congress Card No.:
applied for

British Library Cataloguing-in-Publication Data
A catalogue record for this book is available from the British Library.

Bibliographic information published by the Deutsche Nationalbibliothek
The Deutsche Nationalbibliothek lists this publication in the Deutsche Nationalbibliografie; detailed bibliographic data are available on the Internet at <http://dnb.d-nb.de>.

Print ISBN: 978-3-527-34620-2
ePDF ISBN: 978-3-527-82219-5
ePub ISBN: 978-3-527-82221-8
oBook ISBN: 978-3-527-82220-1

Typesetting SPi Global, Chennai, India
Printing and Binding Markono Print Media Pte Ltd, Singapore

Printed on acid-free paper

10 9 8 7 6 5 4 3 2 1

Contents

Preface

Robots are playing an increasingly bigger role in society. We must consider the symbiotic relationship between humans and robots, as robots may help to improve our lives in the near future. However, conventional robots composed of metals have the disadvantage of being rigid and heavy. Soft robots made of organic materials have attracted much attention recently, as they tend to be soft and light and, therefore, suitable for daily interactions with humans.

Mechanically responsive materials that can move macroscopically by external stimuli, such as light, heat, electricity, chemical reactions, and others, have been studied extensively in materials chemistry over the past two decades. Many mechanically responsive materials such as crystals, polymers, gels, and composites have been developed. The next step is the practical application of these mechanical materials. Specifically, mechanical materials that can move autonomously by external stimuli are promising for soft robots with improved safety and comfort. Arguably, soft robots may be the best application of mechanically responsive materials.

The purpose of this book is to bring readers to the forefront of the current status of mechanically responsive materials for soft robotics. This book consists of three parts: mechanically responsive crystals (Part I), mechanically responsive polymers and composites (Part II), and the application of mechanically responsive materials to soft robotics (Part III). Despite the fact that the history of research on mechanical molecular crystals is as short as 10 years, approximately, many excellent mechanical crystals that show bending, twisting, rotation, jumping, locomotion, self-healing, and shape memory have been developed, as described in Chapters 1–6 of Part I. Although currently limited to basic research, practical application to soft robots is expected in the near future. In contrast, research on mechanical polymer materials precedes crystals and has been conducted for several decades. Recently, mechanical polymer materials have evolved into artificial muscles, photomobile materials, bioinspired soft actuators, inorganic–organic hybrid materials, multi-responsive composite materials, and strain sensor materials, as discussed in Chapters 7–12 of Part II. The application of mechanical materials to soft robots is just the beginning. In Chapters 13–16 of Part III, challenging and versatile applications, such as soft microrobots made from photoresponsive elastomers, four-dimensional printing for assembling soft robots, self-growing of soft robots like plants, and biohybrid robots using muscle tissue, are presented. The history, development, and practical application of the research areas described are expected to be of great interest to readers.

Many people, including robotics and materials researchers, as well as industry and others in the scientific community, are very excited about the recent advances in soft robotics. However, further advances in this field require a hybrid understanding of soft robotics and mechanical materials. It is our hope that this book will provide a bridge between these two research areas for academia and industry, enabling continued development of this exciting field.

Hideko Koshima
Tokyo, Japan

March 2019

Part I

Mechanically Responsive Crystals

1

Photomechanical Behavior of Photochromic Diarylethene Crystals

Seiya Kobatake and Daichi Kitagawa

Department of Applied Chemistry, Graduate School of Engineering, Osaka City University, 3-3-138 Sugimoto, Sumiyoshi-ku, Osaka 558-8585, Japan

1.1 Introduction

Photochromism is defined as a reversible transformation reaction between two isomers having different absorption spectra, which is induced in one or both directions by photoirradiation [1]. Among many photochromic compounds, diarylethenes with heteroaryl groups including thiophene, furan, thiazole, and oxazole rings have excellent properties, such as thermal stability of both isomers, fatigue resistance, high coloration quantum yield, rapid response, and high reactivity even in the crystalline phase [2]. Such diarylethenes have potential applications in ultraviolet (UV) sensors, photoswitches, displays, optical waveguides, optical memories, holographic recording media, nonlinear optics, and actuators. Upon UV light irradiation, diarylethenes exhibit color changes because of a molecular structure change from the open-ring isomer form to the closed-ring isomer form. The colors remain stable in the dark at room temperature. The colored isomers revert to their original colorless isomer forms by irradiation with visible light. The reversible color changes can be repeated many times.

Photochromic compounds that undergo a photochromic reaction in the crystalline phase are known for paracyclophanes, triarylimidazole dimer, diphenylmaleronitrile, aziridines, 2-(2,4-dinitrobenzyl)pyridine, *N*-salicylideneanilines, triazenes, and diarylethenes. The large change in geometrical structures prohibits photochromic reactions in the crystalline phase. Even in the crystalline phase, diarylethenes can undergo thermally irreversible and fatigue-resistant photochromic reactions when diarylethene molecules are fixed in the antiparallel conformation and the distance between the reactive carbons is less than 4.2 Å [3]. The photocyclization reaction results in a color change in the crystals from colorless to yellow, red, blue, or green, as shown in Figure 1.1. The color of the crystals can be maintained if they are stored in the dark. The colored crystals return to the initial colorless ones by irradiation with visible light. In the crystalline phase, the photocyclization quantum yield is close to unity and the coloration/decoloration cycles can be repeated more than 10^4

Mechanically Responsive Materials for Soft Robotics, First Edition. Edited by Hideko Koshima.

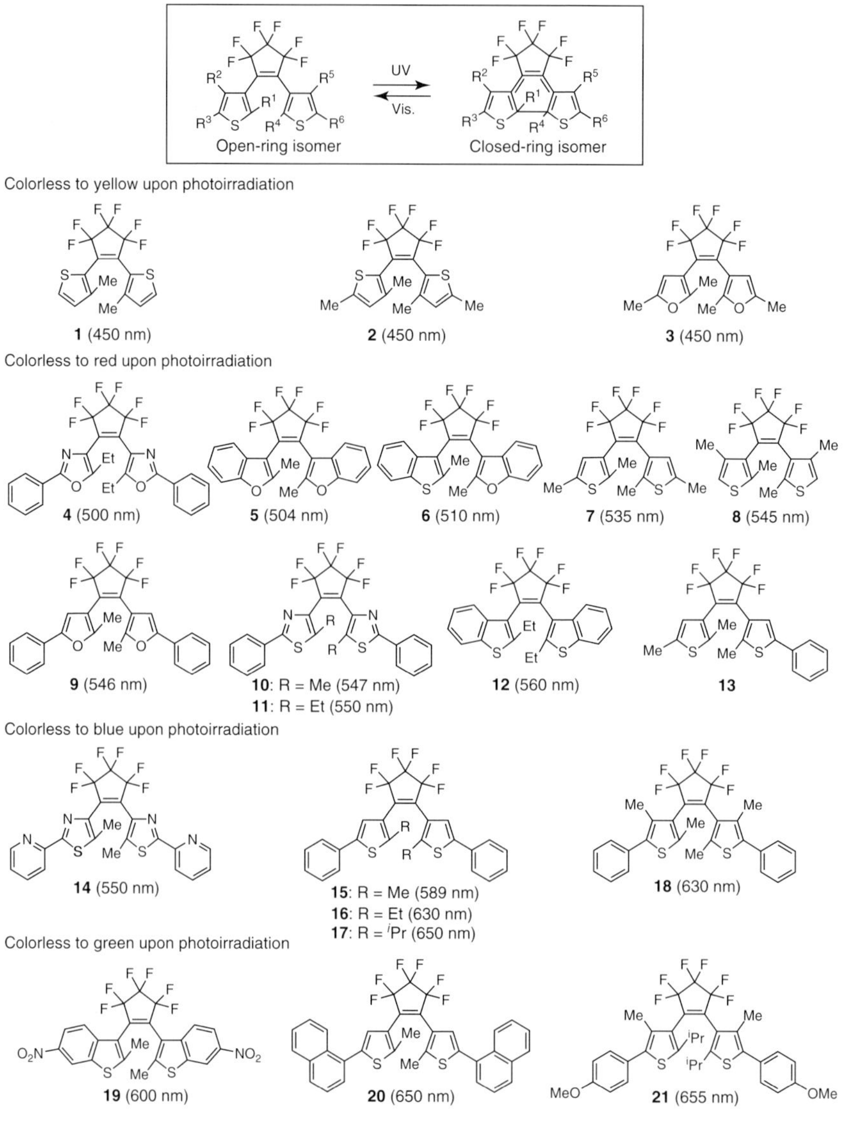

Figure 1.1 Typical examples of diarylethenes that underwent photochromism in the single crystalline phase. Maximum absorption wavelength of the photogenerated closed-ring isomers in crystals is shown in parentheses. When exposed to UV radiation crystals **1–3** turned to yellow, crystals **4–13** to red, crystals **14–16** to blue, and crystals **19–21** to green.

times [2]. There are many studies describing the photochromism of diarylethene crystals, including investigations that report multicolor photochromism [4], dichroism under polarized light [5], fluorescence [6], three-dimensional optical memory [7], diastereoselective cyclization [8], selective photochromic reaction under polarized light [9], theoretical studies [10], Raman spectroscopic studies [11], nanostructures [12], supramolecular architectures [13], nanocrystals [14],

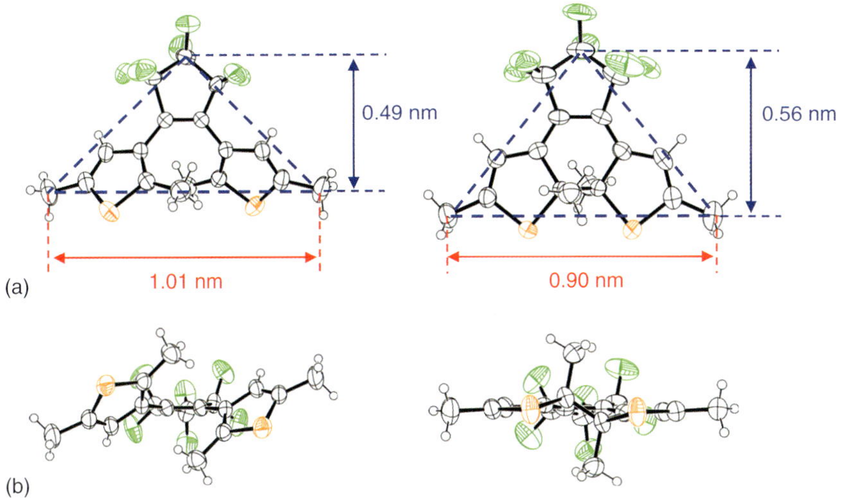

Figure 1.2 (a) Top and (b) side views of the geometrical structures of the open- and closed-ring isomers of 1,2-bis(2,5-dimethyl-3-thienyl)perfluorocyclopentene (**7**) in crystals. The two isomers were isolated and independently recrystallized. Source: Irie et al. 2014 [2b]. Adapted with permission from American Chemical Society.

polymorphism [9a, 15], phase transitions [15b, c], surface wettability [15a, 16], and molecular motion observed by X-ray crystallography [17]. The research on molecular motion observed by X-ray crystallography demonstrated that photochromic reactions of diarylethene molecules in the crystals are accompanied by a change in the unit cell dimensions because of a decrease in the molecular volume resulting from photoisomerization of the open-ring isomer to yield the closed-ring isomer as shown in Figure 1.2 [2b]. The height of the triangle shape increases from 0.49 to 0.56 nm and the base width decreases from 1.01 to 0.90 nm. The side view indicates that the thickness of the molecule is reduced. The change in the geometrical structure of diarylethene molecules plays an important role in photomechanical phenomena.

In 2001, the crystal surface of diarylethene **18** was found to exhibit a photoreversible surface morphology change [18]. The flat crystal surface formed a step with a height of approximately 1 nm upon UV light irradiation. The step was erased by irradiation with visible light. The crystal thickness decreased as a result of the photochromic isomerization of the open-ring isomer to yield the closed-ring isomer. Another surface, which is perpendicular to the surface that formed the step, exhibited a photoreversible valley formation. These reversible surface morphology changes are ascribed to photoinduced contraction in the direction of the long axis of each diarylethene molecule regularly packed within the single crystal. These results indicate that the molecular-scale structural change of individual molecules may induce the macroscopic mechanical movement of materials.

In this chapter, recent developments in the light-driven actuators based on photochromic diarylethene crystals are described.

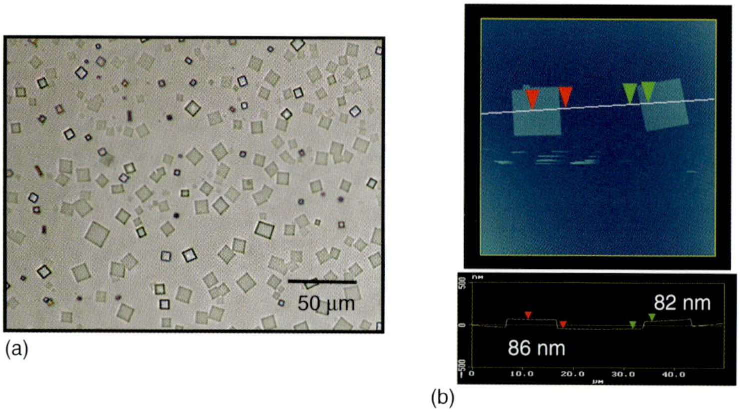

Figure 1.3 (a) Digital microscopic and (b) atomic force microscope (AFM) images for thin microcrystals of **16**.

1.2 Crystal Deformation Exhibiting Expansion/Contraction upon Photoirradiation

A first example of photoreversible macroscopic crystal deformation was a thin microcrystal of 1,2-bis(2-ethyl-5-phenyl-3-thienyl)perfluorocyclopentene (**16**) [19]. The microcrystals were prepared by sublimation on a thin glass plate under atmospheric pressure at 144 °C. A photograph of microcrystals composed of **16** is shown in Figure 1.3. The crystals have several tens of micrometers square in size with thickness of a few hundred nanometers. Upon irradiation with 365 nm light, the crystal turned blue and the blue-colored crystal returned to the initial colorless crystal form after irradiation with visible light. The conversion ratio in the crystal from the open- to the closed-ring isomers was followed by an infrared (IR) absorption microspectroscopy. IR absorption spectra for thin single crystal **16** were taken under polarized IR light to avoid an overlap of peaks. Figure 1.4 shows the IR spectral changes of crystal **16** upon irradiation with 365 nm light. The open-ring isomer in the crystal has two characteristic bands at 1260 and 1350 cm^{-1}. The band at 1350 cm^{-1} was split into two peaks upon UV light irradiation, whereas the band at 1260 cm^{-1} monotonously decreased. The closed-ring isomer has no absorption around 1260 cm^{-1}. The conversion ratio from the open- to the closed-ring isomers can be determined from the decrease of the band at 1260 cm^{-1}. Almost 70% conversion was observed at the photostationary state under irradiation with 365 nm light.

In general, crystals of different molecules have different unit cell parameters, space group, and packing in the lattice. Figure 1.5 shows the photographs of crystals for diarylethenes **16**, **11**, and **17** [19, 20]. A single crystal of **16** with a thickness of 570 nm was reversibly changed from a square-like shape with corner angles of 88° and 92° to a lozenge-like shape with corner angles of 82° and 98° upon alternating irradiation with UV and visible light. The photochromic reaction took place homogeneously in crystals because of their thin crystallized forms with a thickness of several hundred nanometers. Crystals of **11** and **17** have unit cell

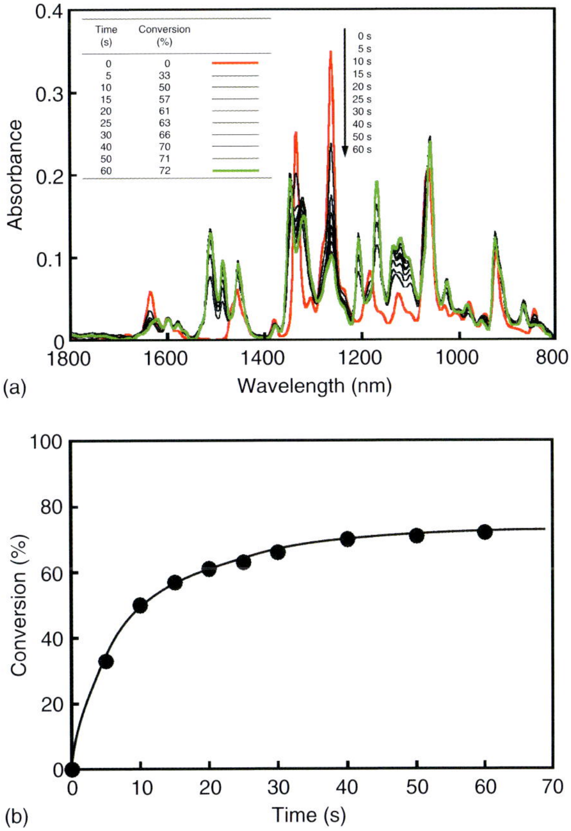

Figure 1.4 (a) IR spectral and (b) conversion change for thin microcrystal of **16** upon irradiation with 365 nm light. IR spectra were detected under polarized IR light. The polarization direction was set to the short axis in the molecule. Source: Kobatake et al. 2007 [19]. Adapted with permission of Springer Nature.

parameters similar to **16**. These diarylethenes have the same space group, i.e. *Pbcn*, as shown in Table 1.1. The molecular packings of **11** and **17** are also similar to that of **16**, as shown in Figure 1.6. Crystal **11** changed its color from colorless to red and its corner angles from 90° and 90° to 86° and 94°, and hence its shape from square to lozenge. Crystal **17** changed its color from colorless to blue and the corner angles from 83° and 97° to 81° and 99° upon irradiation with UV light. Crystals of **16**, **11**, and **17** exhibited a similar crystal shape deformation. These results indicate that the crystal shape deformation depends on the molecular packing of the molecules in the crystal and was not affected by difference in molecular structure.

1.3 Photoresponsive Bending

A first example of photoresponsive bending of diarylethene crystals was a rodlike crystal of 1,2-bis(5-methyl-2-phenyl-4-thiazolyl)perfluorocyclopentene (**10**) [19]. The crystal bent toward the incident UV light source when irradiated with UV light. The bending is attributed to a contraction in the direction of the crystal long axis and a gradient in the extent of the photoisomerization reaction at the crystal surface caused by high light absorbance. The rodlike crystal bent fast in

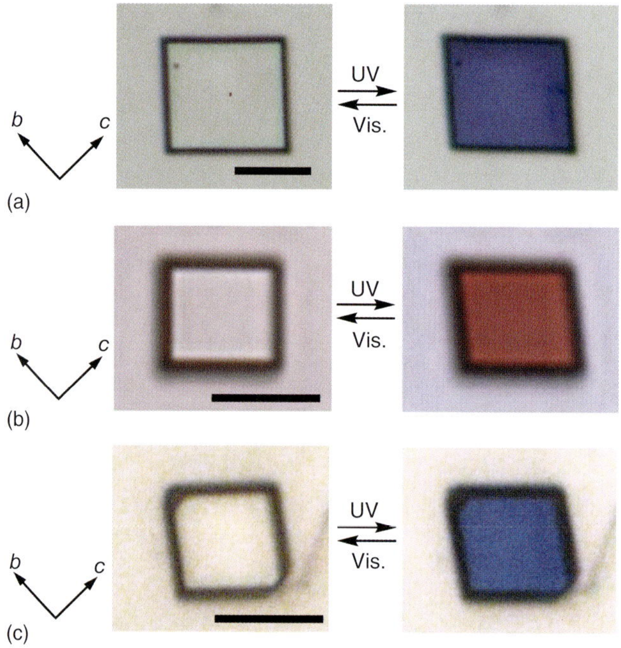

Figure 1.5 Photoinduced crystal shape deformation of crystals (a) **16**, (b) **11**, and (c) **17**. Scale bar is 10 μm. Source: Kobatake et al. 2007 [19, 20]. Adapted with permission of Springer Nature and the European Society for Photobiology, the European Photochemistry Association, and The Royal Society of Chemistry.

a few seconds timescale and could move a gold micro-particle that was 90 times heavier than the single crystal. Moreover, it could launch a tiny silica particle like a tennis ball. Thus, scaling down the crystal size allows achieving macroscopic mechanical responses.

The rodlike crystal of **10** (53 μm × 3 μm × 3 μm) was also used for measuring the response time of the bending because even low power single pulsed laser can induce the bending of the rodlike crystal [19]. As a light source, the pulsed laser of the third harmonics of Nd-YAG laser (355 nm, pulse width = 8 ns, power = 60 mJ/pulse) was used. The bending behavior induced by the single pulsed laser was measured using a high-speed camera with an image intensifier. The exposure time of each frame was 25 μs (40 000 frames s^{-1}). After irradiation with the single pulsed laser, the straight rodlike crystal bent, and the bending process was almost completed in one frame. This means that the response time of the bending shape change is around 25 μs or faster.

Various types of rodlike crystals of diarylethenes have been reported on photoresponsive bending. There are two types of the photoinduced bending behavior: bending away from the incident light and bending toward the incident light [21]. Figure 1.7 shows the molecular structures of crystals that exhibit photoinduced bending [19, 22–31]. When the long axis of the rodlike crystal expands upon UV light irradiation, the crystal bends away from the incident light. On the other hand, when the long axis of the rodlike crystal contracts upon UV light irradiation, the crystal bends toward the incident light.

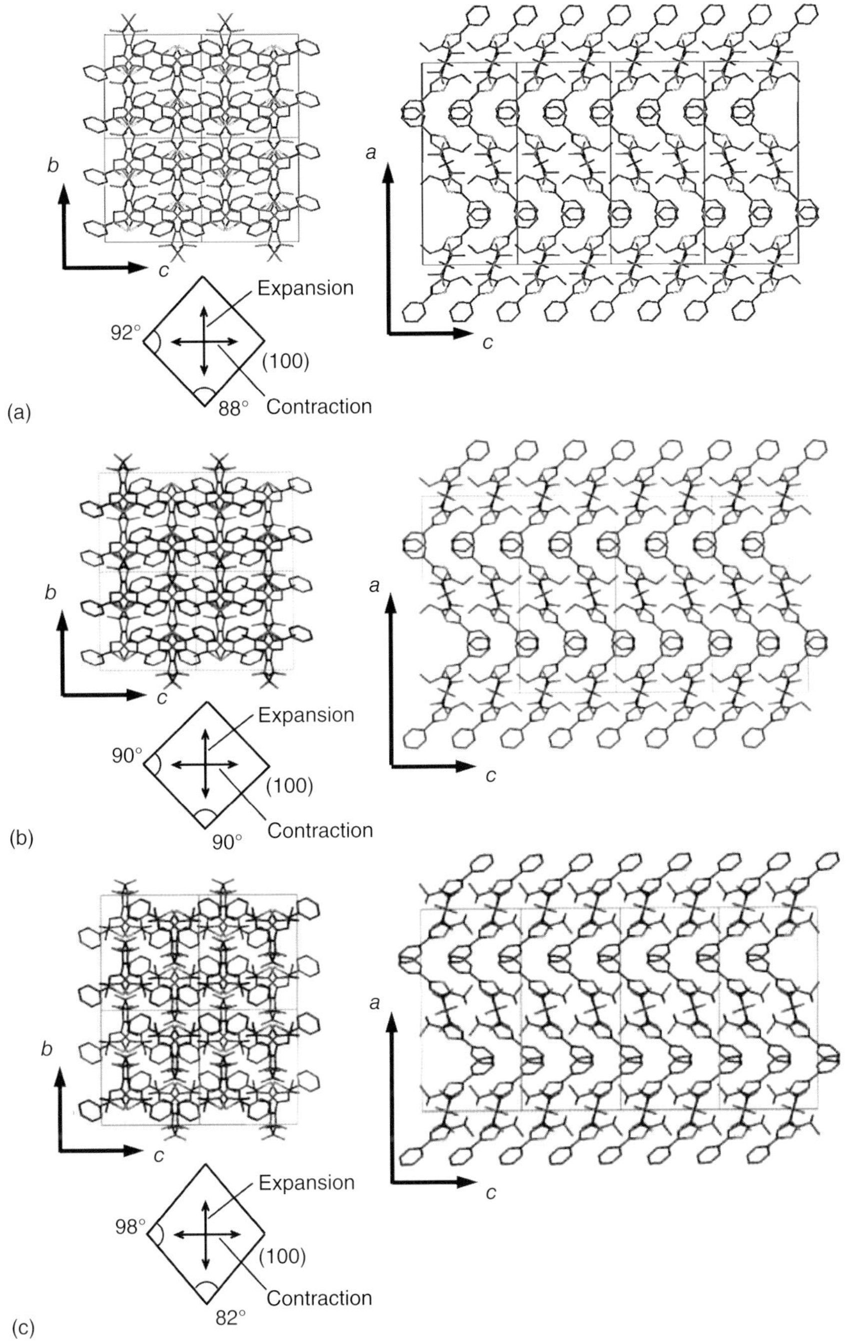

Figure 1.6 Molecular packing of crystals (a) **16**, (b) **11**, and (c) **17** viewed from (100) (left) and (010) faces (right). The arrows indicate the direction of contraction and expansion of the crystals upon UV irradiation. Source: Kobatake et al. 2007 [19, 20]. Adapted with permission of Springer Nature and the European Society for Photobiology, the European Photochemistry Association, and The Royal Society of Chemistry.

Table 1.1 X-ray crystallographic data for **16**, **11**, and **17**.

	16	11	17
Empirical formula	$C_{29}H_{22}F_6S_2$	$C_{27}H_{20}F_6N_2S_2$	$C_{31}H_{26}F_6S_2$
Formula weight	548.61	550.59	576.66
T (K)	123	123	123
Crystal system	Orthorhombic	Orthorhombic	Orthorhombic
Space group	*Pbcn*	*Pbcn*	*Pbcn*
a (Å)	22.332(5)	21.5461(15)	21.307(6)
b (Å)	10.991(2)	10.8096(8)	12.193(4)
c (Å)	10.601(2)	10.8098(8)	10.669(3)
V (Å^3)	2602.0(9)	2517.7(3)	2771.8(14)
Z	4	4	4
Density(cal.) (g cm^{-3})	1.400	1.453	1.382
Corner angles before UV irradiation (°)	88, 92	90, 90	83, 97
Corner angles after UV irradiation (°)	82, 98	86, 94	81, 99

The bending velocity of the photoresponsive bending crystals depends on the molecular structure and the crystal structure. Rodlike crystals of 1,2-bis(2-methyl-5-(4-(1-naphthoyloxymethyl)phenyl)-3-thienyl)perfluorocyclopentene (**22**) bends away from the incident light upon UV light irradiation [22]. It was found that the bending velocity depended on crystal faces subjected to UV light irradiation. Figure 1.8 shows the photoinduced bending behavior of crystal **22** upon irradiation with 365 nm light. When the (001) face was irradiated, the crystal bent slowly. In contrast, when the (010) face was irradiated, the crystal bent significantly fast. This is ascribed to the difference in the absorption coefficients and the photoreacted thickness of the different faces. The depth of the photoreaction in the vicinity of the crystal surface affects the bending velocity. Therefore, this result suggested that crystal thickness is a very important factor in bending behavior.

The dependence of the bending velocity on UV irradiation power was examined using a diarylethene crystal of **22** [23]. For their rodlike crystals, it was revealed that the initial velocity of curvature change (V_{init}) increased in proportion to the power of the incident UV light as shown in Figure 1.9. This result also suggests that the local strain owing to the individual diarylethene molecules structural changes acts cumulatively for the photomechanical bending behavior.

To understand the dependence of the bending velocity on crystal thickness, the bending behaviors of crystals **22**, **24**, and **34** with different thicknesses were investigated [23]. Figure 1.10 shows V_{init} as a function of the crystal thickness. The crystal bent significantly when it was thin. However, when the crystal thickness was 0.62 μm, the crystal could not bend. To explain this relationship, Timoshenko's bimetal model was introduced. The relation between V_{init} and the crystal

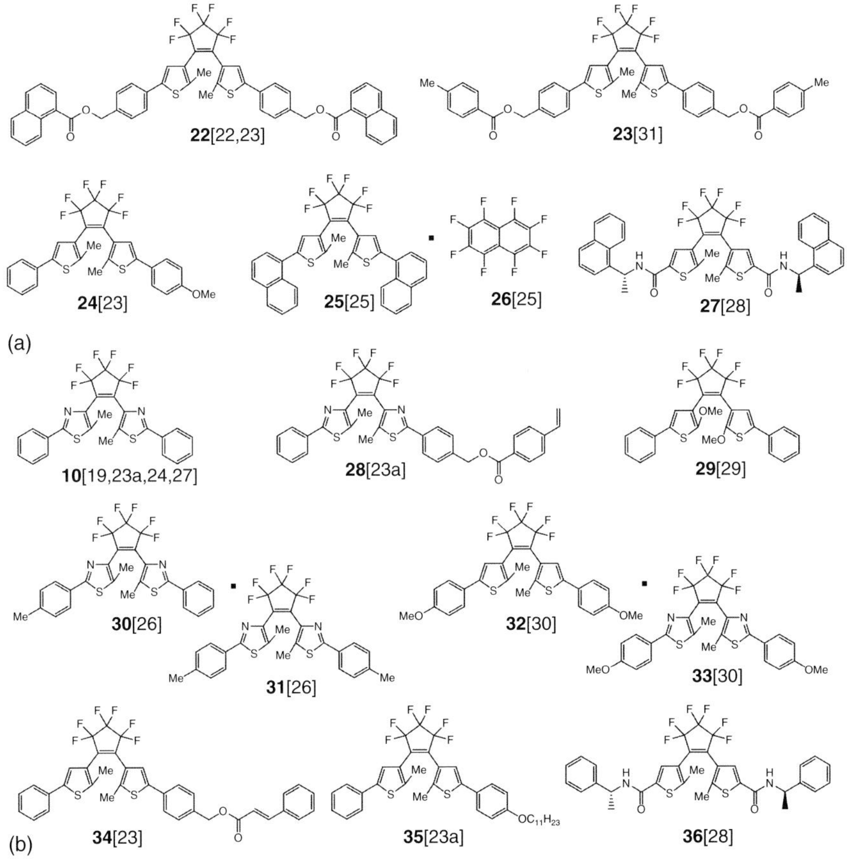

Figure 1.7 Photochromic diarylethene derivatives that exhibit the photoinduced crystal bending: (a) bending away from the incident UV light, (b) bending toward the incident UV light.

thickness was well explained by Timoshenko's bimetal model using two parameters. The depth of the photocyclization reaction in the crystal was defined to be h_2. The initial change in the actuation strain per second in a UV light intensity under the same condition was expressed as $\alpha_{2,\text{init}}$. Timoshenko's bimetal model was applied to various diarylethene crystals and enabled quantitative evaluation of the potential photoinduced bending velocity among different crystals. Thus, it was revealed that h_2 and $\alpha_{2,\text{init}}$ played important roles in determining the bending velocity.

1.4 Dependence of Bending Behavior on Irradiation Wavelength

Utilizing UV light with a different wavelength is expected to change h_2. To assess the effect of a change in h_2 for the photomechanical behavior, a rodlike crystal

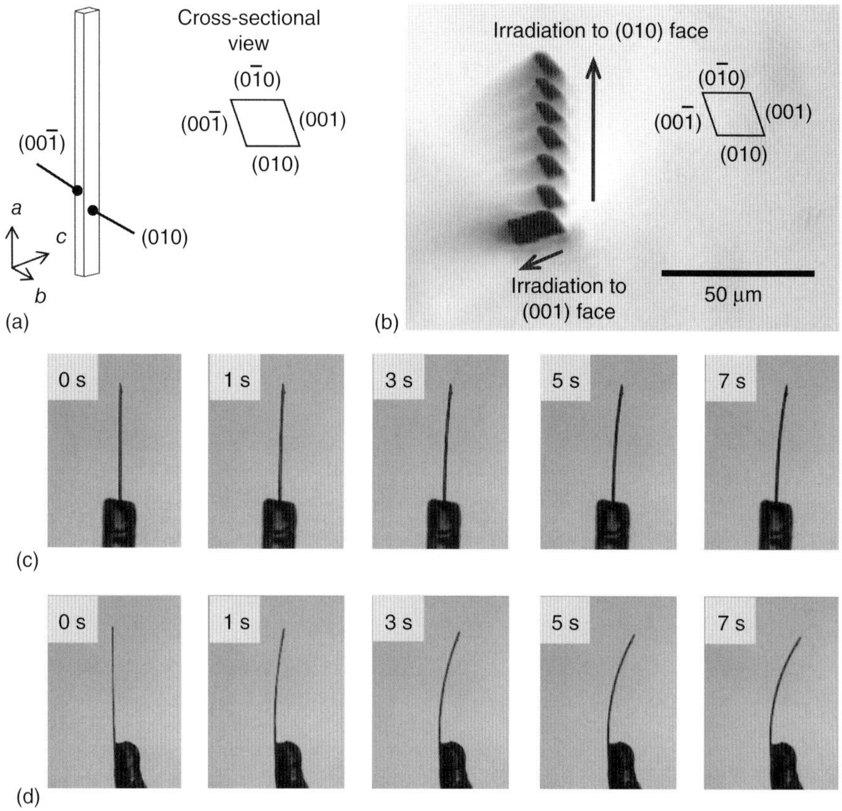

Figure 1.8 (a) Crystal shape and (b) photoinduced bending behavior of crystal **22** upon irradiation with 365 nm light to (c) the (001) face and (d) the (010) face. (b) Superimpose of the photographs observed from tip of the crystal. Source: Kitagawa and Kobatake 2013 [22]. Reproduced with permission of American Chemical Society.

of **10** was irradiated with different wavelengths of 365 and 380 nm light [24]. Figure 1.11 shows photographs of the bending behavior of the crystal. When irradiated with 365 nm light, the crystal bent toward the UV light source. In contrast, when irradiated with 380 nm light, the crystal first bent away from the incident light and then bent toward the light source. This is ascribed to the difference in the depth of the photochromic reaction from the crystal surface. When irradiated with 365 nm light, the photoisomerization of diarylethene molecules took place on only the crystal surface because of its high absorbance capacity. This indicates that the depth of photoisomerization from the crystal surface is very small relative to the crystal thickness and that the conversion of the photoreacted layer increases significantly. In contrast, when irradiated with 380 nm light, the photoisomerization of diarylethene molecules took place deep within the crystal because diarylethene has a low absorption at 380 nm compared with that at 365 nm. This indicates that, in this case, the depth of photoisomerization from the crystal surface was close to the crystal thickness, and the conversion of the photoreacted layer increased slowly. Upon UV light irradiation, photoisomerization

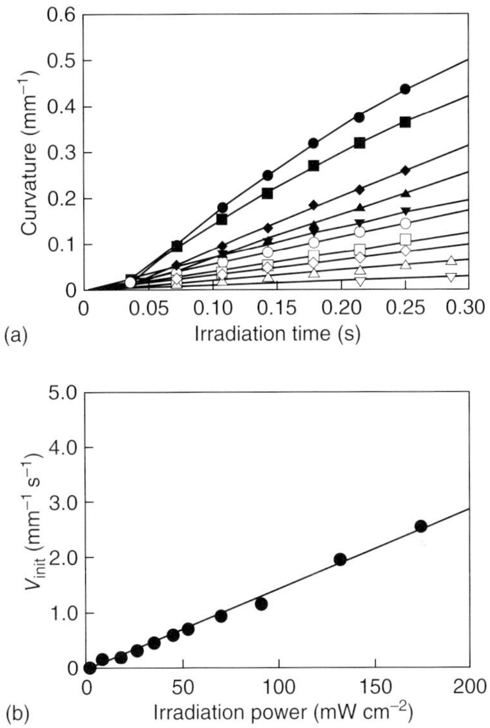

Figure 1.9 (a) Curvature change against UV irradiation time and (b) the initial velocity of curvature change (V_{init}) against power of the incident UV light for crystals **22** (crystal thickness: 6.7 μm). Power of the incident UV light is 174 (●), 132 (■), 91.0 (◆), 70.2 (▲), 52.7 (▼), 44.8 (○), 35.1 (□), 26.3 (◇), 17.6 (△), and 8.17 mW cm^{-2} (▽) for (a). Source: Hirano et al. 2017 [23b]. Reproduced with permission of American Chemical Society.

of diarylethene molecules in the crystalline phase from the open-ring isomer to the closed-ring isomer occurs randomly at first. As this occurs, there could be repulsion between the open-ring isomer and the photogenerated closed-ring isomer, which induces the expansion of the photoreacted layer. As the photoconversion of the diarylethene molecules increases, this repulsion disappears and van der Waals interactions between the closed-ring isomers induce the contraction of the photoreacted layer. When the depth of photochromic reaction is very small relative to the crystal thickness, the crystal cannot bend to a significant extent. In contrast, when the depth of the photochromic reaction is approximately half of the crystal thickness, the crystal can bend to a large extent. Therefore, upon irradiation with 365 nm light, the initial bending away from the light source could not be observed. On the other hand, upon irradiation with 380 nm light, the initial bending away from the light source was clearly observed. The proposed mechanism, shown in Figure 1.12, was well supported by the experimental results with respect to the crystal thickness and the change in the unit cell dimensions associated with the photochromic reaction.

1.5 Photomechanical Work of Diarylethene Crystals That Exhibit Bending

As mentioned previously, how to apply the photomechanical behavior to practical use turns out to be an important task. In this section, several

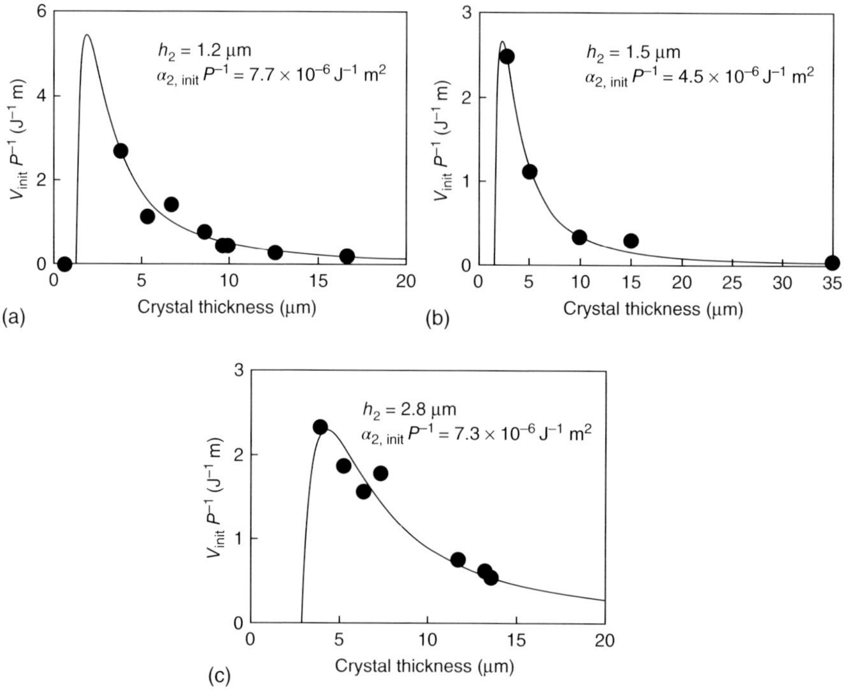

Figure 1.10 Initial velocity of curvature change (V_{init}) relative to the crystal thickness upon irradiation with UV light for diarylethene crystals (a) **22**, (b) **24**, and (c) **34**. The initial velocity is normalized according to the power of the light (*P*). Source: Hirano et al. 2017 [23b]. Reproduced with permission of American Chemical Society.

demonstrations using photomechanical behavior of diarylethene crystals were described.

Cocrystals composed of 1,2-bis(2-methyl-5-(1-naphthyl)-3-thienyl)perfluorocyclopentene (**25**) and perfluoronaphthalene (**26**) with a length of 1–5 mm exhibited photoreversible bending behavior over 250 times upon alternating irradiation with UV and visible light [25]. The crystals bent away from the incident UV light as a result of the elongation of the UV-irradiated thin surface layer, which corresponded to a change in the geometrical structure of the short axis of the diarylethene molecules from the open-ring isomer to the closed-ring isomer. The crystals were able to lift a heavy metal that was 200–600 times heavier than the crystals, as shown in Figure 1.13. The maximum stress in the crystals by photoirradiation was estimated to be 44 MPa, which is 100 times larger than that of muscle (~0.3 MPa). The Young's modulus of the crystal was measured to be 11 GPa. The relatively large Young's modulus enabled the crystals to carry out large mechanical work. Thus, the crystals could be used as "molecular crystal cranes."

Mixed crystals composed of 1-(5-methyl-2-phenyl-4-thiazolyl)-2-(5-methyl-2-*p*-tolyl-4-thiazolyl)perfluorocyclopentene (**30**) and 1,2-bis(5-methyl-2-*p*-tolyl-4-thiazolyl)perfluorocyclopentene (**31**) also exhibited more than 1000 reversible bending cycles upon alternating irradiation with UV and visible light without

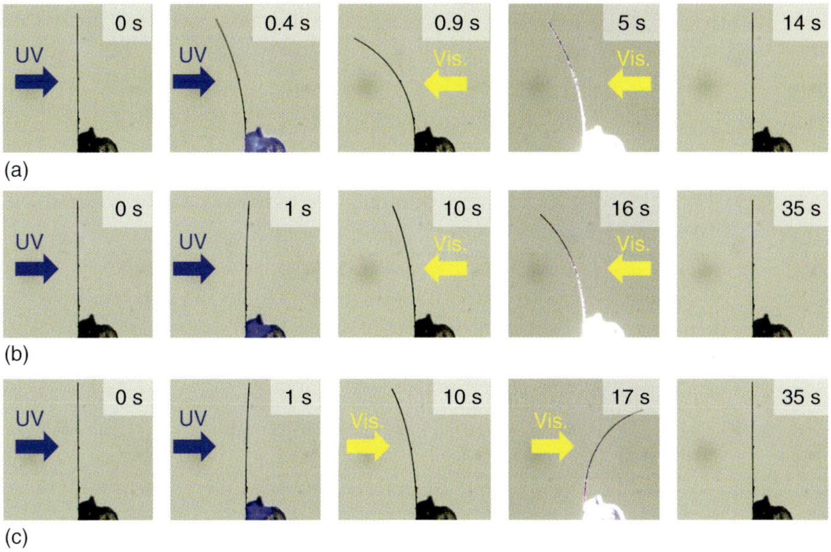

Figure 1.11 Photoinduced bending behavior of diarylethene crystal **10** upon irradiation with 365 (a) and 380 nm light (b and c) from the left side. The visible light was irradiated from the right side (a and b) and the left side (c). Source: Kitagawa et al. 2015 [24]. Reproduced with permission of the Royal Society of Chemistry.

any crystal destruction [26]. The crystal exhibited a bending motion across a wide temperature range from 4.6 to 370 K and even in water. The rodlike crystal bent toward the light source irrespective of the irradiation direction. The edge of the rodlike crystal could be rotated when the intensity of the UV and visible lights was controlled, which resulted in actual mechanical work to rotate a gearwheel, as shown in Figure 1.14. Thus, diarylethene crystals were used as "photoactuators."

The bending behavior of a gold-coated crystal of **10** was investigated [27]. As a result of the large Young's modulus of gold (83 GPa) compared with that of the diarylethene crystals, the gold-coated crystal tended not to bend as the thickness of the gold layer was increased. The relationship between the initial speed of the curvature change and the thickness of the gold-coated crystal was well explained by the extended bimetal model. When the thickness of the gold layer was 14 nm, the velocity of the bending motion was reduced by half. Furthermore, the photoreversible current ON/OFF switching of an electric circuit by the gold-coated diarylethene crystal was demonstrated, as shown in Figure 1.15. The current was able to flow in a stable manner even when the current value was over 30 mA; the current switching could be repeated over 10 cycles. Thus, gold-coated diarylethene crystals could be used as real "photoswitches" in the mesoscopic region.

1.6 New Types of Photomechanical Motion

The crystal shape change is ascribed to photogenerated anisotropic strain within the crystals. In most cases, the local shape change of each diarylethene molecule

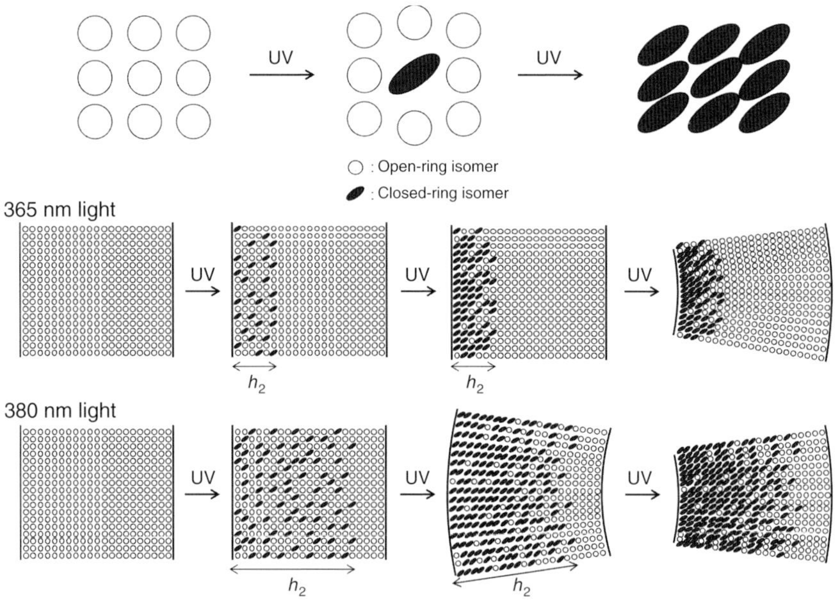

Figure 1.12 Schematic illustration of the mechanism for photoinduced crystal bending upon irradiation with 365 and 380 nm light. Source: Kitagawa et al. 2015 [24]. Reproduced with permission of the Royal Society of Chemistry.

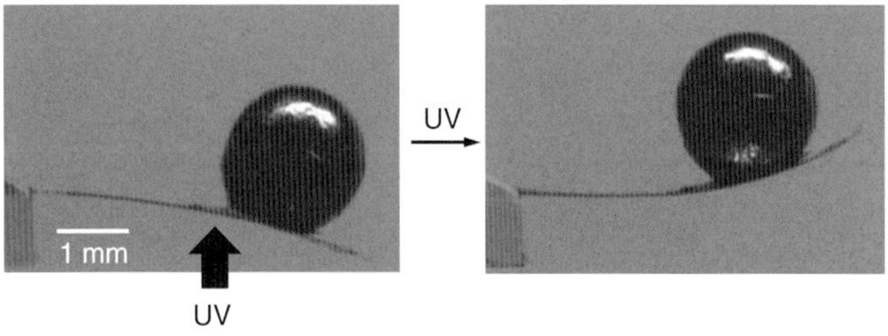

Figure 1.13 Photomechanical work of molecular crystal cantilevers made of **25/26** cocrystals. UV light was irradiated from the lower side of the crystal: 2 mm lead ball, 46.77 mg; crystal cantilever, 0.17 mg. Source: Morimoto and Irie 2010 [25]. Reproduced with permission of American Chemical Society.

contributes additively to the strain of the photoirradiated crystal [30]. However, a cooperative phase transition or domain formation induced by the photoisomerization may produce new types of photomechanical motion. Although the bending behavior is the most common photomechanical motion, the development of photoactuators exhibiting complex motion requires the investigation of new types of photomechanical motion. In the course of evaluating the initial bending speed of various diarylethene crystals, needle-like crystals of 1,2-bis(2-methyl-5-(4-(*p*-toluyloxymethyl)phenyl)-3-thienyl)perfluorocyclopentene (**23**) was found to exhibit unusual stepwise bending behavior, in which the bending rate changed significantly during continuous UV light irradiation,

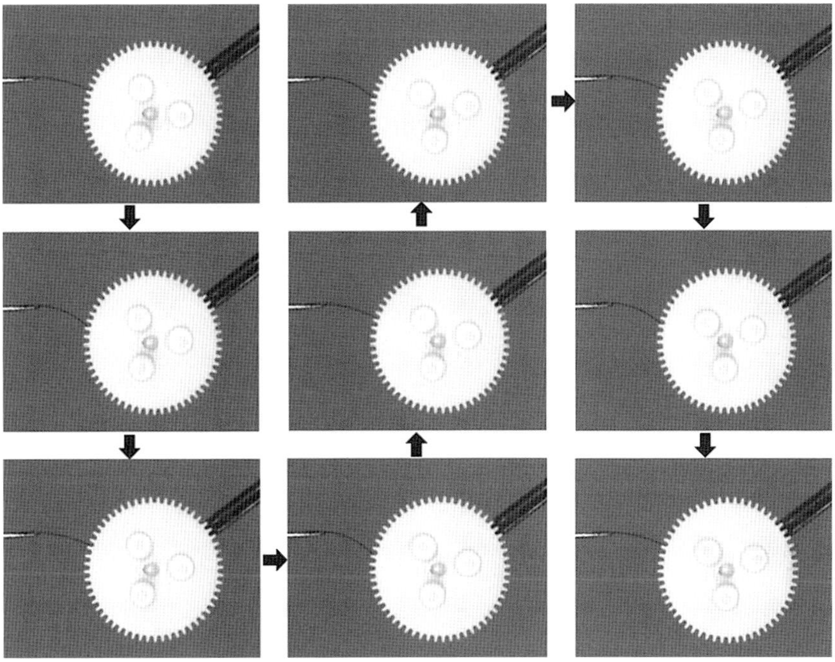

Figure 1.14 Gearwheel rotation operated by a light-driven molecular crystal actuator. The two-component mixed crystal containing **30** and **31** (**30** : **31** = 63/37) (1.3 mm × 60 μm × 12 μm) was fixed on the tip of a metal needle. The gear (diameter: 3.2 mm) was rotated by the crystal, which exhibited reversible bending upon alternating irradiation with UV (365 nm) and visible (>500 nm) light. Source: Terao et al. 2012 [26]. Reproduced with permission of John Wiley & Sons.

as shown in Figure 1.16 [31]. For the bending during the first step, it could be repeated over 100 cycles upon alternating irradiation with UV and visible light. On the other hand, the bending in the second step exhibited a different behavior after several cycles. Furthermore, delayed bending behavior was observed, in which the bending motion was continued even after the UV light was turned off during the second step. These results suggest that the bending during the second step may be ascribed to a change in the crystal packing, such as a phase transition. In fact, the powder X-ray diffraction pattern before and after many cycles of alternating irradiation with UV and visible light was different from each other. In this case, the phase transition was irreversible.

In contrast, crystals of 1,2-bis(2-methyl-5-(4-octyloxyphenyl)-3-thienyl)perfluorocyclopentene (**37**) were found to undergo a reversible thermodynamic single-crystal-to-single-crystal phase transition accompanying a change in crystal length, which was clarified by differential scanning calorimetry measurement, X-ray crystallographic analysis, and direct microscopic observation of the crystal length [32]. When UV irradiation was conducted on the rodlike crystal, the crystal exhibited an unusual photomechanical behavior. The crystal slowly bent away from the incident light at first and quickly bent away from the light source subsequently. After continuous UV light irradiation, the crystal suddenly

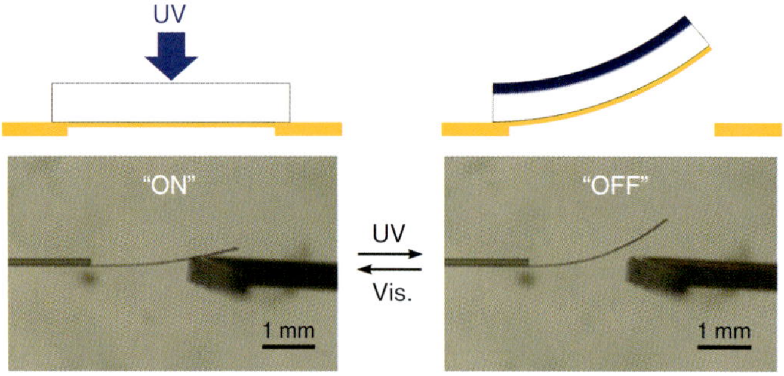

Figure 1.15 Photoreversible current switching using a gold-coated crystal of **10** upon alternating irradiation with UV and visible light. The crystal thickness is 6.2 mm. The gold thickness is 21 nm. Source: Kitagawa and Kobatake 2015 [27]. Reproduced with permission of The Royal Society of Chemistry.

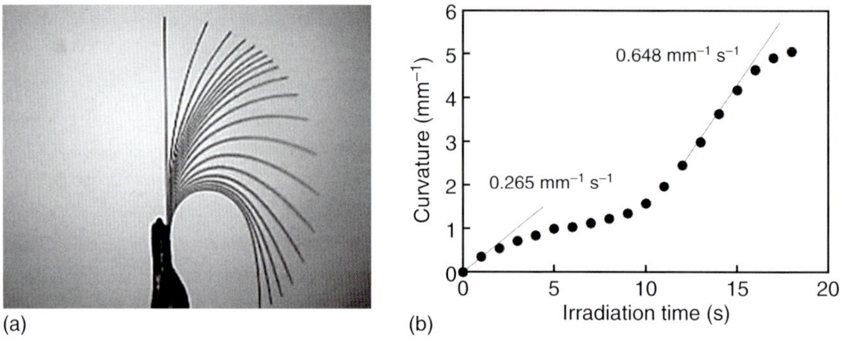

Figure 1.16 Stepwise photoinduced crystal bending of diarylethene **23** upon irradiation with UV light. (a) The photograph is a superimposition of 20 frames taken at intervals of one seconds. (b) Change in curvature relative to irradiation time with UV light. Source: Kitagawa et al. 2016 [31]. Reproduced with permission of The Royal Society of Chemistry.

returned to the initial straight shape as shown in Figure 1.17. From this detailed investigation, it was revealed that the mechanism of the unusual bending is based on the combination of the photochromic reaction and the phase transition. The photoirradiated crystal exhibited thermal bending behavior by changing temperature as shown in Figure 1.18. This is a new type of reversible mechanical material.

As a novel type of photomechanical behavior, ribbon-like crystals of 1-(2-methyl-5-(4-(1-naphthoyloxymethyl)phenyl)-3-thienyl)-2-(2-methyl-5-phenyl-3-thienyl)perfluorocyclopentene (**38**) were found to twist upon UV light irradiation, as shown in Figure 1.19 [33]. The twisted crystal retained its shape in the dark and returned to the initial straight shape upon irradiation with visible light. The photoreversible twisting could be repeated for more than 30 cycles by alternating irradiation with UV and visible light. Both types of twisting, which yield either left-handed or right-handed helices, occurred in almost equal amounts and the twisting direction depended on the face irradiated with UV light. When the $(0\bar{1}0)$ face was irradiated with UV light, the crystal twisted into

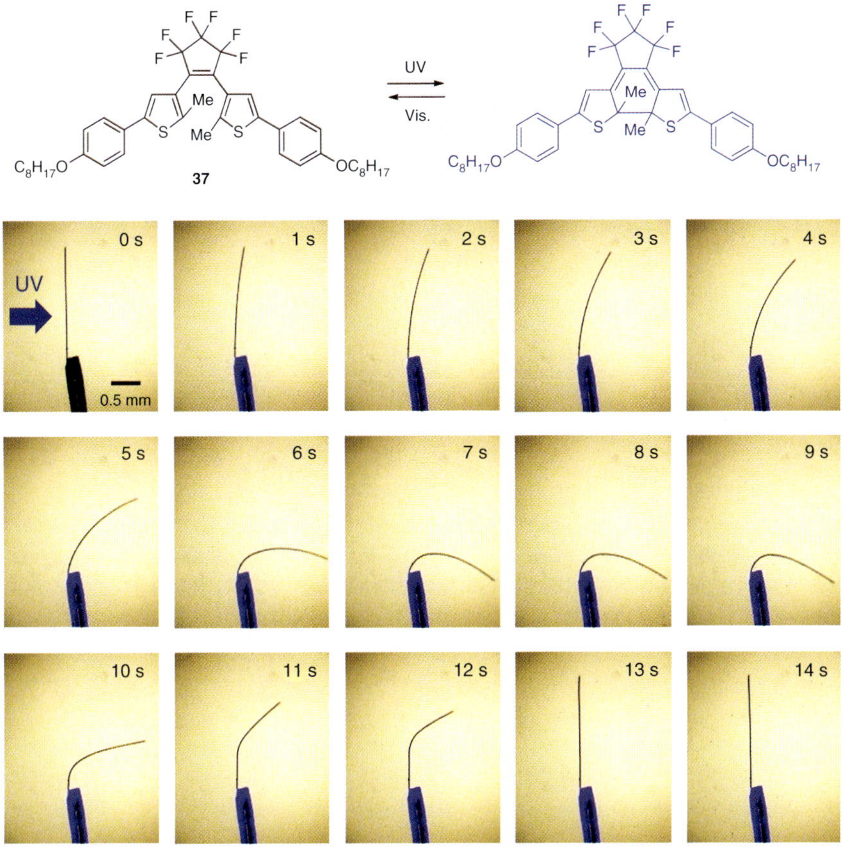

Figure 1.17 Photomechanical behavior of diarylethene **37**. UV irradiation was conducted from left side. Source: Kitagawa et al. 2017 [32]. Reproduced with permission of American Chemical Society.

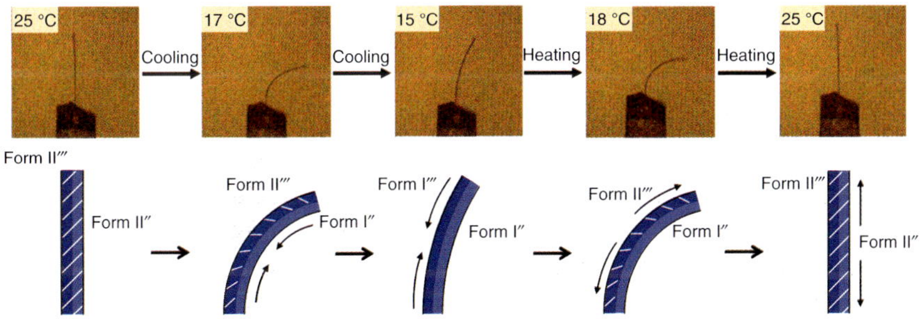

Figure 1.18 Bending behavior of the photoirradiated crystal **37** by changing temperature. The initial crystal was prepared by irradiation with UV light from left side. Source: Kitagawa et al. 2017 [32]. Reproduced with permission of American Chemical Society.

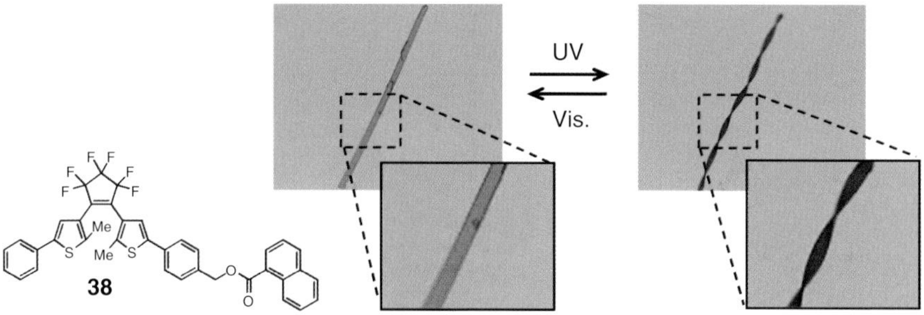

Figure 1.19 Photoinduced crystal twisting of diarylethene **38**. Source: Kitagawa et al. 2013 [33]. Reproduced with permission of John Wiley & Sons.

a right-handed helix. On the other hand, irradiation of the (010) face with UV light resulted in the crystal twisting into a left-handed helix. This is ascribed to the contraction of the crystal in the diagonal direction and a gradient in the photocyclization conversion that occurred below the crystal surface.

Moreover, it was found that the illumination direction affects the mode of the photomechanical twisting motion. The ribbon crystal of diarylethene **15** can be prepared by a sublimation method, while the recrystallization from organic solvents yields the block crystals. For the ribbon crystal, the effect of the illumination direction on the mode of photomechanical motion was investigated as shown in Figure 1.20 [34]. The definition of the illumination direction is depicted in the upper right side of the image. Upon UV light irradiation from the tip of the crystal (incident light angle = 0°), the crystal twisted into a helicoid shape. On the other hand, when UV irradiation was incident on the ribbon at larger angles, it gradually transformed into a cylindrical helix shape. Eventually, when the ribbon crystal was irradiated with an angle of 90°, the crystal exhibited bending rather than twisting. Thus, the incident light angle determines the mode of photomechanical motion. This might be due to the preferential excitation of differently oriented molecules within the crystal. In other words, the photoinduced strain tensor in the crystal, thus the mode of photomechanical deformation, can be controlled by exciting differently oriented molecules. This result suggests that the use of illumination angle to tune the mechanical response of photomechanical molecular crystals provides unique opportunities for the control of their motion.

1.7 Photosalient Effect

In addition to these photomechanical motions such as contraction, expansion, bending, and twisting, a photosalient behavior such as crystal jumping similar to the popping of corn has been also reported for some diarylethene crystals. First example of photosalient effect for diarylethene crystals has been reported for 1,2-bis(5-methyl-2-(2-pyridyl)-4-thiazolyl)perfluorocyclopentene (**14**) and 1,2-bis(2-ethyl-5-phenyl-3-thienyl)perfluorocyclopentene (**16**) [35]. Microcrystals of diarylethene **14** underwent jumping accompanied with photocyclization upon UV light irradiation. When the energy absorbed by the crystal reached

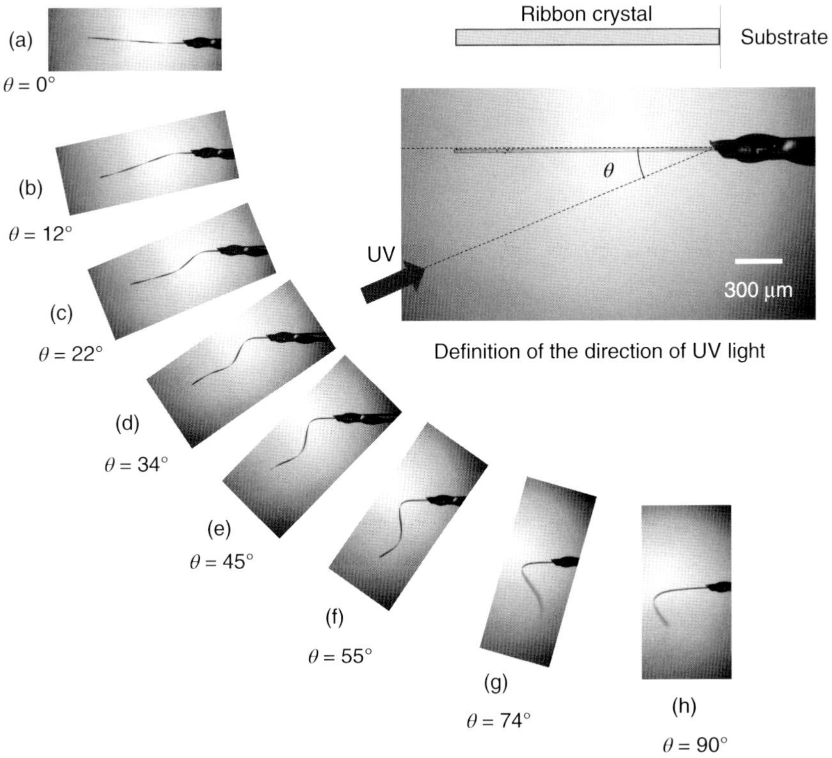

Figure 1.20 Different twisting motions, ranging from a helicoid to a cylindrical helix, of the ribbon crystal **15** depending on the angle of the incident light. Source: Kitagawa et al. 2018 [34]. Reproduced with permission of American Chemical Society.

about 10 μJ, the uniaxial stress induced in the crystal lattice relaxed through directional jumping. If the crystal was prevented from jumping, parallel and equidistant cracks appeared on crystal surfaces. These photomechanical effects could result from a Grinfeld surface instability [36].

Crystals of diarylethenes, 1-(2-methyl-5-(4-(phenylcarbarylmethyl)phenyl)-3-thienyl)-2-(2-methyl-5-phenyl-3-thienyl)perfluorocyclopentene (**39**) and 1-(2-methyl-5-(4-(benzylcarbarylmethyl)phenyl)-3-thienyl)-2-(2-methyl-5-phenyl-3-thienyl)perfluorocyclopentene (**40**), having urethane bonding unit in the molecular structure exhibited the photosalient effect by UV irradiation as shown in Figure 1.21 [37]. From X-ray crystallographic analysis, both crystals have intermolecular hydrogen bonding network at their urethane bonding units in the crystal, which suggests that the diarylethene molecules strongly interact with each other. The influence of crystal characters, such as size (area), thickness, and volume, on the initial speed of a fragment was not observed. The average speed and the kinetic energy for a crystal fragment were estimated to be $4.7\,\mathrm{m\,s^{-1}}$ and 3.2×10^{-12} J for **39** and $1.9\,\mathrm{m\,s^{-1}}$ and 0.50×10^{-12} J for **40**, respectively. The difference in the speed and energy might be due to the difference in the power of intermolecular interaction.

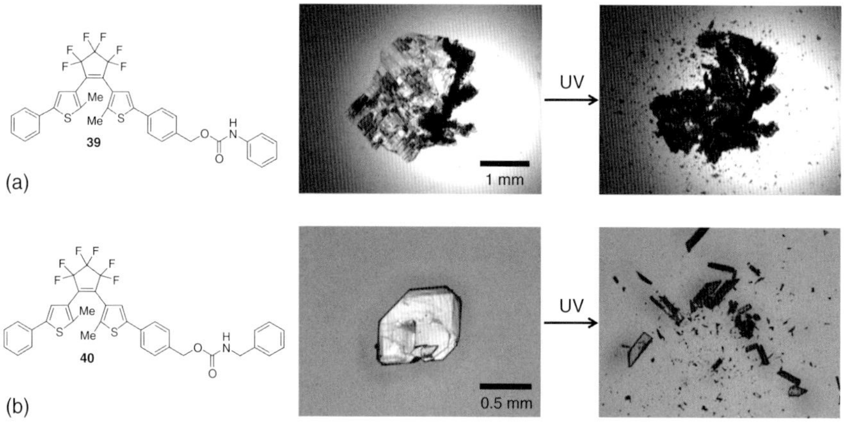

Figure 1.21 Optical photographs of diarylethene crystals (a) **39** and (b) **40** before and after UV irradiation. Source: Kitagawa et al. 2016 [37]. Reproduced with permission of American Chemical Society.

As another example, the photosalient effect of 1,2-bis(5-methyl-2-phenyl-4-thiazolyl)perfluorocyclohexene (**41**) was reported [38]. Crystals of **41**, prepared by a sublimation method under normal pressure, were found to have a hollow structure. Upon UV irradiation, the hollow crystal exhibited the photosalient effect, which burst the crystal into small pieces. The speed at which the broken crystal pieces travelled after the explosion was measured as 2.67 m s^{-1}. To mimic the behavior of "*Impatiens*," small fluorescent beads were inserted into the hollow crystal and exposed to UV light, which resulted in the scattering of the beads by deformation and bursting of the hollow structure as shown in Figure 1.22. This is the first example of the utilization of the photosalient phenomena.

1.8 Summary

In this chapter, recent developments in the light-driven actuators based on photochromic diarylethene crystals have been described. Molecular-scale structural changes of diarylethenes can induce macroscopic mechanical movement of the bulk crystals. The thin micro-sized diarylethene crystals exhibit contraction or expansion upon photoirradiation as a result of structural isomerization of the open-ring isomer to yield the closed-ring isomer. The rodlike crystals can bend away from the incident light or toward the incident light. This behavior occurs because of a gradient in the degree of photochemical conversion within the crystal caused by the high absorbance capacity of its surface. There are various types of photomechanical motions, such as contraction, expansion, bending, twisting, and jumping. These photomechanical behaviors are based on geometrical structure changes in the crystalline phase and can be applied to macro-sized light-driven actuators. In order to apply photoresponsive organic crystals as photomechanical materials to soft robots, microcrystals can be used as part of mechanical elements. Furthermore, combining existing materials with

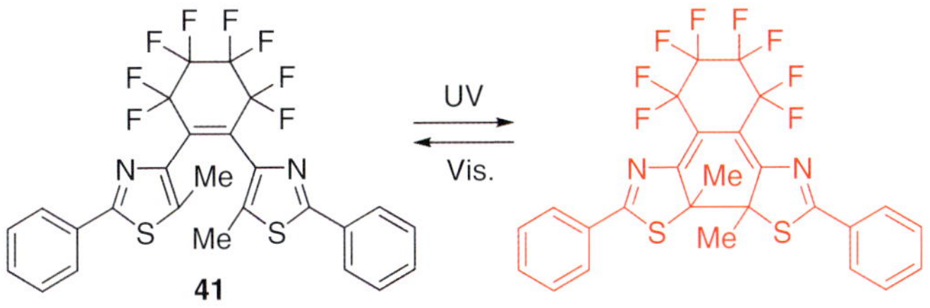

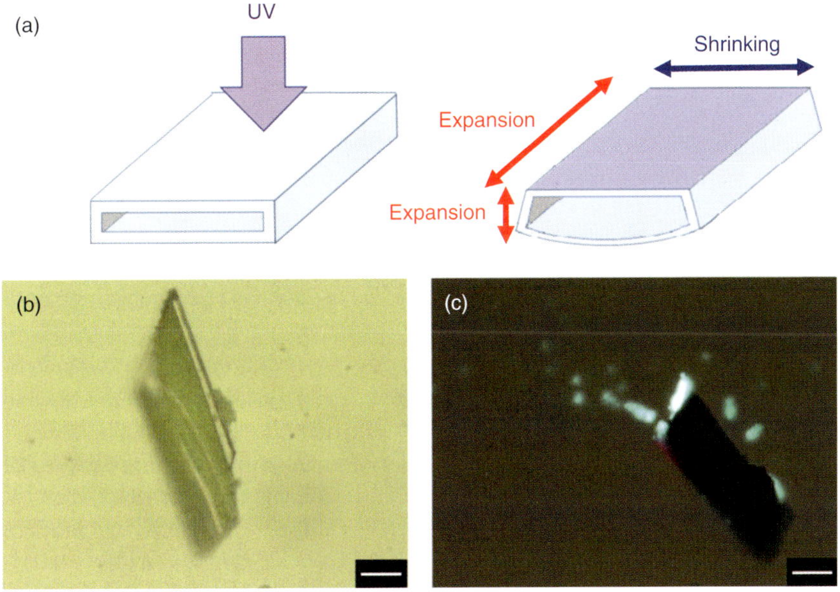

Figure 1.22 UV light-induced photosalient effect of a hollow crystal, mimicking *impatiens*: (a) deformation of the hollow crystal upon UV irradiation, (b) a hollow crystal packed with fluorescent polystyrene beads with 1 mm diameter, (c) UV-induced scattering of fluorescent beads. Scale bars = 20 μm. Source: Hatano et al. 2017 [38b]. Reproduced with permission of John Wiley & Sons.

photomechanical materials will create new functionality. It is expected to be applied to real soft robot parts by developing such materials in the future.

References

1 Dürr, H. and Bouas-Laurent, H. (eds.) (2003). *Photochromism: Molecules and Systems*. Amsterdam: Elsevier.

2 (a) Irie, M. (2000). Diarylethenes for memories and switches. *Chem. Rev.* 100 (5): 1685–1716. (b) Irie, M., Fukaminato, T., Matsuda, K., and Kobatake, S. (2014). Photochromism of diarylethene molecules and crystals: memories, switches, and actuators. *Chem. Rev.* 114 (24): 12174–12277.

3 Kobatake, S., Uchida, K., Tsuchida, E., and Irie, M. (2002). Single-crystalline photochromism of diarylethenes: reactivity-structure relationship. *Chem. Commun.* (23): 2804–2805.

4 (a) Yamada, T., Kobatake, S., and Irie, M. (2002). Single-crystalline photochromism of diarylethene mixtures. *Bull. Chem. Soc. Jpn.* 75 (1): 167–173. (b) Morimoto, M., Kobatake, S., and Irie, M. (2002). Multi-colored photochromic crystals of diarylethene mixture. *Adv. Mater.* 14 (15): 1027–1029. (c) Morimoto, M., Kobatake, S., and Irie, M. (2003). Multicolor photochromism of two- and three-component diarylethene crystals. *J. Am. Chem. Soc.* 125 (36): 11080–11087. (d) Takami, S., Kuroki, L., and Irie, M. (2007). Photochromism of mixed crystals containing bisthienyl-, bisthiazolyl-, and bisoxazolylethene derivatives. *J. Am. Chem. Soc.* 129 (23): 7319–7326.

5 (a) Kobatake, S., Yamada, T., Uchida, K. et al. (1999). Photochromism of 1,2-bis(2,5-dimethyl-3-thienyl)perfluorocyclopentene in a single crystalline phase. *J. Am. Chem. Soc.* 121 (11): 2380–2386. (b) Kobatake, S., Yamada, M., Yamada, T., and Irie, M. (1999). Photochromism of 1,2-bis(2-methyl-6-nitro-1-benzothiophen-3-yl)perfluorocyclopentene in a single-crystalline phase: dichroism of the closed-ring form isomer. *J. Am. Chem. Soc.* 121 (37): 8450–8456. (c) Kobatake, S., Yamada, T., and Irie, M. (2000). Photochromism of diarylethenes in single-crystalline phases. *Mol. Cryst. Liq. Cryst.* 344: 185–190.

6 Fukaminato, T., Kawai, T., Kobatake, S., and Irie, M. (2003). Fluorescence of photochromic 1,2-bis(3-methyl-2-thienyl)ethene. *J. Phys. Chem. B* 107 (33): 8372–8377.

7 Fukaminato, T., Kobatake, S., Kawai, T., and Irie, M. (2001). Three-dimensional erasable optical memory using a photochromic diarylethene single crystal as the recording medium. *Proc. Jpn. Acad., Ser. B* 77: 30–35.

8 (a) Kodani, T., Matsuda, K., Yamada, T. et al. (2000). Reversible diastereoselective photocyclization of a diarylethene in a single-crystalline phase. *J. Am. Chem. Soc.* 122 (40): 9631–9637. (b) Uchida, K., Walko, M., de Jong, J.J.D. et al. (2006). Diastereoselective cyclization of a dithienylethene switch through single crystal confinement. *Org. Biomol. Chem.* 4 (6): 1002–1006. (c) Kobatake, S., Kuma, S., and Irie, M. (2007). Single-crystalline photochromism of diarylethene dimers bridged by a spiro structure. *J. Phys. Org. Chem.* 20 (11): 960–967.

9 (a) Morimoto, M., Kobatake, S., and Irie, M. (2003). Polymorphism of 1,2-bis(2-methyl-5-*p*-methoxyphenyl-3-thienyl)perfluorocyclopentene and photochromic reactivity of the single crystals. *Chem. Eur. J.* 9 (3): 621–627. (b) Kobatake, S., Kuma, S., and Irie, M. (2004). Single-crystalline photochromism of diarylethenes. *Bull. Chem. Soc. Jpn.* 77 (2): 945–951.

10 (a) Asano, Y., Murakami, A., Kobayashi, T. et al. (2003). Theoretical study on novel quantum yields of dithienylethenes cyclization reactions in crystals. *J. Mol. Struct. THEOCHEM* 625 (1–3): 227–234. (b) Kobatake, S., Morimoto, M., Asano, Y. et al. (2002). Absorption spectra of colored isomer of diarylethene in single crystals. *Chem. Lett.* (12): 1224–1225.

11 Saita, K., Kobatake, S., Fukaminato, T. et al. (2008). Raman spectroscopic study on isomers of photochromic 1,2-bis(2,5-dimethyl-3-thienyl)-perfluorocyclopentene in crystal and stability of the closed-ring forms in the open-ring forms. *Chem. Phys. Lett.* 454 (1–3): 42–48.

12 (a) Morimoto, M., Kobatake, S., and Irie, M. (2003). Aryl-perfluoroaryl interaction in photochromic diarylethene crystals. *Cryst. Growth Des.* 3 (5): 847–854. (b) Morimoto, M., Kobatake, S., and Irie, M. (2004). Crystal engineering of photochromic diarylethene single crystals. *Chem. Rec.* 4 (1): 23–38. (c) Morimoto, M., Kobatake, S., and Irie, M. (2003). Photochromism of diarylethenes in nanolayers of a single crystal. *Photochem. Photobiol. Sci.* 2 (11): 1088–1094. (d) Kobatake, S., Matsumoto, Y., and Irie, M. (2005). Conformational control of photochromic reactivity in a diarylethene single crystal. *Angew. Chem. Int. Ed.* 44: 2148–2151. (e) Kobatake, S., Morimoto, M., and Irie, M. (2005). Nanolayered structures in photochromic crystal of 1,2-bis(2-methyl-5-*p*-methoxyphenyl-3-thienyl)perfluorocyclopentene. *Mol. Cryst. Liq. Cryst.* 431: 223–228. (f) Morimoto, M., Kobatake, S., and Irie, M. (2005). Crystal engineering of photochromic diarylethene derivatives by aryl-perfluoroaryl interaction. *Mol. Cryst. Liq. Cryst.* 431: 229–234.

13 Yamamoto, S., Matsuda, K., and Irie, M. (2003). Photochromism of diarylethenes linked by hydrogen bonds in the single-crystalline phase. *Chem. Eur. J.* 9 (20): 4878–4886.

14 (a) Tagawa, N., Masuhara, A., Kasai, H. et al. (2010). Monodispersed and size-controlled diarylethene nanoparticles fabricated by the reprecipitation method. *Mol. Cryst. Liq. Cryst.* 520: 245–250. (b) Tagawa, N., Masuhara, A., Onodera, T. et al. (2011). Polystyrene-encapsulated diarylethene nanocrystals by soap-free emulsion polymerization. *J. Mater. Chem.* 21 (22): 7892–7894. (c) Tagawa, N., Masuhara, A., Onodera, T. et al. (2011). Nanocrystallization process of diarylethene. *Mol. Cryst. Liq. Cryst.* 539: 45–49. (d) Sanz-Menez, N., Monnier, V., Colombier, I. et al. (2011). Photochromic fluorescent diarylethene nanocrystals grown in sol gel thin film. *Dyes Pigm.* 89 (3): 241–245. (e) Lim, S.J., An, B.K., Jung, S.D. et al. (2004). Photoswitchable organic nanoparticles and a polymer film employing multifunctional molecules with enhanced fluorescence emission and bistable photochromism. *Angew. Chem. Int. Ed.* 43 (46): 6346–6350.

15 (a) Kitagawa, D., Yamashita, I., and Kobatake, S. (2011). Control of surface wettability and photomicropatterning with a polymorphic diarylethene crystal upon photoirradiation. *Chem. Eur. J.* 17 (35): 9825–9831. (b) Kitagawa, D. and Kobatake, S. (2014). Thermodynamic phase transition through crystal-to-crystal process of photochromic 1,2-bis(5-phenyl-2-propyl-3-thienyl)perfluorocyclopentene. *Chem. Asian J.* 9 (1): 289–293. (c) Iwaihara, C., Kitagawa, D., and Kobatake, S. (2015). Polymorphic crystallization and thermodynamic phase transition between the polymorphs of a photochromic diarylethene. *Cryst. Growth Des.* 15 (4): 2017–2023.

16 (a) Kitagawa, D. and Kobatake, S. (2012). Morphology, wettability and photomicropatterning of superhydrophobic surface with high adhesive force by crystal growth of a photochromic diarylethene. *Chem. Sci.* 3 (5): 1445–1449. (b) Kitagawa, D., Yamashita, I., and Kobatake, S. (2010). Photoinduced micropatterning by polymorphic crystallization of a photochromic diarylethene in a polymer film. *Chem. Commun.* 46 (21): 3723–3725. (c)

Uchida, K., Izumi, N., Sukata, S. et al. (2006). Photoinduced reversible formation of microfibrils on a photochromic diarylethene microcrystalline surface. *Angew. Chem. Int. Ed.* 45 (39): 6470–6473. (d) Uchida, K., Nishikawa, N., Izumi, N. et al. (2010). Phototunable diarylethene microcrystalline surfaces: lotus and petal effects upon wetting. *Angew. Chem. Int. Ed.* 49 (34): 5942–5944. (e) Uyama, A., Yamazoe, S., Shigematsu, S. et al. (2011). Reversible photocontrol of surface wettability between hydrophilic and superhydrophobic surfaces on an asymmetric diarylethene solid surface. *Langmuir* 27 (10): 6395–6400. (f) Nishikawa, N., Kiyohara, H., Sakiyama, S. et al. (2012). Photoinduced formation of superhydrophobic surface on which contact angle of a water droplet exceeds 170° by reversible topographical changes on a diarylethene microcrystalline surface. *Langmuir* 28 (51): 17817–17824. (g) Nishikawa, N., Sakiyama, S., Yamazoe, S. et al. (2013). Photoinduced self-epitaxial crystal growth of a diarylethene derivative with antireflection moth-eye and superhydrophobic lotus effects. *Langmuir* 29 (25): 8164–8169. (h) Nishikawa, N., Mayama, H., Nonomura, Y. et al. (2014). Theoretical explanation of the photoswitchable superhydrophobicity of diarylethene microcrystalline surfaces. *Langmuir* 30 (35): 10643–10650.

17 (a) Yamada, T., Kobatake, S., and Irie, M. (2000). X-ray crystallographic study on single-crystalline photochromism of 1,2-bis(2,5-dimethyl-3-thienyl)perfluorocyclopentene. *Bull. Chem. Soc. Jpn.* 73 (10): 2179–2184. (b) Yamada, T., Kobatake, S., Muto, K., and Irie, M. (2000). X-ray crystallographic study on single-crystalline photochromism of bis(2,5-dimethyl-3-thienyl)perfluorocyclopentene. *J. Am. Chem. Soc.* 122 (8): 1589–1592. (c) Hamazaki, T., Matsuda, K., Kobatake, S., and Irie, M. (2007). Photostimulated crystal lattice change induced by the photochemical ring-opening reaction of diarylethene molecules. *Bull. Chem. Soc. Jpn.* 80 (2): 365–370.

18 Irie, M., Kobatake, S., and Horichi, M. (2001). Reversible surface morphology changes of a photochromic diarylethene single crystal by photoirradiation. *Science* 291 (5509): 1769–1772.

19 Kobatake, S., Takami, S., Muto, H. et al. (2007). Rapid and reversible shape changes of molecular crystals on photoirradiation. *Nature* 446 (7137): 778–781.

20 Kuroki, L., Takami, S., Yoza, K. et al. (2010). Photoinduced shape changes of diarylethene single crystals: correlation between shape changes and molecular packing. *Photochem. Photobiol. Sci.* 9 (2): 221–225.

21 Kobatake, S. and Kitagawa, D. (2015). Photoinduced mechanical motion of photochromic crystalline materials. In: *Advances in Organic Crystal Chemistry* (eds. R. Tamura and M. Miyata), 533–547. Tokyo: Springer.

22 Kitagawa, D. and Kobatake, S. (2013). Crystal thickness dependence of photoinduced crystal bending of 1,2-bis(2-methyl-5-(4-(1-naphthoyloxymethyl)phenyl)-3-thienyl)perfluorocyclopentene. *J. Phys. Chem. C* 117 (40): 20887–20892.

23 (a) Kitagawa, D., Iwaihara, C., Nishi, H., and Kobatake, S. (2015). Quantitative evaluation of photoinduced bending speed of diarylethene crystals. *Crystals* 5 (4): 551–561. (b) Hirano, A., Hashimoto, T., Kitagawa, D. et al. (2017). Dependence of photoinduced bending behavior of diarylethene crystals on ultraviolet irradiation power. *Cryst. Growth Des.* 17 (9): 4819–4825.
24 Kitagawa, D., Tanaka, R., and Kobatake, S. (2015). Dependence of photoinduced bending behavior of diarylethene crystals on irradiation wavelength of ultraviolet light. *Phys. Chem. Chem. Phys.* 17 (41): 27300–27305.
25 Morimoto, M. and Irie, M. (2010). A diarylethene cocrystal that converts light into mechanical work. *J. Am. Chem. Soc.* 132 (40): 14172–14178.
26 Terao, F., Morimoto, M., and Irie, M. (2012). Light-driven molecular-crystal actuators: rapid and reversible bending of rodlike mixed crystals of diarylethene derivatives. *Angew. Chem. Int. Ed.* 51 (4): 901–904.
27 Kitagawa, D. and Kobatake, S. (2015). Photoreversible current ON/OFF switching by photoinduced bending of gold-coated diarylethene crystals. *Chem. Commun.* 51 (21): 4421–4424.
28 Uchida, K., Sukata, S.I., Matsuzawa, Y. et al. (2008). Photoresponsive rolling and bending of thin crystals of chiral diarylethenes. *Chem. Commun.* (3): 326–328.
29 Kobatake, S., Hasegawa, H., and Miyamura, K. (2011). High-convertible photochromism of a diarylethene single crystal accompanying the crystal shape deformation. *Cryst. Growth Des.* 11 (4): 1223–1229.
30 Ohshima, S., Morimoto, M., and Irie, M. (2015). Light-driven bending of diarylethene mixed crystals. *Chem. Sci.* 6 (10): 5746–5752.
31 Kitagawa, D., Tanaka, R., and Kobatake, S. (2016). Photoinduced stepwise bending behavior of photochromic diarylethene crystals. *CrystEngComm* 18 (38): 7236–7240.
32 Kitagawa, D., Kawasaki, K., Tanaka, R., and Kobatake, S. (2017). Mechanical behavior of molecular crystals induced by combination of photochromic reaction and reversible single-crystal-to-single-crystal phase transition. *Chem. Mater.* 29 (17): 7524–7532.
33 Kitagawa, D., Nishi, H., and Kobatake, S. (2013). Photoinduced twisting of a photochromic diarylethene crystal. *Angew. Chem. Int. Ed.* 52 (35): 9320–9322.
34 Kitagawa, D., Tsujioka, H., Tong, F. et al. (2018). Control of photomechanical crystal twisting by illumination direction. *J. Am. Chem. Soc.* 140 (12): 4208–4212.
35 Colombier, I., Spagnoli, S., Corval, A. et al. (2007). Diarylethene microcrystals make directional jumps upon ultraviolet irradiation. *J. Chem. Phys.* 126 (1): 011101.
36 Grinfeld, M.A. (1986). Instability of the interface between a non-hydrostatically stressed elastic body and melts. *Dokl. Akad. Nauk. SSSR* 290: 1358–1363.

37 Kitagawa, D., Okuyama, T., Tanaka, R., and Kobatake, S. (2016). Photoinduced rapid and explosive fragmentation of diarylethene crystals having urethane bonding. *Chem. Mater.* 28 (14): 4889–4892.
38 (a) Hatano, E., Morimoto, M., Hyodo, K. et al. (2016). Photosalient effect of a diarylethene with a perfluorocyclohexene ring. *Chem. Eur. J.* 22 (36): 12680–12683. (b) Hatano, E., Morimoto, M., Imai, T. et al. (2017). Photosalient phenomena that mimic impatiens are observed in hollow crystals of diarylethene with a perfluorocyclohexene ring. *Angew. Chem. Int. Ed.* 56 (41): 12576–12580.

2

Photomechanical Crystals Made from Anthracene Derivatives

Fei Tong[1], *Christopher J. Bardeen*[1], *and Rabih O. Al-Kaysi*[2]

[1] *University of California Riverside Department of Chemistry, 501 Big Springs Road, Riverside CA, 92521, USA*
[2] *King Saud bin Abdulaziz University for Health Sciences, and King Abdullah International Medical Research Center, Ministry of National Guard Health Affairs Riyadh Department of Basic Sciences, College of Science and Health Professions Riyadh 11426, Saudi Arabia*

2.1 Introduction

In 1966 Isaac Asimov published his science fiction novel *Fantastic Voyage* [1], which was later made into a movie with the same title. The dream of injecting autonomous machines smaller than human cells inside a living body to perform a function, deliver therapeutics, and cure disease has been feeding the imagination of aspiring scientists ever since [2, 3]. Nowadays, with the development of micro transistors and nanolithography [4], submillimeter-scale autonomous machines are slowly becoming a reality. Unfortunately, developing nanobatteries with a large power to size ratio to fuel such machines for long durations is still a challenge that stifles further innovations in that field. Not to mention the challenge of remotely guiding such micromachines via radio frequency signals further complicates matters. Inspired by biological microorganisms such as bacteria and their flagellum, one can engineer micromachines that spontaneously channel chemical fuel from the surrounding environment into a nanoengine. Building such complex micromachines is still beyond our technological reach. Janus nanowire and nanoparticles made by fusing different metals can decompose hydrogen peroxide in aqueous solutions forming bubbles and concentration gradients that propel them in solution [5–7]. These rudimentary motors are rigid but can perform simple translational motion. The need for a form of energy/fuel that can be transduced through space and is independent of the surrounding medium is a far better option. Magnetic fields and acoustic waves are one option that has been used to cause translational motion of magnetic nano- and micromotors in vivo [8–10]. While magnetic fields die off rapidly with distance and acoustic waves are stochastic, light is the ideal fuel for powering nano- and microscale, noncontact actuators, since photons travel fast, cover long distances without losing energy, can penetrate into a wider variety of media, transport both energy and information, and are not subject to economic tariffs. Our motivation is to develop materials that can directly convert light into mechanical work. Some crystals

Mechanically Responsive Materials for Soft Robotics, First Edition. Edited by Hideko Koshima.

[11–13] and polymers [14–18] that are mostly organic in nature are able to absorb photons to initiate a photochemical reaction that will reconfigure the internal structure of the polymer or crystal. When the reconfiguration is done in a controlled manner, useful work in the form of bending, twisting, rolling, jumping, or coiling can be harnessed from these systems. In this chapter, we will survey photomechanical crystals made from a specific type of light absorbing chromophore, anthracene. These derivatives can be grown into crystals with different shapes and sizes leading to a variety of photomechanical responses.

2.2 Elements of Photomechanical Molecular Crystals

Historically, one could argue that the field of molecular crystal photomechanics is c. 100 years old [13, 19–21]. A systematic study of crystal packing effects on photochemical outcomes was pioneered by Schmidt and Ginsburg [20], who coined the term "topochemistry" [22–25]. Earlier accounts of photomechanical phenomenon associated with molecular crystals were disregarded and considered more of a nuisance than as a feature. In many cases, the photoreaction generates phase-separated regions composed of photoproduct and reactant molecules within the reacting crystal [26]. Nonelastic deformation caused by internal strain resulting from the interface between different phases often leads to fracture and disintegration of the original crystal [27–29]. In general, a photomechanical crystal has the following: molecules that can photochemically react in condensed media and an ordered matrix to amplify the work generated from the photochemical reaction. Some molecules can convert light to motion through photochemistry [30, 31]. A list of some of these photochemically reactive molecules is provided in Figure 2.1. Individually, these photochemical reactions produce very small displacement on the order of a few angstroms (Å). To amplify the work output and displacement from these individual photochemical reactions on the order of microns or millimeters, the molecules must be ordered or aligned so they all "push" in one direction. A common strategy is to covalently attach these photochemical units in a liquid crystal elastomer matrix [16, 32]. In other cases we allow the photochemical units to self-assemble into a single crystal. Unfortunately, designing a photomechanical crystal is limited by many factors, most important are the ability of the chromophore to perform a photochemical reaction in the solid state and maintaining the structural integrity of the crystal post photochemical reaction. Conserving the structural integrity of the crystal post a photomechanical action is a crucial requirement if these crystals are to be used in the real world as actuators or parts of a larger micromachine. In many cases, photochemically reactive molecular crystals generate forces that can shatter the crystal while forming photoproducts. Our group has shown that although millimeter-size crystals composed of the anthracene derivative 9-*tert*-butyl-anthracene ester (**9TBAE**) shattered under UV illumination, 200 nm diameter crystalline nanowires composed of the same molecule can expand up to 15% without breaking when irradiated with 365 nm light [33]. These photomechanical nanowires were driven by a crystal-to-crystal $4\pi+4\pi$ photodimerization (Figure 2.2).

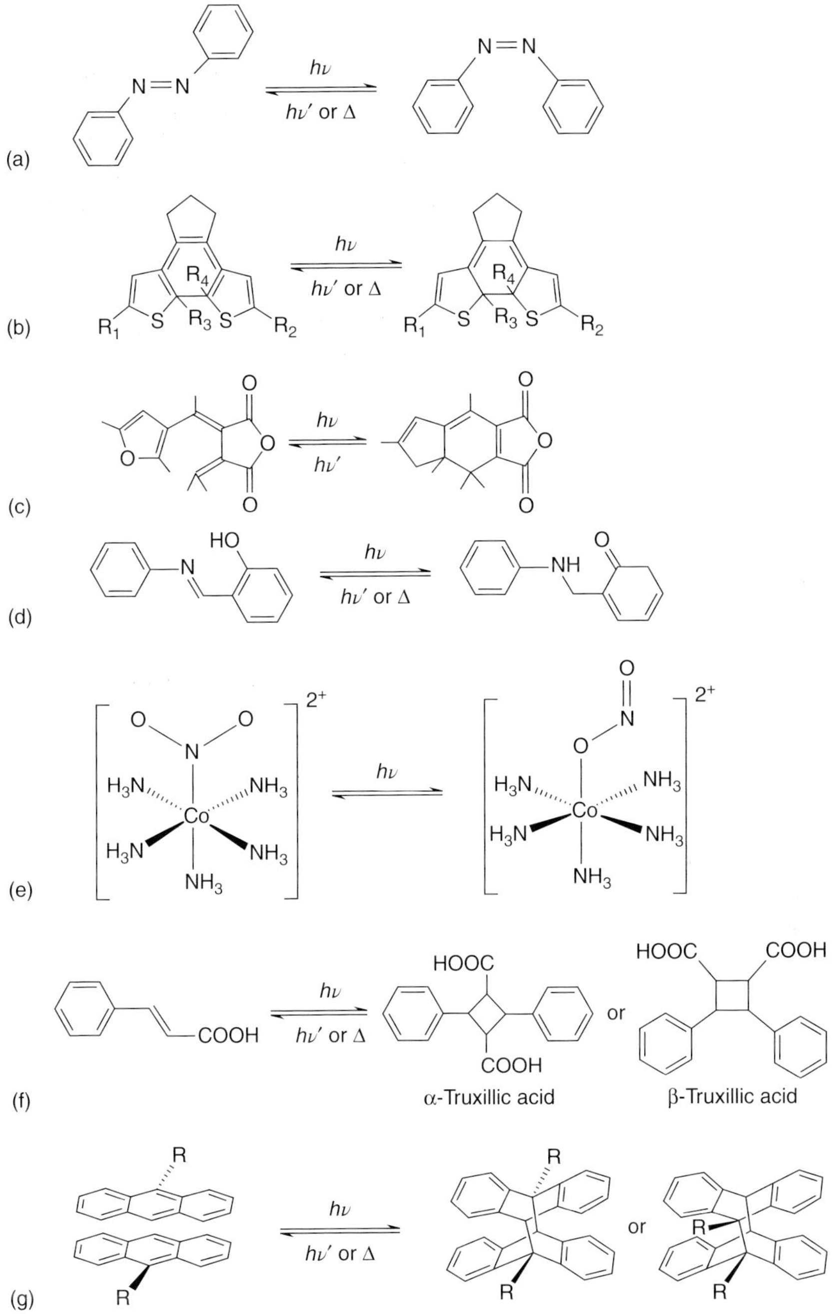

Figure 2.1 Some reversible photochromic reactions used in photomechanical crystals. (a) trans-to-cis photoisomerization of azobenzenes; (b) electrocyclic ring formation and cleavage reaction of diarylethenes; (c) electrocyclic ring formation and cleavage isomerization of furylfulgide; (d) intramolecular hydrogen transfer reaction of salicylideneanilines; (e) intramolecular bond linkage isomerization of a nitropentaamminecobalt(III) complex; (f) 2π+2π cycloaddition reaction; and (g) 4π+4π cycloaddition reaction. Source: White 2017 [13]. Reprinted with permission of John Wiley & Sons.

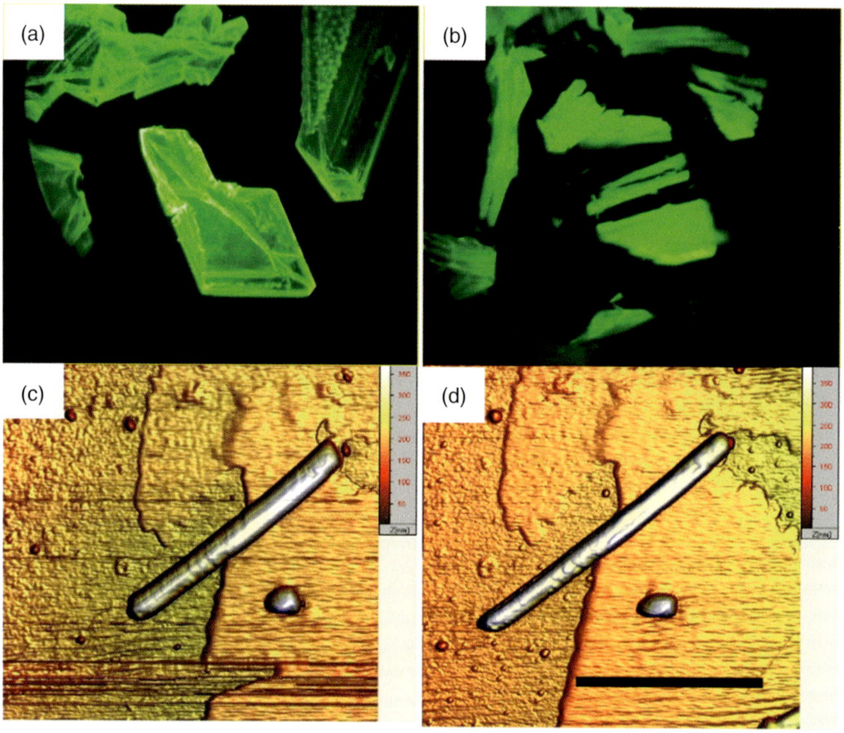

Figure 2.2 Optical microscope images of millimeter-size **9TBAE** crystals before (a) and after (b) 365 nm irradiation. The crystal disintegrates during irradiation. (c) Atomic force microscope (AFM) scan of a crystalline **9TBAE** nanowire before illumination and (d) after illumination with 365 nm. Scale bar is 6 μm. Source: Al-Kaysi et al. 2006 [33]. Reprinted with permission of ACS.

To prevent photomechanical crystals from disintegrating due to build up of interfacial strain, we perfected methods to grow these crystals into nanowires via solvent annealing inside anodic aluminum oxide (AAO) templates [34, 35], or into microplates, microwires, microblocks, or microsheaf by coprecipitation or acid-catalyzed hydrolysis in aqueous surfactants [36–38]. These nano- and microcrystals alleviate strain buildup during a photochemical reaction due to the placement of more molecules on the surface relative to the bulk. An added advantage for using small or thin crystals is to increase the photochemical yield by minimizing the internal filter effect in cases when the photoproduct absorbs the incident radiation. An important factor that could affect crystal survival is the fact that the mechanical properties of a molecular microcrystal can change in nontrivial ways during photoreaction, perhaps making the crystal more plastic and better able to withstand internal stresses [39]. The photomechanical crystal community is still young; nonetheless it seems that they all agree on classifying photomechanical crystals into two major groups: P-type (photon-reversible) and T-type (thermally reversible) [40–45]. In this chapter, we will introduce a third type of photomechanical crystals, the nonreversible type (NR-type). In P-type photomechanical crystals, the photoproducts can be reverted to reactants by absorbing a different wavelength of light that typically matches the maximum absorption of the generated photoproduct. This reversibility can cause mechanical photo-actuation in the crystal up to several hundred or thousand times, depending on the material used. Typical photomechanical crystals of this type are made from diarylethenes [39] and in some cases of spiropyrans [11, 46] or azobenzenes [47]. In T-type crystals, the photochemical product is thermally labile and can be decomposed at ambient temperature to regenerate the reactant. In theory, any unimolecular photoreaction can be thermally reverted to its reactant if provided with enough thermal energy. But when dealing with photomechanical crystals, the applied thermal energy must not exceed the crystal-binding energy that might cause melting or chemical decomposition of the crystal. T-type photochemical crystals are typically allowed to reset at room temperature over a varied period depending on the stability of the photoproduct [48–50]. NR-type crystals can't survive either thermal or photochemical reversibility. Applying thermal energy or very short wavelength (<254 nm) may cause the photoproduct to revert to the reactant but not without imparting significant damage to the crystal matrix (melting or sublimation in the case of thermal energy) or secondary photoreactions with pendent functional groups in the case of irradiating with a short wavelength.

2.3 The Advantage of Using Anthracene Derivatives in Photomechanical Crystals

Typically, photomechanical crystals are composed of diarylethene [45, 51–54] and azobenzene [55–58] derivatives. Throughout the evolution of the field, researches discovered that other types of photochemical reactions could be amplified in the crystalline phase to yield photomechanical crystals. Crystalline anthracene is photochemically stable, giving off faint blue fluorescence when irradiated with UV light. The arrangement of anthracene units in the crystal matrix topologically disfavors $4\pi+4\pi$ photodimerization [59]. On the other hand, many anthracene derivatives are photochemically reactive in the solid state [60, 61]. For the past 15 years, our group has synthesized numerous anthracene derivatives that show photochemical reactivity in the solid state. In Figure 2.3 we see the general molecular formula of some anthracene derivatives that are photochemically reactive in the solid state.

Using anthracene derivatives as the photomechanical "engine" has several advantages. First, the photochemistry and photophysics of anthracenes are well established, which provides us with the necessary tools to understand the mechanism involved. Second, anthracene derivatives absorb strongly in the UV (365 nm) region and their absorption spectrum can be easily extended to the visible region by simple conjugation with double bonds at the C-9 or central carbon. Third, $4\pi+4\pi$ photodimerization provides a relatively large displacement/absorbed photon, which translates to more force/photon. Fourth, anthracene derivatives are relatively stable to oxygen with very few side products possible. Fifth, derivatives of anthracene are mainly solid at room temperature, with a good solubility in organic solvents, hence they can be easily grown into

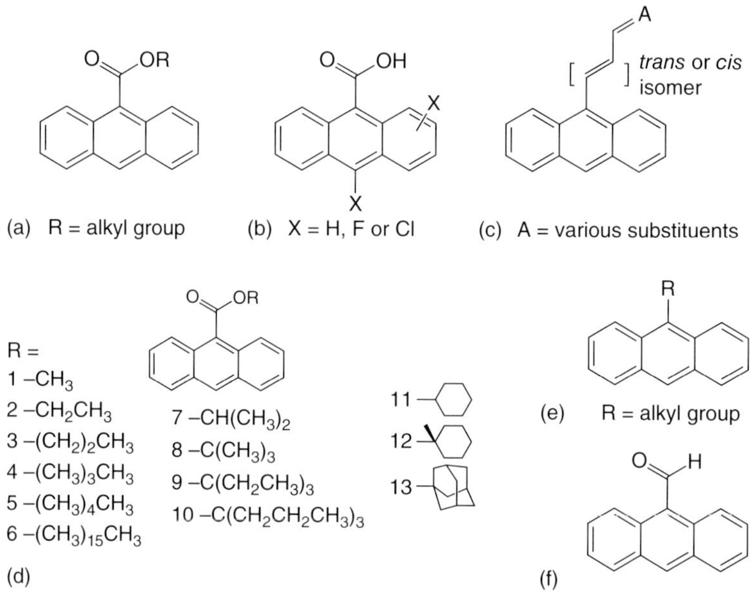

Figure 2.3 Structure of some photomechanical anthracene derivatives. (a) 9-Anthracene carboxylate esters, (b) derivatives of 9-anthracene carboxylic acid, (c) conjugated anthracenes, (d) 9-anthracene carboxylate esters with different R groups, (e) 9-alkyl anthracene derivatives, and (f) 9-anthraldehyde.

crystalline ribbons, nanowires, or other shapes. Sixth, synthesis of anthracene derivatives, in many cases, does not involve complex synthetic methodology.

2.4 Types of Anthracene Photomechanical Crystals

As mentioned earlier, photomechanical crystals fall under three types – P-type, T-type, and NR-type – with each family of derivatives leaning heavily toward a preferred type (Figure 2.1). Diarylethenes, for example, are mainly P-type, while azobenzene derivatives (especially when incorporated inside liquid crystalline elastomers (LCE)) are mostly T-type, and P-type. In the case of anthracene derivatives, we find that they tend to be diverse with types spanning NR-, T-, and P-types.

2.4.1 NR-Type Anthracene Derivatives

2.4.1.1 9-Anthracene Carboxylate Ester Derivatives

The reaction of 9-anthracene carboxylic acid (**9AC**) with an aliphatic alcohol in the presence of an acylating agent such as trifuloroacetic anhydride in dry benzene yields the corresponding carboxylate ester in relatively high yields [33, 60, 61]. A series of ester derivatives (Figure 2.3a,d) were synthesized by our group and assayed for photochemical reactivity in the solid state. All of the ester derivatives were photoreactive in the solid state with a clean 4π+4π photodimerization photochemistry between parallel-stacked anthracene units <3.8 Å apart arranged in an anti-conformation. Using light with $\lambda < 405$ nm the photochemical yield was near 100%. Unfortunately, stresses generated during the photoreaction due to the formation of a bimorph phase (monomer/photodimer) caused large crystals to disintegrate and in many cases in a photosalient [62–64] way during irradiation (Figure 2.2a,b). To prevent crystal disintegration these derivatives were grown into crystalline nanowires by solvent annealing inside AAO templates [34, 35, 53, 65, 66]. Nanowires made from these derivatives expanded along their long axis from 7% to 15% when irradiated with 365 nm light (Figure 2.2c,d). Detailed investigation of these derivatives revealed a unique aspect of their solid-state photochemistry [61]. The crystal structure of the photodimer produced via the solid-state photochemistry has a metastable structure different from what formed when the dimers are crystallized from liquid solution. The metastable crystal structure slowly converts into the lower energy solution-grown form over the course of months. Since the photomechanical response is dictated by the metastable crystal structure, this research illustrated that the photomechanical response can arise from the formation of nonequilibrium structures and can be difficult to predict based on equilibrium crystal structures. This system demonstrates that the ~100% conversion strategy can lead to large shape changes without bimorph formation. Thus our studies of the 9-anthracene carboxylate esters illustrated several general concepts in the field of photomechanical crystals: the importance of crystal size and shape, the role of molecular packing, and the ability to generate significant motion even in the absence of a bimorph. In order to fully understand the mechanism of photomechanical expansion of these nanowires, a representative derivative of these class of compounds, **9TBAE** (Figure 2.3d, R = $C(CH_3)_3$), is being thoroughly studied in an attempt to get a full understanding of the mechanism [67]. When a methyl bridge tethers two 9-anthracene carboxylate esters, an A–A type monomer is formed. Nanowires made from this monomer can photopolymerize (Scheme 2.1), with slight photomechanical expansion, to form tough and highly crystalline polymer nanowires [68]. Different diameter nanowire photopolymers were made (35 and 200 nm). These photopolymer nanowires can be thermally unzipped to regenerate the monomer; unfortunately, the unzipped nanowires quickly melt in the process.

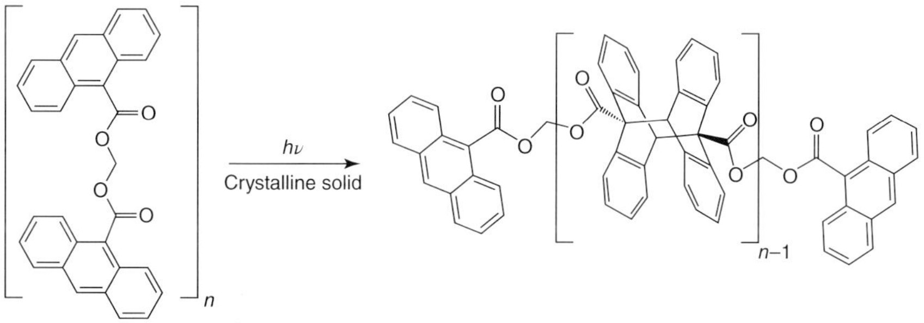

Scheme 2.1 Photopolymerization of an A–A type anthracene monomer in the solid state to form crystalline polymer chains.

2.4.1.2 9-Methylanthracene

To further understand the relationship between photomechanical behavior and crystal structure, 9-methylanthracene (**9MA**) 4π+4π solid-state photochemistry was chosen as our model system since there is no difference between the crystal structure of the solid-state photoproduct and the solution-grown crop. **9MA** photodimerization reaction serves as an ideal model system to illustrate how heterogeneous reaction kinetics and crystal shape can be used as design elements for the development of new photomechanical materials [69]. The amenability of **9MA** to form different crystal shapes was helpful in understanding the mechanism involved. Although nanowires of **9MA** were photomechanically responsive (bending), nonetheless they showed very little photomechanical expansion compared to **9TBAE**. By varying the recrystallization conditions we were able to form different crystal shapes of **9MA** with the same basic crystal structure. By using the floating drop method microneedles and microribbons of **9MA** were obtained. The microribbons twisted under irradiation, while the microneedles bent. Examples of these shape changes are shown in Figure 2.4a,b.

The net shape change after a long period of irradiation was very small with slight elongation (<3%) along the long axis, but at intermediate times the crystals were highly deformed. Maximum crystal deformation was achieved at intermediate stages during irradiation were photodimerization is halfway complete. At 100% conversion, the crystal returns to its original shape [70]. Given the success of utilizing **9MA** to study the photomechanical behavior

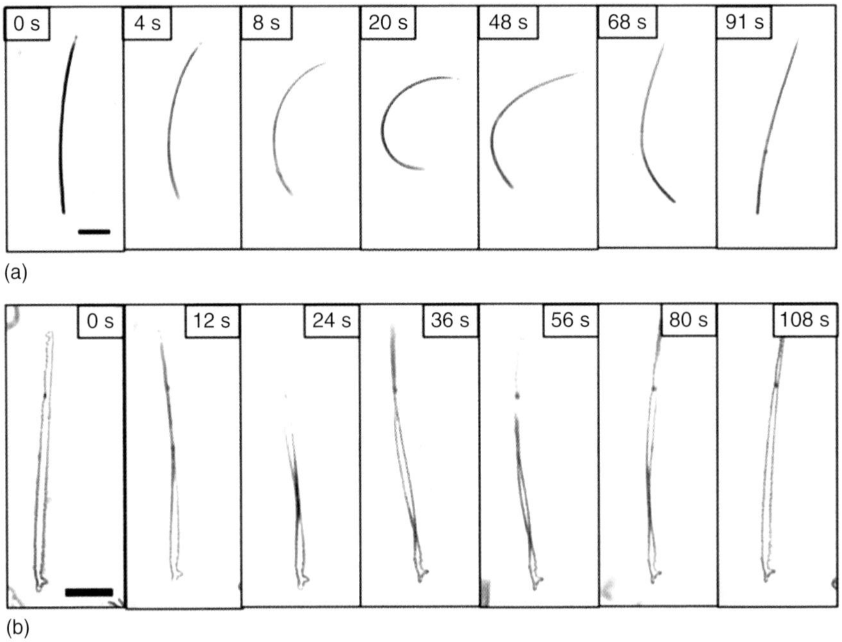

Figure 2.4 (a) Optical microscope images of bending and unbending of a **9MA** microneedle. (b) Twisting and untwisting of a **9MA** microribbon. Light source is 365 nm UV irradiation. Scale bars: 20 μm. Source: Kim et al. 2014 [70]. Reprinted with permission of ACS.

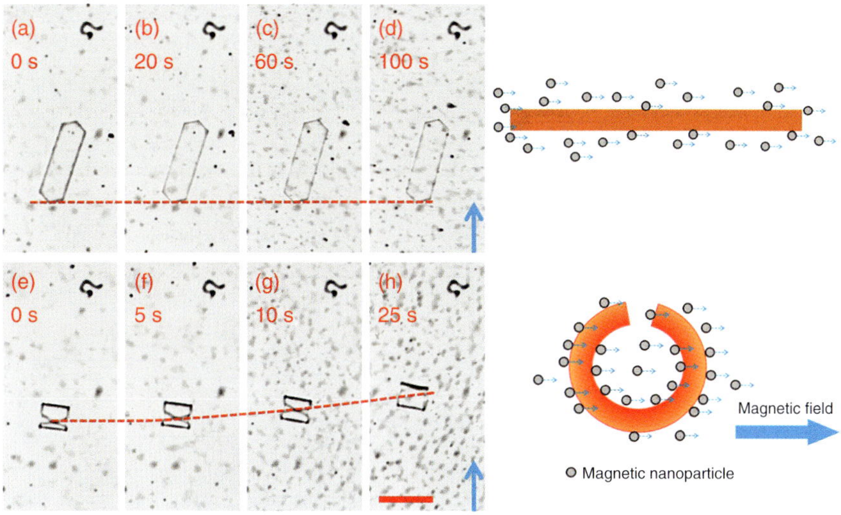

Figure 2.5 (a)–(d) An elongated **9MA** hexagonal microplate without UV light irradiation. As an external magnetic field is applied (cyan arrow), the superparamagnetic CNCs aggregate and migrate in the direction of the field, but the flat platelet is unaffected. (e)–(h) The microplate from panels (a)–(d) is exposed to a pulse of 405 nm UV light causing it to curl up into a semicylinder. When the magnetic field is applied again, the CNCs carry the crystal in the direction of the field. Scale bar: 30 μm. The diagrams at right show a side view of the **9MA** microplate trapping the superparamagnetic CNCs after UV irradiation, which then carries the cylinder in the direction of the magnetic field. Source: Tong et al. 2018 [38]. Reprinted with permission of John Wiley & Sons.

of these crystals, we investigated the photomechanics of other **9MA** crystal habits grown using other methods. When a concentrated solution of **9MA** in *N*,*N*-dimethylformamide (DMF) was slowly injected in an aqueous sodium dodecyl sulfate (SDS) solution in water, a metastable supersaturated solution of the **9MA** inside the SDS micelles was formed. Precipitation of the supersaturated solution via seeded growth generates elongated hexagonal platelets with **9MA** pairs oriented perpendicular to the crystal long axis. Exposure to visible or UV light causes photoinduced rolling up and unrolling of these 700 nm thick microplates. The rolled up cylindrical shape of these microcrystal plates can trap superparamagnetic nanoparticles, which enables it to be dragged along in a magnetic field gradient. This novel photoinduced shape change, made possible by surfactant-assisted crystal growth method, opens up the possibility of using light to modulate the crystal translational motion of the microribbon (Figure 2.5). This was a proof-of-concept of interfacing magnetic field with photomechanical crystals to perform translational motion [38].

2.4.1.3 9-Cyanoanthracne, 9-Anthealdehyde, and 9,10-Dinitroanthracene

Our group has pioneered the use of molecular crystals made from anthracene derivatives to study photomechanical behavior. G. Kaupp et al. made significant observations in solid-state photochemistry and its effect on the surface topology of molecular crystals using atomic force microscopy on molecular crystals made

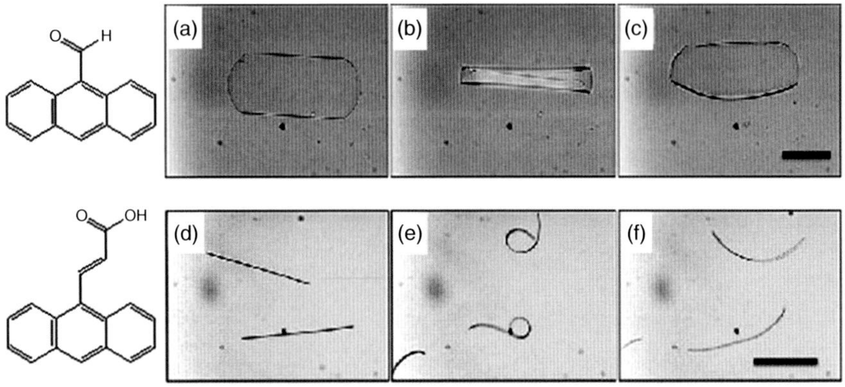

Figure 2.6 Optical microscope image of 9-anthraldehyde microplate folding and unfolding when exposed to 475 nm light: (a) before light exposure; (b) during light exposure; and (c) after light exposure. Scale bar = 50 μm for panels. Optical microscope image of **9-AYAA** microwire (~60 μm long) coiling and uncoiling when exposed to 475 nm light: (d) before light exposure; (e) during light exposure; and (f) after light exposure. Scale bar = 30 μm. Source: Al-Kaysi et al. 2015 [36]. Reproduced with permission of The Royal Society of Chemistry.

from anthracene derivatives [28, 29, 71, 72]. Nonetheless, the focus of their research was on observing phase rebuilding on the surface of these crystals during solid-state reactions rather than the effect of a photochemical reaction on the entire bulk of the crystal. Another example where photochemistry was the focus rather than the photomechanics is the photodecomposition of 9,10-dinitroanthracene crystals to yield anthraquinone with the release of NO gas [73]. Another solid-state photoreaction that was extensively studied is that of 9-cyanoanthracene [74–76]. Microneedles of 9-cyanoanthracene bend and disintegrate upon excitation with 405 nm light revealing the potential use of this derivative as a photomechanical crystal. A promising derivative that can be used as photochemical crystals is 9-anthraldehyde (Figure 2.3f) with a well-established photochemistry [19]. The literature does not report any study related to the photomechanics of 9-anthraldehyde molecular crystals. Our group was able to grow pancake-like crystals of 9-anthraldehyde using acid-catalyzed hydrolysis of anthracen-9-yl methylene diacetate precursor in aqueous surfactant [36] (Figure 2.6a–c). These crystals could be rolled and unrolled like a tortilla using visible light; unfortunately, the transformation was crystal-to-amorphous, which made mechanistic investigation using X-ray diffraction techniques impractical.

2.4.1.4 Conjugated Anthracene Derivatives with Trans-to-Cis Photochemistry

In order to shift the absorption spectrum of anthracene derivatives to the visible, conjugation with double bonds extending from the central or 9-carbon did the job. Nanowires made from 3-anthracen-9-yl acrylic acid (**9-AYAA**) with a single double bond extension showed photomechanical responses such as bending, coiling, and uncoiling when exposed to visible light (475 nm). More prominent display of photomechanical activity was observed from ultra-long microwires made from the same derivative grown via a chemical reaction method in aqueous surfactant. The microwires could photomechanically coil forming closed loops, then uncoil when the photochemical reaction reached completion [36] (Figure 2.6d–f). Unfortunately, the transition was a crystal-to-amorphous, which rendered further investigation into the mechanism of coiling not trivial. To complicate matters, the solid-state photochemistry is not unimolecular, thus multiple photoproducts were observed and photoreversibility could not be achieved. Alkyl ester derivatives of **9-AYAA** were not photoreactive in the solid state, which made further investigation unattractive. To push the absorption spectrum of anthracene derivatives further into the visible regions, derivatives with the general molecular formula depicted in Figure 2.3c were synthesized. The absorption spectra of the *E* isomers extend beyond 550 nm and a reversible unimolecular *E*-to-*Z* photoisomerization in solution. A library of these derivatives is listed in Figure 2.7. Nanowires made from either the *E*- or *Z*-dimethyl-2(3-anthracen-9-yl)allylidene)malonate (**DMAAM**) (Figure 2.7a; with $R_3 = R_4 = H$; $R_1 = R_2 = COO\text{-}Me$) do not show the typical photomechanical expansion exhibited by **9TBAE** and other ester analogs. Nanowires suspended in a dilute surfactant solution tend to spontaneously coil to a "dot" upon a brief pulse with 475 nm light [77] (Figure 2.8). The rate of coiling is accelerated by a factor of 10 in the presence of a surfactant such as SDS or Cetyltrimethylammonium Bromide (CTAB) [78]. This synergy between the surfactant and photomechanical nanowires was further evident with microblocks composed of *Z*-**DMAAM** that were grown by seeded coprecipitation from a saturated solution of *Z*-**DMAAM** in SDS surfactant. When these microblocks were pulsed with 365 nm light, an amorphous layer composed of a mixture of *Z*- and *E*-**DMAAM** slowly peels off after 15–20 seconds. The peeling process could be repeated until the entire microblock is peeled off [79, 80] (Figure 2.9). Molecules made from (*E*)-3-(3-(anthracene-9-yl)allylidene)-1,5-dioxaspiro[5.5]undecane-2,4-dione (Figure 2.7c, $n = 4$) can be grown into red elongated microsheets by gently agitating a supersaturated solution of the compound in aqueous SDS at 45 °C. The microsheets are around 700 nm thick with variable lengths. They can be tightly rolled when briefly exposed to visible light. Continued exposure to visible light causes the crystals to unroll and photoisomerize to the *Z* isomer (Figure 2.10).

2.4.2 T-Type Photomechanical Crystals Based on Reversible 4π+4π Photodimerization

Photoproducts formed from the photodimerization of **9MA** or 9-anthracene carboxylate esters can, in principle, be reversed by heating or irradiation at 254 nm. In practice, high temperature thermal "unzipping" is detrimental to the crystal structure and 254 nm irradiation cannot achieve 100% back reaction due to the formation of a photostationary state and secondary photoreactions with pendent functional groups. To make a T-type photomechanical crystal that can be cycled multiple times the reverse reaction must spontaneously reset at room temperature. **9AC** crystallizes in a head-to-head "syn" arrangement [81], rather than the head-to-tail "anti" arrangement common in most 9-substituted anthracenes [75]. Although the syn arrangement is often assumed to disfavor 4π+4π photocycloaddition reaction due to electronic and steric factors,

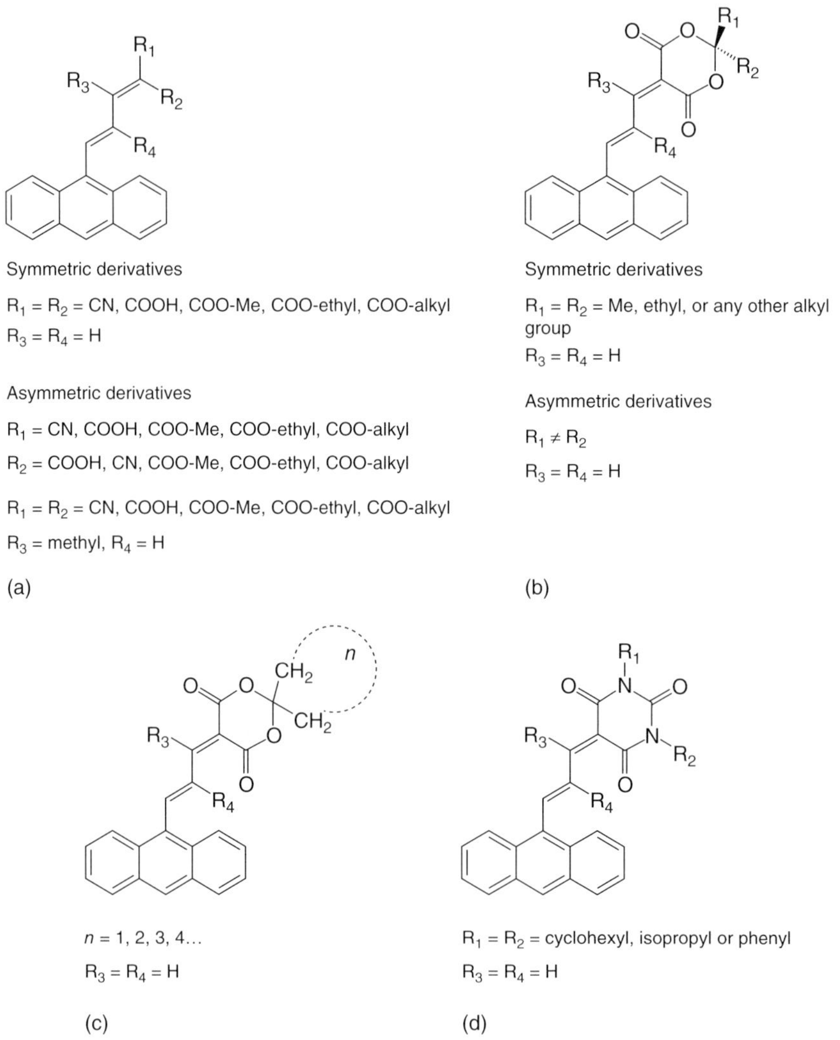

Figure 2.7 Molecular structure of some conjugated anthracene derivatives.

solid-state NMR measurements showed that **9AC** does, in fact, undergo the $4\pi+4\pi$ photocycloaddition reaction characteristic of anthracenes in the solid state [82]. Thus, solid-state photodimerization of **9AC** under UV light yields a thermally unstable photodimer that spontaneously reverts back to the monomer state within a few minutes at room temperature. Reversible expansion and contraction (actuation) of individual **9AC** nanowire was monitored using atomic force microscopy [83]. After UV light exposure (365 nm) and ~100% conversion of an individual **9AC** nanowire to photodimer, it briefly expands by 1–3% but then returns to its original shape after 10 minutes at room temperature. To induce actuation on a single nanowire, we irradiated its central region. The nanowire instantly bends under the influence of the UV light (Figure 2.11).

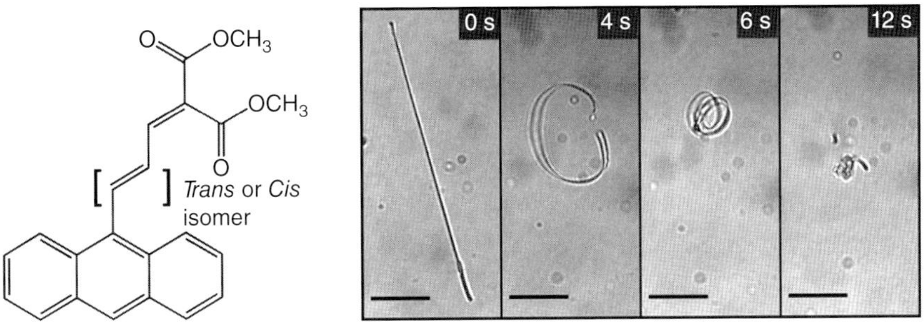

Figure 2.8 Optical microscope mages of a coiling nanowire, with a 200 nm diameter, composed of (*E*)-**DMAAM** after 1 second of light exposure at time 0 second. Coiling occurs after 10–20% of the *trans* isomer has been converted to the *Z* isomer. Scale bars: 10 μm. Source: Kim et al. 2013 [77]. Reprinted with permission of John Wiley & Sons.

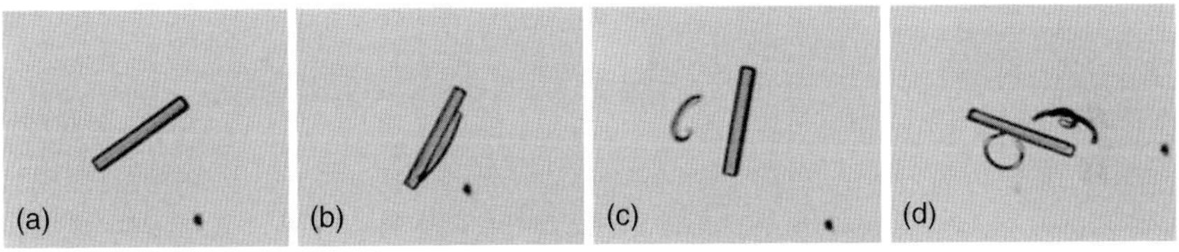

Figure 2.9 Microblocks composed of *Z*-**DMAAM** peeling after subjected to a brief pulse of UV light. (a) Initial, (b) after first pulse, (c) after second pulse, and (d) after the third pulse, notice the accumulation of peels.

Allowing the nanowire to reset to its original shape for five minutes in the dark, the actuation cycle could be repeated multiple times. Secondary photoreactions with oxygen cause the cycling to taper off after 5–10 cycles as the fluorescence of the recovered nanowire slowly dies off. Improving cycling times can, in theory, be achieved by removing oxygen from the surrounding or by coating **9AC** nanowires with a thin layer of silica protective shell without loss of photomechanical activity [65, 66, 84]. To better control the motion of these photomechanical nanowires, we combined **9AC** nanowires with near-infrared two-photon excitation (2PE) technique [85]. We can precisely control the bend angle of the nanowire, reversibly, with the same magnitude and duration even on nanowires as small as 35 nm in diameter. Given the success in making reversible nanowires from **9AC**, we attempted to grow crystalline microribbons from the same material using a modified floating drop method [49]. Instead of bending or expanding under uniform UV light illumination, these microribbons reversibly twist (twisting actuation), as previously observed in the **9MA** microribbons [70]. The reset time for **9AC** photomechanical crystals of 10 minutes is impractical for many applications; the need to modify the molecule was imperative in order to obtain shorter reset times. But to maintain the T-type identity of the **9AC** crystals we must retain the H-bond in the carboxylic acid group. In an attempt to accelerate the photomechanical response by increasing the rate of the spontaneous backward dimer dissociation reaction at room temperature, we systematically modified the **9AC** molecule by replacing

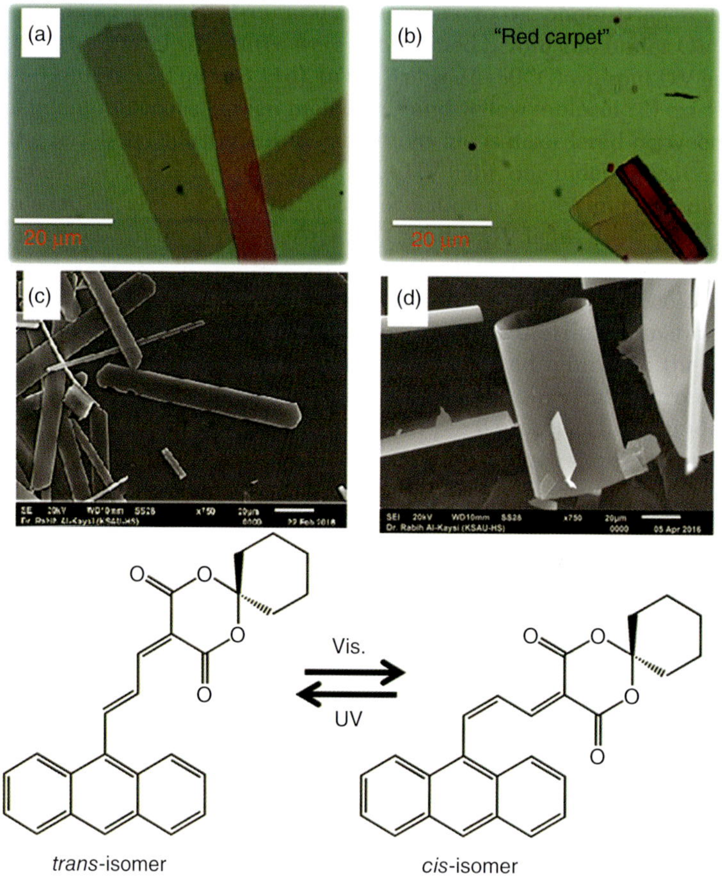

Figure 2.10 Microsheets of (*E*)-3-(3-(anthracene-9-yl)allylidene)-1,5-dioxaspiro[5.5]undecane-2,4-dione. (a) Before exposure to light made of *trans* isomer, (b) after exposure to visible light made of a mixture of *trans* and *cis*, (c) SEM image of unrolled sheets, and (d) SEM image of a tightly coiled sheet.

the H at the C-10 position, directly across from the COOH group, with CH_3, F, Cl, Br, and even another COOH [81]. Unfortunately, none of the derivatives succeeded in increasing the rate of the backward reaction due to topologically unfavorable packing in the crystal. Only the **10-F-9AC** derivative crystallized in head-to-head stacks that exhibited the same reversible photochemistry as **9AC** but with a much slower reset time due to an increased stability of the photoproduct. Our next attempt was to modify the side rings of the anthracene molecule. We synthesized four fluorinated derivatives of 9**AC** and characterized their properties [48]. The spectroscopic behavior and crystal structures of 4-fluoro-9-anthracene carboxylic acid (**4-F-9AC**), 2-fluoro-9-anthracene carboxylic acid (**2-F-9AC**), 10-fluoro-9-anthracene carboxylic acid (**10-F-9AC**), and 2,6-difluoro-9-anthracene carboxylic acid (**2,6d-F-9AC**) are all very similar to those of **9AC**. However, their photomechanical properties varied widely. UV light induces a 4π+4π photodimerization and a mechanical response in

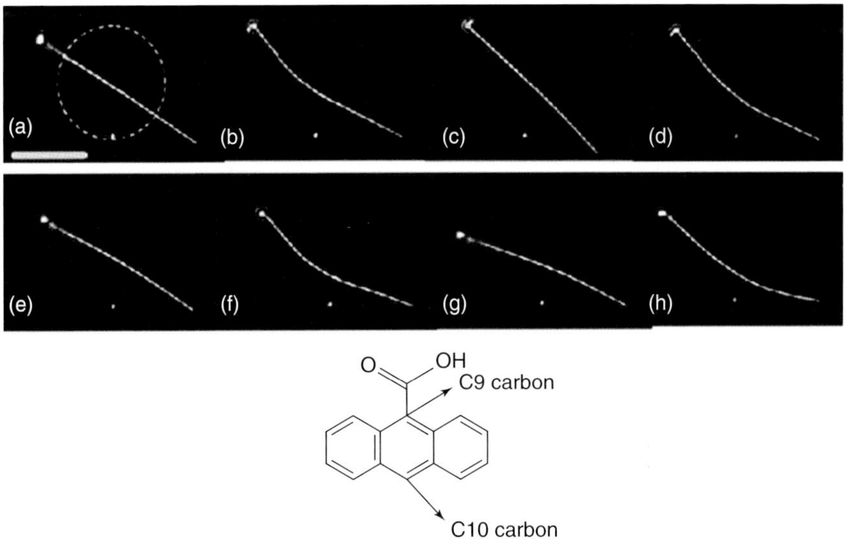

Figure 2.11 A 200 nm diameter nanowires made from **9AC** (~60 μm long) briefly exposed to 365 nm light. Dotted circle shows the illuminated region (35 μm in diameter). The nanowires repeatedly flex back and forth (after UV illumination = panels b, d, f, h; after dark period = panels a, c, e, g). The time required to revert is around 5 minutes at room temperature. Scale bar = 20.7 μm. Source: Al-Kaysi and Bardeen 2007 [83]. Reprinted with permission of John Wiley & Sons.

crystalline microneedles and ribbons. Both the photodimer dissociation rate and the mechanical recovery varied by more than 1 order of magnitude, with **4-F-9AC** exhibiting the most rapid recovery time, on the order of 30 seconds at room temperature. Surprisingly millimeter-size crystals of **4-F-9AC** did not disintegrate upon photoconversion. Microneedles could undergo more than 100 photomechanical bending cycles without any signs of fatigue or deterioration and microribbons could twist and untwist rather rapidly as well (Figure 2.12). Nanoindentation measurements show that this crystal has a slightly reduced elastic modulus and a significantly reduced hardness, making it less brittle than the **9AC** crystal. Given the similarity of the crystal packing in all five molecules, the improved photomechanical properties must arise from subtle changes in intermolecular interactions. These results demonstrate that it is possible to significantly improve the properties of photomechanical materials through small modifications of the molecular structure.

Motivated by the superior photomechanical behavior of **4-F-9AC** we tried different crystallization techniques to grow more complex crystal shapes. When the pH of an aqueous solution of **4-F-9AC** is slowly decreased, mostly X-shaped microcrystals with vertical asymmetry crystallized out. Top or bottom irradiation of these crystals with UV light caused them to rotate either clockwise or counterclockwise simulating a ratchet-like motion (Figure 2.13) [50]. Acid-catalyzed hydrolysis of the *tert*-butyl ester of **4-F-9AC** (*tert*-butyl 4-fluoroanthracene-9-carboxylate) in the presence of an anionic surfactant such as SDS coupled with gentle agitation generated sheaf-like microcrystals

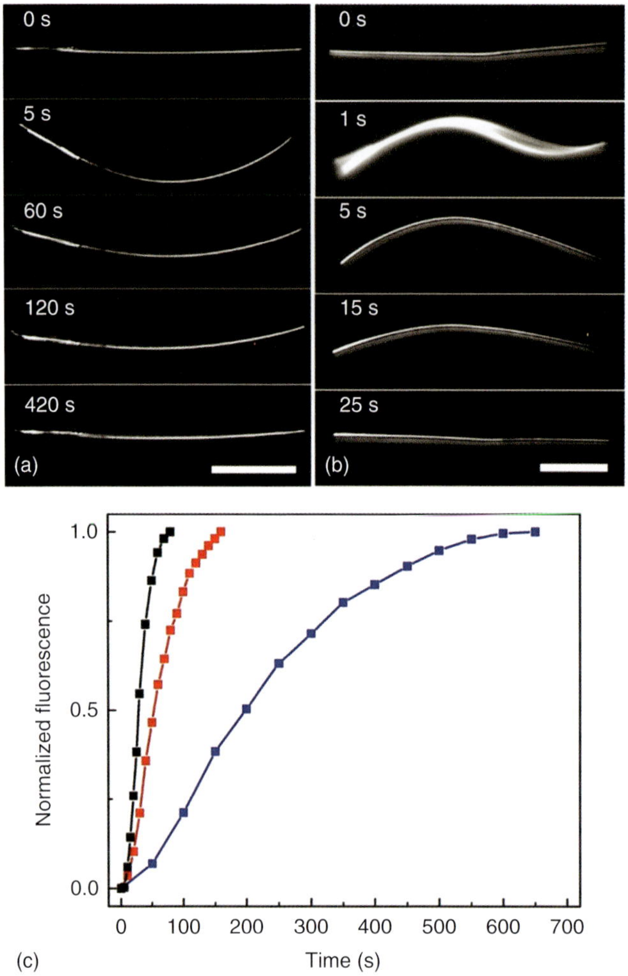

Figure 2.12 Series of dark field optical microscopy images of (a) **9AC** and (b) **4-F-9AC** microcrystals after a 1 second exposure to 405 nm light causes them to deform. **4-F-9AC** crystal reset in 25 seconds, while the **9AC** crystal requires c. 7 minutes to reset. Both scale bars are 50 μm. (c) Fluorescence recovery curves of **4-F-9AC** (black), **2-F-9AC** (red), and **9AC** (blue) showing the different photodimer dissociation rates for the different crystals. Source: Zhu et al. 2014 [48]. Reprinted with permission of ACS.

of **4-F-9AC**. These highly branched "microbrushes" can be actuated and reset repeatedly while in suspension (T-type). The rapid sweeping motion of the arms was able to sweep silica microspheres and manipulate larger objects in the proximity of the arms (Figure 2.14).

2.4.3 P-Type Anthracene Derivatives

An effective strategy in designing P-type anthracene photomechanical crystals is to synthesize anthracene chromophore with *E*-to-*Z* photochemistry

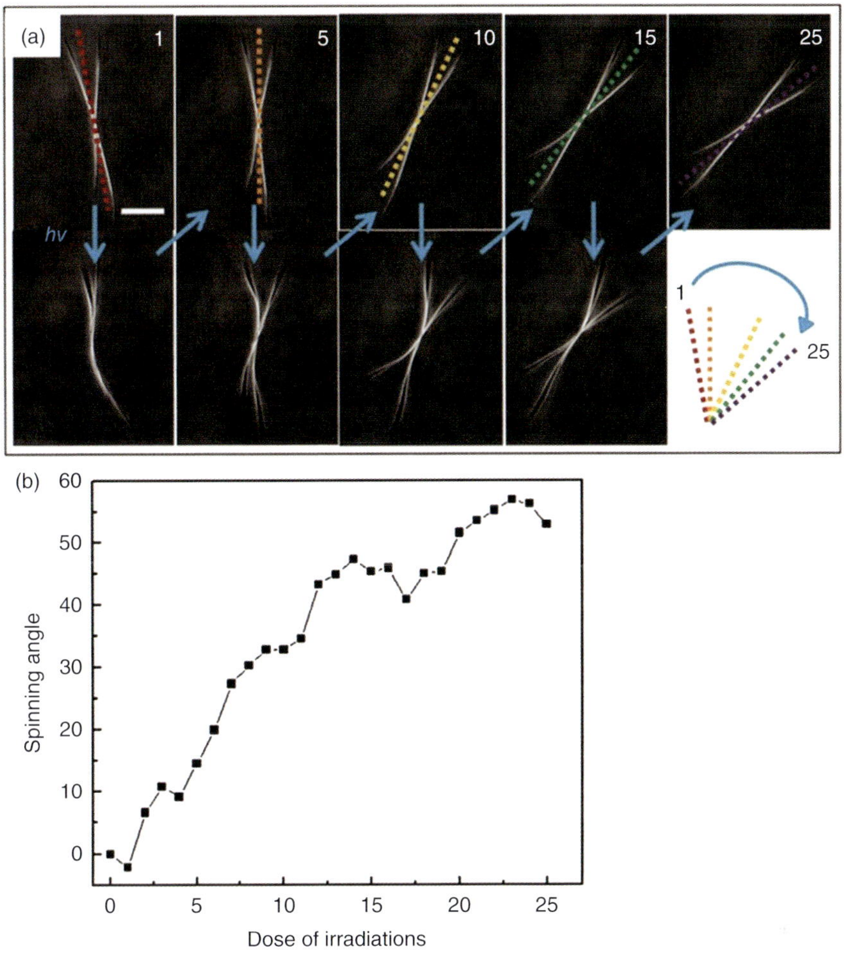

Figure 2.13 Optical microscope images of X-shaped **4-F-9AC** crystal rotating clockwise after each irradiation dose. (a) A total of five cycles of the X-shaped crystal before and after irradiation are illustrated. Scale bar: 20 mm. (b) Plot of the rotation angle versus number of UV light doses. Source: Zhu et al. 2016 [50]. Reprinted with permission of John Wiley & Sons.

that can be reversed using a wavelength greater than 254 nm. Reaction of malononitrile with (*E*)-3-(anthracen-9-yl) acrylaldehyde formed ruby-red crystals (*E*-2-(3-anthracen-9-yl-allylidene)-malononitrile (**DVAM**), Figure 2.7a; $R_1 = R_2 = CN$, $R_3 = R_4 = H$) or nanowires (after solvent annealing inside AAO templates) that can be actuated with different wavelengths of light [43]. Since the reverse reaction could not be regenerated 100% via irradiation at any wavelength, this type of photomechanical motion should be classified as a *pseudo*-P-type instead of a pure P-type typically reserved for diarylethenes. The broad absorption spectrum due to intramolecular charge-transfer formation between the electron donor anthracene and the electron acceptor malononitrile head group gives this derivative its unique photophysical and photomechanical

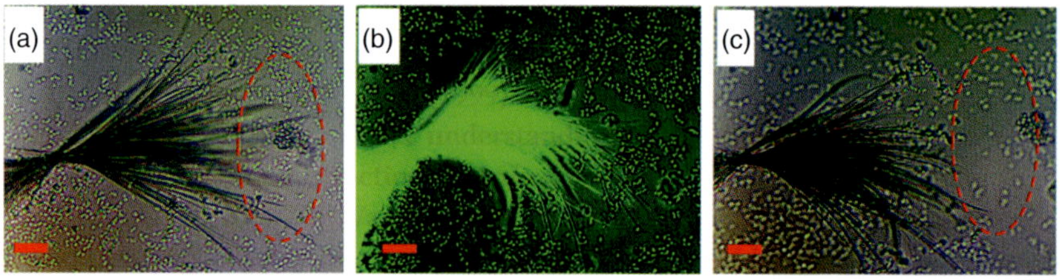

Figure 2.14 (a) Suspended branched microcrystals of **4-F-9AC** with microspheres before UV exposure. (b) During UV (365 nm) light exposure the branches curl in the same direction and sweep the microspheres toward the middle. (c) After several cycles, the microspheres were brushed away. Scale bar = 25 μm. Source: Al-Kaysi et al. 2017 [37]. Reproduced with permission of The Royal Society of Chemistry.

characteristics. Millimeter long microwires made from *Z*-**DVAM** do not show photomechanical activity when deposited on a glass slide. On the other hand, when these microwires are freely suspended or completely immersed in an aqueous surfactant solution such as SDS, they can perform continuous and autonomous wiggling and translational motion, reminiscent of nematode motility, when irradiated with intense visible light [79]. Mechanism of this motion is currently under investigation.

2.5 Synthesis of Anthracene Derivatives

Difficulty in predicting the crystal structure of anthracene derivatives plunged most of our work on anthracene photomechanical crystals in the experimental realm. We synthesized hundreds of anthracene derivatives, cherry-picking those that show photochemical reactivity in the solid state. A good example of this unpredictable nature was the case of C-10-substituted **9AC** derivatives mentioned in Section 2.4.2, and the aliphatic ester derivatives of **9-AYAA** in Section 2.4.1. Another example of anthracene derivatives that fail to show photochemical reactivity in the solid state are the 9-substituted alkyl anthracene derivatives (Figure 2.3e) with the exception of **9MA**. From a synthetic point of view, we rarely used anthracene as a starting point for the synthesis of our anthracene derivatives. Chemical supply companies provide commercially available anthraquinone derivatives, 9-anthraldehyde, and (*E*)-3-(anthracen-9-yl)acrylaldehyde that serve as a starting point in our synthesis. For the synthesis of 10-substituted-9AC derivatives we start by preparing **9AC** by reacting 9-anthraldehyde with aqueous sodium hydroxide, Cannizzaro reaction, followed by separation and workup. Chloro and bromo C-10 derivatives of **9AC** were prepared by direct halogenation. For the F-substituted **9AC** derivatives we start with the amine-substituted anthraquinone followed by fluorination via the Balz–Schiemann reaction [86] (Scheme 2.2). The final product is obtained by lithiation followed by bubbling CO_2 gas and workup to form the free acid.

For the conjugated derivatives (Figure 2.7a–d), the major reaction is the Knoevenagel condensation of the (*E*)-3-(anthracen-9-yl)acrylaldehyde with the appropriate malonate derivative (Scheme 2.3). Of course, not all the derivatives

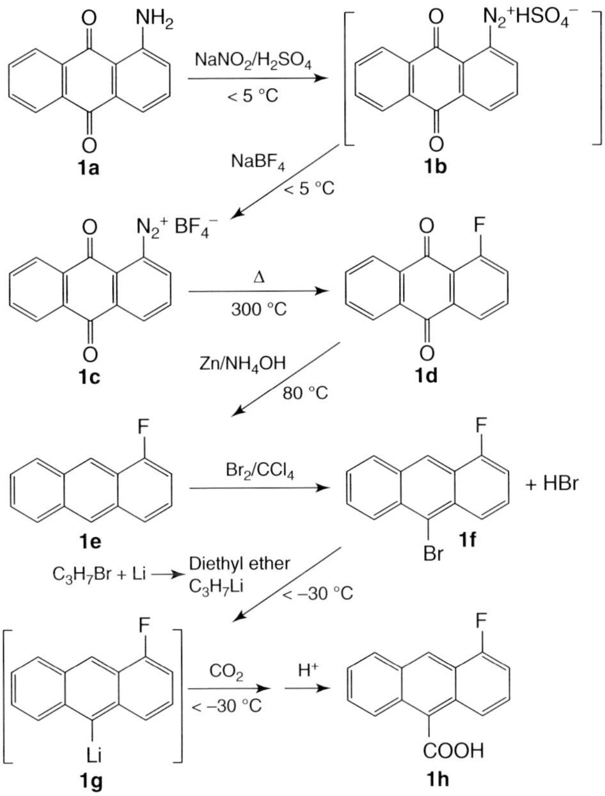

Scheme 2.2 Multistep synthesis of 4-fluoro-9-anthracene carboxylic acid (**4-F-9AC**).

synthesized required simple synthetic procedures, in some cases, the synthesis was more involved with the use of hazardous reagents such as P_2O_5 to make 9-*tert*-butylanthracene, which unfortunately did not show any photomechanical behavior in the solid state but had other interesting properties nonetheless [87].

2.6 Future Direction and Outlook

2.6.1 Modeling Reaction Dynamics in Molecular Crystals

Although many of the topochemical laws governing solid-state photochemical reactions were laid down by Shmidt and coworkers, there are still some unanswered questions regarding the mechanism that affects the entire crystal structure in a solid-state photochemical reaction [20, 22–25]. In photomechanical materials, the photoreaction of individual monomers making up the crystal is the primary step. The rate and direction of the photomechanical response of the crystal are affected by the overall rate of the individual monomer photoreaction and the spatial distribution of the photoproducts. It is still mechanistically vague whether photochemical reactions in a crystal are homogeneously or heterogeneously distributed throughout the crystal. It would be interesting to know

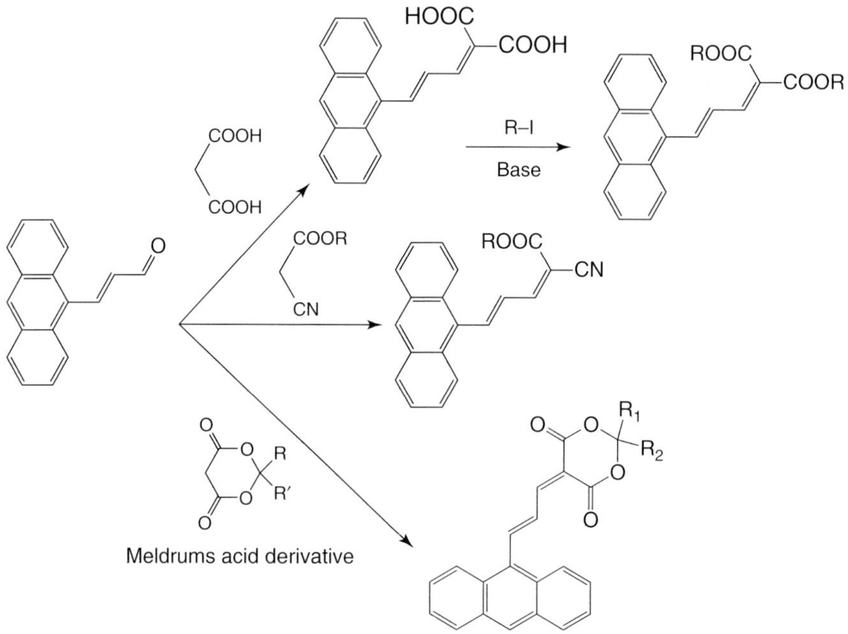

Scheme 2.3 General synthetic scheme for the synthesis of some conjugated anthracene derivatives.

how quickly motion on the molecular scale is translated to motion of the entire photomechanical crystal. An interesting question is how large or what is the size threshold of a photomechanical crystal before fracturing instead of bending or coiling [88]? How can we relate the photomechanical crystal's Young's modulus to its photomechanical behavior? We need to achieve a predictive understanding of crystal structure and packing using X-ray and electron diffraction and how it affects the photomechanical response. In the long run, a quantitative understanding of how micron-scale motions arise from molecule-scale geometry changes will be crucial for the rational design of new photomechanical materials. In order to advance the use of these photomechanical microcrystals in in vitro or in vivo applications, we need to further understand the behavior of such microstructures in low Reynold's number environments and their ability to swim.

2.6.2 New Anthracene Derivatives and Crystal Shapes

Organic synthesis provides us with the tools necessary to make new photochemical engines and improve existing ones with chemical modifications. In general, an ideal photochemical engine should have some of the following properties: (i) high speed of photochemical reaction with high quantum yields; (ii) large structural change, which implies a large force; (iii) can be cycled many times without photodegrading to secondary products. To improve the efficiency of a photomechanical engine is to increase the overall yield of the reaction via a self-propagating chain reaction with an effective quantum yield >1 [89]. Another important factor is molecular photostability over long cycling or exposure times. One way to tackle this problem is to increase the oxidation potential of the molecules by replacing H's with fluorine atoms. In addition to making new molecules, we must consider crystal shape and morphology. In previous Section (2.4.1.2), we have shown how changing the morphology of photochemically reactive crystals changes the overall motion of the photomechanical crystals as seen in the example of bending **9MA** microneedles versus twisting **9MA** microribbons. Our work has focused on the use of nano- and micro-structured molecular crystals as photomechanical elements for two reasons. First, as described above, these structures are more robust and are able to survive photochemical reactions that lead to the disintegration of larger crystals. Second, we think that the advantages of photomechanical structures are most pronounced for small-scale applications where electrical connections are impractical, for example for structures functioning inside biological cells. The key to this approach is to develop ways to grow small molecular crystals with reasonably well-defined shapes and sizes. We have used solvent annealing in AAO templates to fabricate molecular crystal nanowires and controlled growth on flat substrates to generate microneedles and ribbons [49, 50, 70]. In this latter approach, however, the crystals still have a wide size distribution. One can tune the crystal growth conditions to obtain new shapes and better size distributions, but this is typically an empirical approach. For example, we have used acid-catalyzed hydrolysis of precursor molecules in aqueous surfactants like SDS to generate uniform microwires or microplates of photoactive molecules that could undergo photoinduced coiling and bending [36, 37]. We have utilized the seeded growth method from a supersaturated solution of photoreactive precursor molecules from aqueous surfactants to generate flat microribbons that show the different orientation of monomers inside the crystal [38]. While there has been encouraging progress in the creation of molecular crystal arrays in which both the crystal size and orientation can be controlled to some degree [90–93], there is still a need for general methods that can generate molecular crystals with uniform shapes and sizes.

2.6.3 Interfacing Photomechanical Molecular Crystals with Other Materials

The real challenge for these molecular crystals is how can they survive the elements and incorporate them with more complex composite structures. The integration of organic photomechanical structures into larger assemblies is a largely unexplored area. The first challenge is to figure out ways to protect these molecular crystals from the elements such as oxygen or organic solvent vapors. In one of our earlier attempts, we coated nanowires made from **9AC** with a thin layer or silica using solvent annealing inside a previously silica coated AAO template [65]. Recently, Bardeen and coworkers have shown that it is possible to coat perylene organic crystals with a layer of graphene [94]. The coated crystals are resistant to dissolution in solvent vapors. This promising result could lead to techniques for coating photomechanical crystals that can be used in environments that will otherwise dissolve the uncoated crystals such as using **4-F-9AC** in a high pH medium. Experiments showing that some photomechanical crystals can turn gears and move microspheres [51, 52], all provide preliminary demonstrations that it is possible to do useful work with these materials. But there is no general strategy for incorporating photomechanical elements as actuators in a mechanical device like a walker or swimmer.

Another attractive strategy is to incorporate nanometer- to micrometer-sized molecular crystals into a polymer matrix to make a composite material that can be processed more easily [95–97]. Interfacing of actuating photomechanical molecular crystals like diarylethene (P- or T-type) with various metals was demonstrated by Tsujioka et al. [98, 99].

2.7 Conclusion

The field of photomechanical molecular crystals and polymers is growing exponentially despite a slow beginning. We hope we were able to highlight the importance and diversity of using anthracene derivatives as a photomechanical engine on par with other widely used chromophores such as diarylethenes and azobenzenes. Using photomechanical crystals made from anthracene derivatives we can access three different types of photomechanical crystals such as NR-, T-, and P-types. The mode of photomechanical response from such crystals can be influenced by the illumination conditions and shape of the crystal. These derivatives can be grown into nanowires, microribbons, microsheaf, microblocks, or microneedles. The high Youngs modulus of these crystals gives them a rapid response time and stiffness, making them excellent candidates for actuator application. More derivatives are being synthesized and studied every day leading to a better understanding of the photomechanical mechanism involved.

Acknowledgments

This research was supported by the National Science Foundation grant DMR-1508099 (C.J.B.) and KSAU-HS/KAIMRC through grant RC10/104 (R.O.K.).

References

1 Asimov, I. (2011). *Fantastic Voyage: A Novel*. Bantam Books.
2 Sengupta, S., Ibele, M.E., and Sen, A. (2012). Fantastic voyage: designing self-powered nanorobots. *Angew. Chem. Int. Ed.* 51 (34): 8434–8445.
3 Kostarelos, K. (2010). Nanorobots for medicine: how close are we? *Nanomedicine* 5 (3): 341–342.
4 Seisyan, R.P. (2011). Nanolithography in microelectronics: a review. *Tech. Phys.* 56 (8): 1061.
5 Sánchez, S., Soler, L., and Katuri, J. (2015). Chemically powered micro-and nanomotors. *Angew. Chem. Int. Ed.* 54 (5): 1414–1444.
6 Ma, X., Hahn, K., and Sanchez, S. (2015). Catalytic mesoporous Janus nanomotors for active cargo delivery. *J. Am. Chem. Soc.* 137 (15): 4976–4979.

7 Wang, W., Duan, W., Ahmed, S. et al. (2013). Small power: autonomous nano-and micromotors propelled by self-generated gradients. *Nano Today* 8 (5): 531–554.
8 Li, J., Li, T., Xu, T. et al. (2015). Magneto–acoustic hybrid nanomotor. *Nano Lett.* 15 (7): 4814–4821.
9 Wang, J. and Gao, W. (2012). Nano/microscale motors: biomedical opportunities and challenges. *ACS Nano* 6 (7): 5745–5751.
10 Gao, W. and Wang, J. (2014). Synthetic micro/nanomotors in drug delivery. *Nanoscale* 6 (18): 10486–10494.
11 Kim, T., Zhu, L., Al-Kaysi, R.O., and Bardeen, C.J. (2014). Organic photomechanical materials. *ChemPhysChem* 15 (3): 400–414.
12 Naumov, P.P., Chizhik, S., Panda, M.K. et al. (2015). Mechanically responsive molecular crystals. *Chem. Rev.* 115 (22): 12440–12490.
13 White, T.J. (2017). *Photomechanical Materials, Composites, and Systems: Wireless Transduction of Light into Work*. Wiley.
14 Yu, Y. and Ikeda, T. (2006). Soft actuators based on liquid-crystalline elastomers. *Angew. Chem. Int. Ed.* 45 (33): 5416–5418.
15 Yu, Y., Nakano, M., and Ikeda, T. (2003). Photomechanics: directed bending of a polymer film by light. *Nature* 425 (6954): 145.
16 Ikeda, T., Mamiya, J., and Yu, Y. (2007). Photomechanics of liquid-crystalline elastomers and other polymers. *Angew. Chem. Int. Ed.* 46 (4): 506–528.
17 Zeng, H., Wani, O.M., Wasylczyk, P., and Priimagi, A. (2018). Light-driven, caterpillar-inspired miniature inching robot. *Macromol. Rapid Commun.* 39 (1): 1700224.
18 Zeng, H., Wasylczyk, P., Wiersma, D.S., and Priimagi, A. (2018). Light robots: bridging the gap between microrobotics and photomechanics in soft materials. *Adv. Mater.* 30 (24): 1703554.
19 Schmidt, G.M.J. (1971). Photodimerization in the solid state. *Pure Appl. Chem.* 27 (4): 647–678.
20 Schmidt, G.M.J. and Ginsburg, D. (1976). *Solid state photochemistry*. Verlag Chemie.
21 Sun, S.-S. and Dalton, L.R. (2016). *Introduction to Organic Electronic and Optoelectronic Materials and Devices*. CRC Press.
22 Cohen, M.D. (1975). The photochemistry of organic solids. *Angew. Chem. Int. Ed. Engl.* 14 (6): 386–393.
23 McBride, J.M., Segmuller, B.E., Hollingsworth, M.D. et al. (1986). Mechanical stress and reactivity in organic solids. *Science* 234 (4778): 830–835.
24 Ramamurthy, V. and Venkatesan, K. (1987). Photochemical reactions of organic crystals. *Chem. Rev.* 87 (2): 433–481.
25 Luty, T. and Eckhardt, C.J. (1995). General theoretical concepts for solid state reactions: quantitative formulation of the reaction cavity, steric compression, and reaction-induced stress using an elastic multipole representation of chemical pressure. *J. Am. Chem. Soc.* 117 (9): 2441–2452.
26 Keating, A.E. and Garcia-Garibay, M.A. (1998). Photochemical solid-to-solid reactions. *Mol. Supramol. Photochem.* 2: 195–248.

27 Kaupp, G. (1992). Photodimerization of cinnamic acid in the solid state: new insights on application of atomic force microscopy. *Angew. Chem. Int. Ed. Engl.* 31 (5): 592–595.
28 Kaupp, G. (1992). Photodimerization of anthracenes in the solid state: new results from atomic force microscopy. *Angew. Chem. Int. Ed. Engl.* 31 (5): 595–598.
29 Kaupp, G. and Plagmann, M. (1994). Atomic force microscopy and solid state photolyses: phase rebuilding. *J. Photochem. Photobiol., A* 80 (1–3): 399–407.
30 Irie, M. (2000). Photochromism: memories and SwitchesIntroduction. *Chem. Rev.* 100 (5): 1683–1684.
31 Bouas-Laurent, H. and Dürr, H. (2003). *Photochromism: Molecules and Systems*. Elsevier.
32 White, T.J. and Broer, D.J. (2015). Programmable and adaptive mechanics with liquid crystal polymer networks and elastomers. *Nat. Mater.* 14 (11): 1087.
33 Al-Kaysi, R.O., Müller, A.M., and Bardeen, C.J. (2006). Photochemically driven shape changes of crystalline organic nanorods. *J. Am. Chem. Soc.* 128 (50): 15938–15939.
34 Al-Kaysi, R.O., Ghaddar, T.H., and Guirado, G. (2009). Fabrication of one-dimensional organic nanostructures using anodic aluminum oxide templates. *J. Nanomater.* 2009: 1–14.
35 Al-Kaysi, R.O. and Bardeen, C.J. (2006). General method for the synthesis of crystalline organic nanorods using porous alumina templates. *Chem. Commun.* 3 (11): 1224–1226.
36 Al-Kaysi, R.O., Zhu, L., Al-Haidar, M. et al. (2015). Chemical reaction method for growing photomechanical organic microcrystals. *CrystEngComm* 17 (46): 8835–8842.
37 Al-Kaysi, R.O., Tong, F., Al-Haidar, M. et al. (2017). Highly branched photomechanical crystals. *Chem. Commun.* 53 (17): 2622–2625.
38 Tong, F., Xu, W., Al-Haidar, M. et al. (2018). Photomechanically induced magnetic field response by controlling molecular orientation in 9-methylanthracene microcrystals. *Angew. Chem. Int. Ed.* 57 (24): 7080–7084.
39 Karunatilaka, C., Bučar, D.-K., Ditzler, L.R. et al. (2011). Softening and hardening of macro- and nano-sized organic cocrystals in a single-crystal transformation. *Angew. Chem. Int. Ed.* 50 (37): 8642–8646.
40 Bleger, D. and Hecht, S. (2015). Visible-light-activated molecular switches. *Angew. Chem. Int. Ed.* 54 (39): 11338–11349.
41 Irie, M., Fukaminato, T., Matsuda, K., and Kobatake, S. (2014). Photochromism of diarylethene molecules and crystals: memories, switches, and actuators. *Chem. Rev.* 114 (24): 12174–12277.
42 Kobatake, S. and Kitagawa, D. (2015). Photoinduced mechanical motion of photochromic crystalline materials. In: *Advances in Organic Crystal Chemistry* (eds. R. Tamura and M. Miyata), 533–547. Springer.
43 Zhu, L., Tong, F., Zaghloul, N. et al. (2016). Characterization of a P-type photomechanical molecular crystal based on the $E \rightarrow Z$ photoisomerization of 9-divinylanthracene malonitrile. *J. Mater. Chem. C* 4 (35): 8245–8252.

44 Towns, A. (2014). Colorant, photochromic. In: *Encyclopedia of Color Science and Technology*, 1–9. Springer.
45 Kitagawa, D. (2014). Photochromic reaction behavior and solid state property changes of diarylethene crystals. PhD Dissertation. Osaka City University.
46 Abendroth, J.M., Bushuyev, O.S., Weiss, P.S., and Barrett, C.J. (2015). Controlling motion at the nanoscale: rise of the molecular machines. *ACS Nano* 9 (8): 7746–7768.
47 García-Amorós, J. and Velasco, D. (2012). Recent advances towards azobenzene-based light-driven real-time information-transmitting materials. *Beilstein J. Org. Chem.* 8: 1003.
48 Zhu, L., Tong, F., Salinas, C. et al. (2014). Improved solid-state photomechanical materials by fluorine substitution of 9-anthracene carboxylic acid. *Chem. Mater.* 26 (20): 6007–6015.
49 Zhu, L., Al-Kaysi, R.O., and Bardeen, C.J. (2011). Reversible photoinduced twisting of molecular crystal microribbons. *J. Am. Chem. Soc.* 133 (32): 12569–12575.
50 Zhu, L., Al-Kaysi, R.O., and Bardeen, C.J. (2016). Photoinduced Ratchet-like rotational motion of branched molecular crystals. *Angew. Chem. Int. Ed.* 55 (25): 7073–7076.
51 Kobatake, S., Takami, S., Muto, H. et al. (2007). Rapid and reversible shape changes of molecular crystals on photoirradiation. *Nature* 446 (7137): 778–781.
52 Terao, F., Morimoto, M., and Irie, M. (2012). Light-driven molecular-crystal actuators: rapid and reversible bending of rodlike mixed crystals of diarylethene derivatives. *Angew. Chem. Int. Ed.* 51 (4): 901–904.
53 Tong, F., Kitagawa, D., Dong, X. et al. (2018). Photomechanical motion of diarylethene molecular crystal nanowires. *Nanoscale* 10 (7): 3393–3398.
54 Morimoto, M. and Irie, M. (2010). A diarylethene cocrystal that converts light into mechanical work. *J. Am. Chem. Soc.* 132 (40): 14172–14178.
55 Koshima, H., Ojima, N., and Uchimoto, H. (2009). Mechanical motion of azobenzene crystals upon photoirradiation. *J. Am. Chem. Soc.* 131 (20): 6890–6891.
56 Bushuyev, O.S., Tomberg, A., Friščić, T., and Barrett, C.J. (2013). Shaping crystals with light: crystal-to-crystal isomerization and photomechanical effect in fluorinated azobenzenes. *J. Am. Chem. Soc.* 135 (34): 12556–12559.
57 Koshima, H. and Ojima, N. (2012). Photomechanical bending of 4-aminoazobenzene crystals. *Dyes Pigm.* 92 (2): 798–801.
58 Bushuyev, O.S., Singleton, T.A., and Barrett, C.J. (2013). Fast, reversible, and general photomechanical motion in single crystals of various azo compounds using visible light. *Adv. Mater.* 25 (12): 1796–1800.
59 Cruickshank, D.W.J. (1956). A detailed refinement of the crystal and molecular structure of anthracene. *Acta Crystallogr.* 9 (11): 915–923.
60 Heller, E. and Schmidt, G.M.J. (1971). Topochemistry. Part XXXIII. The solid-state photochemistry of some anthracene derivatives. *Isr. J. Chem.* 9 (4): 449–462.

61 Zhu, L., Agarwal, A., Lai, J. et al. (2011). Solid-state photochemical and photomechanical properties of molecular crystal nanorods composed of anthracene ester derivatives. *J. Mater. Chem.* 21 (17): 6258.
62 Naumov, P., Sahoo, S.C., Zakharov, B.A., and Boldyreva, E.V. (2013). Dynamic single crystals: kinematic analysis of photoinduced crystal jumping (the photosalient effect). *Angew. Chem. Int. Ed.* 52 (38): 9990–9995.
63 Medishetty, R., Husain, A., Bai, Z. et al. (2014). Single crystals popping under UV light: a photosalient effect triggered by a [2+2] cycloaddition reaction. *Angew. Chem. Int. Ed.* 53 (23): 5907–5911.
64 Medishetty, R., Sahoo, S.C., Mulijanto, C.E. et al. (2015). Photosalient behavior of photoreactive crystals. *Chem. Mater.* 27 (5): 1821–1829.
65 Al-Kaysi, R.O., Dillon, R.J., Zhu, L., and Bardeen, C.J. (2008). Template assisted synthesis of silica-coated molecular crystal nanorods: from hydrophobic to hydrophilic nanorods. *J. Colloid Interface Sci.* 327 (1): 102–107.
66 Abumaree, M.H., Zhu, L., Bardeen, C.J. et al. (2011). Fabrication of biologically active surface-modified Taxol nanowires using anodic aluminum oxide templates. *RSC Adv.* 1 (5): 884.
67 Yang, C., Zhu, L., Kudla, R.A. et al. (2016). Crystal structure of the meta-stable intermediate in the photomechanical, crystal-to-crystal reaction of 9-*tert*-butyl anthracene ester. *CrystEngComm* 18 (38): 7319–7329.
68 Al-Kaysi, R.O., Dillon, R.J., Kaiser, J.M. et al. (2007). Photopolymerization of organic molecular crystal nanorods. *Macromolecules* 40 (25): 9040–9044.
69 Takegoshi, K., Nakamura, S., and Terao, T. (1998). Solid-state photodimerization of 9-methylanthracene as studied by solid-state 13C NMR. *Solid State Nucl. Magn. Reson.* 11 (3–4): 189–196.
70 Kim, T., Zhu, L., Mueller, L.J., and Bardeen, C.J. (2014). Mechanism of photoinduced bending and twisting in crystalline microneedles and microribbons composed of 9-methylanthracene. *J. Am. Chem. Soc.* 136 (18): 6617–6625.
71 Kaupp, G. (2002). Solid-state reactions, dynamics in molecular crystals. *Curr. Opin. Solid State Mater. Sci.* 6: 131–138.
72 Kaupp, G. (2016). Organic solid-state reactions. In: *Encyclopedia of Physical Organic Chemistry*, 1–80.
73 Salzillo, T., Bilotti, I., Della Valle, R.G. et al. (2012). Crystal-to-crystal photoinduced reaction of dinitroanthracene to anthraquinone. *J. Am. Chem. Soc.* 134 (42): 17671–17679.
74 Cohen, M.D., Ludmer, Z., Thomas, J.M., and Williams, J.O. (1971). The role of structural imperfections in the photodimerization of 9-cyanoanthracene. *Proc. R. Soc. Lond. A* 324 (1559): 459–468.
75 Stevens, B., Dickinson, T., and Sharpe, R.R. (1964). Photodimerization in crystalline 9-cyanoanthracene. *Nature* 204 (4961): 876.
76 Salzillo, T., Zaccheroni, S., Della Valle, R.G. et al. (2014). Micro Raman investigation of the photodimerization reaction of 9-cyanoanthracene in the solid state. *J. Phys. Chem. C* 118 (18): 9628–9635.
77 Kim, T., Al-Muhanna, M.K., Al-Suwaidan, S.D. et al. (2013). Photoinduced curling of organic molecular crystal nanowires. *Angew. Chem. Int. Ed.* 52 (27): 6889–6893.

78 Tong, F., Liu, M., Al-Kaysi, R.O., and Bardeen, C.J. (2018). Surfactant-enhanced photoisomerization and photomechanical response in molecular crystal nanowires. *Langmuir* 34 (4): 1627–1634.

79 Al-Kaysi, R.O. (2019). Spontaneous peeling of tetragonal microcrystals with short pulses of UV-light. US Patent 10/179,857, filed 23 October 1980 and issued 15 January 2019.

80 Tong, F., Al-Haidar, M., Zhu, L. et al. (2019). Photoinduced peeling of molecular crystals. *Chem. Commun.* 55: 3709–3712.

81 Zhu, L., Al-Kaysi, R.O., Dillon, R.J. et al. (2011). Crystal structures and photophysical properties of 9-anthracene carboxylic acid derivatives for photomechanical applications. *Cryst. Growth Des.* 11 (11): 4975–4983.

82 Ito, Y. and Fujita, H. (1996). Formation of an unstable photodimer from 9-anthracenecarboxylic acid in the solid. *J. Org. Chem.* 98 (9): 5677–5680.

83 Al-Kaysi, R.O. and Bardeen, C.J. (2007). Reversible photoinduced shape changes of crystalline organic nanorods. *Adv. Mater.* 19 (9): 1276–1280.

84 Abumaree, M.H., Al-Suwaidan, S., and Al-Kaysi, R.O. (2012). Effect of surface-modified paclitaxel nanowires on U937 cells in vitro: a novel drug delivery vehicle. *J. Nanomater.* 2012: 1–9.

85 Good, J.T., Burdett, J.J., and Bardeen, C.J. (2009). Using two-photon excitation to control bending motions in molecular-crystal nanorods. *Small* 5 (24): 2902–2909.

86 Al-Kaysi, R.O. (2016). Alternative synthesis of 1,8-difluoroanthracene via the Balz–Schiemann reaction. *Adv. Chem.* 2016: 1–5.

87 Tong, F., Cruz, C.D., Jezowski, S.R. et al. (2014). Pressure dependence of the forward and backward rates of 9-*tert*-butylanthracene Dewar isomerization. *J. Phys. Chem. A* 118 (28): 5349–5354.

88 Zaziski, D., Prilliman, S., Scher, E.C. et al. (2004). Critical size for fracture during solid–solid phase transformations. *Nano Lett.* 4 (5): 943–946.

89 Kuzmanich, G., Gard, M.N., and Garcia-Garibay, M.A. (2009). Photonic amplification by a singlet-state quantum chain reaction in the photodecarbonylation of crystalline diarylcyclopropenones. *J. Am. Chem. Soc.* 131 (Step 3): 11606–11614.

90 Briseno, A.L., Mannsfeld, S.C.B., Ling, M.M. et al. (2006). Patterning organic single-crystal transistor arrays. *Nature* 444 (7121): 913.

91 Liu, S., Mannsfeld, S.C.B., Wang, W.M. et al. (2008). Patterning of α-sexithiophene single crystals with precisely controlled sizes and shapes. *Chem. Mater.* 21 (1): 15–17.

92 Odom, T.W., Thalladi, V.R., Love, J.C., and Whitesides, G.M. (2002). Generation of 30–50 nm structures using easily fabricated, composite PDMS masks. *J. Am. Chem. Soc.* 124 (41): 12112–12113.

93 Minemawari, H., Yamada, T., Matsui, H. et al. (2011). Inkjet printing of single-crystal films. *Nature* 475 (7356): 364.

94 Li, W., Tierce, N.T., Bekyarova, E., and Bardeen, C.J. (2018). Protection of molecular microcrystals by encapsulation under single-layer graphene. *ACS Omega* 3 (7): 8129–8134.

95 Koshima, H., Matsudomi, M., Uemura, Y. et al. (2013). Light-driven bending of polymer films in which salicylidenephenylethylamine crystals are aligned magnetically. *Chem. Lett.* 42 (12): 1517–1519.
96 Lan, T. and Chen, W. (2013). Hybrid nanoscale organic molecular crystals assembly as a photon-controlled actuator. *Angew. Chem. Int. Ed.* 52 (25): 6496–6500.
97 Sahoo, S.C., Nath, N.K., Zhang, L. et al. (2014). Actuation based on thermo/photosalient effect: a biogenic smart hybrid driven by light and heat. *RSC Adv.* 4 (15): 7640–7647.
98 Tsujioka, T., Sesumi, Y., Takagi, R. et al. (2008). Selective metal deposition on photoswitchable molecular surfaces. *J. Am. Chem. Soc.* 130 (32): 10740–10747.
99 Kitagawa, D. and Kobatake, S. (2015). Photoreversible current ON/OFF switching by the photoinduced bending of gold-coated diarylethene crystals. *Chem. Commun.* 51 (21): 4421–4424.

3

Mechanically Responsive Crystals by Light and Heat

Hideko Koshima[1], *Takuya Taniguchi*[2], *and Toru Asahi*[1,3]

[1] *Waseda University, Research Organization for Nano and Life Innovation, 513 Wasedatsurumaki-cho, Shinjuku, Tokyo 162-0041, Japan*
[2] *Waseda University, Center for Data Science, 27 Waseda-cho, Shinjuku, Tokyo, 162-0042, Japan*
[3] *Waseda University, Department of Life Science and Medical Bioscience, Graduate School of Advanced Science and Engineering, 2-2 Wakamatsu-cho, Shinjuku, Tokyo 162-8480, Japan*

3.1 Introduction

A crystal is the ultimate solid material, in which molecules are arranged three dimensionally. Contrary to their appearance as hard, fragile objects, molecular crystals are attractive as materials mechanically responsive to external stimuli. A number of mechanical crystals have been developed over the past decade. Macroscopic crystal motion is based on solid-state photoreactions and phase transitions, which give rise to molecular motions and thereby deform the bulk crystal. The following provides an overview of the transition of our studies from solid-state organic photoreactions in crystals to photomechanical motion of bulk crystals, i.e. from the microscopic motion of molecules to macroscopic crystal motion.

Solid-state organic photochemistry originated from the first report of topochemical [2+2] photocycloaddition of *trans*-cinnamic acids in crystals, by Cohen and Schmidt in 1964 [1]. Since then, considerable advances in the development of solid-state photoreactions for various applications have occurred [2–4]. Solid-state photoreactions offer the advantage of high product selectivity, due to the low molecular mobility in the crystal lattice. However, the reactions have been limited mostly to unimolecular (intramolecular) reactions in one-component crystals, such as Norrish type II hydrogen abstraction of carbonyl compounds [5, 6]. Over the last three decades, we have developed various bimolecular (intermolecular) photoreactions in two-component molecular crystals (cocrystals) composed of different molecules [7–21], with reaction types including reduction, decarboxylation, cyclization, condensation, and addition reactions. One of the most important advantages of utilizing a cocrystal as a reaction medium is that even if a component molecule has no photoreactivity, new reactivity can be induced, e.g. via photoinduced electron transfer by combining an electron donor and acceptor species in a cocrystal [7, 10, 12–18]. Ultimately, the occurrence of a stereoselective reaction results in successful

Mechanically Responsive Materials for Soft Robotics, First Edition. Edited by Hideko Koshima.

absolute asymmetric synthesis, using a chiral crystal formed spontaneously from achiral molecules [13, 19].

Molecules making up the crystal move microscopically during reactions. This suggests that bulk single crystals should also move during reactions on a macroscopic scale. Thus, over the past two decades, our research has focused on photoreaction-induced macroscopic mechanical motion of bulk single crystals. Our work has shown that the surface morphologies of crystals change as the single-crystal-to-single-crystal (SCSC) photoreaction progresses [20, 22–24]. However, the difficulty lies in resolving how these morphological changes are converted into macroscopic mechanical motion of a bulk single crystal. In 2007, Irie and coworkers reported rapid, reversible bending in rod microcrystals of photochromic diarylethene under photoirradiation, thereby succeeding in linking molecular-level shape changes to macroscale mechanical motion in crystals [25]. Since this report, we have uncovered new photomechanical crystals that exhibit bending motion [26–34]. These crystals are typically made from photoisomerizable (photochromic) molecules, such as azobenzene, salicylideneaniline, fulgide, and benzobarrelene (Figure 3.1), in which the macroscopic motions are based on trans–cis photoisomerization, enol–keto photoisomerization (photo-tautomerization), photo-electrocyclization, and Norrish type II hydrogen abstraction, respectively. Our research showed that autonomous motion of a bulk single crystal could be induced via a reversible photoreaction (photoisomerization). In Section 3.2, photomechanical motion (mainly bending) by photoisomerization is introduced.

The thermal SCSC phase transitions of crystals also cause crystal bending [35], expansion/contraction [36, 37], and jumping [38]. However, with the exception of crawling motion with repeated crystal-to-melt phase changes under light irradiation, the locomotion of crystals has been limited [39]. Despite

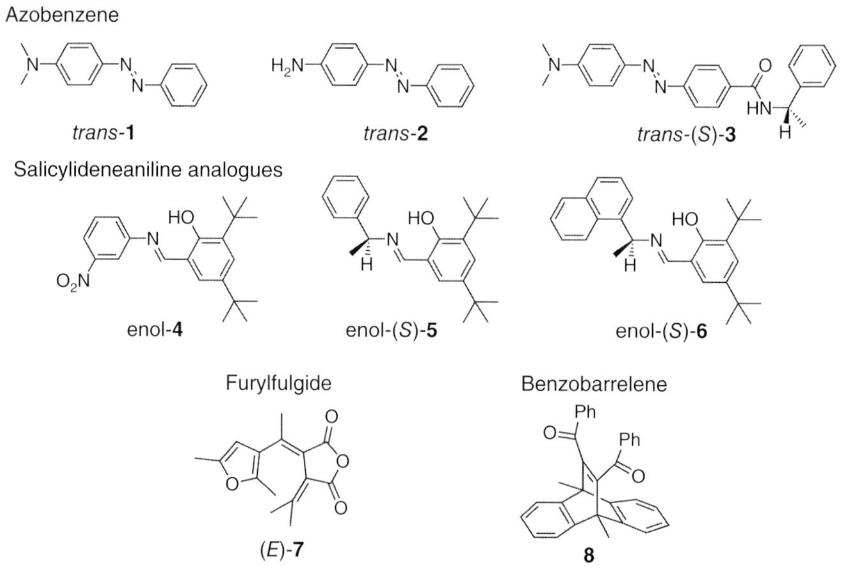

Figure 3.1 Compounds exhibiting mechanical motion in single crystals.

the present limitations, molecular crystals are expected to play important roles as locomotive materials and are being researched extensively in the field of soft robotics [40]. In our research, we have observed inchworm-like "walking" and fast "rolling" locomotion of chiral azobenzene crystals, induced thermally by a reversible SCSC phase transition [41]. In Section 3.3, crystal locomotion is discussed in more detail.

Recently, we discovered a new phase transition in a photochromic chiral salicylideneamine crystal, which we refer to as a "photo-triggered phase transition" [42]. The crystal, which exhibits a thermal SCSC phase transition reversible by heating and cooling, transforms into the identical phase upon light irradiation at temperatures lower than the thermal transition temperature. This photo-triggered phase transition enables thin crystals to bend stepwise via light irradiation, and has the potential to extend the functions of photo-responsive solid materials. In Section 3.4, the manifestation mechanism of the photo-triggered phase transition is examined.

Numerous mechanical crystals have been developed over the past decade and are summarized in several reviews [43–46]. Such mechanically responsive crystals are important for future applications in sensors, switches, actuators, artificial muscles, and soft robotics. However, to date, there have been few studies on organic crystals as potential mechanical materials for the aforementioned applications. Instead, much of the focus has been on polymers. We speculate that preconception of crystals as being hard and fragile has discouraged research in this area; this situation is similar to that of glass in its rigid form, as opposed to fiberglass, which is flexible enough to bend freely. Finally, we discuss why crystals are important as mechanical (actuation) materials, based on their mechanical properties, including Young's modulus.

3.2 Photomechanical Bending of Crystals by Photoreactions

3.2.1 Azobenzene

Azobenzene is a well-known chromophore that undergoes trans–cis photoisomerization. A large number of photo-functional polymers, elastomers, and gels modified with azobenzene have been reported, due to the ease of modification. When *trans*-azobenzene molecules are isomerized to the cis-form upon photoirradiation, a large conformational change accompanies the excited state, leading to a shape change or mechanical motion of polymer materials. Many photomechanical azobenzene-containing elastomers and gels have been developed to date [47–49]. In contrast, it has been thought that trans–cis photoisomerization of azobenzene cannot occur in the crystalline state. Nevertheless, we demonstrated that an azobenzene crystal can be bent via photoisomerization near the surface of the microcrystal [26].

3.2.1.1 Bending

Platelike crystals of *trans*-4-(dimethylamino)azobenzene (*trans*-**1**), with a thickness of several micrometers, were prepared by sublimation in an atmosphere slightly lower than the melting point. Upon irradiation with ultraviolet

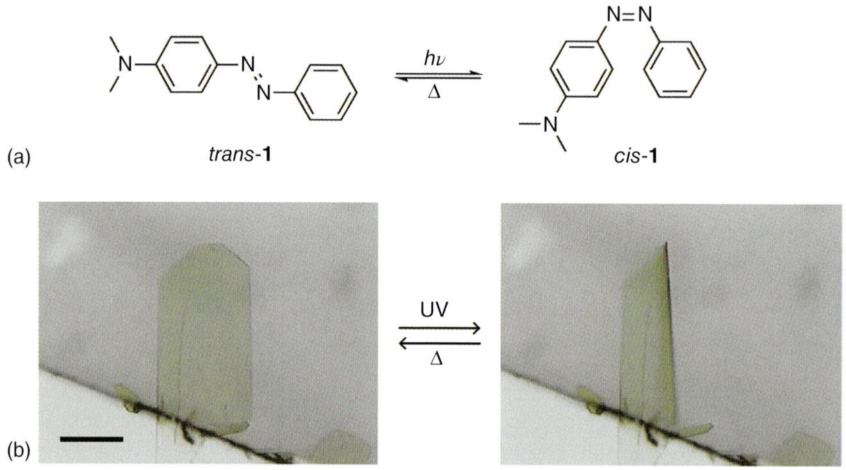

Figure 3.2 (a) Trans–cis photoisomerization of azobenzene *trans*-**1**. (b) Photomechanical bending of a platelike crystal of azobenzene *trans*-**1** upon ultraviolet (UV) light irradiation at 365 nm from the right rear. The scale bar is 200 μm. Source: Koshima et al. 2009 [26]. Reproduced with permission of the American Chemical Society.

(UV) light at 365 nm, a thin platelike crystal (thickness: 5 μm) of *trans*-**1** quickly bent away from the incident light to reach a maximum deflection angle of 180°, forming a semicircle (Figure 3.2) [26]. When the UV illumination was removed, the crystal returned slowly to its initial flat shape after tens of seconds. The mechanical bending was reversible over 100 cycles of UV irradiation.

This novel photomechanical bending response of the azobenzene crystal is especially exciting, having the potential to usher in new fields of study based on photomechanical phenomena in crystals. Previously, it had been thought that the trans–cis photoisomerization of azobenzene could not occur due to the large geometric changes that would be required in such a densely packed crystal lattice. Unambiguous evidence of trans–cis photoisomerization in the crystalline state was obtained from proton nuclear magnetic resonance spectra of *trans*-**1** microcrystals after irradiation, as evident in d_6-benzene peaks of the cis-isomer. However, UV–visible (UV–Vis) diffuse reflectance spectra did not show significant changes after UV light irradiation, most likely due to the very small trans–cis photoisomerization response. After photoisomerization, the *cis*-**1** product formed an amorphous phase, accompanied by a concomitant small change in the roughness of the crystal surface, as determined by atomic force microscopy. In summary, the bending effect is attributable to a gradient in the extent of UV-induced trans–cis photoisomerization as a function of light penetration, causing the expansion of the irradiated crystal surface to result in a bent macrostructure.

An analogous amino-substituted azobenzene (*trans*-**2**) crystal also demonstrates photomechanical bending [27]. The thin platelike crystal quickly bent away from the light upon UV light irradiation. Removing the illumination resulted in the gradual return of the crystal to its initial linear shape after a few minutes, by thermal cis-to-trans back-isomerization. Under visible light irradiation at 530 nm, the crystal recovered more quickly (within one minute) to its initial straight form, due to cis-to-trans back-photoisomerization, revealing that both the bending and straightening of the *trans*-**2** crystals can be controlled photochemically by UV and visible light.

Crystals of azo dyes, such as Methyl Red and Disperse Red 13, in which a strong electron-withdrawing group is combined with a strong donor, bend under visible light irradiation [50]. The crystals and cocrystals of perhalogenated *cis*-azobenzenes bent irreversibly by crystal-to-crystal cis-to-trans photoisomerization into a polycrystalline aggregate of its trans-isomer, and did not return to their initial shape [51–53]. Recently, multidirectional deformation of azobenzene molecular crystals was also reported upon UV irradiation, in the form of bending toward and away from a light source, as well as bidirectional motion [54]. A light-driven crystalline nanowire actuator with an azobenzene derivative was developed by employing a direct writing method. Single nanowires displayed photoinduced reversible bending; the bending behavior followed the first-order kinetics associated with azobenzene photoisomerization [55].

3.2.1.2 Twisted Bending

Most photomechanical crystals exhibit bending motion. In contrast, crystals showing twisted motion are limited to anthracene [56, 57] and diarylethene [58, 59]. In this case, chiral crystals are expected to be especially receptive to induced twisting motion, as they typically maintain a chiral form during photochemical reactions, allowing an enantioselective reaction to be achieved. The photomechanical motion of chiral crystals of *trans*-azobenzene derivative [*trans*-(*S*)-**3**] with an (*S*)-phenylethylamide group has been compared with the racemic crystal [28].

UV–Vis absorption spectra of powdered chiral crystals upon UV irradiation measured with an optical waveguide spectrometer changed due to trans-to-cis photoisomerization (Figure 3.3a, b). Changes in the spectra after stopping UV light revealed the half-life (12.4 seconds) of cis-to-trans thermal back-isomerization of the chiral crystal (Figure 3.3c), which was slightly faster than that of racemic crystals (18.7 seconds). Under UV irradiation, the long platelike chiral microcrystal bent away from the light source with a twisting motion (Figure 3.3d), compared with only slight bending of the platelike racemic microcrystal. The possible mechanism of bending and twisting motion was discussed based on the optimized cis-conformer from calculations, as crystallographic analysis after UV irradiation was not possible (Figure 3.3e). The cis-conformer has torsional conformation due to the repulsion of two phenyl rings. Thus, the bending motion with twisting is caused by elongation along the *b*-axis and shrinkage along the *a*-axis (Figure 3.3f).

3.2.2 Salicylideneaniline and Analogues

Salicylideneaniline and salicylideneamine (Schiff bases) are typical photochromic compounds that exhibit enol-to-keto photoisomerization in the crystalline state [60]. One of the features of these compounds is that the keto-to-enol back-isomerization of the many derivatives undergoes photochemical, as well as

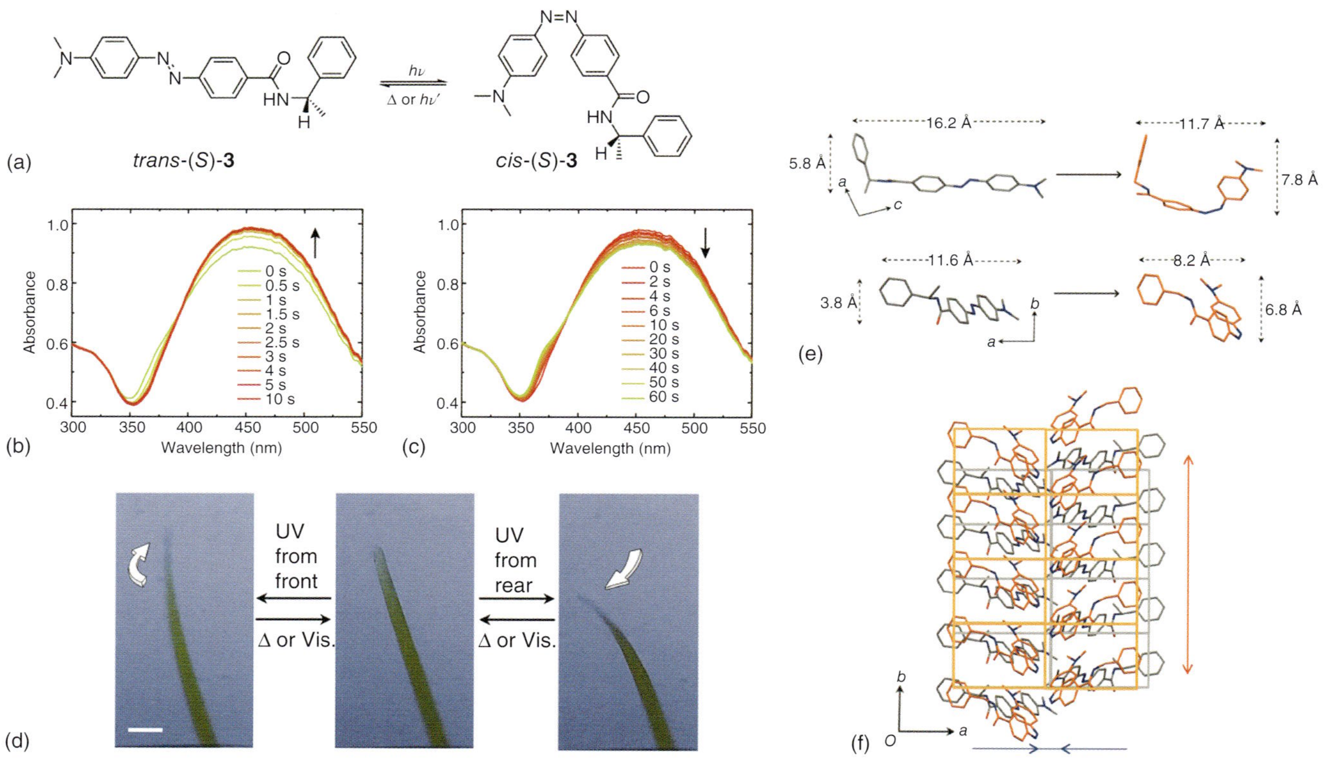

Figure 3.3 (a) Trans–cis photoisomerization of chiral azobenzene *trans*-(*S*)-**3**. Ultraviolet–visible (UV–Vis) absorption spectra (b) under ultraviolet (UV) light irradiation, and (c) after stopping UV light. (d) Photomechanical bending with twisting of a platelike crystal of *trans*-(*S*)-**3** upon UV light irradiation. The scale bar is 100 μm. (e) Molecular conformations of *trans*-(*S*)-**3** (gray) and optimized *cis*-(*S*)-**3** (orange) on the (010) and $(00\bar{1})$ faces. (f) Overlaid molecular arrangements on the (001) face of *trans*-(*S*)-**3** (gray) and *cis*-(*S*)-**3** (orange). Source: Taniguchi et al. 2016 [28]. Reproduced with the permission of John Wiley & Sons.

thermal changes. A wide variety of derivatives of salicylideneaniline are easily prepared as crystals. Thus, the Schiff bases are considered to be promising photomechanical crystal candidates.

3.2.2.1 Bending and the Mechanism

A thin (1 μm) platelike crystal of *N*-3,5-di-*tert*-butylsalicylidene-3-nitroaniline in enol form (enol-**4**) bent away from UV light (365 nm) irradiation (Figure 3.4b) [29]. The bending motion was accompanied by a color change from pale yellow to reddish-orange, due to the formation of *trans*-keto-**4** by photoisomerization (Figure 3.4a). Subsequent illumination with visible light (530 nm) returned the bent shape to its initial straight form by keto-to-enol back-photoisomerization, with fading of the reddish-orange color to the initial pale yellow. The bend angle depends on the intensity and UV irradiation time. The recovery to a straight form was slow (1400 seconds) when UV illumination was removed, due to the long thermal *trans*-keto-to-enol back-isomerization (half-life: 42 days) [61]. However, the recovery time was dramatically shortened to 43 seconds by back-photoisomerization with visible light irradiation. This minimal movement allows for excellent repeatability of the reversible bending, over as many as 200 cycles in these studies (Figure 3.4c).

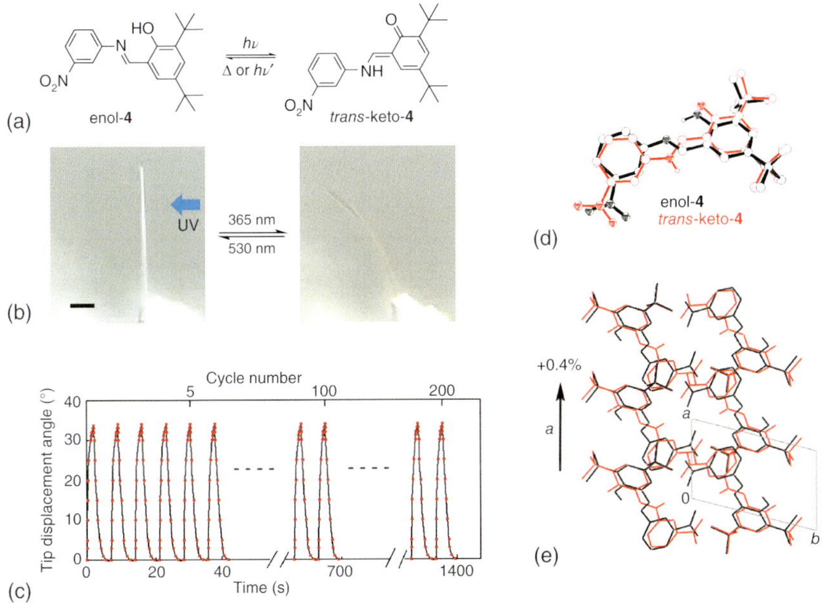

Figure 3.4 (a) Enol–keto photoisomerization of salicylideneaniline enol-**4**. (b) Photomechanical bending of a platelike crystal of enol-**4** upon UV light irradiation from the right. The scale bar is 10 μm. (c) The repeatability of the reversible bending can be observed over as many as 200 cycles. (d) ORTEP drawings show the disordered structures of enol-**4** (black) and *trans*-keto-**4** (red) isomers after UV light irradiation at the 25% probability level. (e) Molecular arrangements on the (001) face. Source: Koshima et al. 2011 [29]. Reproduced with permission of the Royal Society of Chemistry.

The manifestation mechanism of photomechanical bending can be elucidated by crystallographic analysis of a thin enol-**4** microcrystal before and after UV light irradiation. After UV irradiation, the crystal exhibited a disordered structure of the enol and the *trans*-keto isomers, at a ratio of 0.9 : 0.1 (Figure 3.4d). The length of the *a*-axis in the unit cell increased by 0.42% (Figure 3.4e). Hence, photoisomerization elongated the crystal length along the *a*-axis near the (001) crystal surface. In contrast, given that no photoisomerization occurred in the absence of light, the unit cell dimensions remained constant at the non-illuminated surface, resulting in the bending of the microcrystal away from the light source. When UV or visible (530 nm) irradiation was stopped, reverse isomerization from *trans*-keto-**4** to enol-**4** brought the crystal back to its initial straight shape.

3.2.2.2 Comparison of Chiral and Racemic Crystals

Solid-state photoreactivity in chiral crystals usually differs from that in achiral crystals, because the motion of molecules is restricted in the crystal lattice and the chiral environment is retained during reactions [62]. This suggests that photomechanical motion should be different between chiral and achiral crystals. The photomechanical bending behavior of chiral crystals composed of *S*-enantiomer of photochromic *N*-3,5-di-*tert*-butylsalicylidene-1-phenylethylamine in enol form [enol-(*S*)-**5**] has been compared with that of both achiral crystals of the racemic compound [enol-(*rac*)-**5**] of *S*- and *R*-enantiomers (Figure 3.5) [30, 31]. Both platelike chiral and achiral crystals exhibited similar reversible bending upon alternate irradiation with UV and visible light (Figure 3.5b, e). The reversible bending of the achiral enol-(*rac*)-**5** crystal was repeated over 100 times, while the tip displacement angles of the chiral enol-(*S*)-**5** crystal gradually decreased with repeated cycles (Figure 3.5c, f). In situ X-ray measurements revealed that the bending motion of the chiral and achiral crystals was caused by a slight shrinkage of the long axis of the platelike crystals, where the shrinkage was induced by the shape change of salicylidenephenylethylamine molecules upon photoinduced proton transfer.

Measurement and evaluation of mechanical properties are important for crystalline materials. Young's modulus of the enol-(*S*)-**5** crystal is 0.8 GPa, which is smaller than that of the enol-(*rac*)-**5** crystal (2.6 GPa). We measured also the photomechanical work of platelike crystals while lifting a metal weight under UV irradiation (Figure 3.5g). The crystal cantilevers, made of enol-(*S*)-**5** and enol-(*rac*)-**5**, lifted metal rings that were up to 50 and 300 times heavier, respectively, than the cantilevers upon UV irradiation (Figure 3.5h). The lifting work (614 nJ mg^{-1} crystal) per milligram of the achiral enol-(*rac*)-**1** crystal was twofold larger than that of the chiral enol-(*S*)-**1** crystal (343 nJ mg^{-1} crystal). Overall, the achiral enol-(*rac*)-**5** crystal was found superior to the chiral enol-(*S*)-**5** crystal as a light-driven actuator.

3.2.3 Fulgide

Fulgides constitute a class of photochromic compounds that undergo reversible electrocyclic ring-closure and ring-opening reactions [63]. The furylfulgide

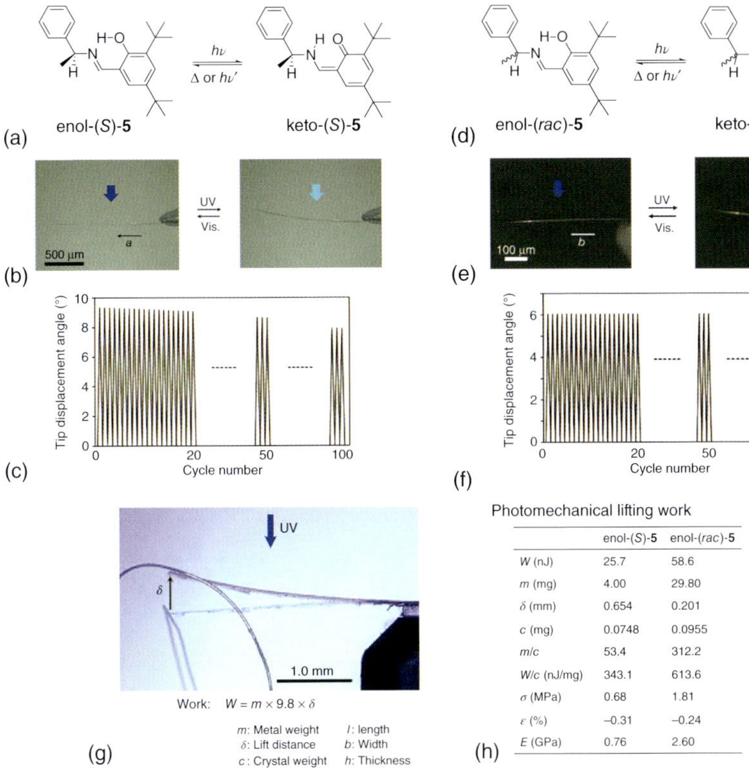

	enol-(*S*)-**5**	enol-(*rac*)-**5**
W (nJ)	25.7	58.6
m (mg)	4.00	29.80
δ (mm)	0.654	0.201
c (mg)	0.0748	0.0955
m/*c*	53.4	312.2
W/*c* (nJ/mg)	343.1	613.6
σ (MPa)	0.68	1.81
ε (%)	−0.31	−0.24
E (GPa)	0.76	2.60

Figure 3.5 Comparison of bending and mechanical properties of chiral and racemic salicylideneamine crystals of enol-(*S*)-**5** and enol-(*rac*)-**5**. Enol–keto photoisomerization of (a) enol-(*S*)-**5** and (d) enol-(*rac*)-**5**. Bending of thin platelike crystals of (b) enol-(*S*)-**1** and (e) enol-(*rac*)-**5** upon UV light irradiation from the upper side. Repeatability of the reversible bending of (c) enol-(*S*)-**1** and (f) enol-(*rac*)-**5** crystals over as many as 100 cycles. (g) Photomechanical lifting work of the crystal cantilever of enol-(*S*)-**1**. (h) Maximum strain and stress of enol-(*S*)-**1** and enol-(*rac*)-**1** crystals. Source: Koshima et al. 2013 [30]. Reproduced with permission of the American Chemical Society.

(*E*)-2-[1-(2,5-dimethyl-3-furyl)ethylidene]-3-isopropylidenesuccinic anhydride in *E*-form [(*E*)-**7**] exhibits photochromism, resulting in a closed isomer [(*C*)-**7**] in the crystalline state (Figure 3.6a) [64–66].

A thin hexagonal crystal (thickness: 2 μm) was irradiated from the diagonal underside with UV light at 365 nm for one second. The crystal curled from the right upper corner toward the light, reaching a maximum twisted curl after two seconds, with a color change from pale yellow to red due to the formation of the closed (*C*)-**7** isomer (Figure 3.6b) [32]. On the other hand, a narrow plate-like microcrystal (width: 6 μm) bent toward the light upon UV light irradiation, reaching a maximum tip displacement angle of 9° after two seconds (Figure 3.6c). Subsequent illumination with visible light at >390 nm returned the crystal to its initial straight shape after 30 seconds. This reversible bending was observed over 200 cycles of alternating irradiation with UV (2 seconds) and visible light (30 seconds).

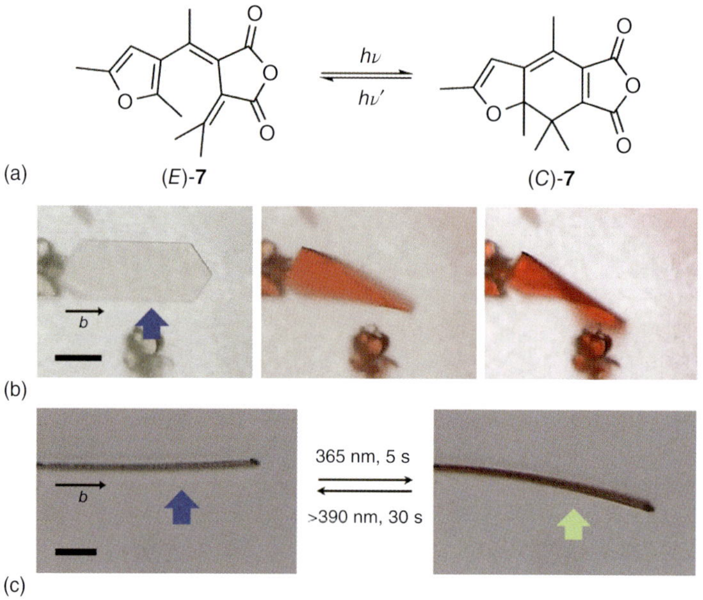

Figure 3.6 (a) Photo-electrocyclization of furylfulgide (*E*)-**7**. (b) Curling of a thin hexagonal crystal upon UV light irradiation. The scale bar is 50 μm. (c) Reversible bending of a narrow, platelike crystal upon alternate irradiation with UV and visible light. The scale bar is 20 μm. Source: Koshima et al. 2011 [32]. Reproduced with permission of the Chemical Society of Japan.

Crystallographic analysis of (*E*)-7 after two-photon excitation at 742 nm by pulsed laser light showed 0.13% shrinkage of the *b*-axis length of the unit cell, providing direct evidence for the photomechanical bending mechanism of the platelike crystal toward the light source [67]. There exists a disordered structure with contributions from both the (*E*)-7 and the (*C*)-7 isomers at a ratio of 0.945 : 0.055. The (*E*)-7 molecules have torsional conformation between the succinic anhydride portion and the furan ring, and are arranged at the (101) face as a twofold screw along the *b*-axis [64]. In contrast, the (*C*)-7 isomer is nearly planar [66]. Therefore, the ring-closure photoisomerization from the torsional (*E*)-7 isomer to the nearly planar (*C*)-7 isomer shrinks the crystal length near the irradiated surface. In contrast, at the back surface, the crystal length remains the same, as photoisomerization does not occur in the absence of light. Thus, the unit cell dimensions remain constant for the non-illuminated surface, causing the crystal to bend toward the light source.

3.2.4 Carbonyl Compounds

Norrish type II hydrogen abstraction of carbonyl compounds is one of typical photoreactions. It is known that isopropylbenzophenone derivatives undergo Norrish type II hydrogen abstraction, followed by photocyclization, to give the cyclobutenols in the crystals upon UV irradiation [5]. Highly enantioselective photocyclization often occurs in the salt crystals of carboxylic acids with chiral and achiral amines [8, 11, 18–24]. Most of the crystals react without cracking or breaking, due to small changes in the crystal structure resulting from SCSC transformation. The surface morphologies of the single crystals become rough as photocyclization proceeds; after completion of the reaction, the crystals return to their initial smooth surfaces [22, 23]. The surface morphology changes are induced by changes in the molecular structures from the reactant molecule to the product molecule, resulting in molecular motion at the crystal surface.

Dibenzobarrelene derivatives are known to cause various photochemical reactions in the crystalline state, such as tri-π-methane rearrangement, di-π-methane rearrangement, Norrish type II reactions, and [2+2] cycloaddition [68, 69]. Dibenzobarrelene, such as 11,12-dibenzoyl-9,10-dihydro-9,10-dimethyl-9,10-ethenoanthracene (**8**), exhibits photochromism in the crystal, which is caused by a long-lived triplet biradical (**9**), formed from a Norrish type II intramolecular hydrogen abstraction initiated by UV light irradiation (Figure 3.7a) [70].

When a platelike crystal was irradiated with UV light at 365 nm, two-step bending was observed. In the first step, the crystal quickly bent away from the light source, with an accompanying crystal color change from colorless to purple derived from the triplet biradical (Figure 3.7b) [34]. The bending returned gradually and partially after stopping UV irradiation, probably due to the rearrangement and molecular packing of photoproducts. Upon a third dose of UV exposure, the crystal bent and then broke into two pieces in the next moment.

In the second step, under prolonged UV light exposure, the bending returned slowly and then the crystal bent up toward the opposite direction, accompanied by an additional color change to light yellow (Figure 3.7c). Spectroscopic measurements and X-ray crystallographic analysis suggested that a long-lived biradical species is generated immediately upon UV light irradiation via Norrish type II intramolecular hydrogen abstraction, and then the final photoproducts form under continuous UV exposure. X-ray crystallographic analysis before and after UV light irradiation for a few seconds revealed that the longitudinal axis (*a*-axis) of the crystal became elongated slightly after irradiation, which is consistent with the direction of first-step bending. Based on these results, we propose that first-step bending may be induced by a biradical species, generated via a Norrish type II intramolecular hydrogen abstraction, whereas second-step bending is believed to originate from the formation of a mixture of final photoproducts under prolonged light irradiation.

3.3 Locomotion of Crystals by Thermal Phase Transition

Over the course of research on chiral azobenzene crystals of *trans*-(*S*)-**3**, we unexpectedly found that the *trans*-(*S*)-**3** crystal exhibited a reversible SCSC phase transition at 145 °C (Figure 3.8a) [41]. X-ray crystallographic analysis revealed that the lattice constants changed discontinuously due to the phase transition from the low-temperature (LT) to the high-temperature (HT) phase. The length of the *b*-axis shortened by 0.30% due to the LT-to-HT phase transition. Crystal structures at both the LT and HT phases were disordered due to a pedal motion

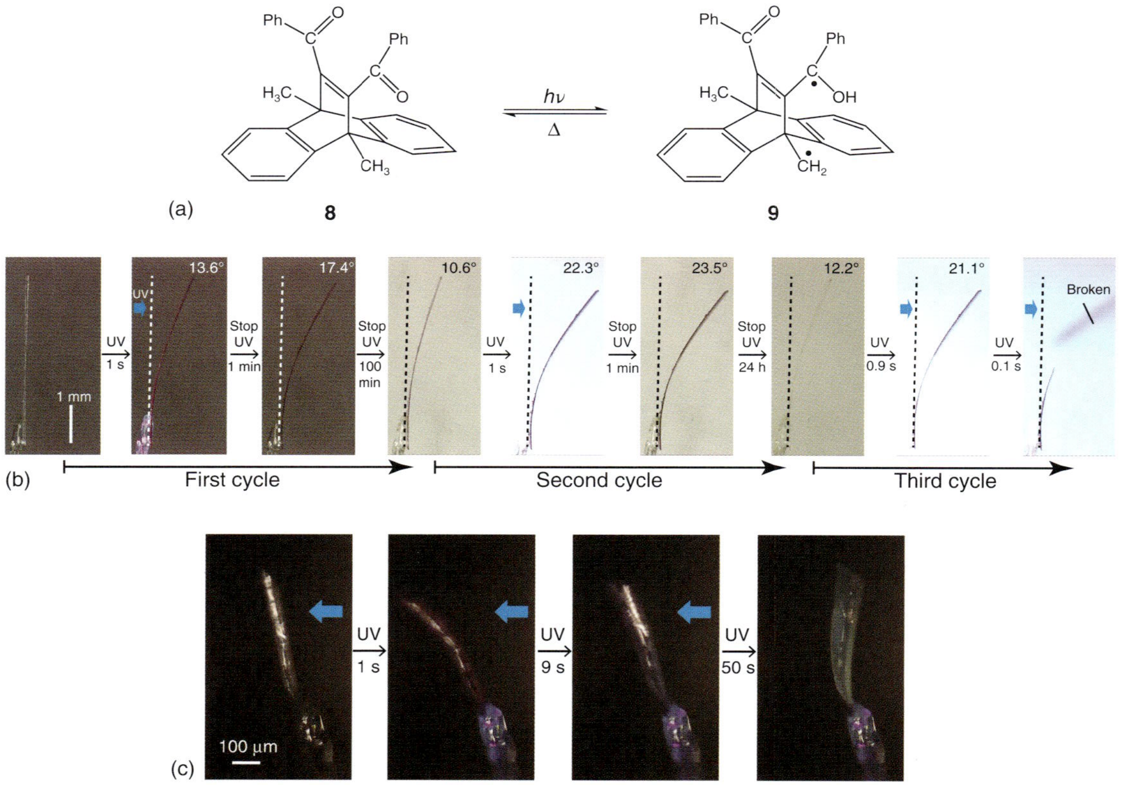

Figure 3.7 (a) Photochromism of dibenzobarrelene **8**. (b) First step of photomechanical bending of a platelike crystal upon UV light irradiation from the left for one second. The angle at the upper right corner in each panel shows the tip displacement angle. Dotted lines indicate the initial position at the tip displacement angle (0°). (c) Two-step bending of a platelike crystal under continuous UV light exposure for one minute. Source: Taniguchi et al. 2018 [34]. Reproduced with permission of the Royal Society of Chemistry.

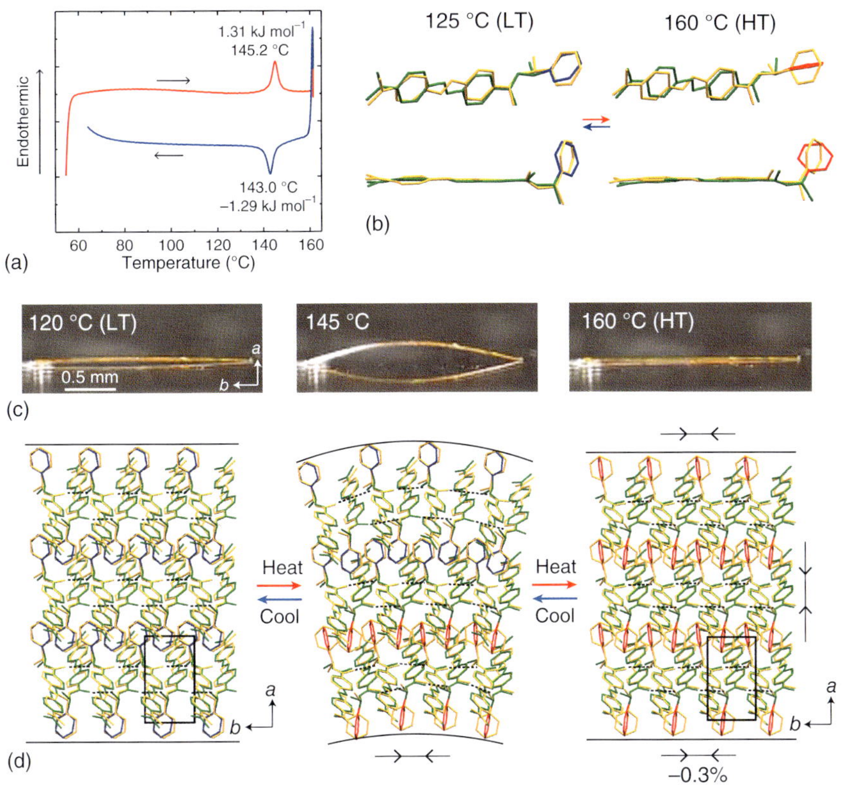

Figure 3.8 (a) Differential scanning calorimetry (DSC) curve of the chiral azobenzene *trans*-(*S*)-**3** measured between 60 and 160 °C upon heating and then cooling. (b) Molecular conformation at 125 and 160 °C. (c) Bending motion due to the phase transition. (d) Molecular packing on (001) face. Black dotted lines represent NH—O=C hydrogen bonds along the *b*-axis. Hydrogen atoms are omitted for clarity. Source: Taniguchi et al. 2018 [41]. Reproduced with permission of Springer Nature. https://creativecommons.org/licenses/by/4.0/.

of the azobenzene moiety (Figure 3.8b). Here, the dihedral angle can be defined by the angle between the azobenzene plane and the phenyl ring of the side chain. The dihedral of one conformer changed by 25° due to the phase transition from LT to HT, by rotating the phenyl ring. The small conformational change allows surrounding molecules to condense along the *b*-axis, which is the longitudinal direction of the crystal.

When a thin platelike crystal was heated with one edge fixed, the crystal bent along the *b*-axis, and then became straight with slight shortening of the length (Figure 3.8c). On cooling, the crystal bent again and then returned to its initial shape. This bending motion originates from a temperature gradient in the thickness direction. On heating, the phase transition from the LT to HT phase was initiated from the lower surface of the crystal. Thus, the lower part of the crystal shrank along the *b*-axis, inducing the bending motion (Figure 3.8d). Then, the crystal straightened when the phase transition to the HT phase proceeded through the entire crystal. On subsequent cooling, the reverse phase transition

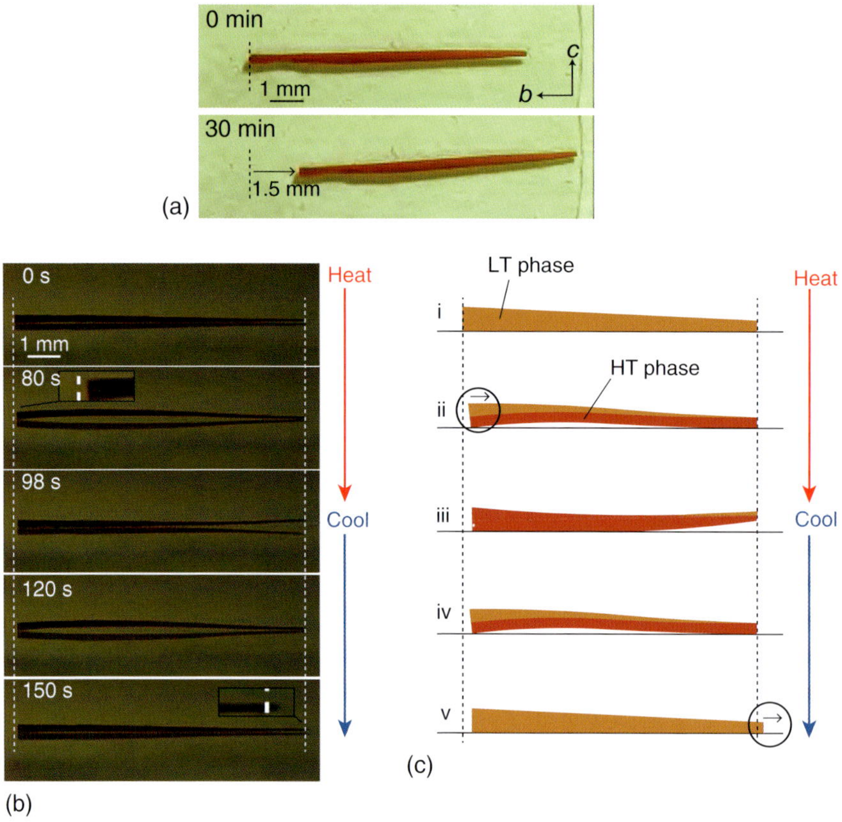

Figure 3.9 (a) Inchworm-like walking of the crystal *trans*-(*S*)-**3**. (b) Sideview of the inchworm-like walking in a heat and cool cycle. Dotted lines represent the initial position of each edge. (c) Possible mechanism of the inchworm-like walking. Source: Taniguchi et al. 2018 [41]. Reproduced with permission of Springer Nature. https://creativecommons.org/licenses/by/4.0/.

from the HT to LT phase began at the upper surface, due to cooling by the surrounding air, and the crystal bent due to elongation of the length. The crystal returned to its initial length as the entire crystal became the LT phase.

3.3.1 Inchworm-Like Walking

When a long platelike crystal with a thickness gradient in the longitudinal direction was repeatedly heated and cooled between 139 and 154 °C (heating rate: 0.6 $°C\,s^{-1}$; cooling rate: 1.0 $°C\,s^{-1}$) on silanized glass, surprisingly, the crystal walked slowly to the right along the *b*-axis, like an inchworm (Figure 3.9a). The walking distance reached 1.5 mm after 30 minutes (walking speed: 3.0 $mm\,h^{-1}$). The average stroke on heating and cooling was 34 μm, that is 0.4% of the crystal length, which is consistent with the length change of the *b*-axis at the phase transition.

When the crystal was viewed from the side, the thickness gradient was clearly visible; the left edge was thicker, and the right edge was thinner (Figure 3.9b).

Upon heating between 132 and 150 °C at a rate of 0.2 °C s^{-1}, the crystal remained straight until the substrate temperature reached 140 °C (0 second). Then, the crystal started to bend gradually, and the thicker left edge moved slightly (47 μm) to the right, as shown in the enlarged picture (at 80 seconds). The bending ended suddenly at 98 seconds, and simultaneously the right edge sprang up (at 98 seconds), retaining its lifted shape until reaching 150 °C.

Subsequent cooling between 150 and 132 °C at a speed of 0.4 °C s^{-1} completed the lifting-up behavior of the right edge at 118 seconds, and then crystal bending began again (at 120 seconds). Crystal bending decreased gradually with the temperature. The crystal finally returned to its initial straight shape with movement (42 μm) of the right edge to the right after cooling (at 150 seconds). A schematic of a possible mechanism for the inchworm-like behavior is given in Figure 3.9c. On heating, a platelike crystal (i) with a thickness gradient bends due to the phase transition from the LT to the HT phase from the lower part (ii). The larger bending in the thicker left part leads to movement to the right. When the phase transition to the HT phase proceeds throughout the entire crystal, the thicker left part becomes suddenly straight and, simultaneously, the thinner right edge is lifted up like a spring, probably due to relaxation of the strain generated by the abrupt end of bending (iii). On subsequent cooling, the thicker part starts to bend again due to the phase transition from the HT to the LT phase, with cooling by the surrounding air (iv). Once the phase transition to the LT phase is complete, the thinner right edge moves largely to the right, by slipping on a silanized glass surface (v), as the thinner right part is lighter with less friction than the thicker left part. Thus, the unsymmetrical crystal shape with a thickness gradient is the driving force of inchworm-like walking.

3.3.2 Fast Rolling Locomotion

Surprisingly, some crystals rolled very fast under one process, i.e. heating or cooling. When a long, thin platelike crystal with a width gradient was heated on silanized glass from 134 to 147 °C (heating rate: 1 °C s^{-1}), the crystal began to bend with fluctuations at ~137 °C. Then, the crystal rolled with repeating flips at 141 °C, and finally stopped at ~142 °C (Figure 3.10a). The locomotion distance reached 3.7 mm in only 0.8 seconds (speed: 4.6 mm s^{-1}). The rolling locomotion was much faster than inchworm-like walking. Rolling was also observed during the cooling process.

This rolling locomotion is due to an unbalanced shape during bending motion (Figure 3.10b). When a long, thin platelike crystal of the LT phase is heated, the crystal bends with some tilt, due to the phase transition to the HT phase. The tilting occurs autonomously via an unsymmetrical crystal shape with a width gradient, in the direction in which the angle of the crystal corner becomes obtuse, leading to a loss of balance and the first flip. The incline/lowering of the center of gravity induces the driving force of flipping. During the first flip, the whole crystal returns to the LT phase due to cooling by the surrounding air. Then, the crystal turns inside out, is again heated, and bends with tilting. This time, the bending occurs due to the phase transition to the HT phase, and is in a direction in which the angle of the crystal corner is acute (probably due to the remaining

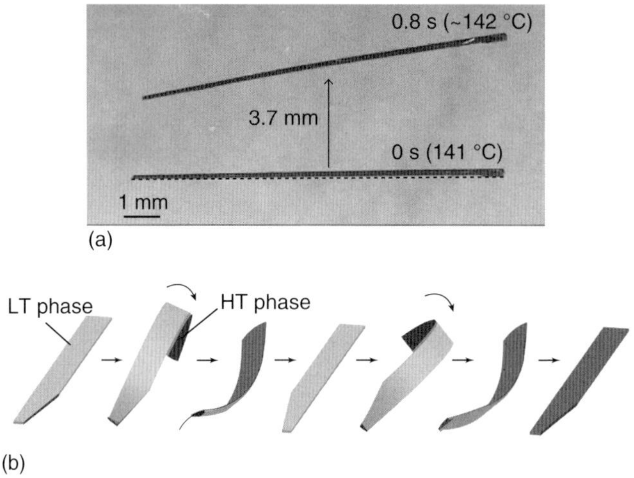

Figure 3.10 (a) Fast rolling locomotion of the crystal *trans*-(*S*)-**3**. (b) Possible mechanism of the rolling due to an unsymmetrical shape. Source: Taniguchi et al. 2018 [41]. Reproduced with permission of Springer Nature. https://creativecommons.org/licenses/by/4.0/.

momentum), leading to the second flip. When the entire crystal reaches the HT phase with further heating, the locomotion stops. Under cooling, the same explanation can be applied to the rolling locomotion. Thus, the driving force of fast rolling is generated from the unsymmetrical shape of the crystal.

3.4 Diversification of Mechanical Motion by Photo-triggered Phase Transition

3.4.1 Discovery and the Mechanism of Photo-triggered Phase Transition

The discovery of locomotion with chiral azobenzene *trans*-(*S*)-**3** crystals prompted the question: How does the mechanical motion change if both photoisomerization and thermal phase transitions occur in one crystal? To this end, we explored several photochromic compounds and found that photochromic chiral salicylideneamine [enol-(*S*)-**6**] (Figure 3.11a) exhibits two thermal phase transitions near −80 and 40 °C due to crystal phase changes $\alpha \leftrightarrow \beta$ and $\beta \leftrightarrow \gamma$, respectively (yellow circles in Figure 3.11b) [42]. As a representative, the α angle changed from 94° to 84° due to the phase transition $\alpha \rightarrow \beta$, and then became 90° by the thermal phase transition $\beta \rightarrow \gamma$ due to the crystal system change from triclinic to monoclinic.

Surprisingly, the unit cell constants of the β phase measured under UV light converted to that of the γ phase at temperatures of −50, −20, 0, and 20 °C (orange circles in Figure 3.11b). In contrast, the α angle measured at −120 °C (α phase) and 60 °C (γ phase) did not change under UV irradiation. This new phase transition $\beta \rightarrow \gamma$ by UV irradiation is referred to as the photo-triggered phase transition (Figure 3.11c).

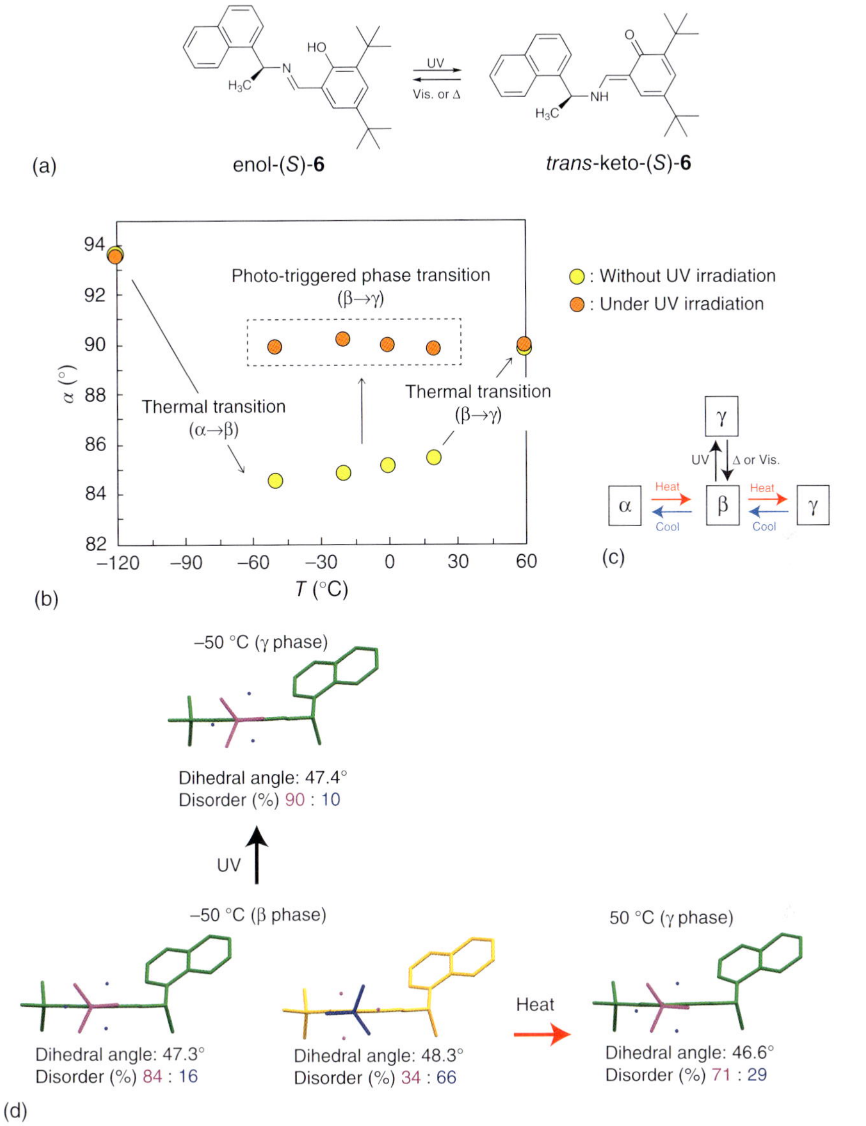

Figure 3.11 (a) Enol–keto photoisomerization of chiral salicylideneamine enol-(*S*)-**6**. (b) Photo-triggered phase transition of the enol-(*S*)-**6** based on the α angle change. (c) Diagram of the phase change due to temperature change and photoirradiation. (d) Crystal structure change by the thermal phase transition and the photo-triggered phase transition. Two independent molecules (1 and 2) at β phase are drawn in green and yellow. Disordered *tert*-butyl substituents are drawn in magenta and blue; the solid lines correspond to major orientation, and dots to minor orientation of the disorder. Hydrogen atoms are omitted for clarity. Source: Taniguchi et al. 2019 [42]. Reproduced with permission of Springer Nature. https://creativecommons.org/licenses/by/4.0/.

For elucidation of structural changes due to the photo-triggered phase transition, the crystal structures were determined before and under UV irradiation at −50 and 50 °C (Figure 3.11d). In the β phase (−50 °C) without UV light, two independent molecules (green molecule 1 and yellow molecule 2) exist in the lattice. The *tert*-butyl substituent in each molecule is disordered between the two orientations. At 50 °C (γ phase) without UV light, one independent molecule exists in a lattice, and the molecular conformation is more similar to molecule 1 than molecule 2 in the β phase. At −50 °C under UV light, the crystal phase became γ phase identically due to the photo-triggered phase transition. Here, photoisomerization underwent in the crystal but the *trans*-keto molecule was not found, probably due to insufficient quantities for structural determination.

The photo-triggered γ phase is metastable and returns to the stable β phase after stopping UV light due to thermal back-isomerization from *trans*-keto to enol form. The reverse phase transition γ → β proceeds faster under visible light because half-life of back-photoisomerization is much shorter than that of thermal back-isomerization.

Thus, the production of *trans*-keto molecules is essential for the photo-triggered phase transition. Namely, the photo-triggered phase transition should originate from the strain of *trans*-keto molecules generated by enol–keto photoisomerization. Before photoirradiation, the crystal is in the β phase (Figure 3.12a). Upon UV irradiation, *trans*-keto molecules are produced due to photoisomerization at the irradiated surface, generating strain in the crystal (Figure 3.12b). The strain due to *trans*-keto molecules induces conformational changes of the surrounding molecules, which further influence the adjacent surrounding molecules like a domino effect, triggering crystal structure change to the γ phase (Figure 3.12c). When *trans*-keto molecules are produced at a certain low ratio, the whole crystal completes the phase transition β → γ. When the light irradiation is stopped, thermal back-isomerization gradually occurs in the crystal, triggering the reverse phase transition γ → β.

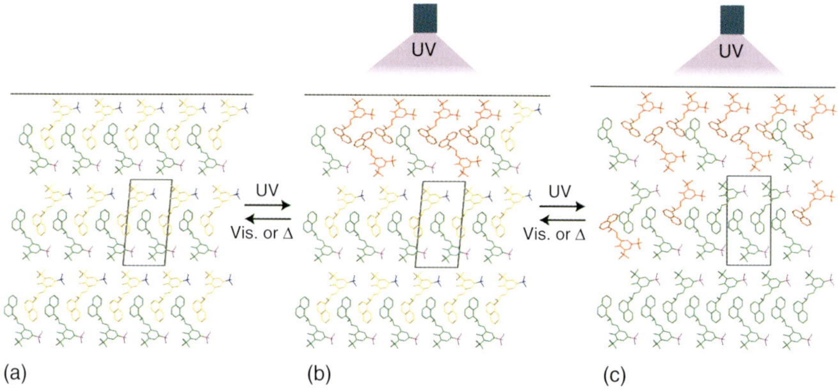

Figure 3.12 Mechanism of photo-triggered phase transition of the crystal enol-(*S*)-**6**. (a) β phase before irradiation. (b) Photoisomerization at the surface of β phase crystal. (c) Photo-triggered phase transition β → γ due to strain of *trans*-keto molecules. *Trans*-keto molecules are drawn in red. Source: Reproduced with permission of Taniguchi et al. 2019 [42], licensed under Creative Commons Attribution 4.0 International License.

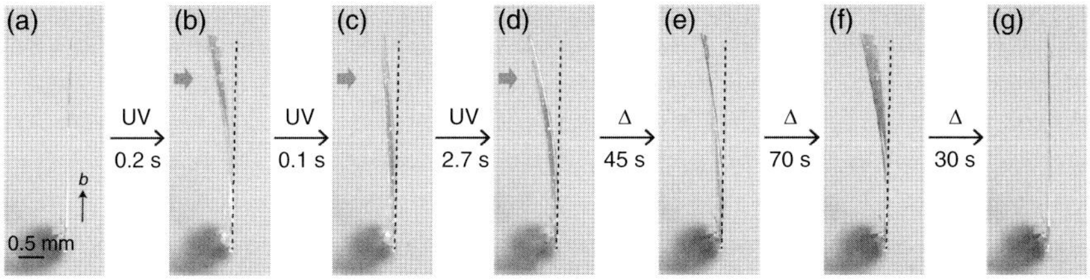

Figure 3.13 (a) Initial shape before irradiation of the crystal enol-(*S*)-**6**. (b) Bending with twist due to photoisomerization at β phase. (c) Sudden disappearance of the twist due to photo-triggered phase transition β → γ. (d) Bending without twist due to photoisomerization at γ phase. (e) Bending relaxation due to back-isomerization at γ phase. (f) Gradual appearance of twist due to phase transition γ → β. (g) Return to initial shape due to back-isomerization at β phase. Source: Reproduced with permission of Taniguchi et al. 2019 [42], licensed under Creative Commons Attribution 4.0 International License.

3.4.2 Stepwise Bending

From the mechanical function perspective, the photo-triggered phase transition leads to stepwise bending of the crystal enol-(*S*)-**6** (Figure 3.13). When the (001) face of a thin platelike crystal in the β phase was irradiated with UV light (365 nm, 60 mW cm^{-2}) at room temperature, the crystal bent toward the light source with a twisting motion in 0.2 seconds (a, b). In the next moment, the twisted shape suddenly disappeared (c). Then, the crystal continued to bend without twisting under prolonged UV irradiation (d). After stopping the UV light, the bending gradually returned without twisting, within 45 seconds (e). The bent crystal started to twist slowly, and then exhibited the most twisted shape at 70 seconds (f). The bending with a twist gradually relaxed, and then the crystal returned to its initial straight shape (g).

The bending behavior during the photo-process can be divided into three steps: (i) bending with twisting due to enol–keto photoisomerization at the β phase, (ii) disappearance of the twist due to the photo-triggered phase transition β → γ, and (iii) bending without twisting due to enol–keto photoisomerization in the γ phase. Relaxation process can be explained by the reverse process. Initially, bending relaxes due to thermal back-isomerization in the γ phase. Then, twisted shape gradually appears due to the progression of the phase transition γ → β triggered by thermal back-isomerization. The initial unbent shape is then restored due to thermal back-isomerization in the β phase.

The creation of crystal motion by light has thus far been possible only by photoisomerization; however, motion via photo-triggered phase transitions are also possible. Thus, the photo-triggered phase transition leads to a new strategy to diversify the mechanical motion of photo-responsive crystals.

3.5 Why Crystals?

Finally, we discuss why crystals are important as mechanically responsive materials. As mechanical properties of materials, Young's modulus and strain are fundamental parameters used to correlate the motion and actuation force of

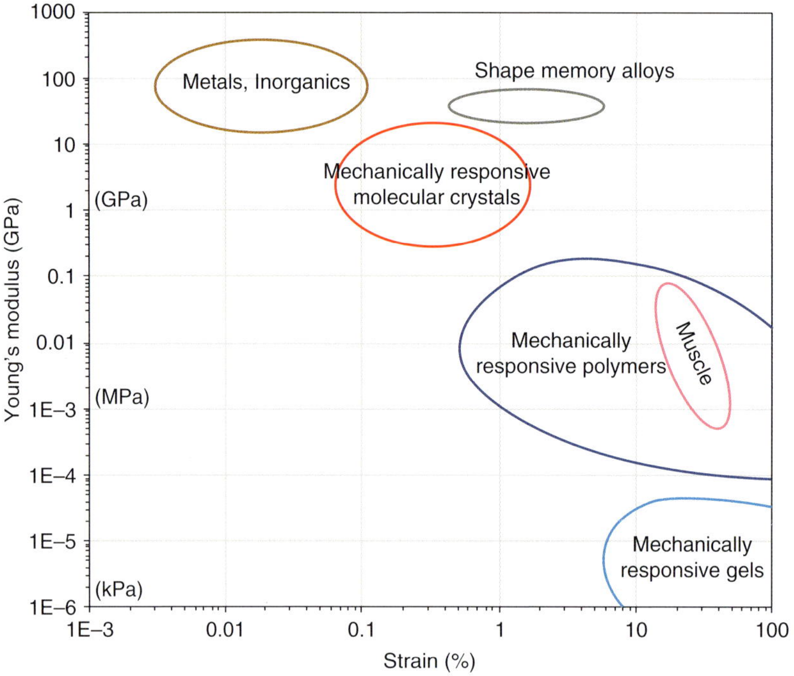

Figure 3.14 Relationship between Young's modulus and strain of actuation materials. The ranges are drawn based on the values for molecular crystals [30, 41, 71], gels [49], polymers [72], metals and alloys [73].

materials. Figure 3.14 shows the correlation between strain and Young's modulus of actuation materials, including molecular crystals. Metals and alloys actuating by thermal expansion, piezoelectricity, magnetostriction, and the shape memory effect have relatively large Young's modulus values, in the range of 10–500 GPa; the strain of these materials ranges from 0.01% to 10%. Mechanically responsive polymers typically have smaller Young's modulus values of 1–100 MPa and larger strain, sometimes more than 100%. The mechanical properties of polymers are nearly equivalent to those of human skeletal muscle. In the case of gels, Young's modulus values are much smaller (~50 kPa) than that of polymers. Rubbers also have a small Young's modulus (~100 kPa), comparable to those of gels and soft polymers.

On the other hand, photo- and thermo-mechanical molecular crystals generally have Young's modulus values of 1–10 GPa and strain of 0.1–1%. The mechanical properties of mechanically responsive molecular crystals are located between metallic inorganic materials and organic polymers with respect to Young's modulus and strain. Thus, mechanically responsive molecular crystals effectively fill the gap between inorganic materials and organic polymers, and, as such, are expected to show new applications in the future with further development.

However, it may be impractical or difficult to apply only one single crystal as an actuator for soft robotics. Hybridization of single crystals with polymers provides the opportunity to create actuators with more realistic movements. We have reported a hybrid crystal assembly, in which platelike salicylideneamine crystals are aligned in silicone polymer films by a magnetic field and undergo reversible photomechanical bending upon alternate irradiation with UV and visible light [74]. Hybrid photomechanical membranes, in which nanorod crystals are assembled with poly(vinylidene fluoride-*co*-hexafluoropropylene) polymers, have also been reported [75].

3.6 Summary and Outlook

In this chapter, we presented recent developments in mechanical molecular crystals responsive to light and heat. The photomechanical bending motions of various crystals, such as azobenzene, salicylideneaniline, fulgide, and benzobarrelene, originate from trans–cis photoisomerization, enol–keto photoisomerization, photo-electrocyclization, and Norrish type II hydrogen abstraction, respectively, as typical photoreactions. The research results suggest the possibility of finding new photomechanical crystals via other photoreactions. Salicylideneaniline derivatives are considered the most promising as mechanical crystal candidates, due to their ease of synthesis and the occurrence of both photochemical and thermal back-isomerization.

The thermal phase transitions of crystals are useful for the creation of mechanical motion. Chiral azobenzene crystals exhibit inchworm-like walking and fast rolling locomotion on heating and cooling near the phase transition point. Such crystal locomotion behavior can potentially lead to a new field of study: crystal robotics.

A new photo-triggered phase transition, discovered in a photochromic chiral salicylideneamine crystal, is a phenomenon in which the crystal with the thermal phase transition transforms into an identical phase upon light irradiation at temperatures lower than the thermal transition temperature. The photo-triggered phase transition has the potential to extend the functions of photo-responsive solid materials.

Crystals have the advantage of a larger Young's modulus than polymers. Hybridization of crystals with polymers provides the opportunity to create actuators with more realistic movements. The next step in the study of mechanically responsive crystals is their practical application in creating soft robots.

References

1 Cohen, M.D. and Schmidt, G.M.J. (1964). Topochemistry. Part I. A survey. *J. Chem. Soc.*: 1996–2000.
2 Desiraju, G.R. (ed.) (1987). *Organic Solid State Chemistry*. Amsterdam: Elsevier.
3 Ohashi, Y. (1993). *Reactivity in Molecular Crystals*. Weinheim: Wiley-VCH.
4 Toda, F. (ed.) (2000). *Organic Solid-State Reactions*. Dordrecht: Kluwer Academic Publishers.

5 Ito, Y. and Matsuura, T. (1988). Efficiency for solid-state photocyclization of 2,4,6,-triisopropylbenzophenones. *Tetrahedron Lett.* 29: 3087–3090.
6 Ihmels, H. and Scheffer, J.R. (1999). The Norrish type II reaction in the crystalline state: toward a better understanding of the geometric requirements for γ-hydrogen atom abstraction. *Tetrahedron* 55: 885–907.
7 Koshima, H., Yao, X., Wang, H. et al. (1994). Preparation, crystal structure and photochemistry of a novel molecular compound between duroquinone and durene. *Tetrahedron Lett.* 35: 4801–4804.
8 Koshima, H., Maeda, A., Masuda, N. et al. (1994). Ionic chiral handle-induced asymmetric synthesis in a solid state Norrish type II photoreaction. *Tetrahedron Asymmetry* 5: 1415–1418.
9 Koshima, H., Ding, K., and Matsuura, T. (1994). Stoichiometrically sensitized decarboxylation occurring in a molecular crystal composed of phenanthridine and 3-indoleacetic acid. *J. Chem. Soc. Chem. Commun.*: 2053–2054.
10 Koshima, H., Ding, K., Chisaka, Y., and Matsuura, T. (1995). Chirality memory controlled by the crystal lattice of a molecular compound. *Tetrahedron Asymmetry* 6: 101–104.
11 Hirotsu, K., Okada, K., Mizutani, H. et al. (1996). X-ray study of asymmetric photoreaction of 2,4,6-triisopropyl-4′-carboxybenzophenone. *Mol. Cryst. Liq. Cryst.* 277: 99–106.
12 Koshima, H., Ding, K., Chisaka, Y. et al. (1996). Solid-state photodecarboxylation induced by exciting the CT bands of the complexes of arylacetic acids and 1,2,4,5-tetracyanobenzene. *J. Org. Chem.* 61: 2352–2357.
13 Koshima, H., Ding, K., Chisaka, Y., and Matsuura, T. (1996). Generation of chirality in a two-component molecular crystal of acridine and diphenylacetic acid and its absolute asymmetric photodecarboxylating condensation. *J. Am. Chem. Soc.* 118: 12059–12065.
14 Koshima, H., Wang, Y., Matsuura, T. et al. (1997). Photoreactivities of two kinds of bimolecular crystals formed from acridine and phenothiazine. *J. Chem. Soc. Perkin Trans.* 2: 2033–2038.
15 Koshima, H., Ding, K., Chisaka, Y. et al. (1997). Stoichiometrically sensitized decarboxylation occurring in the two-component molecular crystals of aza aromatic compounds and aralkyl carboxylic acids. *J. Am. Chem. Soc.* 119: 10317–10324.
16 Koshima, H., Nakagawa, T., and Matsuura, T. (1997). Enantioselective photoreaction occurring in a chiral bimolecular crystal formed from acridine and *R*-(−)- or *S*-(+)-phenylpropionic acid. *Tetrahedron Lett.* 38: 6063–6066.
17 Koshima, H., Hayashi, E., Shirafuji, K. et al. (1997). Photoreaction in two-component molecular crystals of α-hydroxycarboxylic acids and aza aromatic compounds. *J. Photochem. Photobiol., A* 129: 57–66.
18 Koshima, H., Matsushige, D., and Miyauchi, M. (2001). Enantiospecific single crystal-to-single crystal photocyclization of 2,5-diisopropyl-4′-carboxybenzophenone in the salt crystals with (*S*)- and (*R*)-phenylethylamine. *CrystEngComm* 3: 141–143.
19 Koshima, H., Kawanishi, H., Nagano, M. et al. (2005). Absolute asymmetric photocyclization of isopropylbenzophenone derivatives using a cocrystal

approach involving single-crystal-to-single-crystal transformation. *J. Org. Chem.* 70: 4490–4497.

20 Koshima, H., Fukano, M., and Uekusa, H. (2007). Diastereospecific photocyclization of a isopropylbenzophenone derivative in crystals and the morphological changes. *J. Org. Chem.* 72: 6786–6791.

21 Koshima, H., Ide, Y., Fukano, M., and Uekusa, H. (2008). Single-crystal-to-single-crystal photocyclization of 4-(2,4,6-triisopropylbenzoyl)benzoic acid in the salt crystal with (*S*)-phenylethylamine. *Tetrahedron Lett.* 49: 4346–4348.

22 Koshima, H., Ide, Y., and Ojima, N. (2008). Surface morphology changes of a salt crystal of 4-(2,5-diisopropylbenzoyl)benzoic acid with (*S*)-phenylethylamine via single-crystal-to-single-crystal photocyclization. *Cryst. Growth Des.* 8: 2058–2060.

23 Koshima, H., Ide, Y., Yamasaki, S., and Ojima, N. (2009). Changes in the surface morphology of salt crystals of 4-(2,5-diisopropylbenzoyl)benzoic acid with amines via single-crystal-to-single-crystal photocyclization. *J. Phys. Chem. C* 113: 11683–11688.

24 Koshima, H., Fukano, M., Ojima, N. et al. (2014). Absolute asymmetric photocyclization of triisopropylbenzophenone derivatives in crystals and their morphological changes. *J. Org. Chem.* 79: 3088–3093.

25 Kobatake, S., Takami, S., Muto, H. et al. (2007). Rapid and reversible shape changes of molecular crystals on photoirradiation. *Nature* 446: 778–781.

26 Koshima, H., Ojima, N., and Uchimoto, H. (2009). Mechanical motion of azobenzene crystals upon photoirradiation. *J. Am. Chem. Soc.* 131: 6890–6891.

27 Koshima, H. and Ojima, N. (2012). Photomechanical bending of 4-aminoazobenzene crystals. *Dyes Pigments* 92: 798–801.

28 Taniguchi, T., Fujisawa, J., Shiro, M. et al. (2016). Mechanical motion of chiral azobenzene crystals with twisting upon photoirradiation. *Chem. Eur. J.* 22: 7950–7958.

29 Koshima, H., Takechi, K., Uchimoto, H. et al. (2011). Photomechanical bending of salicylideneaniline crystals. *Chem. Commun.* 47: 11423–11425.

30 Koshima, H., Matsuo, R., Matsudomi, M. et al. (2013). Light-driven bending crystals of salicylidenephenylethylamines in enantiomeric and racemate forms. *Cryst. Growth Des.* 13: 4330–4337.

31 Takanabe, A., Tanaka, M., Johmoto, K. et al. (2016). Optical activity and optical anisotropy in photomechanical crystals of chiral salicylidenephenylethylamines. *J. Am. Chem. Soc.* 138: 15066–15077.

32 Koshima, H., Nakaya, H., Uchimoto, H., and Ojima, N. (2011). Photomechanical motion of furylfulgide crystals. *Chem. Lett.* 41: 107–109.

33 Koshima, H., Uchimoto, H., Taniguchi, T. et al. (2016). Mechanical motion of molecular crystals induced by [4+4] photodimerisation. *CrystEngComm* 18: 7305–7310.

34 Taniguchi, T., Kubota, A., Moritoki, T. et al. (2018). Two-step photomechanical motion of a dibenzobarrelene crystal. *RSC Adv.* 8: 34314–34320.

35 Shima, T., Muraoka, T., Hoshino, N. et al. (2014). Thermally driven polymorphic transition prompting a naked-eye-detectable bending and straightening motion of single crystals. *Angew. Chem. Int. Ed.* 53: 7173–7178.

36 Yao, Z.S., Mito, M., Kamachi, T. et al. (2014). Molecular motor-driven abrupt anisotropic shape change in a single crystal of a Ni complex. *Nat. Chem.* 6: 1079–1083.

37 Liu, G., Liu, J., Liu, Y., and Tao, X. (2014). Oriented single-crystal-to-single-crystal phase transition with drastic changes in the dimensions of crystals. *J. Am. Chem. Soc.* 136: 590–593.

38 Panda, M.K., Runčevski, T., Sahoo, S.C. et al. (2014). Colossal positive and negative thermal expansion and thermosalient effect in a pentamorphic organometallic martensite. *Nat. Commun.* 5: 4811.

39 Uchida, E., Azumi, R., and Norikane, Y. (2015). Light-induced crawling of crystals on a glass surface. *Nat. Commun.* 6: 7310.

40 Rus, D. and Tolley, M.T. (2015). Design, fabrication and control of soft robots. *Nature* 521: 467–475.

41 Taniguchi, T., Sugiyama, H., Uekusa, H. et al. (2018). Walking and rolling of crystals induced thermally by phase transition. *Nat. Commun.* 9: 538.

42 Taniguchi, T., Sato, H., Hagiwara, Y. et al. (2019). Photo-triggered phase transition of a crystal. *Commun. Chem.* 2: 19.

43 Irie, M., Yokoyama, Y., and Seki, T. (eds.) (2013). *New Frontiers in Photochromism*. Tokyo, Japan: Springer.

44 Irie, M., Fukaminato, T., Matsuda, K., and Kobatake, S. (2014). Photochromism of diarylethene molecules and crystals: memories, switches, and actuators. *Chem. Rev.* 114: 12174–12277.

45 Kim, T., Zhu, L., Al-Kaysi, R.O., and Bardeen, C.J. (2014). Organic photomechanical materials. *ChemPhysChem* 15: 400–414.

46 Naumov, P., Chizhik, S., Panda, M.K. et al. (2015). Mechanically responsive molecular crystals. *Chem. Rev.* 115: 12440–12490.

47 Yu, Y., Nakano, M., and Ikeda, T. (2003). Photomechanics: directed bending of a polymer film by light. *Nature* 425: 145.

48 White, T.J., Tabiryan, N.V., Serak, S.V. et al. (2008). A high frequency photo-driven polymer oscillator. *Soft Matter* 4: 1796–1798.

49 Takashima, Y., Hatanaka, S., Otsubo, M. et al. (2012). Expansion–contraction of photoresponsive artificial muscle regulated by host–guest interactions. *Nat. Commun.* 3: 1270.

50 Bushuyev, O.S., Singleton, T.A., and Barrett, C.J. (2013). Fast, reversible, and general photomechanical motion in single crystals of various azo compounds using visible light. *Adv. Mater.* 25: 1796–1800.

51 Bushuyev, O.S., Tomberg, A., Friščić, T., and Barrett, C.J. (2013). Shaping crystals with light: crystal-to-crystal isomerization and photomechanical effect in fluorinated azobenzenes. *J. Am. Chem. Soc.* 135: 12556–12559.

52 Bushuyev, O.S., Tomberg, A., Vinden, J.R. et al. (2016). Azo· · ·phenyl stacking: a persistent self-assembly motif guides the assembly of fluorinated cis-azobenzenes into photo-mechanical needle crystals. *Chem. Commun.* 52: 2103–2106.

53 Bushuyev, O.S., Friščić, T., and Barrett, C.J. (2016). Photo-induced motion of azo dyes in organized media: from single and liquid crystals, to MOFs and machines. *CrystEngComm* 18: 7204–7211.

54 Hao, Y., Huang, S., Guo, Y. et al. (2019). Photoinduced multi-directional deformation of azobenzene molecular crystals. *J. Mater. Chem. C* 7: 503–508.
55 Lee, J., Oh, S., Pyo, J. et al. (2015). A light-driven supramolecular nanowire actuator. *Nanoscale* 7: 6457–6461.
56 Zhu, L., Al-Kaysi, R.O., and Bardeen, C.J. (2011). Reversible photoinduced twisting of molecular crystal microribbons. *J. Am. Chem. Soc.* 133: 12569–12575.
57 Kim, T., Zhu, L., Mueller, L.J., and Bardeen, C.J. (2014). Mechanism of photoinduced bending and twisting in crystalline microneedles and microribbons composed of 9-methylanthracene. *J. Am. Chem. Soc.* 136: 6617–6625.
58 Kitagawa, D., Nishi, H., and Kobatake, S. (2013). Photoinduced twisting of a photochromic diarylethene crystal. *Angew. Chem. Int. Ed.* 52: 9320–9322.
59 Kitagawa, D., Tsujioka, H., Tong, F. et al. (2018). Control of photomechanical crystal twisting by illumination direction. *J. Am. Chem. Soc.*: 4208–4212.
60 Amimoto, K. and Kawato, T. (2005). Photochromism of organic compounds in the crystal state. *J. Photochem. Photobiol. C* 6: 207–226.
61 Johmoto, K., Sekine, A., Uekusa, H., and Ohashi, Y. (2009). Elongated lifetime of unstable colored species by intermolecular hydrogen bond formation in photochromic crystals. *Bull. Chem. Soc. Jpn.* 82: 50–57.
62 Inoue, Y. and Ramamurthy, V. (2002). *Chiral Photochemistry*. New York, NY: Marcel Dekker.
63 Yokoyama, Y. (2000). Fulgides for memories and switches. *Chem. Rev.* 100: 1717–1740.
64 Yoshioka, Y., Tanaka, T., Sawada, M., and Irie, M. (1989). Molecular and crystal structures of *E*-and *Z*-isomers of 2,5-dimethyl-3-furylethylidene (isopropylidene)-succinic anhydride. *Chem. Lett.* 18: 19–22.
65 Kobatake, S. and Irie, M. (2004). Photochromism of furylfulgide in a single-crystalline phase. *Chem. Lett.* 33: 904–905.
66 Khedhiri, L., Corval, A., Casalegno, R., and Rzaigui, M. (2004). Molecular conformation and structural changes in crystalline photochromism of 3-furylfulgide. *Chem. A Eur. J.* 108: 7473–7478.
67 Harada, J., Nakajima, R., and Ogawa, K. (2008). X-ray diffraction analysis of photochromic reaction of fulgides: crystalline state reaction induced by two-photon excitation. *J. Am. Chem. Soc.* 130: 7085–7091.
68 Pokkuluri, P.R., Scheffer, J.R., and Trotter, J. (1990). Novel photorearrangements of bridgehead-substituted dibenzobarrelene derivatives in solution and the solid state. *J. Am. Chem. Soc.* 112: 3676–3677.
69 Chen, J., Scheffer, J.R., and Trotter, J. (1992). Differences in photochemical reactivity of 9,10-ethenoanthracene derivatives in liquid and crystalline media. *Tetrahedron* 48: 3251–3274.
70 Sajimon, M.C., Ramaiah, D., Suresh, C.H. et al. (2007). Photochromic dibenzobarrlenes: long-lived triplet biradical intermediates. *J. Am. Chem. Soc.* 129: 9439–9445.
71 Morimoto, M. and Irie, M. (2010). A diarylethene cocrystal that converts light into mechanical work. *J. Am. Chem. Soc.* 132: 14172–14178.

72 Montero de Espinosa, L., Meesorn, W., Moatsou, D., and Weder, C. (2017). Bioinspired polymer systems with stimuli-responsive mechanical properties. *Chem. Rev.* 117: 12851–12892.

73 Liu, K., Cheng, C., Cheng, Z. et al. (2012). Giant-amplitude, high-work density microactuators with phase transition activated nanolayer bimorphs. *Nano Lett.* 12: 6302–6308.

74 Koshima, H., Matsudomi, M., Uemura, Y. et al. (2013). Light-driven bending of polymer films in which salicylidenephenylethylamine crystals are aligned magnetically. *Chem. Lett.* 42: 1517–1519.

75 Lan, T. and Chen, W. (2013). Hybrid nanoscale organic molecular crystals assembly as a photon-controlled actuator. *Angew. Chem. Int. Ed.* 52: 6496–6500.

4

Crawling Motion of Crystals on Solid Surfaces by Photo-induced Reversible Crystal-to-Melt Phase Transition

Yasuo Norikane and Koichiro Saito

Electronics and Photonics Research Institute, National Institute of Advanced Industrial Science and Technology (AIST), Central 5, Higashi 1-1-1 Tsukuba, Ibaraki 305-8565, Japan

4.1 Introduction

The motion of natural and artificial objects has attracted significant attention from the fundamental and practical points of view. Nature has developed fascinating molecular systems that inspire scientists to attempt the design and fabrication of novel mimicking systems that produce externally powered or self-propelling motion [1–10]. Molecular-sized to centimeter-sized objects have been proposed, whose motion is produced by diverse energy sources, such as chemical, thermal, electric, electromagnetic, and light energies. Among the various types of motions such as translational, rotational, and deformational, the translational motion has been of interest to researchers, because this motion is essential for living organisms for finding food and escaping from an enemy. In addition, perfecting this motion for an artificial object can facilitate the design of a carrier that can transport drugs, biomaterials, such as DNA and cells, and parts of electronic devices.

Light is a powerful energy source and an excellent tool for triggering and controlling the direction of motion. Optical manipulation, such as optical trapping and transporting, is a successful and important tool in many research areas in optics, physics, biological sciences, and chemistry [11]. In optical manipulation, optical forces from laser beams are utilized in the application of optical tweezers and optical torque, from the microscale to the nanoscale. The motion and position of a target can be precisely controlled by this method. However, it is difficult to move a large number of objects simultaneously, and a special optical apparatus is required. On the other hand, light energy has also been used to initiate the photothermal effect, photocatalytic reactions, photolytic decomposition, and photochromic reactions to create the motion of materials and objects [2]. In these materials, light energy is absorbed by the active materials, and the energy is converted to kinetic energy. In principle, the driving force for the motion is the formation of an asymmetric field of chemical products or energy. In particular, photochromic reactions have been incorporated in many active materials.

Mechanically Responsive Materials for Soft Robotics, First Edition. Edited by Hideko Koshima.

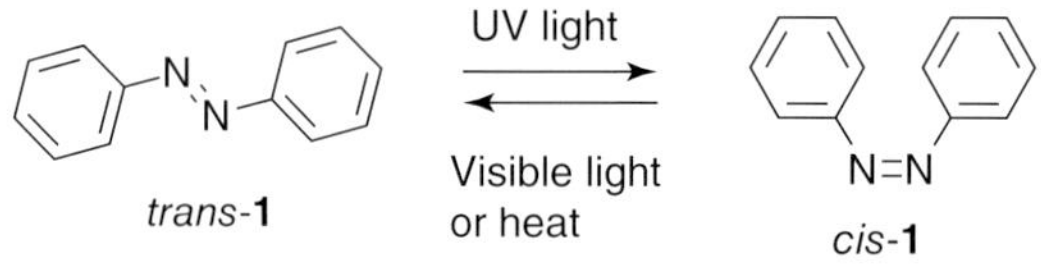

Figure 4.1 The isomerization of azobenzene.

This chapter focuses on the recent progress in photo-induced crystal-to-melt (liquid) phase transition and the crawling motion of crystals. Both phenomena are caused by the photoisomerization of azobenzene (**1**) (Figure 4.1), a key photochromic reaction, and thus, the mechanism of the crawling motion is strongly related to the phase transition behavior of the crystals, although the mechanism of the crawling crystal is still under study. Together with the basic knowledge of the photo-induced crystal-to-liquid phase transition, its potential application is reviewed because the phase transition phenomena extends the research field, as it has been recognized for its versatility in potential applications.

4.2 Isomerization of Azobenzene

Azobenzene is one of the well-known photochromic compounds. It undergoes photo-induced reversible changes in its molecular structure between the trans and cis isomers, accompanied by changes in absorption spectra. In the trans form, two phenyl rings are at opposite positions with respect to the N=N bond, while they face the same side in the cis form (Figure 4.1). Generally, the photochemical trans-to-cis and cis-to-trans isomerizations of azobenzene take place upon irradiation with UV and visible (Vis, blue) light, respectively. The cis isomer is metastable due to the steric repulsion of the two benzene rings, and it undergoes thermal cis-to-trans isomerization in the dark. Azobenzene has been widely used in various research fields because of its high photoswitching ability and its ease in chemical modification. There are many excellent reviews concerning the synthesis [12], basic photochemistry [13], and optical applications [14, 15] of azobenzene, as well as photoswitching biomolecules [16] and photo-induced mechanical motions [17, 18].

As described above, azobenzene changes its absorption spectra by photoirradiation. The absorption spectra of azobenzene are shown in Figure 4.2. There are two absorption maxima in the trans isomer: one in the visible, and the other in the UV region. The intense band around 320 nm is attributed to the $\pi \rightarrow \pi^*$ transition, while the weaker band around 430 nm is due to the forbidden $n \rightarrow \pi^*$ transition. The molar extinction coefficients (ε) of *trans*-azobenzene in hexane are reported to be 22 300 and 400 $l\,mol^{-1}\,cm^{-1}$ at 318 and 432 nm, respectively [19]. The cis isomer, on the other hand, exhibits absorption peaks at 250 and 440 nm. Accordingly, there is almost no absorption band around 300–350 nm, where the trans isomer exhibited the strong absorption. Instead, the intensity of the $n \rightarrow \pi^*$ transition is stronger than that of the trans isomer. The reported molar extinction coefficients at the absorption maxima are 11 500 and 1100 $l\,mol^{-1}\,cm^{-1}$ at 247 and 437 nm, respectively [20].

Azobenzene exhibits efficient reversible photoisomerization between the trans and cis isomers. From the trans to the cis isomer, the quantum yields ($\Phi_{t\rightarrow c}$, in

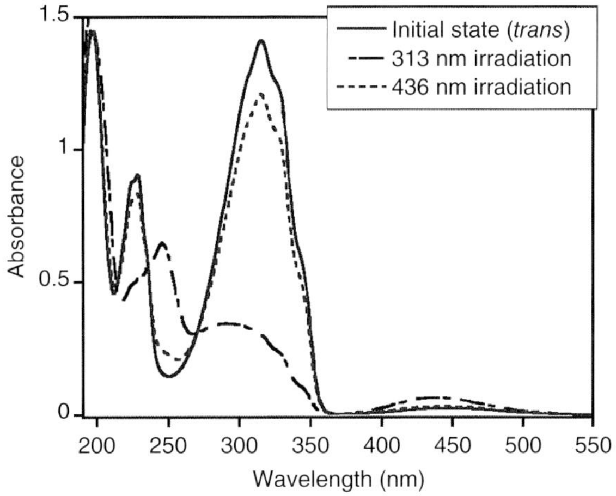

Figure 4.2 Changes in the absorption spectrum of azobenzene: (**1**) in *n*-hexane upon irradiation with 313 and 436 nm light. The light irradiation was carried out until the photostationary state was established.

hexane) are 0.11 and 0.25, upon irradiation at 317 and 439 nm, respectively [21]. The isomerization from the cis to the trans isomer, on the other hand, exhibits quantum yields ($\Phi_{c\rightarrow t}$) of 0.44 [22] and 0.56 [21] upon irradiation at 317 and 439 nm, respectively. There is a slight solvent effect for those quantum yield values [13, 21, 22]. The excitation wavelength dependence of the quantum yields is due to the different mechanisms in each electronic state: a rotation mechanism in the $\pi \rightarrow \pi^*$ (S_2) excited state and an inversion mechanism in the $n \rightarrow \pi^*$ (S_1) excited state. The inversion mechanism is a flipping of the bending angle of the C—N=N bond, while the rotation mechanism is a twisting of the dihedral angle of the C—N=N—C moiety. Those mechanisms were simply proposed based on the experimental observations in azobenzene derivatives with bulky substituents [19] and macrocyclic structure [23]. However, recent studies, including ultrafast spectroscopy techniques and molecular simulations, have suggested that the azobenzene isomerization mechanism is not as simple as described above. The discussion about the reaction mechanism based on the recent studies can be found in a review [13].

Because of the large difference in the absorption spectra between the trans and cis isomer of azobenzene, the isomer ratio in the photostationary state (pss) can be drastically "switched" by selecting the irradiation wavelength. Therefore, it can be said that the trans-to-cis and cis-to-trans photoisomerizations of azobenzene take place upon irradiation with UV and visible (blue) light, respectively. A more detailed expression of the isomer ratio in the pss is given by

$$\frac{[\text{cis}]_{\text{pss}}}{[\text{trans}]_{\text{pss}}} = \frac{\varepsilon_t \times \Phi_{t\rightarrow c}}{\varepsilon_c \times \Phi_{c\rightarrow t}} \tag{4.1}$$

where $[\text{cis}]_{\text{pss}}$ and $[\text{trans}]_{\text{pss}}$ are the concentrations of the cis and trans isomers at the pss, respectively and parameters ε and Φ are the molar extinction coefficient and the quantum yield, respectively, at a particular irradiation wavelength. According to this equation, the enrichment of either isomer concentration can be controlled by selecting the wavelength where the other isomer does not relatively absorb. For example, enrichment of the cis isomer concentration can be achieved by UV light irradiation, such as 365 nm. On the other hand, a high concentration of trans isomer is established by the irradiation of visible light, such as 436 nm. Therefore, generally speaking, most studies have applied UV (313 and 365 nm) and visible light (436 nm) in studies regarding azobenzene and its derivatives.

As described, it should be noted that the cis isomer of azobenzene is metastable, and it reverts to the original trans isomer, even in the dark. This thermal isomerization is observed, for example, with a rate constant (k) of $2.48 \times 10^{-6}\ s^{-1}$ in hexane at 20 °C [24]. Thus, the lifetime, which is defined as $\tau = 1/k$, of *cis*-azobenzene is 4.7 days. An Eyring analysis gives activation parameters, such as the activation enthalpy ($\Delta H^{\ddagger}$) and the activation entropy ($\Delta S^{\ddagger}$) of thermal isomerization. For example, analyzing the same solution gives $\Delta H^{\ddagger}$ and $\Delta S^{\ddagger}$ values of 21.2 kcal mol^{-1} and −12.0 cal $mol^{-1}\ K^{-1}$, respectively [24]. Although there are only a few reports on the effect of solvents on the thermal isomerization of simple azobenzene, it is indicated that polar solvents enhance the reaction rate [24, 25]. The relationship between those activation parameters has been used to discuss the reaction mechanism of the thermal isomerization mechanism, especially through the use of isokinetic (or compensation) plots [24–28].

Since the isomerization of azobenzene involves a large change in the molecular structure, this process is affected by the molecular environment, steric distortion, and/or crystal packing to which the azobenzene is exposed. For example, in glassy polymer matrices, photo- and thermal isomerizations of azobenzene are affected by the free volume in the matrices [29–31]. In the crystal phase, isomerization rarely occurs [32], and only limited examples of the isomerization of simple azobenzene can be found in the literature. In these cases, the isomerization takes place only at the crystal surface [33, 34].

4.3 Phase Transitions in Liquid Crystals (Liquid-Crystal-to-Isotropic)

Azobenzene exhibits efficient photoisomerization in solution and liquid-crystal (LC) media, because their highly fluidic property provides a large degree of free volume to the azobenzene molecule. In particular, there are numerous reports concerning the photochemical modulation of the properties of LCs and polymers, such as phase, morphology, and optical properties. For example, the molecular orientation of LC materials can be tuned in out-of-plane and in-plane directions by using a command surface [35–38]. The phase transition from LC to isotropic phases can be induced by the photoisomerization of azobenzene, when azobenzene derivatives are dissolved in LCs [39–42].

As for polymeric materials, the surface is drastically modulated, and surface relief gratings can be produced by laser irradiation [43, 44]. In addition, photomechanical behavior such as bending motion has been reported, and it has been utilized in actuators for soft robotics [45–47]. Although the details of photoresponsive polymeric actuators are described in the other chapters, some examples of polymer films that exhibit photomechanical bending behavior are described in the following section.

4.4 Phase Transitions in Crystal Phase (Crystal-to-Melt)

4.4.1 Characteristics of the Crystal-to-Melt Phase Transition

As described above, isomerization rarely occurs in the crystal phase of simple azobenzene. For example, grounded crystals of azobenzene was analyzed by IR spectroscopy, and it was reported that photoisomerization from trans to cis isomer does not occur in the solid state [32]. However, the photoresponsivity is drastically enhanced by its molecular design, and the crystal of an azobenzene derivative becomes photochemically active to exhibit the phase transition from the crystal to the liquid phase by light irradiation. The schematic concept of the phase transition is shown in Figure 4.3. One can readily imagine that there are potential applications in diverse fields, if the state of matter (solid or liquid) of a material can be switched by light without changing the temperature. Importantly, this sort of material possesses reversibility, a key property for materials developed to reduce the environmental footprint.

The crystal-to-liquid phase transition was first reported in azobenzene derivatives having a cyclic skeleton with long alkoxy chains (**2**, Figure 4.4) [48–50]. The cyclic azobenzene **2** has a melting point of 100 °C, and it shows an LC phase above this temperature up to c. 120 °C. Notably, this compound melts upon UV light irradiation at room temperature, as shown in Figure 4.5. The dark area in Figure 4.5 corresponds to the liquid phase because the microscopic photograph was taken under a crossed-polarizer configuration. This phenomena is in contrast with the fact that simple azobenzene does not show significant change in the crystal upon UV light irradiation, although the melting point of azobenzene (68 °C) [32] is much lower than that of compound **2**. To elucidate the molecular mechanism of the phase transition, the single crystal structure was determined [50]. It was revealed that the crystal shows the characteristic coexistence of

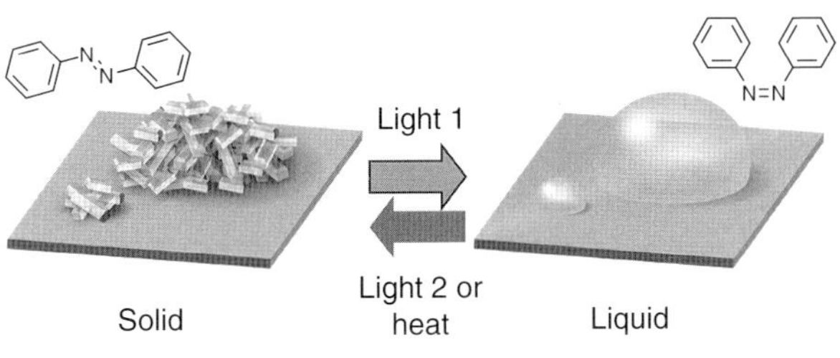

Figure 4.3 Schematic representation showing the concept of the photo-induced crystal-to-melt (liquid) phase transition by photoisomerization of azobenzene.

Figure 4.4 Macrocyclic azobenzene (**2**) showing the photo-induced crystal-to-liquid phase transition.

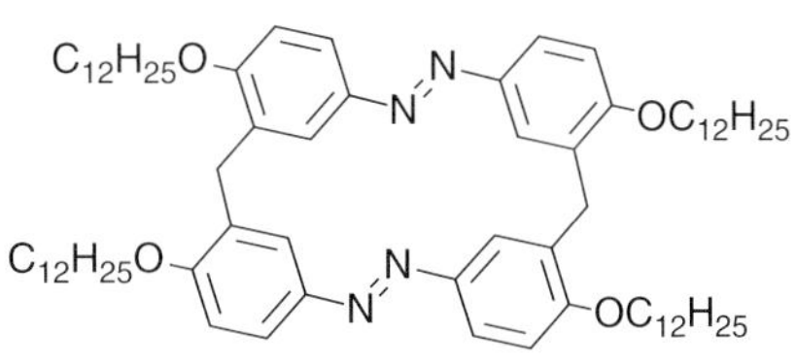

2

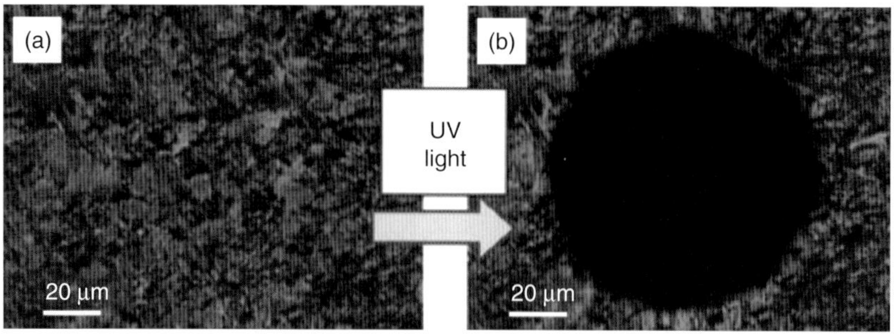

Figure 4.5 Polarizing optical microscope photographs of **2** at 25 °C before (a) and after (b) UV light irradiation for five seconds. Source: Norikane et al. 2011 [48]. Reproduced with permission of RSC.

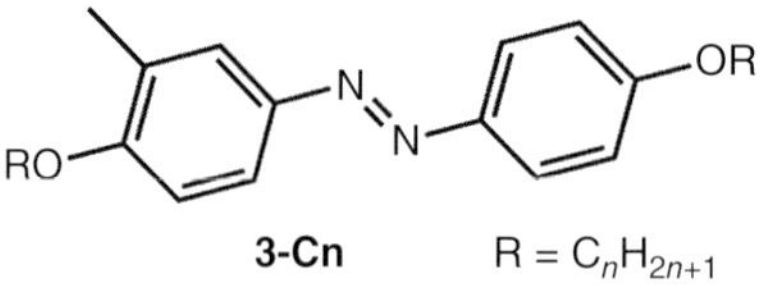

Figure 4.6 Simplified molecular design of azobenzene (**3-Cn**) showing the photo-induced crystal-to-liquid phase transition.

structurally ordered and disordered portions. Azobenzene chromophores form a one-dimensional ordered network of π–π interactions, while highly disordered alkyl chains exist, even at room temperature, between the ordered networks. The disordered structure of the crystal plays an important role in providing the free volume for azobenzene moieties, which can be important in dynamic motion, such as isomerization.

To simplify the molecular structure for investigation of the structure–property relationship of the crystal-to-liquid phase transition properties, a simplified molecular design without a cyclic skeleton (**3-Cn**) was proposed, as shown in Figure 4.6. A simple molecular design enabled facile synthesis, and it was revealed that acyclic azobenzenes with a methyl group at the 3 position exhibit phase transition [51, 52]. This simple modification of the introduction of a single methyl group onto one of the two benzene rings of azobenzene drastically changes the photo-induced photoresponsivity. In sharp contrast to this molecular design, the azobenzene containing two methyl groups at 3- and 3′ positions and that without the methyl groups did not exhibit phase transition. For instance, the azobenzene crystal, **3-C6**, with this molecular design strategy shows a photo-induced phase transition and exhibits the liquid state. The melted state showed a viscosity of 460 mPa s, which remained for about one hour at room temperature in the dark. The melting temperature of the melted state was −6 °C, while that of the original trans state was 87 °C. Since the molecular design of the simple substitution of a single methyl group was successful, the effect of the chain length of the alkoxy groups at the 4 and 4′ positions was studied [52]. It was found that the alkyl chains with $C_n = 6–10$ showed a relatively fast phase transition out of $C_n = 1–18$ (Figure 4.7).

Regarding the effect of the heat produced by the photoirradiation during the phase transition of the crystal, the photoexcited molecules release energy as heat

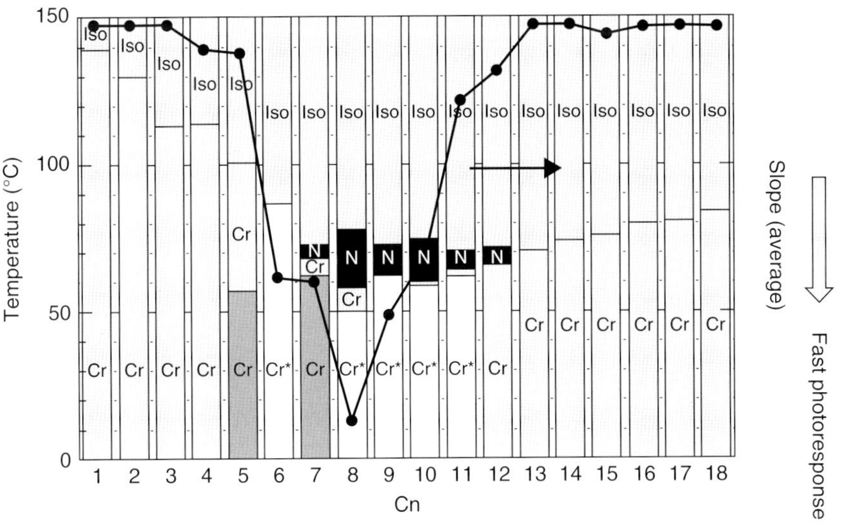

Figure 4.7 Thermal phase transition temperature (bar graphs) of **3-Cn** measured by differential scanning calorimetry at second heating process, and a plot of relative speed of photo-induced crystal-to-liquid phase transition (line graph). Cr, Cr*, N, and Iso denote crystal, metastable crystal, nematic LC, and isotropic (liquid) phases, respectively. Source: Norikane et al. 2016 [52]. Reproduced with permission of SPST.

through the deactivation process, and an increase in temperature is unavoidable. Actually, a temperature increase in the crystal powder of the **3-Cn** series was approximately 3 K, upon an irradiation at 200 mW cm^{-2} [51]. The crystals of **3-C6** were irradiated by different intensities of UV light (5–408 mW cm^{-2}). As a result, it was found that the photo-induced phase transition did not depend on light intensity, and melting was observed for the same total photon dose [52]. Therefore, it can be concluded that the heating effect is not the main reason for the photo-induced crystal-to-liquid phase transition, but the photoisomerization reaction of azobenzene is the driving force for the phase transition.

The crystal-to-liquid phase transition can be readily observed by eye [51], X-ray diffraction [49, 50], optical and polarizing optical microscopy [48, 49, 51], and spectroscopically [52]. The phase transition occurs at the crystal surface as suggested [50], and the irradiation time required for complete liquefaction strongly depends on the sample thickness and light intensity. Particularly, during irradiation, the melted state covers the remaining crystal and absorbs the incoming photons, resulting in an additional effect to slow down the phase transition.

4.4.2 Potential Applications of Crystal-to-Melt Transition

The concept of photo-induced crystal-to-liquid (or melt) transition has impacted not only the solid-state photochemistry field but also the wide fields of polymer science, soft materials, and smart materials. If the state of matter, such as the solid or liquid state of a material, can be switched by light without changing its temperature, it can be applied to various fields. For example, there already exists a variety of proof-of-concept studies, such as photoresists [51], smart reversible adhesives [52–54], gas storage materials [55], solar thermal fuels [56], self-healing materials [57], and photoswitching coating layer for biodegradable polymers [58]. In addition, novel mechanical motions by using materials showing a crystal-to-liquid phase transition have been found: crawling [59], swimming [60], and bending motions [61, 62]. Here, some examples of compounds that exhibit a crystal-to-liquid phase transition, with demonstrations suggesting potential application, are explained.

A rod-shaped amphiphilic azobenzene (**4**, Figure 4.8), having a hydrophilic tetra(ethylene glycol) chain and two ethyl groups, exhibits a crystal-to-liquid phase transition [63]. This compound shows light-directed dynamic spheres that change their morphology between spherical and disassembled states. The melting temperature of the trans isomer is 50 °C, and the two ethyl groups contribute to weaken the intermolecular π–π interactions. Star-shaped tetra(azobenzene)methane compounds (**5**, Figure 4.8), consisting of four azobenzene units covalently connected to a tetrahedral carbon atom, form porous crystals [55]. Irradiation of these crystals by UV light converts them into the nonporous liquid phase. By using these compounds, it was demonstrated that CO_2 gas can be adsorbed/released by phase transition. Notably, the melting temperature of these compounds are above 300 °C, although they exhibit a photo-induced phase transition at room temperature. An ionic azobenzene possessing a quaternary nitrogen atom (**6**, Figure 4.8) exhibits a photo-induced crystal-to-ionic liquid phase transition. It was demonstrated that this compound can be used as a solar thermal fuel that stores photon energy and releases energy as heat.

Conventional photoresists are generally not reusable because irreversible reactions, such as photopolymerization, degradation, or cross-linking reactions, are applied. In addition, in photolithography processes, significant amounts of acids and/or bases are used to remove the photoresists for development or washing. Materials that can be easily reused and reduce the environmental footprint are desired. An attempt has been made to use azobenzene as a photoresist, to make a solid/liquid pattern on surfaces by patterned irradiation [51]. The developmental process using azobenzene was much simpler than the conventional method: the melted azobenzene layer can be removed by wiping, blowing, or washing by solvents without the use of acids or bases. In particular, the azobenzene **3-C10** has good film-forming property on solid surfaces by spin-coating methods. This compound has been used for photolithography processes to fabricate copper [51] as well as for the coating layer to switch the biodegradable properties of polymers [58]. Biodegradable polymers are environmentally friendly materials, and their degradation takes place through the action of hydrolytic enzymes of microorganisms. From the application viewpoint, these polymers should maintain their functions during use and start degrading just after disposal. The use of azobenzene **3-C10** as a coating layer demonstrated that the enzymatic degradation of poly(L-lactide) and poly(ε-caprolactone) is switched on by UV irradiation [58]. The solid layer of the azobenzene prevents enzymes (proteinase K and lipase) from degrading the polymer. On the other hand, the enzymes started to degrade the polymer surfaces after irradiation.

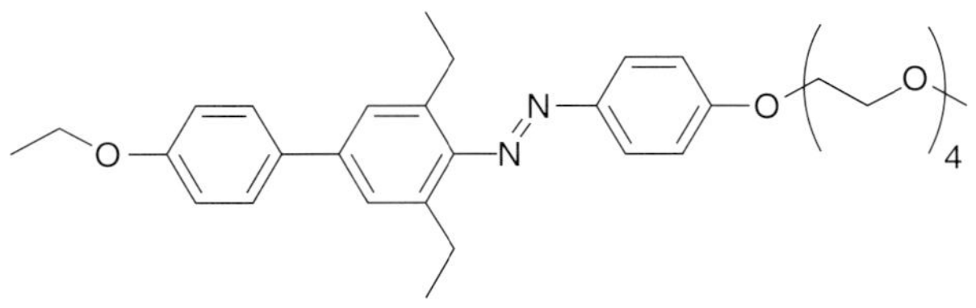

4

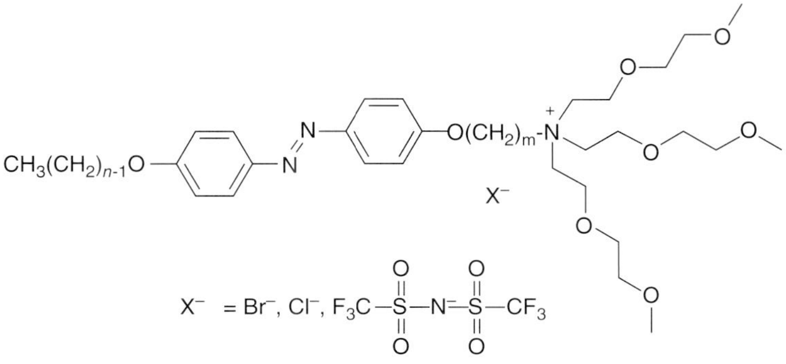

6

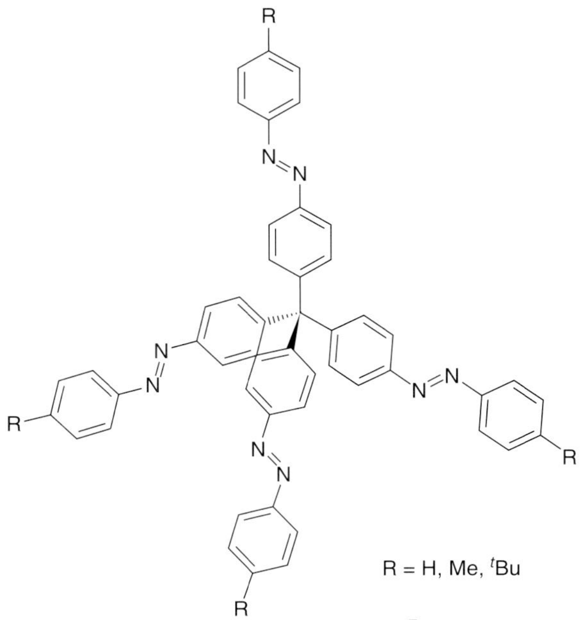

5

Figure 4.8 Molecular structures of azobenzene derivatives (**4–6**) showing the photo-induced crystal-to-liquid phase transition.

The concept of the crystal-to-liquid phase transition has significantly impacted the field of adhesives. If the adhesive force of a material can be switched reversibly by external stimuli such as light, it can be used to recycle expensive or precious parts, and improve manufacturing processes, especially in temporal tacking. For this purpose, different kinds of molecules possessing azobenzene with various molecular weights have been studied. Larger molecules tend to form noncrystalline states, such as glassy LC and high-ordered LC states, which are out of the scope of this chapter, and adhesion properties of those materials were reviewed [64]. As for small-molecular-weight materials, adhesion properties have been demonstrated in the simple azobenzene **3-Cn**, with alkyl chain lengths from C6 to C18 [52]. The adhesion force was determined by a pulling test (the tensile lap-shear adhesive test) of two thermally fused glass plates sandwiching the azobenzene derivatives. The two glass plates were pulled to the opposite direction, and the test was carried out with and without light irradiation. UV irradiation melted the azobenzene, and the tensile shear strength values decreased to nearly zero. Visible-light irradiation, on the other hand, recovered the adhesion strength, up to c. $50\,N\,cm^{-2}$. This value is similar to that of commercially available double-sided tape.

4.4.3 Mechanical Motions Derived from the Crystal-to-Liquid Phase Transition

By utilizing the crystal-to-liquid phase transition found in azobenzene derivatives, novel mechanical motions have been found such as crawling [59], swimming [60], and bending motions [61, 62]. As for the crawling motion, this phenomenon is the main topic of this chapter and is described in detail in the next section. Here, we introduce the swimming and bending motions. When the crystals of 4-methoxyazobenzene are placed on a water surface, the crystals floated. When the crystals were subjected to UV light irradiation, they started moving on the surface, away from the light source (Figure 4.9). On observation, the crystals exhibited various types of motions, such as spinning, splitting, winding, and linear motions. It is notable that the motion continued for a few minutes after the irradiation ceased, indicating that the motion was not caused by the heat produced by the light. The cause of the motion is the photo-induced crystal-to-liquid phase transition. The photochemically produced cis isomer induces the motion by dissolution. A light-powered boat was prepared by soaking filter paper with azobenzene. The boat moved at the speed of c. $5\,cm\,s^{-1}$, while the irradiation of visible light slowed the motion. The boats fuel was stored in an inert solid state, and light can switch the fuel between activated and deactivated states.

As described above, a methyl group at the 3 position of azobenzene drastically enhanced the photo-induced phase transition properties. According to this molecular design strategy, a monomer (**8**) has been developed, and the copolymer consisting of this monomer exhibits photomechanical bending [61]. The azobenzene **8** exhibits a rapid phase transition from the crystal to the liquid phase upon UV light irradiation. In contrast, the azobenzene monomer

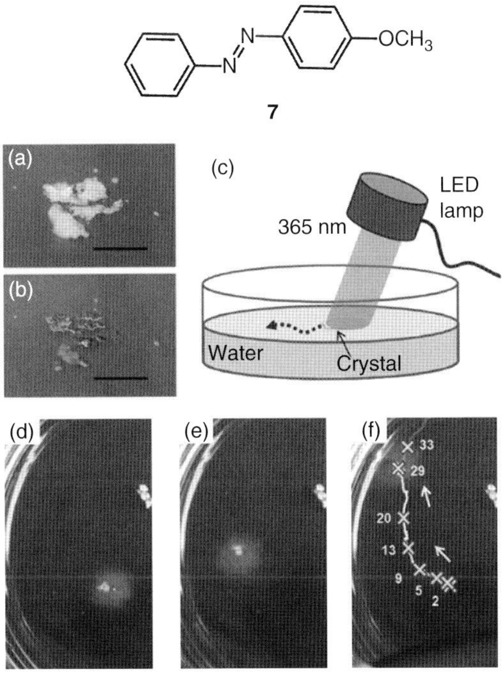

Figure 4.9 Molecular structure of 4-methoxyazobenzene (**7**) (above). Photographs of the crystal powder of **7** (a) before and (b) after UV light irradiation. The scale bars are 5 mm. (c) Schematic of the experiment of crystal motion on a water surface. (d–f) Snapshots of the crystal motion on a water surface. (f) Trajectory of the motion. Cross marks represent the position at every second and the numbers denote the displacement (mm) from the original position. Source: Norikane et al. 2016 [60]. Reproduced with permission from RSC.

without the methyl group at the 3 position did not show the photo-induced phase transition. This fact strongly supports substitution at the 3 position in azobenzene as a robust strategy for molecular design, as it significantly changes the molecular packing. By using the azobenzene **8** as a monomer, a freestanding polymer film, produced by free-radical copolymerization with **9**, having an LC (smectic) property. It was found that the T_g (glass transition temperature) value of the **8**/**9** copolymer decreased upon UV irradiation from 29 to 16 °C, recovering in the dark or through visible-light irradiation. This polymer exhibits photomechanical bending with a bending speed (displacement/time) of 0.4–2 mm s^{-1}, at a light intensity of 3.28–45 mW cm^{-2} (Figure 4.10). Compared to other photomechanical bending films, this polymer has a greater efficiency in causing mechanical motion. The photoswitchable T_g seems to be beneficial for achieving efficient photomechanical actuation in the polymer network.

Another photomechanical bending film has been fabricated by using a low-density polyethylene (LDPE), commercially available wrap film, and an azobenzene derivative having an ethyl group at the 3 position, which exhibits a photo-induced crystal-to-liquid phase transition [62]. The film preparation is simple in that a tetrahydrofuran (THF) solution is dropped onto the LDPE with rubbing pretreatment; then, the coated film is thermally annealed. The film exhibited reversible bending behavior toward the UV light. The microgrooves fabricated by the rubbing process provide templates for the molecular orientation of azobenzene. Along the microgroove, it was suggested that the photo-induced phase transition of azobenzene brings about a contraction force, which originates from the change in the molecular length.

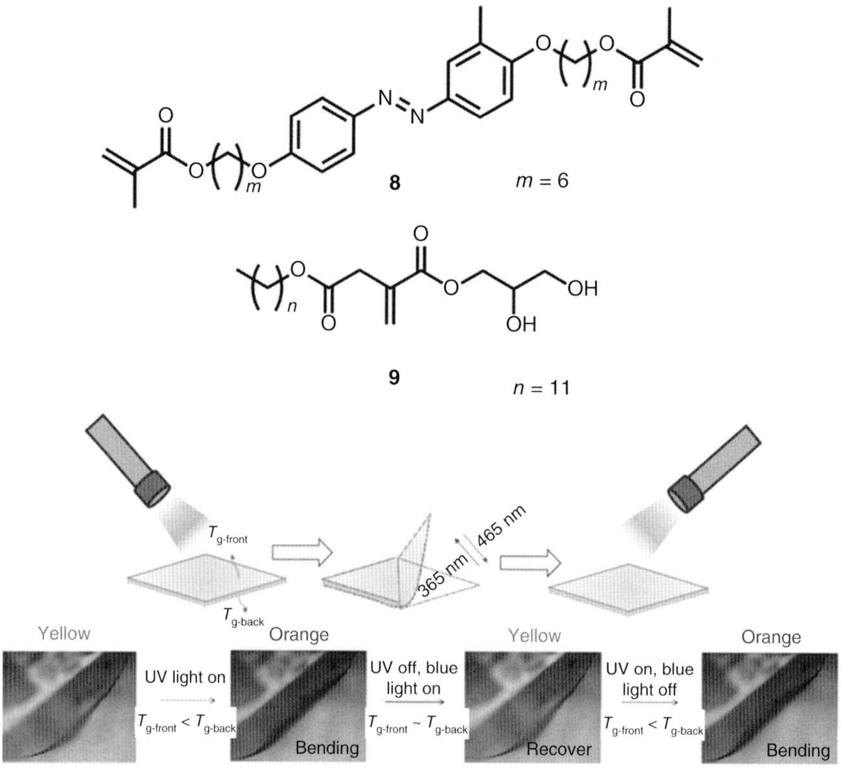

Figure 4.10 Molecular structure of a monomer showing the photo-induced crystal-to-liquid phase transition (**8**) and a liquid-crystalline monomer (**9**) (above). The photomechanical response of the copolymer consisting **8** and **9**. UV light cause bending and visible light straighten the film. Source: Yue et al. 2018 [61]. Reproduced with permission from Springer Nature.

4.5 Photo-induced Crawling Motion of Azobenzene Crystals

4.5.1 Discovery of the Crawling Motion of Crystal on Solid Surface

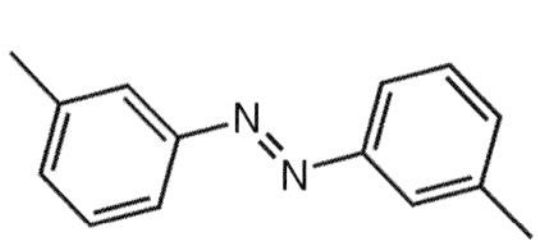

10

Figure 4.11 Molecular structure of 3,3′-dimethylazobenzene (**10**).

Here, we introduce the serendipitous discovery of the directional and continuous translational motion of crystal on a glass surface in the course of the investigation of the photo-induced crystal-to-melt phase transition of azobenzene derivatives. Usually, the experiments of the photo-induced crystal-to-melt phase transition in our laboratory had been in a manner that the photoirradiation was carried out by alternate photoirradiation of UV and visible light. However, we irradiated crystals of 3,3′-dimethylazobenzene (**10**, Figure 4.11) simultaneously with two different wavelengths for other purposes. Surprisingly, those crystals started *crawling* on the surface. They also *climbed* on the vertical surfaces! The motion was directional and crystals moved away from the UV light source. The motion proceeded without changing the crystal orientation, despite the large deformation of the crystal shape. Importantly, this motion can be achieved by simply using light-emitting diodes or Hg lamp as light sources in a fixed position. It is notable that no expensive apparatus, such as laser or positioning device, is required. It is also notable that no special treatment, such as chemical modification [65–67], spatial gradients [68], or application of ratchet potential [69], is necessary for the solid surface.

4.5.2 Characteristics of the Crawling Motion of Crystals

Micro-sized (c. 20–30 μm) crystals of **10** placed on a cover glass were irradiated with 365 nm UV light and 465 nm visible light, as shown in Figure 4.12a. This compound exhibits trans → cis and cis → trans isomerizations by 365 and 465 nm irradiation, respectively. In addition, in the crystal phase, the trans → cis and cis → trans isomerizations induce crystal-to-liquid and liquid-to-crystal phase transitions, respectively. As can be seen in Figure 4.12, the 3,3′-dimethylazobenzene (DMAB) crystals move away from the UV light and toward the visible light. It is quite interesting that the motion occurs regardless of whether the crystals are single crystalline, polycrystalline, or a distorted shape.

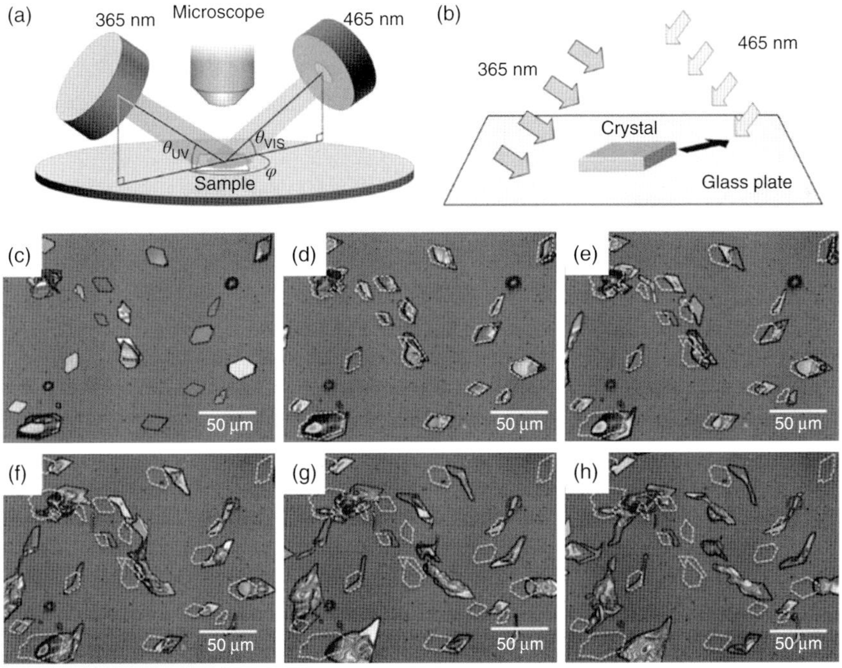

Figure 4.12 (a) Schematic of the experimental setup. (b) Schematic of the crawling motion. (c–h) Microscope images of the motion of crystal of **10** after irradiation for $t = 0$ (c), 3 (d), 6 (e), 10 (f), 15 (g), and 20 minutes (h). Source: Uchida et al. 2015 [59]. Reproduced with permission of Springer Nature.

However, the velocity of the motion varies by crystal. The average speed of the motion was c. 1.5 μm. In contrast, supercooled liquid droplets or already melted droplets by photoirradiation did not move. Those observations clearly indicate that the sample phase should be in the solid phase to achieve the motion.

To optimize the condition for the crawling motion, experiments with various light intensities were carried out at the positions of two light sources at the fixed angles of $\theta_{UV} = \theta_{Vis} = 45°$, $\varphi = 180°$ (Figure 4.13a). When the crystals were irradiated by only UV light, they melted to droplets and did not move. When the crystals were irradiated by only visible light, no morphological change or motion was observed. Therefore, simultaneous irradiation with the two light sources are required to obtain the motion. Since the velocity of the crystals varies by crystal, the motion of the crystals was analyzed statistically, and the average velocity was plotted against the light intensity of each light. This revealed that the balance of the intensities of two light sources are important for moving the crystals. When the intensity of the UV light is much more intense than that of the visible light (Figure 4.13a, the dashed line in the back right corner), the crystal melts and becomes droplets, which do not move. On the other hand, when the UV light is relatively weak (Figure 4.13a, the dashed line in left front), the crystals do not move without changing their shape. The optimized condition for the crawling motion was found at UV and visible light intensities of 200 and 50–60 mW cm^{-2}, respectively, and the average velocity was 2.0 μm min^{-1}.

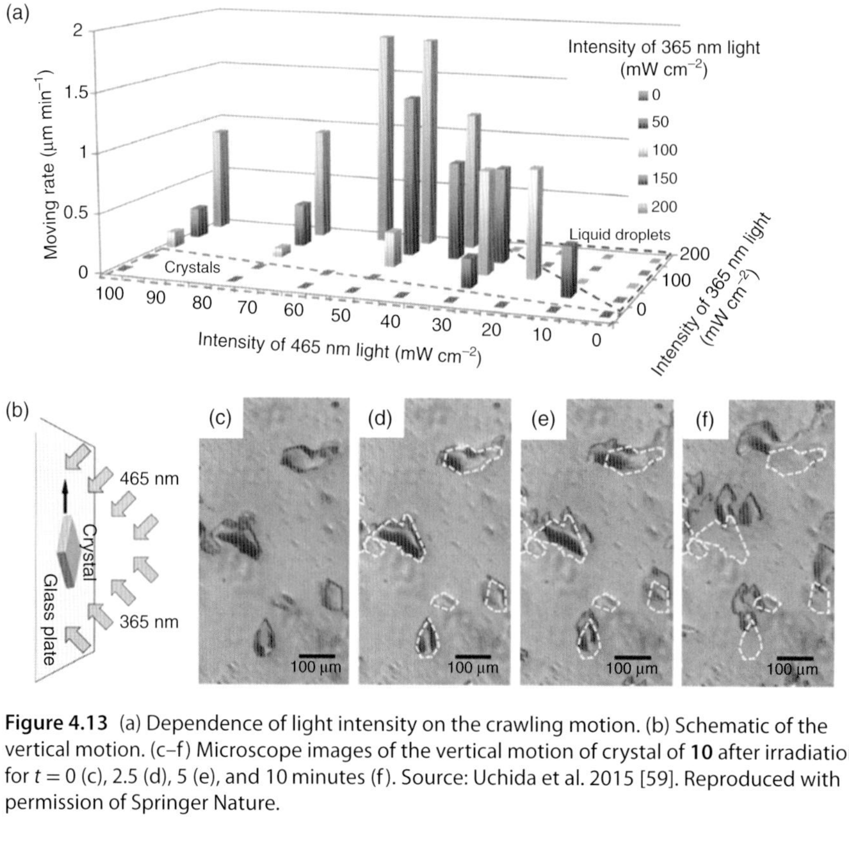

Figure 4.13 (a) Dependence of light intensity on the crawling motion. (b) Schematic of the vertical motion. (c–f) Microscope images of the vertical motion of crystal of **10** after irradiation for $t = 0$ (c), 2.5 (d), 5 (e), and 10 minutes (f). Source: Uchida et al. 2015 [59]. Reproduced with permission of Springer Nature.

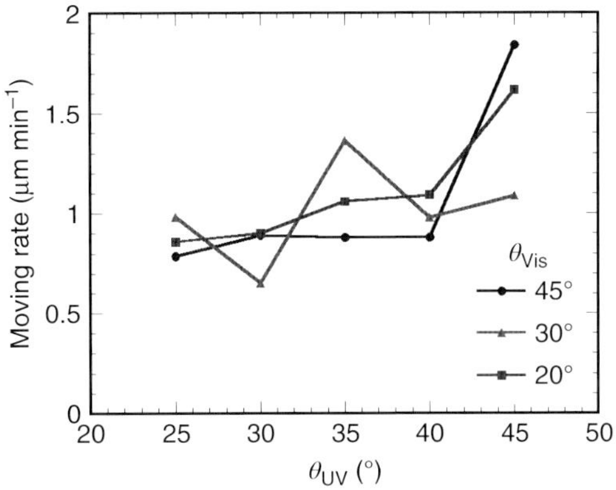

Figure 4.14 Dependence of angle of light source on the crawling motion. Source: Uchida et al. 2015 [59]. Reproduced with permission of Springer Nature.

To investigate the effect of the irradiation angle, irradiations with various angles of light sources were carried out while fixing the intensities of the UV and the visible light at 200 and 60 mW cm^{-2}, respectively. The angles were scanned in the ranges of 25–45° and 20–45° for θ_{UV} and θ_{Vis}, respectively. The results show that the velocity relies more on the UV light angle than on the visible light angle. A higher velocity is obtained at higher angles on θ_{UV}, as shown in Figure 4.14.

It is very interesting that the direction of crystal orientation is maintained while the crystal shape is highly deformed during the motion. When a single crystal, shown in Figure 4.15a, is irradiated from the direction parallel to the longer diagonal axis of the crystal, the whole crystal moved away from the UV light source. During the motion, the front edge of the crystal grows, whereas the rear edge shrinks. On the other hand, when a single crystal is irradiated perpendicular to the longer diagonal axis of the crystal, the crystal shape changes with the elongation along the longer diagonal axis (Figure 4.15d). It should be noted that the angles of the corners of the crystal are maintained at c. 56° and 124° during the motion.

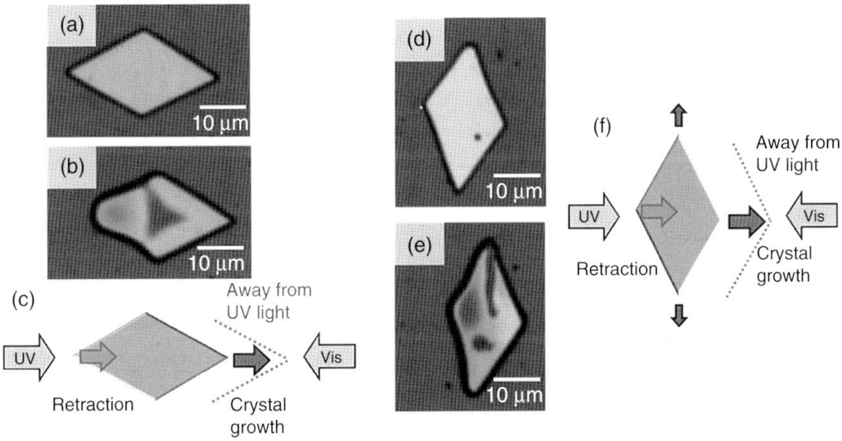

Figure 4.15 (a–c) Motion of a single crystal of **10** when the longer diagonal axis of the crystal is (a–c) parallel and (d–f) perpendicular to the direction of irradiation. Source: Uchida et al. 2015 [59]. Reproduced with permission of Springer Nature.

Three-dimensional measurements by laser microscopy revealed that the crystals of initial thickness of c. 1–4 μm became thinner during the motion, especially at the front edge. Observations by polarizing optical microscopy showed that the direction of optical anisotropy of the crystal was maintained during the motion. This indicates that the crystal motion takes place without changing its molecular orientation. Therefore, from the above results, the motion is caused by retraction (melting) of the rear edge and a crystal growth process at the front edge.

Surprisingly, this crystal motion can be achieved on a wall of glass; the crystals climb vertically, as shown in Figure 4.13b–f. This result shows that the motion is not caused by gravity or other artifacts. These microcrystals are less influenced by gravity than the adsorption forces between a crystal and a glass surface.

4.5.3 Mechanism of the Crawling Motion

Although the mechanism of the crystal motion has not been clearly revealed, it seems that a nonequilibrium condition is produced by the two light sources. It should be noted that the entire sample is uniformly irradiated by two light sources, and the light intensity seems to be constant within the microscopes field of view, because the distances to the light sources are substantially larger than the size of the dimension of the view. On the surface of the crystals, melting and crystallization take place at the same time, and an equilibrium between the solid and liquid phases is established. For the lateral face of the crystal, however, the intensity ratio between UV and visible light is slightly different at the front and rear edges. As a result, the equilibrium is shifted to the opposite direction at the two edges. Therefore, the crystal growth process on the glass surface is important. Actually, when the photoirradiation is carried out from the backside (from the face where the crystals contact the glass), the crystal motion is not observed, although some change in morphology is observed.

4.5.4 Crawling Motion of Azobenzene Crystals

To extend the applicability of the crawling motion observed in DMAB described above, the motion of azobenzene crystals (**1**) was investigated. Under room temperature, no motion was observed when the crystals of **1** were irradiated. However, the crawling motion was observed at 50 °C. This result indicates that the crystal motion is strongly related to the photo-induced crystal-to-melt phase transition, because the **1** crystals do not exhibit a photo-induced phase transition at room temperature, but they do at 50 °C. Most importantly, these results indicate that the crawling motion can be expected in other molecular systems that show reversible crystal-to-melt phase transitions induced by photochromic reactions.

4.6 Conclusion

The photo-induced crystal-to-melt phase transition found in azobenzene derivatives has attracted much attention from vast areas of research, because of its potential application in smart materials through its reusable characteristics.

In addition, the phase transition provides novel mechanical motions, such as crawling, swimming, and bending motions, that are useful for designing active materials and soft robotics.

The crawling motion of the crystal is intriguing because an object (crystal) travels on a solid surface under an ambient atmosphere (at solid/air interface). This is in contrast with numerous examples of stimuli-responsive and self-propelling motion of small objects in solution or at the solid/liquid interface. This motion would be one of candidates for use in a "carriage vehicle" to transport objects/chemicals on a solid surface without a flow channel. Recently we reported a negative phototactic behaviour of crystals of an aminoazobenzene derivative where the crystals on a glass move away from the light source during visible light irradiation [70].

However, so far, the speed of the crawling motion is quite slow (up to $10\,\mu m\,min^{-1}$), and more importantly, the mechanism of the motion is not clear. Understanding the mechanism of the crawling motion is a prerequisite to establishing a system that can accelerate the speed and precisely control the shape and position of the crystals.

References

1 Wang, J. (2013). *Nanomachines*. Weinheim: Wiley-VCH.

2 Xu, L., Mou, F., Gong, H. et al. (2017). Light-driven micro/nanomotors: from fundamentals to applications. *Chem. Soc. Rev.* 46 (22): 6905–6926.

3 Parmar, J., Ma, X., Katuri, J. et al. (2015). Nano and micro architectures for self-propelled motors. *Sci. Technol. Adv. Mater.* 16 (1): 014802.

4 Zhao, G. and Pumera, M. (2012). Macroscopic self-propelled objects. *Chem. Anal. Asian J.* 7: 1994–2002.

5 Mei, Y., Solovev, A.A., Sanchez, S., and Schmidt, O.G. (2011). Rolled-up nanotech on polymers: from basic perception to self-propelled catalytic microengines. *Chem. Soc. Rev.* 40 (5): 2109–2119.

6 Pumera, M. (2010). Electrochemically powered self-propelled electrophoretic nanosubmarines. *Nanoscale* 2 (9): 1643–1649.

7 Sánchez, S. and Pumera, M. (2009). Nanorobots: the ultimate wireless self-propelled sensing and actuating devices. *Chem. Anal. Asian J.* 4: 1402–1410.

8 Wang, J. (2009). Can man-made nanomachines compete with nature biomotors. *ACS Nano* 3 (1): 4–9.

9 Paxton, W.F., Sundararajan, S., Mallouk, T.E., and Sen, A. (2006). Chemical locomotion. *Angew. Chem. Int. Ed. Engl.* 45 (33): 5420–5429.

10 Paxton, W.F., Sen, A., and Mallouk, T.E. (2005). Motility of catalytic nanoparticles through self-generated forces. *Chem. Eur. J.* 11 (22): 6462–6470.

11 Gao, D., Ding, W., Nieto-Vesperinas, M. et al. (2017). Optical manipulation from the microscale to the nanoscale: fundamentals, advances and prospects. *Light Sci. Appl.* 6: e17039.

12 Merino, E. (2011). Synthesis of azobenzenes: the coloured pieces of molecular materials. *Chem. Soc. Rev.* 40 (7): 3835–3853.

13 Bandara, H.M.D. and Burdette, S.C. (2012). Photoisomerization in different classes of azobenzene. *Chem. Soc. Rev.* 41 (5): 1809–1825.
14 Delaire, J.a. and Nakatani, K. (2000). Linear and nonlinear optical properties of photochromic molecules and materials. *Chem. Rev.* 100 (5): 1817–1845.
15 Natansohn, A. and Rochon, P. (2002). Photoinduced motions in azo-containing polymers. *Chem. Rev.* 102 (11): 4139–4175.
16 Beharry, A.A. and Woolley, G.A. (2011). Azobenzene photoswitches for biomolecules. *Chem. Soc. Rev.* 40 (8): 4422–4437.
17 Ikeda, T., Mamiya, J., and Yu, Y. (2007). Photomechanics of liquid-crystalline elastomers and other polymers. *Angew. Chem. Int. Ed. Engl.* 46 (4): 506–528.
18 Seki, T. (2018). A wide array of photoinduced motions in molecular and macromolecular assemblies at interfaces. *Bull. Chem. Soc. Jpn.* 91: 1026–1057.
19 Rau, H. and Yu-Quan, S. (1988). Photoisomerization of sterically hindered azobenzenes. *J. Photochem. Photobiol. A Chem.* 42: 321–327.
20 Forber, C.L., Kelusky, E.C., Bunce, N.J., and Zerner, M.C. (1985). Electronic spectra of cis-and trans-azobenzenes: consequences of ortho substitution. *J. Am. Chem. Soc.* 107 (21): 5884–5890.
21 Bortolus, P. and Monti, S. (1979). Cis-trans photoisomerization of azobenzene. Solvent and triplet donors effects. *J. Phys. Chem.* 83 (6): 648–652.
22 Siampiringue, N., Guyot, G., Monti, S., and Bortolus, P. (1987). The cis → trans photoisomerization of azobenzene: an experimental re-examination. *J. Photochem.* 37 (1): 185–188.
23 Rau, H. and Lüddecke, E. (1982). On the rotation-inversion controversy on photoisomerization of azobenzenes. experimental proof of inversion. *J. Am. Chem. Soc.* 104 (6): 1616–1620.
24 Asano, T., Okada, T., Shinkai, S. et al. (1981). Temperature and pressure dependences of thermal cis-to-trans isomerization of azobenzenes which evidence an inversion mechanism. *J. Am. Chem. Soc.* 103 (17): 5161–5165.
25 Otruba, J.P. and Weiss, R.G. (1983). Liquid crystalline solvents as mechanistic probes. 11. The syn ↦ anti thermal isomerization mechanism of some low-"bipolarity" azobenzenes. *J. Org. Chem.* 48 (20): 3448–3453.
26 Kuriyama, Y. and Oishi, S. (1999). Mecchanism of thermal isomerization of azobenzene in zeolite cavities. *Chem. Lett.* 28 (10): 1045–1050.
27 Nagamani, S.A., Norikane, Y., and Tamaoki, N. (2005). Photoinduced hinge-like molecular motion: studies on xanthene-based cyclic azobenzene dimers. *J. Org. Chem.* 70 (23): 9304–9313.
28 Norikane, Y., Kitamoto, K., and Tamaoki, N. (2003). Novel crystal structure, cis-trans isomerization, and host property of meta-substituted macrocyclic azobenzenes with the shortest linkers. *J. Org. Chem.* 68 (22): 8291–8304.
29 Yoshii, K., Machida, S., and Horie, K. (2000). Local free volume and structural relaxation studied with photoisomerization of azobenzene and persistent spectral hole burning in poly(alkyl methacrylate)s at low temperatures. *J. Polym. Sci., Part B Polym. Phys.* 38 (23): 3098–3105.
30 Mita, I., Horie, K., and Hirao, K. (1989). Photochemistry in polymer solids. 9. Photoisomerization of azobenzene in a polycarbonate film. *Macromolecules* 22: 558–563.

31 Victor, J.G. and Torkelson, J.M. (1987). On measuring the distribution of local free volume in glassy polymers by photochromic and fluorescence techniques. *Macromolecules* 20 (9): 2241–2250.

32 Tsuda, M. and Kuratani, K. (1964). Isomerization of cis-azobenzene in the solid phase. *Bull. Chem. Soc. Jpn.* 37: 1284–1288.

33 Nakayama, K., Jiang, L., Iyoda, T. et al. (1997). Photo-induced structural transformation on the surface of azobenzene crystals. *Jpn. J. Appl. Phys.* 36: 3898–3902.

34 Ichimura, K. (2009). Reversible photoisomerisability and particle size changes of mill-dispersed azobenzene crystals in water. *Chem. Commun.*: 1496–1498.

35 Ichimura, K. (2000). Photoalignment of liquid-crystal systems. *Chem. Rev.* 100 (5): 1847–1873.

36 Ichimura, K., Suzuki, Y., Seki, T. et al. (1988). Reversible change in alignment mode of nematic liquid crystals regulated photochemically by "command surfaces" modified with an azobenzene monolayer. *Langmuir* 4 (5): 1214–1216.

37 Fukuhara, K., Nagano, S., Hara, M., and Seki, T. (2014). Free-surface molecular command systems for photoalignment of liquid crystalline materials. *Nat. Commun.* 5: 3320.

38 Seki, T. (2016). Light-directed alignment, surface morphing and related processes: recent trends. *J. Mater. Chem. C* 4 (34): 7895–7910.

39 Tazuke, S., Kurihara, S., and Tomiki, I. (1987). Amplified image recording in iquid crystal media by means of photochemically triggered phase transition. *Chem. Lett.*: 911–914.

40 Ikeda, T. and Tsutsumi, O. (1995). Optical switching and image storage by means of azobenzene liquid-crystal films. *Science* 268 (5219): 1873–1875.

41 Ikeda, T. (2003). Photomodulation of liquid crystal orientations for photonic applications. *J. Mater. Chem.* 13 (9): 2037–2057.

42 Yamamoto, T., Nishiyama, I., and Yokoyama, H. (2007). Novel photoinduced phase transition observed in three-dimensional liquid-crystalline phase of azobenzene compound. *Chem. Lett.* 36 (9): 1108–1109.

43 Kim, D.Y., Tripathy, S.K., Li, L., and Kumar, J. (1995). Laser-induced holographic surface relief gratings on nonlinear optical polymer films. *Appl. Phys. Lett.* 66 (10): 1166–1168.

44 Rochon, P., Batalla, E., and Natansohn, A. (1995). Optically induced surface gratings on azoaromatic polymer films. *Appl. Phys. Lett.* 66 (2): 136–138.

45 Yu, H. and Ikeda, T. (2011). Photocontrollable liquid-crystalline actuators. *Adv. Mater.* 23: 2149–2180.

46 White, T.J. and Broer, D.J. (2015). Programmable and adaptive mechanics with liquid crystal polymer networks and elastomers. *Nat. Mater.* 14 (11): 1087–1098.

47 Bushuyev, O.S., Aizawa, M., Shishido, A., and Barrett, C.J. (2018). Shape-shifting azo dye polymers: towards sunlight-driven molecular devices. *Macromol. Rapid Commun.* 39 (1): 1–14.

48 Norikane, Y., Hirai, Y., and Yoshida, M. (2011). Photoinduced isothermal phase transitions of liquid-crystalline macrocyclic azobenzenes. *Chem. Commun.* 47 (6): 1770–1772.

49 Uchida, E., Sakaki, K., Nakamura, Y. et al. (2013). Control of the orientation and photoinduced phase transitions of macrocyclic azobenzene. *Chem. Eur. J.* 19: 17391–17397.

50 Hoshino, M., Uchida, E., Norikane, Y. et al. (2014). Crystal melting by light: X-ray crystal structure analysis of an azo crystal showing photoinduced crystal-melt transition. *J. Am. Chem. Soc.* 136: 9158–9164.

51 Norikane, Y., Uchida, E., Tanaka, S. et al. (2014). Photoinduced crystal-to-liquid phase transitions of azobenzene derivatives and their application in photolithography processes through a solid–liquid patterning. *Org. Lett.* 16 (19): 5012–5015.

52 Norikane, Y., Uchida, E., Tanaka, S. et al. (2016). Photoinduced phase transitions in rod-shaped azobenzene with different alkyl chain length. *J. Photopolym. Sci. Technol.* 29 (1): 149–157.

53 Akiyama, H. and Yoshida, M. (2012). Photochemically reversible liquefaction and solidification of single compounds based on a sugar alcohol scaffold with multi azo-arms. *Adv. Mater.* 24 (17): 2353–2356.

54 Akiyama, H., Kanazawa, S., Okukyama, Y. et al. (2014). Photochemically reversible liquefaction and solidification of multiazobenzene sugar-alcohol derivatives and application to reworkable adhesives. *ACS Appl. Mater. Interfaces* 6: 7933–7941.

55 Baroncini, M., D'Agostino, S., Bergamini, G. et al. (2015). Photoinduced reversible switching of porosity in molecular crystals based on star-shaped azobenzene tetramers. *Nat. Chem.* 7: 634–640.

56 Ishiba, K., Morikawa, M.-A., Chikara, C. et al. (2015). Photoliquefiable ionic crystals: a phase crossover approach for photon energy storage materials with functional multiplicity. *Angew. Chem. Int. Ed.* 54 (5): 1532–1536.

57 Zhou, H., Xue, C., Weis, P. et al. (2017). Photoswitching of glass transition temperatures of azobenzene-containing polymers induces reversible solid-to-liquid transitions. *Nat. Chem.* 9 (October): 145–151.

58 Kikkawa, Y., Tanaka, S., and Norikane, Y. (2017). Photo-triggered enzymatic degradation of biodegradable polymers. *RSC Adv.* 7 (88): 55720–55724.

59 Uchida, E., Azumi, R., and Norikane, Y. (2015). Light-induced crawling of crystals on a glass surface. *Nat. Commun.* 6: 7310.

60 Norikane, Y., Tanaka, S., and Uchida, E. (2016). Azobenzene crystals swim on water surface triggered by light. *CrystEngComm* 18 (38): 7225–7228.

61 Yue, Y., Norikane, Y., Azumi, R., and Koyama, E. (2018). Light-induced mechanical response in crosslinked liquid-crystalline polymers with photoswitchable glass transition temperatures. *Nat. Commun.* 9 (1): 3234.

62 Hu, J., Li, X., Ni, Y. et al. (2018). A programmable and biomimetic photo-actuator: a composite of a photo-liquefiable azobenzene derivative and commercial plastic film. *J. Mater. Chem. C* 6: 10815–10821.

63 Okui, Y. and Han, M. (2012). Rational design of light-directed dynamic spheres. *Chem. Commun.* 48: 11763–11765.

64 Yamamoto, T., Norikane, Y., and Akiyama, H. (2018). Photochemical liquefaction and softening in molecular materials, polymers, and related compounds. *Polym. J.* 50: 551–562.

65 Ichimura, K., Oh, S.-K., and Nakagawa, M. (2000). Light-driven motion of liquids on a photoresponsive surface. *Science* 288: 1624–1626.

66 Oh, S., Nakagawa, M., and Ichimura, K. (2002). Photocontrol of liquid motion on an azobenzene monolayer. *J. Mater. Chem.* 12: 2262–2269.

67 Berna, J., Leigh, D., Lubomska, M. et al. (2005). Macroscopic transport by synthetic molecular machines. *Nat. Mater.* 4: 704–710.

68 Chaudhury, M.K. and Whitesides, G.M. (1992). How to make water run uphill. *Science* 256: 1539–1541.

69 Astumian, R.D. (1997). Thermodynamics and kinetics of a Brownian motor. *Science* 276: 917–922.

70 Koichiro Saito, Mio Ohnuma, and Yasuo Norikane (2019). Negative phototactic behaviour of crystals on a glass surface. *Chem. Commun.* DOI: 10.1039/c9cc03826d.

5

Bending, Jumping, and Self-Healing Crystals

Panče Naumov[1,2]*, Stanislav Chizhik*[3,4]*, Patrick Commins*[1]*, and Elena Boldyreva*[4,5]

[1] *New York University Abu Dhabi, P.O. Box 129188, Saadiyat Island, Abu Dhabi, United Arab Emirates*
[2] *Radcliffe Institute for Advanced Study Harvard University 10 Garden Street, Cambridge MA 02138, USA*
[3] *Institute of Solid State Chemistry and Mechanochemistry, Group of the Reactivity of Solids, Siberian Branch of Russian Academy of Sciences, ul. Kutateladze, 18, Novosibirsk 630128, Russia*
[4] *Novosibirsk State University, Chair of the Solid State Chemistry, Pirogova Street, 2, Novosibirsk 630090, Russia*
[5] *Boreskov Institute of Catalysis, Siberian Branch of Russian Academy of Sciences, Laboratory of the Advanced Synchrotron Radiation Studies Lavrentieva ave, 5, Novosibirsk 630090, Russia*

5.1 Bending Crystals

5.1.1 General Mechanism of Crystal Bending

Bending of slender crystals irradiated from one side is the simplest, yet remarkable mechanical response of an ordered entity such as a crystal. Large macroscopic changes in the crystal geometry can be achieved in a highly controllable and reproducible manner. The first examples of photoinduced bending were documented as early as in the 1980s [1–5], and the field was expanded in the 2000s, with a number of new and significant reports explaining these fundamental properties of molecular crystals [6–10]. The work in this research field published before 2015 has been summarized in a recent extensive review on mechanically responsive crystals [11]. Since then published publications were included in a more recent paper [12], and in the meantime new, important articles have become available.

Crystal bending results from nonuniformity of its transformation and the corresponding lattice strain that develops across the crystal normal to the irradiated surface and is caused by light absorption. The theory describing the photoinduced bending [11] was recently verified by high-precision, detailed experiments [12]. The model assumes several conditions, which should be met if any photobending crystal is considered suitable for any practical applications, regardless of the underlying chemistry, and they can be derived from a detailed quantitative model describing the process:

(1) *First, pure bending can accompany nonuniform transformation if the symmetry of a crystal is not too low*, i.e. the crystal is at least orthorhombic [11]. We note that pure bending is also possible with monoclinic structures; however,

Mechanically Responsive Materials for Soft Robotics, First Edition. Edited by Hideko Koshima.

only under special conditions that include specifically a certain orientation of the unique symmetry plane/axis with respect to the irradiated crystal face. Generally, for monoclinic and, especially, for triclinic crystals the bending is very likely to be accompanied by some degree of twisting. Although twisting is also of interest, it is rather difficult to be quantified completely and thoroughly, because the irradiation conditions change significantly over time if the geometry of the crystal and the light source is fixed, and the crystal experiences variation in light flux as it twists. Also, the possibilities of practical applications of crystal twisting appear to be limited, and thus it remains underexplored. Therefore, we will not cover the analysis of crystal twisting in this chapter. Qualitative and semiquantitative approaches to the description of the macroscopic twisting were discussed previously [11].

(2) *Second, the transformation should be devoid of structural phase transitions or amorphization.* If the bending is aimed for practical applications in devices the transformation must be not only chemically reversible but also should be rapid. Chemical reversibility facilitates establishing quantitative relations between the macroscopic strain and parameters that characterize the elementary stages of the process, since the same crystal can be repeatedly used in a series of experiments, thereby reducing or even eliminating the data scatter from irreproducibility of the mechanical effects of different crystals.

(3) *Third, the highest efficiency of transformation of the absorbed light energy into work* (as measured by the amplitude and rate of tip deflection) *can be accomplished for thin crystals* when the thickness does not exceed the penetration depth of irradiation causing the transformation more than several times (if all other parameters are identical). Physically, this means that the transformation strains are translated into shape transformation almost completely, leaving only a negligible level of residual stress that does not produce any useful work. *This condition is in line with the general requirement for durability of a bending actuator.* The residual stress facilitates deterioration of crystal quality during bending by increasing the contribution of the plastic component of the deformation, as well as by cracking and fatigue. On the other hand, the overall toughness of the crystals against the reaction force depends on crystal thickness, and it should be higher for thicker crystals. This results in a counter-requirement related to the crystal thickness, which should be sufficient to resist the force. In order to satisfy these conditions for the crystal thickness, either the chemical composition of the crystal or the irradiation wavelength should be selected so that the light penetration depth is close to the required crystal thickness.

The crystal bending is a consequence of the structural strain, which is not uniform in the direction normal to the crystal surface. The strain is largest at the irradiated surface and decreases in the direction from the surface toward the depth of the crystal. These nonuniform strains are usually related to nonuniform distribution of the reaction products (isomers, dimers, etc.). The products act as dilation centers that locally distort the parent structure. The dilation centers are accumulated in the crystal during the irradiation. The rate of accumulation of these products is nonuniform across the crystal bulk, since the intensity of the light that penetrates into the crystal decreases exponentially from its surface toward the interior. The photochemical transformation requires a photon; however, not all absorbed photons can cause a photochemical transformation. Some photons are absorbed by the already formed reaction product. Moreover, not all photons that are absorbed by the reactant cause a reaction, and the probability of such event depends on the quantum yield of the photochemical transformation.

In the course of a photochemical transformation, the crystal goes through a series of states with nonuniform distribution of the product. This manifests as continual change in the macroscopic crystal deformation. If the crystal is irradiated from one side, it usually bends. Another possible outcome, which is perhaps less readily observable but is as equally important is elongation or shortening of the crystal. One of the most difficult yet least studied problems is the extent to which the light absorption and the resulting macroscopic strain are coupled to each other.

The variety of phenomena observed when crystals are exposed to light can be classified based on the delay between the action (excitation by light absorption) and the response (crystal deformation). Typically, bending of a crystal caused by a photochemical transformation occurs in a quasi-static mode when it is not necessary to take into account the dynamic effects (i.e. the kinetic energy gained during bending) as the movements caused by corresponding deformations are much slower compared to the sound propagation in the solid. In a rough approximation, in order to approach deformation rate close to the limit determined by the rate of sound propagation, a measurable bending of a crystal that is 1 mm long would require chemical transformation that propagates within $\sim$1 μs or faster. Taking into account that each transformation act requires at least one photon, this would require absorption by the crystal of several hundred watts of light. Even if a suitable source of light is available for this purpose, such strong light would probably melt, sublime, or decompose the crystal. Therefore, all realistic examples of crystal photobending always correspond to quasi-static deformation.

The above reasoning means that the momentary distribution of the reaction product (i.e. of the dilation centers) through the crystal bulk directly determines the macroscopic deformation. The bending is always coupled to redistribution of the product in the crystalline bulk; however, this does not mean that the bending is necessarily synchronized with the irradiation by light. The latter would be true only if the excited state that is obtained by light absorption is rapidly transformed into a reaction product that does not undergo any further transformations – chemical reactions or physical processes, including slow relaxation processes related to optimization of the molecular conformations or intermolecular bonds in the actual crystalline environment. Then and only then the macroscopic bending would be synchronized with the light absorption – the crystal deformation would be initiated when the light is switched on and terminated immediately when the light is switched off. Hereafter, we shall refer to such photobending mode as *synchronous*.

In an alternative scenario, the primary photochemical transformation can be followed by other chemical or physical processes, which have comparable or lower rates as the rate of the primary photoreaction. In this case, the accumulation of the intermediate or final products/states will no longer be in synchrony with the irradiation by light. Each stage of such complex transformation sequence can be described by an individual kinetic equation, independent of the irradiation. The macroscopic crystal deformation will be delayed with respect to the irradiation. Accordingly, bending can be observed not *during* the irradiation, but with a delay *after* the light has been switched off [13]. This type of bending can be referred to as *asynchronous.*

A photochemical transformation that causes synchronous bending can be expected to be of a single-crystal-to-single-crystal type. It can occur when local distortions that are introduced into the starting crystal structure by the transformed molecules are relatively small and can be accommodated by the lattice environment. If the local perturbations are large, then plastic deformation or structural amorphization become likely. The amorphous state is metastable; the disordered/amorphous structure will further structurally relax and possibly recrystallize to form a new phase. These processes will be inevitably delayed, and the mechanical response in the case of a photoreaction accompanied by plastic deformation or amorphization will be asynchronous. A mechanical response accompanied by disorder/amorphization cannot be considered as reversible, even if the photochemical reaction itself is reversible. Such systems are not likely to be suitable for practical application in devices.

5.1.2 Kinetic Model of the Transformation

Let us consider a model that can describe the kinematics of a synchronous photoinduced bending. We shall assume for clarity that the photochemical transformation induced by light is a reversible photoisomerization. The starting chemical species (**A**) transforms into an excited state (**A***) by absorbing a photon. This excited state transforms with a certain probability into a product species (**B**). This form **B** is thermodynamically less stable than form **A** and transforms thermally back to the starting species **A** (Scheme 5.1).

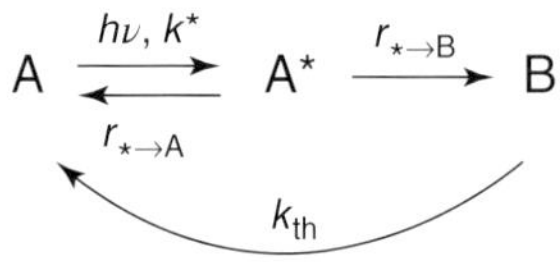

Scheme 5.1 Scheme of the thermoreversible reaction of photoisomerization.

The effective rate constant of formation of the excited state **A*** is determined by the rate of absorption of photons by the reactive species. It is proportional to the normalized light intensity (in photons s^{-1} cm^{-2}) and the light absorption coefficient μ_A (in cm^{-1}) for the species **A**, which are present in the starting crystal with a concentration c_0: $k^* = I\mu_A/c_0$. The absorption coefficient μ_A, which is used in the Lambert–Beer equation, depends on the wavelength, and its maximum value corresponds to the UV–Vis absorption maximum. In the context of the phenomena discussed in this chapter, it is important that molecules in a crystal are oriented in a regular way with respect to the incident light. The value of μ_A depends on the orientation of the molecule with respect to the polarization vector of light.

In order to estimate the value of k^*, one can take into account that the absorption coefficient for a compound in a solution with concentration c is related to the value of extinction ε as $\mu = \varepsilon c \ln 10$. Therefore, $k^* \sim I\varepsilon_A$. For a light source with normalized intensity $\sim 10^{18}$ photons s^{-1} cm^{-2} (typical for light-emitting diodes, LEDs) and extinction ranging from 10^2 (weakly absorbing, almost transparent compounds) to $10^5\,l\,mol^{-1}\,cm^{-1}$ (strongly absorbing dyes) this gives k^* in the range from 0.1 to $100\,s^{-1}$.

Typical lifetimes of excited states that are related to the rate constants r_i in the above reaction scheme can be in the range 10^{-10} to 10^{-8} seconds. Therefore, the values of constants r_i are several orders of magnitude larger than k^*. This condition means that the kinetic equation describing the change in the concentration of $\mathbf{A}^*$ can be considered in a quasistationary approximation when the excited state is in an equilibrium with the original species $\mathbf{A}$

$$[\mathbf{A}^*] = [\mathbf{A}]\frac{k^*}{r_{*\rightarrow A} + r_{*\rightarrow B}} \tag{5.1}$$

The concentration of the excited molecules is negligibly small as compared with the sum of concentrations of the two species, i.e. $[\mathbf{A}] + [\mathbf{B}] \approx c_0$. The kinetic equation for $\mathbf{B}$, in which a dimensionless normalized concentration $c_B = [\mathbf{B}]/c_0$ is used, is:

$$\frac{dc_B}{dt} = k_{ph}(1 - c_B) - k_{th}c_B \tag{5.2}$$

Equation (5.2) takes into account the reverse thermal transformation with a rate constant k_{th}. The rate constant of the phototransformation is defined as $k_{ph} = k^*\varphi$, where k^* is multiplied by the quantum yield of the reaction, $\varphi = r_{*\rightarrow B}/(r_{*\rightarrow A} + r_{*\rightarrow B})$. The value of the quantum yield depends on the ratio of the rates of de-excitation via different channels, and it is always <1. Obviously, the maximum efficiency of conversion of light energy into chemical transformation and the corresponding mechanical effects is the largest for $\varphi \sim 1$. The value of k_{th} can be estimated using equations for a first-order reaction. It depends, among else, on the activation energy and the temperature.

The temperature dependence of the phototransformation rate k_{ph} is not so significant, since it is determined only by a weak $\varphi(T)$ dependence and by $\mu_A(T)$ that results from a temperature shift and broadening of the UV–Vis spectrum. On the contrary, k_{th} depends strongly on temperature. This accounts for the existence of three possible scenarios of the transformation. At low temperatures, when $k_{th} \ll k_{ph}$, the transformation can be considered as a purely photochemical one and can proceed to completion. In the opposite case, when, $k_{th} \gg k_{ph}$, the transformation rate is determined solely by k_{th}, and gives a small stationary degree of transformation $c_B = k_{ph}/k_{th}$. In the intermediate cases, when $k_{th} \sim k_{ph}$, the stationary transformation degree $c_B = k_{ph}/(k_{th} + k_{ph})$ is reached at a rate that depends on $k_{th} + k_{ph}$.

Equation (5.2) does not reveal explicitly the fact that the light intensity decreases in the deeper layers of the crystal due to absorption; the value of k_{ph} decreases correspondingly. This can be taken into account for redefining the photoreaction constant as $k_{ph} = I_0\mu_A\varphi/c_0$. The equation can be then modified to take into account the decrease in light intensity when moving from the surface into the depth of the crystal, and the corresponding spatial distribution of the isomers is nonuniform:

$$\frac{dc_B(x)}{dt} = k_{ph}\frac{I(x)}{I_0}(1 - c_B(x)) - k_{th}c_B(x) \tag{5.3}$$

where I_0 is the initial intensity of light penetrating through the crystal surface (which may differ from the intensity of the incident light since part of light is reflected or scattered). The change in light intensity with depth x is described by the generalized Lambert–Beer law:

$$\frac{I(x)}{I_0} = \exp\left(-\int_0^x \mu(x')dx'\right) \tag{5.4}$$

The total absorption coefficient μ can be represented as a linear combination of absorption by the species **A** and **B** having different absorption coefficients, μ_A and μ_B:

$$\mu(x) = \mu_A(1 - c_B(x)) + \mu_B c_B(x) \tag{5.5}$$

If the two absorption coefficients are equal to each other ($\mu_A = \mu_B$), then Eq. (5.4) is transformed into the usual form of the Lambert–Beer law.

Equations (5.3)–(5.5) define the kinetics of the nonuniform accumulation of product **B** in the crystal bulk in a simple approximation that does not take into account additional possible complications such as change in irradiation conditions due to crystal bending, or a more complex $I(x)$ dependence that can arise because of the light refraction that occurs if light impinges on a convex surface formed by several faces. The problem can be solved analytically only in case when $\mu_A = \mu_B$, or, as an approximation, for low degrees of transformation either at the initial stage, when $(k_{ph} + k_{th})t \ll 1$, or in case of high temperatures, when $k_{th} \gg k_{ph}$, and the equilibrium of the total transformation is shifted strongly to the left.

In the two latter cases the concentration of the product **B** is small, and almost the entire light absorption accounts for absorption by the reactive species **A**: $\mu \approx \mu_A$. Then, at the initial stage of the reaction, when $(k_{ph} + k_{th})t \ll 1$, solving Eqs. (5.3)–(5.5) we obtain:

$$c_B(x, t) = k_{ph}t \exp(-\mu_A x) \tag{5.6}$$

The concentration profile decreases exponentially through the crystal bulk as a function of x and increases linearly with time. Characteristic length characterizing the profile is μ_A^{-1}.

At high temperatures, $k_{th} \gg k_{ph}$, and the solution of the Eqs. (5.3)–(5.5) is:

$$c_B(x, t) = \frac{k_{ph}}{k_{th}} \exp(-\mu_A x)(1 - e^{-k_{th}t}) \tag{5.7}$$

The profile remains exponential, but its amplitude grows similar to the transformation degree of a first-order reaction, which proceeds with a rate constant k_{th} until an equilibrium transformation degree is reached, described by a profile $k_{ph}k_{th}^{-1} \exp(-\mu_A x)$.

Analytical solution of the system of Eqs. (5.3)–(5.5) for the case when $\mu = \mu_A = \mu_B$ allows one to follow the propagation of the reaction till the high degrees of transformation. This solution is given by the equations:

$$c_B(x,t) = F(x)G(x,t)$$
$$F(x) = \frac{k_{ph}\exp(-\mu x)}{k_{ph}\exp(-\mu x) + k_{th}}$$
$$G(x,t) = 1 - \exp\{-t[k_{ph}\exp(-\mu x) + k_{th}]\} \quad (5.8)$$

The concentration profile is given by the product of two functions. $F(x)$ depends solely on depth x and determines the final distribution of the product concentration in the crystal on prolonged irradiation. $G(x, t)$ determines the kinetics of approaching this final state at different depths from the irradiated crystal surface.

Typical appearance of $F(x)$ is shown in Figure 5.1a. The curves calculated for different ratios of the rate constants of the direct (photo) and reverse (thermal) reactions are plotted to show the profile of the maximum attainable transformation degree.

The analysis of $F(x)$ makes it possible to predict when a transformation can be completed (or close to being completed) throughout the crystal bulk. To achieve completeness, the reverse transformation must be significantly slower than the direct photoreaction, that is $k_{th} \ll k_{ph}$. This is related to the slowing down of the photoreaction with depth because of the significant light absorption and decrease in light intensity when it reaches deeper crystal layers. To achieve 99% transformation of crystals with thickness of 2 μ^{-1}, the ratio k_{th}/k_{ph} must be $<3 \times 10^{-3}$. If also $G(x,t)$ function is included in the analysis, the time required to achieve different transformation degrees can be calculated. This function shows that the time required to achieve a high transformation degree, which is predicted by $F(x)$ at $k_{th} \ll k_{ph}$ at large depth, grows exponentially with the crystal depth.

In practice, when kinetics of a reaction is studied experimentally, one usually implicitly assumes the reaction has stopped if its rate has decreased several orders of magnitude relative to the initial rate (for example it has decreased

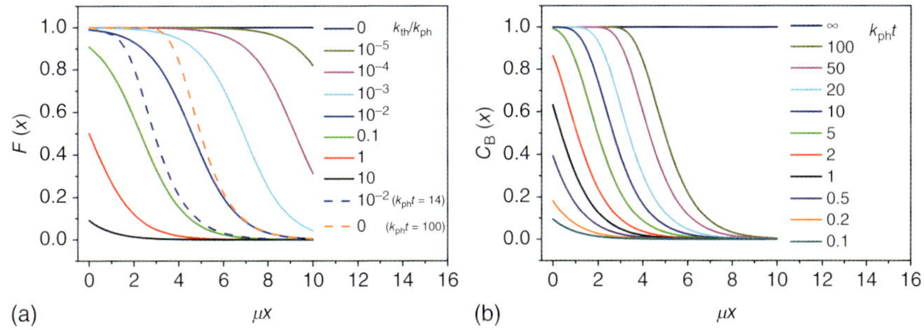

Figure 5.1 (a) Distribution of the maximum attainable transformation degree as a function of the depth from the crystal surface (in dimensionless coordinate μx) for different ratios of rate constants k_{th}/k_{ph} (dashed lines are added for comparison to intermediate transformation profiles reached when the transformation rate decreases a hundred times). (b) Propagation of the profile of the phototransformation into the crystal depth for $k_{th} = 0$ for different $k_{ph}t$.

hundred times, which can be accepted here as low rate criterion). The analysis of Eq. (5.8) shows that at $k_{th}/k_{ph} > 0.1$ the reaction is close to the final stage of the transformation defined by $F(x)$ when it slows down a hundred times. This result is almost independent of the crystal thickness as the reaction naturally stops within a layer $\sim\mu^{-1}$ due to fast reverse transformation. However, even at $k_{th}/k_{ph} = 10^{-2}$ the transformation degree cannot approach $F(x)$ in crystals thicker than $2\,\mu^{-1}$ when the reaction is considered finished by the low rate criterion. It can be seen in Figure 5.1a where the curve $c_B(x)$ corresponding to the criterion is plotted by dashed line ($k_{th}/k_{ph} = 10^{-2}$, $k_{ph}t \approx 14$), which significantly deviates from corresponding $F(x)$ at $\mu x > 2$. Finally, if $k_{th} = 0$ the transformation reaches 99% only within the layer of $\sim 4\,\mu^{-1}$ with the criterion (the dashed line $k_{th}/k_{ph} = 0$, $k_{ph}t \approx 100$ in Figure 5.1a).

Thus, the state with 99% transformation degree can be achieved in practice only with thin crystals, for example with thickness that does not exceed $\sim 2\,\mu^{-1}$. Of course, this limitation can be also overcome; however, the duration of the irradiation must be increased exponentially. The propagation of transformation profile $c_B(x)$ into the crystal depth with increasing time of the irradiation (at $k_{th} = 0$) is plotted in Figure 5.1b. After a complete transformation is achieved at the crystal surface ($k_{ph}t \sim 5$), further transformation is accompanied by a uniform shift of the transformation profile inside the crystal at the distance proportional to $\mu^{-1} \ln(k_{ph}t)$.

5.1.3 Mechanical Response of a Crystal to Irradiation

Mechanical response of the irradiated crystals is caused by deformations that are induced by the photoreaction. The process of product accumulation can be described as formation of a solid solution of the product **B** in the structure of the reactant **A**. The Vegard's law in crystallography relates approximately linearly the unit cell parameters of a crystal structure of a solid solution with the ratio of its components. The change in unit cell parameters can be described in terms of the strains that are caused by the formation of the solid solution of **B** in **A** with concentration c_B. Such strains will depend linearly on the concentration related by some coefficient, which is in a general case an anisotropic tensor. However, when describing the bending of thin crystals, it is sufficient to consider only one of the components of this tensor, namely the component that corresponds to structural strain along the long axis of the crystal. Structural strain in the axial direction ε_{tr}, which is caused by the photochemical transformation, can be described as

$$\varepsilon_{tr}(x) = \varepsilon_0 c_B(x) \tag{5.9}$$

where ε_0 is the linear strain coefficient.

The value ε_{tr} is not strictly speaking the final macroscopic strain of the crystal. Instead, this is the transformation strain that corresponds to real deformation in case it is constant throughout the crystal bulk (i.e. in the case of uniform transformation). In order to find real deformation of a crystal, one needs to solve the equations of the mechanical equilibrium, of the theory of elasticity, similarly to the case of nonuniform strains arising because of the nonuniform distribution of temperature in the bulk of an elastic solid.

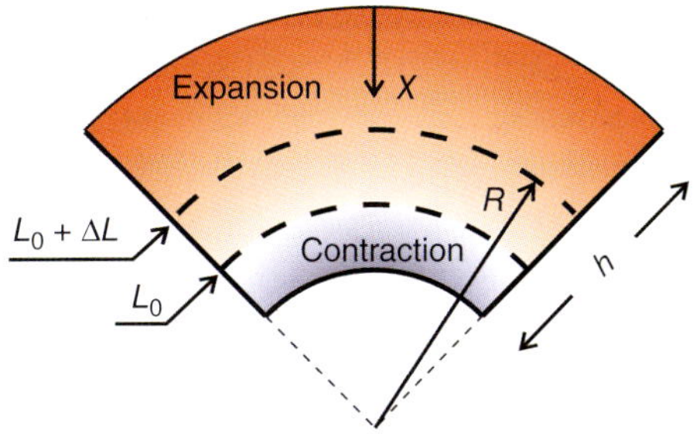

Figure 5.2 Schematic presentation of crystal bending showing the expanded and the contracted regions, with definition of some of the related parameters.

The solution of such a problem for the case of distribution of the transformation strain $\varepsilon_{tr}(x)$, which is uniform along the irradiated surface of a crystal and nonuniform normal to the crystal axis, describes uniform crystal bending combined with its general elongation or shortening, depending on the sign of ε_0. The elastic strain ε_{el}, corresponding to such change of the shape, defined relative to the initial state of a crystal (which is neither stretched nor squeezed during the process) can be presented as a linear function of x:

$$\varepsilon_{el}(x) = \frac{h/2 - x}{R} + \frac{\Delta L}{L_0} \tag{5.10}$$

where R is the curvature radius corresponding to the observed bending, ΔL is a general crystal elongation with respect to its initial length, L_0. Geometrically the deformation described by Eq. (5.10) is explained in Figure 5.2, which schematically represents a longitudinal cross section of a bent crystal. The central line of this cross section has the curvature radius R. The length of the arc at a distance x from the external convex surface depends on x as

$$L(x) = L_0 + \frac{L_0}{R}\left(\frac{h}{2} - x\right) + \Delta L \tag{5.11}$$

The relative elongation $(L - L_0)/L_0$ is equal to the strain ε_{el} in Eq. (5.10). The length of the arc L at the central line, $x = h/2$, exceeds the starting value by ΔL. A so-called neutral line (or a neutral surface in a three-dimensional object such as a crystal) is located at the depth of $x = h/2 + R\Delta L/L_0$. Expansion at this line is equal to zero, and the length of the arc is $L = L_0$. In a general case, this neutral line can be virtual and be located outside the crystal.

The longitudinal expansion ε_{el} results from the deformation of the transformation $\varepsilon_{tr}(x)$, which is nonuniformly distributed through the crystal bulk. The ε_{el} must compensate for $\varepsilon_{tr}(x)$ at its best.

The parameters of strain, which can be derived from the solution of the elasticity problem, are as follows:

$$\frac{1}{R} = \frac{12}{h^3}\int_0^h \varepsilon_{tr}(x)\left(\frac{h}{2} - x\right)\mathrm{d}x = \frac{12\varepsilon_0}{h^3}\int_0^h c_B(x)\left(\frac{h}{2} - x\right)\mathrm{d}x$$

$$\frac{\Delta L}{L_0} = \frac{1}{h}\int_0^h \varepsilon_{tr}(x)\mathrm{d}x = \frac{\varepsilon_0}{h}\int_0^h c_B(x)\mathrm{d}x \tag{5.12}$$

The sign of the R^{-1} (oftentimes referred to as "curvature," κ) depends on the sign of ε_0. It determines the bending direction: the product of transformation is

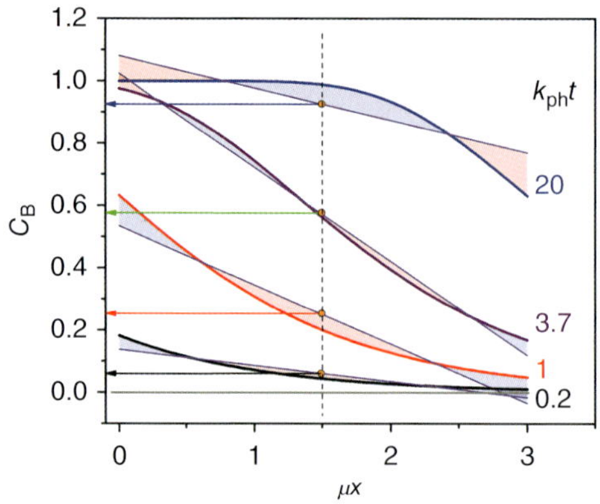

Figure 5.3 Consecutive product concentration profiles $c_B(x)$ in a crystal with the thickness 3 μ^{-1} calculated using Eq. (5.8) for different $k_{ph}t$ and in the absence of a reverse thermal reaction ($k_{th} = 0$) and their corresponding approximations by linear functions.

at a convex side of the crystal if the structure expands and at the concave side if it shrinks. This solution corresponds to the condition that no external forces act on the crystal. Physically, it means that a crystal adopts the shape that corresponds to the minimum residual strains in its bulk. The residual strain is the transformation strain that does not disappear after the crystal shape has changed. It is determined by the difference between the finite macroscopic deformations and the transformation strain: $\varepsilon_{el}(x) - \varepsilon_{tr}(x)$. Since the elastic energy is proportional to the square of the strain, Eq. (5.10) can be considered a least-squares linear approximation of the transformation strain $\varepsilon_{tr}(x)$. The value of the curvature R^{-1} corresponds to the average slope of $\varepsilon_{tr}(x)$, whereas $\Delta L/L_0$ is the value of the strain $\varepsilon_{tr}(x)$ averaged over the depth. Several consecutive $c_B(x)$ in a crystal with thickness 3 μ^{-1} calculated using Eq. (5.8) for $k_{th} = 0$ and approximated by linear functions are plotted in Figure 5.3. The approximating lines are up to a coefficient ε_0 equal to the strain functions in Eq. (5.10).

The slope of these curves is equal to $(R\varepsilon_0)^{-1}$. It first increases, until it reaches the maximum value (in this case at $k_{ph}t \approx 3.7$), and subsequently decreases (to zero for an infinite process). Average degree of transformation (indicated by arrows pointing at the ordinate axis) corresponds to the values $\Delta L/(L_0\varepsilon_0)$. It continuously grows with time, reaching unity at the infinity. The colored regions between a curve and its linear approximation correspond to the relative residual strains $(\varepsilon_{el}(x) - \varepsilon_{tr}(x))/\varepsilon_0$. These are the regions that correspond to the non-compensated expansion and compression of the crystal relative its non-strained state. The magnitude of the residual strains changes with time as the photoreaction takes place and the crystal bends – it increases starting from zero at the beginning to a maximum value achieved at a certain time moment. It then falls down to zero at the end (in case, when $k_{th} = 0$), and the reaction can be completed. The maximum attainable values of the residual strains strongly depend on the ratio of the crystal thickness to the characteristic absorption depth (Figure 5.4). At $\mu h < 1$ they are negligibly small. As the crystal thickness increases, they grow proportionally to the square of μh. This means that the maximum density of the elastic energy of the residual strains in the crystal bulk grows proportionally to $(\mu h)^4$, whereas the energy normalized to the thickness, which is important for achieving critical states that can cause plastic deformation or fragmentation, grows as $(\mu h)^5$. For

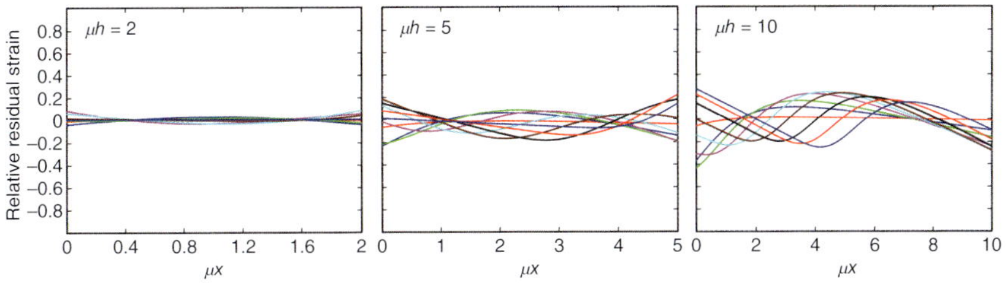

Figure 5.4 The profiles of the residual strains $(\varepsilon_{el}(x) - \varepsilon_{tr}(x))/\varepsilon_0$ in the crystals of different thickness at different time moments, $k_{ph}t$ from 0.1 to 200.

example for the crystals of $[Co(NH_3)_5NO_2]Cl(NO_3)$ crystal fragmentation was observed if their thickness exceeded ~50 μm (about $5\,\mu^{-1}$ for this compound) [12]. The critical value μh depends on both ε_0 and the mechanical strength of a material. As a preliminary estimate of the value of μh, above which irreversible mechanical relaxation becomes highly likely, is close to $\mu h \sim 5$ also for other compounds. After having reached this limit, the residual strains grow very fast.

In order to use thick crystals in a precritical mode (i.e. by avoiding plastic deformation or fragmentation), it is necessary to stop the photoreaction at low degrees of transformation.

As it can be concluded from Eq. (5.12), the time-dependent spatial distribution of the product concentration c_B in the crystal bulk corresponds to two synchronously changing characteristics of the macroscopic crystal deformation, the curvature R^{-1}, and the relative elongation $\Delta L/L_0$. Usually, in the experiments dedicated to photobending only the bending itself is followed, since this is the most apparent effect. The small crystal elongation is often neglected, since it is not always easy to measure it reliably or to even notice it. However, if the length change reaches 1% or more, the value becomes measurable reliably. These measurements can give useful information that is complementary to the measurements of the curvature, since the dependences of the two parameters, the curvature and length change, on the product concentration profile $c_B(x)$ are different.

The analysis of the model described above can provide a set of general qualitative characteristics of the mechanical response of a crystal to photoirradiation. One can obtain simple formula that can be used to calculate important values, such as the initial bending rate, maximum curvature, characteristic time required to achieve the maximum curvature, and the stationary curvature that can be achieved on prolonged irradiation, which are measurable experimentally.

If $k_{th} \ll k_{ph}$, then the crystal curvature increases with time, reaches a maximum value, and then decreases steadily to zero, as the transformation approaches completion. The initial rate of the change of curvature can be calculated as

$$\left(\frac{dR^{-1}}{dt}\right)_{ini} = k_{th}\varepsilon_0\mu_A\left[\frac{6(\mu_A h - 2 + (\mu_A h + 2)\exp(-\mu_A h))}{(\mu_A h)^3}\right] \tag{5.13}$$

The function of $\mu_A h$ in the square brackets is equal to 1 at $\mu_A h = 0$ and decreases monotonically as $\mu_A h$ increases (approximately as $\exp(-0.45\mu_A h)$, if $\mu_A h < 5$, and as $6/(\mu_A h)^2$, if $\mu_A h > 10$). The maximum curvature that can be achieved at an

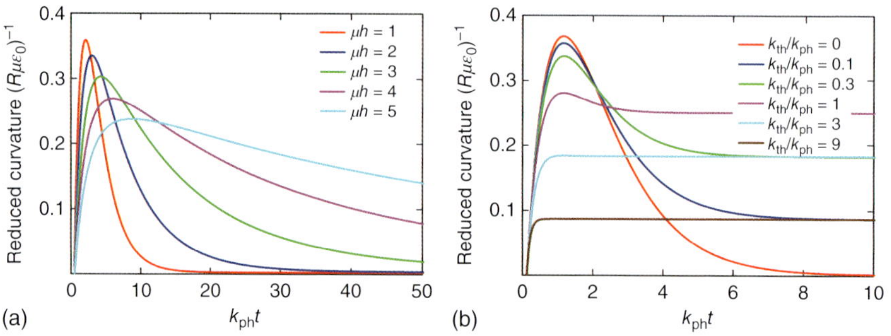

Figure 5.5 Kinetics of changing the curvature of the crystals: (a) crystals with different thickness at $k_{th} = 0$; (b) thin crystals with $\mu h \ll 1$, different ratios of the rate constants of the (direct) photoreaction and (reverse) thermal reaction, k_{th}/k_{ph}.

intermediate degree of transformation can be estimated for thin crystals (with $\mu h \ll 1$) for a constant averaged absorption coefficient $\mu = \mu_A = \mu_B$. The estimate gives

$$(R^{-1})_{max} = \varepsilon_0 \mu / e \tag{5.14}$$

and it is reached at a time moment corresponding to $k_{ph}t = 1$. For thicker crystals the value of the maximum curvature somewhat decreases, namely to ~0.65 of the maximum value given by Eq. (5.14) at $\mu h = 5$. The time that is necessary to achieve maximum curvature steadily grows with μh and can be well approximated as $k_{ph}t \approx \exp(0.45\mu h)$. As a crystal gets thicker, the initial rate of increasing the crystal curvature becomes lower (Eq. (5.13)), the value of the maximum curvature decreases (Eq. (5.14)), and can be achieved later (Figure 5.5a).

If the rate constant of the reverse thermal reaction is comparable with k_{ph}, then the transformation in the whole crystal bulk can never approach a completion (see Figure 5.1). This means that the crystal never unbends completely even at very prolonged irradiation, but its curvature reaches some final stationary value. For thin crystals with $\mu h < 1$ this stationary curvature can be approximated as

$$\left(\frac{1}{R}\right)_{fin} = \varepsilon_0 \mu \frac{k_{th}/k_{ph}}{(1 + k_{th}/k_{ph})^2} \tag{5.15}$$

The value of the stationary curvature is maximal at $k_{th} = k_{ph}$, when it is equal to $\varepsilon_0 \mu / 4$. At $k_{th} > k_{ph}$ the dependence of the crystal curvature versus irradiation time does not have an intermediate maximum: the curvature of the crystal simply reaches the stationary value determined by Eq. (5.15) (Figure 5.5b).

5.1.4 A Case Study, Linkage Isomerization of $[Co(NH_3)_5NO_2]Cl(NO_3)$

As an example, a very detailed quantitative study of crystal bending on irradiation by light has been carried out for the linkage isomerization of $[Co(NH_3)_5NO_2]Cl(NO_3)$ [12]. Experimental data obtained from the analysis of a series of images of bending crystals under variable temperature conditions (Figure 5.6) could be quantitatively described with a high precision using the mathematical model discussed above (Figure 5.7).

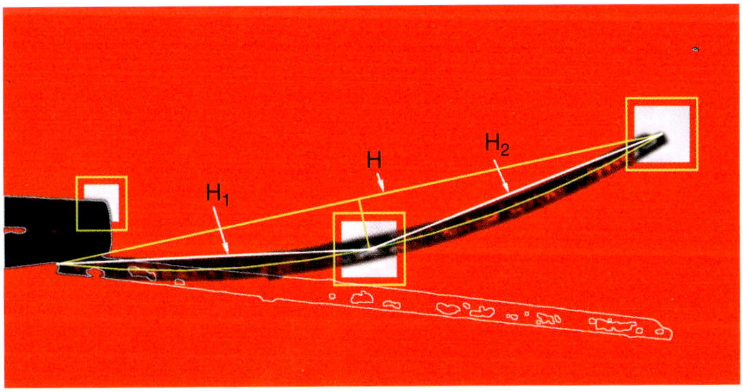

Figure 5.6 A method of measuring the parameters of crystal deformation based on the automatic recognition of the shape of the bent crystal.

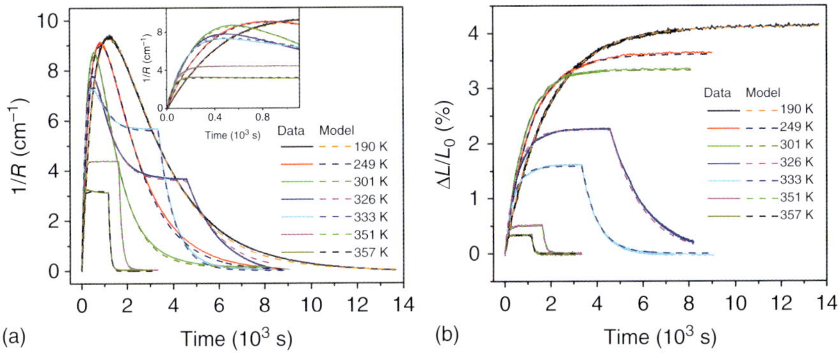

Figure 5.7 Kinetics of change in the crystal curvature (a) and relative crystal elongation (b) of the crystals of $[Co(NH_3)_5NO_2]Cl(NO_3)$ for crystal bending observed on irradiation with blue light ($\lambda = 465$ nm) at different temperatures: experimental data (solid lines) and the dependencies calculated based on the mathematical model described above (dashed lines).

5.1.5 Concluding Remarks

The modeling of photoinduced crystal bending described in this chapter is very important for not only developing the theory of this phenomenon, but also practical applications. Using the model, one can determine a series of parameters that cannot be measured experimentally at all or with such a high precision. These parameters include the absorption coefficients of the parent species **A** and the product **B** (μ_A and μ_B) in a given crystalline environment and in a given orientation with respect to the incident light, the quantum yield of the solid-state photochemical reaction (φ), and the rate constant of the reverse solid-state thermal reaction (k_{th}). The possibilities of obtaining these parameters simply by measuring the crystal bending are rather unique because, in contrast to reactions in solution, it is usually very difficult to obtain these parameters for a solid-state transformation. For practical applications it is also important that this modeling provides the opportunity to analyze the compounds that are potentially interesting for applications as photocontrollable actuators. The model can predict the optimum conditions to obtain the best mechanical response without destroying the crystal or deteriorating its properties.

5.2 Salient Crystals

5.2.1 Salient Effects

When some crystals are heated, cooled, irradiated with light, or even simply pricked with a hard object they are able to acquire a momentum and move. This process is usually (but not always) accompanied with disintegration whereupon the crystals move but they also split or blast. These dynamic phenomena are some of the most exciting recent developments in solid-state chemistry [14]. They are generally referred to as *salient effects*, and the crystals that are capable of these effects are called thermosalient, photosalient, and mechanosalient crystals when the mechanical response is induced by heat, light, or force, respectively [10, 15, 16]. The thermosalient phenomenon was first reported in 1983 [17], but the term was coined in 1987 [18]. In 2010, some of us briefly reported the photosalient effect in a crystal of a small molecule that is a model for the green fluorescent protein (GFP) chromophore [19], and we also reported the first detailed study of the thermosalient crystal oxitropium bromide [20]. The latter work built upon the PhD work of S. Zamir [21] and a brief conference proceedings report on the same compound [22]. Mechanistically the thermosalient effect is the simplest and it is also well explored, and most of the discussion here will be focused on this phenomenon. It is now well established, with over 30 documented compounds (for recent examples see Refs. [23–35]), with increasing reports on disintegrative phenomena induced by light (for examples see Refs. [36–40]) and it has become apparent that they are more common than it has been thought in the past (Figure 5.8).

(a)
(b)
(c)
(d)

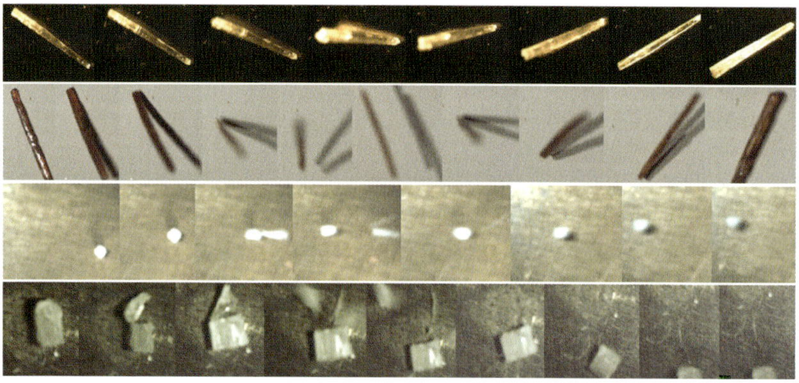

Figure 5.8 Motility in thermosalient crystals recorded with high-speed camera: 1,2,4,5-tetrabromobenzene (TBB, 2; a), (phenylazophenyl)palladium hexafluoroacetylacetonate (PHA, 1; b), *N′*-2-propylidene-4-hydroxybenzohydrazide (PHBH, 9; c), oxitropium bromide (OXTB, 4; d). The snapshots were taken at intervals of 1 ms. Source: Sahoo et al. 2013 [41]. Reproduced with permission of American Chemical Society.

The salient crystals can be considered as a subclass of a broader class of crystalline materials whose structure changes over time, known as *dynamic crystals* [11, 42, 43]. A unique and very important asset of the salient crystals is their ability to convert disordered motion in their environment (heat) or photons into ordered motion (work) on a macroscopic scale [41]. As discussed subsequently, perhaps even more important aspect of this phenomenon is that this occurs on very fast time scales with mechanisms considered to be analogous to that of martensitic transitions, and thus they are occasionally classified as *organic-based analogues of the inorganic martensites.* Together with other exotic mechanical properties of molecular crystals, such as other organic martensites [44–46], elastically and plastically reconfigurable crystals [42], shape-memory and self-healing crystals [47], some of which are the topic of other chapters in this book, these phenomena have brought some of the most exciting recent developments in solid-state chemistry, mechanochemistry, and crystal engineering, and have received an international acclaim. It should be noted, however, that the use of the term martensitic transition has been debated in the 1970s, as it cannot explain all details of these phase transitions in molecular crystals.

The motion and explosion of the crystals is visually appealing, and it is sometimes used for practical demonstrations to capture students' attention. This is particularly effective with compounds that have thermosalient transition temperature close to room temperature, such as 1,2,4,5-tetrabromobenzene (TBB) [48] and their motion can be induced by light heating. The observation of the violent shattering of the crystals of oxitropium bromide under microscope using halogen lamp was also the reason [20] for our motivation to study the mechanism of this effect. Although it is very difficult to quantify the motion of thermosalient (TS) crystals, and normally qualitative descriptions such as "violent," "strong," and "vigorous" are commonly used to describe these phenomena, in some cases the explosion of the crystals is particularly violent and crystals can leap over distances from several millimeters to several tens of centimeters. One of the strongest motions was observed with the compound named *im*ine of *acet*one (IMACET) where crystals were found up to 1 m from their original position [49]. As described subsequently even such qualitative description can sometimes be correlated with the structure, particularly in the case of closely related chemical structures.

It is important to note that not all disintegrative processes can be classified as thermo/photosalient effect. Crystals oftentimes disintegrate during decomposition or desolvation, releasing stable gaseous products, such as dioxygen, dinitrogen, carbon dioxide, water, nitrogen oxides, and other gases that can commonly cause disintegration, sometimes with ballistic outcome. Stoichiometric or occluded solvents such as water or organic solvents that were incorporated in the lattice during crystallization or other surface impurities can be overheated during the heating and can also contribute to either slow or very fast disintegration processes. However, these processes that are based on decomposition should not be regarded as salient effects, they are a mere consequence of the evaporation of the chemical products during decomposition. The term thermosalient effect is reserved for rapid mechanical motion or violent disintegration of crystals *during a phase transition*. A notable example that was reported recently was qualified as thermosalient, but does not appear to undergo a phase transition [50], although the reason for the motion of these crystals remains elusive. Slow expansion of the unit cell during heating cannot be taken as a reason for crystal motion because the crystal motility requires a momentum, which can only be acquired by springing off a solid surface, and that cannot be accomplished by slow expansion.

5.2.2 Mechanism of the Thermosalient Transition

The thermosalient transition is known to occur, upon either heating or cooling from the room temperature, to a phase that is structurally very similar to the original phase. Qualitatively, this phenomenon can be analyzed as a three-stage process:

(1) accumulation of strain in the lattice without change of the crystal packing,
(2) initiation of the phase transition by a defect or at the corner of the crystal,
(3) rapid propagation of the phase front throughout the crystal with the release of the mechanical strain by self-propulsion of the crystal.

The first two stages are the cause, while the third stage is the consequence of what is visually observed as a thermosalient effect [11]. The accumulation of strain occurs during the heating or cooling (depending on the phase transition temperature), and the crystal can be thought of as being able to absorb the strain due to its "softness." The intermolecular vibrations (phonons) that are responsible for the transition have been thoroughly studied in the case of TBB (for selected recent studies, see refs [51–53].). In *N′*-2-propylydene-4-hydroxybenzohydrazide, the transition was also attributed to spring-like phonons that soften by increasing the temperature and assist the phase transition [54]. In a nutshell, a low-frequency phonon, which is responsible for the transition softens upon transitioning, connects two minima that are very close in energy and are structurally also similar. In case of TBB, the critical vibration consists of very slight tilting of the aromatic rings [53], and reflects faithfully the small structural difference between the two phases [48]. Therefore, the softness could mean that the crystal can absorb the thermal energy, which is effectively distributed among the lattice vibrational modes, until one vibration takes the system over the barrier and to the new phase that is structurally very similar to the original phase. This scenario is in line with the structural similarity between the two phases, which in all but several cases [55–58], are isomoprhous and they are structurally very similar [41].

The first order of the thermosalient transitions and the rapid progression of the phase front throughout the crystal resembles in many ways the mechanism of the martensitic transitions observed with metals [59]. These are first-order, diffusionless "military" transitions. The TS crystals therefore are one of the subclasses of molecular solids that undergo such transitions, although other examples of martensitic transitions in molecular solids were reported recently. The rapid progression of the phase transition induces acoustic emission which can be detected on the surface of the crystal [20, 60].

Detailed variable temperature studies of the unit cells of several thermosalient compounds [47, 59, 61, 62] have shown that these compounds exhibit strong and anisotropic expansion before they undergo the thermosalient phase transition. Although this conclusion cannot be generalized for all thermosalient compounds and probably there are exceptions, the compounds that have been studied in detail thus far show anisotropic expansion (positive and negative unidirectional or bidirectional expansion), and some of them exhibit remarkably large expansion. For instance, between 223 and 348 K the alpha form of the palladium complex (phenylazophenyl)palladium hexafluoroacetylacetonate (PHA) [59] undergoes very strong positive thermal expansion along the *a*-axis of $260.4(3) \times 10^{-6}\ K^{-1}$ and much smaller expansion along the *b*-axis, $39.4(4) \times 10^{-6}\ K^{-1}$ (Figure 5.9). The strong expansion along the *a*-axis is compensated by a considerably strong negative thermal expansion along the *c*-axis of $-79.9(2) \times 10^{-6}\ K^{-1}$. In case of another thermosalient solid, the pyroglutamic acid [61], form α shows stronger than normal thermal expansion along the *a* and *c* axes, with thermal expansion coefficients $43.4(7) \times 10^{-6}\ K^{-1}$ and $23.7(4) \times 10^{-6}\ K^{-1}$, and considerably strong expansion along the *b*-axis, $105.5(5) \times 10^{-6}\ K^{-1}$. Form β shows a peculiar thermal expansion where two axes undergo negative thermal expansion with $-54.8(8) \times 10^{-6}\ K^{-1}$ and $-3.62(8) \times 10^{-6}\ K^{-1}$ while the *b*-axis undergoes exceptionally strong uniaxial positive thermal expansion with $303(1) \times 10^{-6}\ K^{-1}$. Another compound, *N′*-2-propylydene-4-hydroxybenzohydrazide, an IMACET [49, 62] was reported to undergo thermal expansion of $368(10) \times 10^{-6}\ K^{-1}$ [54]. The relationship between the thermal expansion during heating or cooling and the thermosalient effect warrants further attention and should be explored in detail.

The phase transitions associated with the thermosalient effect appear to be the fastest transitions in molecular crystals. The observed motion or disintegration of the crystals is the consequence of the very rapid progression of the transformation throughout the crystal, which has been directly observed by using high-speed video recording [59, 61]. These results provide evidence that the phase transition does not proceed simultaneously, but rather in a spatially resolved manner; however, the progression is very fast. Perhaps the best example that illustrates the difference in rate between the thermosalient and non-thermosalient transition is that of PHA [59]. In this material, which is pentamorphic and all five phases can be obtained by heating or cooling, only one of the phase transitions is thermosalient, that is between phases α and γ. From the two transitions at high temperature, α-to-γ and γ-to-β, the first one (342.7–354.5 K) is thermosalient while the latter (368.8 K) is not, providing opportunity to directly compare the mechanisms of the two transitions. High-speed recordings clearly show the difference in the mechanism: the thermosalient transition occurs starting from one end of the crystal, which acts as a defect, and occurs very rapidly toward the other end, with a rate of $0.54\ m\ s^{-1}$. The other, non-thermosalient transition is initiated at multiple nucleation points, from where it spreads radially within the crystal bulk at a rate $4.34 \times 10^{-5}\ m\ s^{-1}$. This is a striking example that shows the difference between TS and non-TS process in the same material that are only about 20 K apart [59].

In all cases, regardless whether it occurs above or below room temperature, the thermosalient effect is reversible. However, due to disintegrative processes such as explosion or fragmentation, the debris generated in the aftermath of the

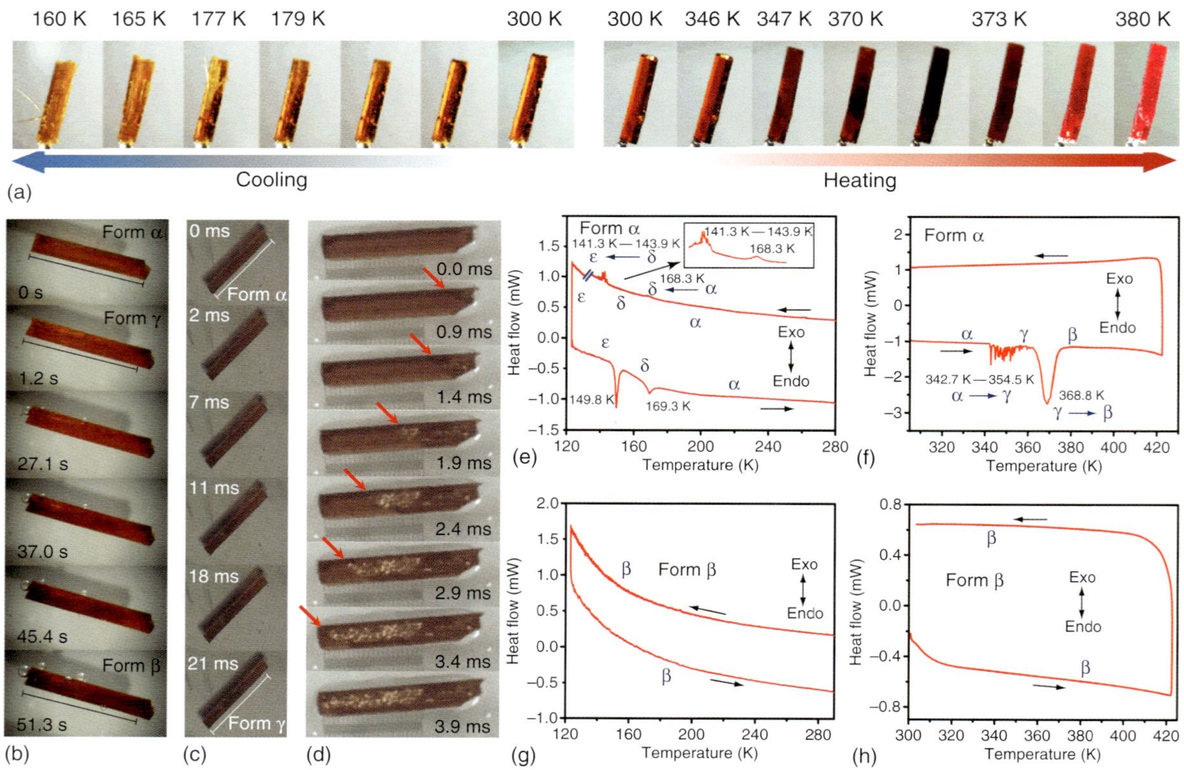

Figure 5.9 Crystal habit, thermochromic, and thermal profiles of PHA crystals. (a) Morphology and color of single crystals of PHA that were cooled and heated in a stream of nitrogen gas (the temperature values are uncorrected). Note the expansion of the crystal at high temperature and the splitting into fibrous microcrystals at low temperature. The overall brightness of the photographs was enhanced uniformly for clarity. (b–d) High-speed snapshots of the migration of the habit plane and color change during the phase transition. The edge of the habit plane is marked with red arrow in panel (d). (e–h) Heat effects and phase transitions upon thermal cycling of PHA monitored by differential scanning calorimetry (heating/cooling rate: 10 K min^{-1}). Source: Panda et al. 2014 [59]. Reproduced with permission of Springer Nature.

transition remains dynamic and it can move if it is taken over the phase transition, although usually its propensity for motion is much less pronounced. The motion of the smaller pieces that may have separated is also usually less pronounced, presumably due to accumulation of defects.

5.2.3 Thermal Signature of the Thermosalient Effect

The thermal profile of the thermosalient transition of unground crystals is characterized by a spiky ("saw-tooth") appearance [15, 16, 41]. By a series of experiments using both single crystals and multiplets (twins) it has been confirmed that – at least when the material is transferred over the phase transition for the first time – each of the individual peaks corresponds to transition of a single crystal [41]. When the transition is induced by heating, all crystals eventually transition up to a threshold temperature. It was also found that twinned crystals usually undergo the transition at lower temperature than the single crystals, as expected from the additional strain in their structure at the interface between the two components (Figure 5.10) [41, 48].

When the crystals are grinded following the usual procedure for preparation for differential scanning calorimetry (DSC), as expected, the individual peaks

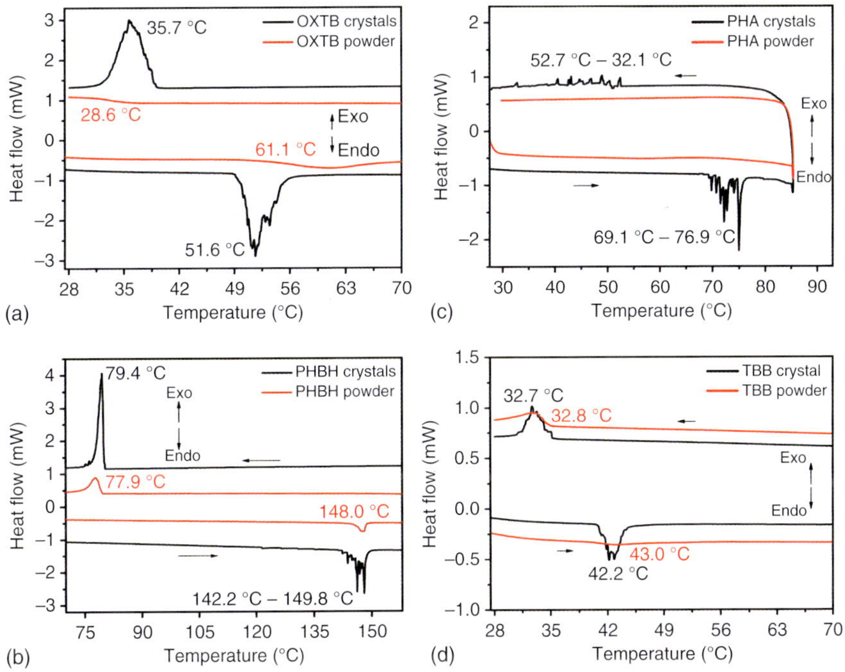

Figure 5.10 Thermal signature on heating and cooling of crystals (black curve) and powders (red curve) of thermosalient (jumping) solids: (phenylazophenyl)palladium hexafluoroacetylacetonate (PHA; b), 1,2,4,5-tetrabromobenzene (TBB; d), oxitropium bromide (OXTB; a), and *N*′-2-propylidene-4-hydroxybenzohydrazide (PHBH; c). Source: Sahoo et al. 2013 [41]. Reproduced with permission of American Chemical Society.

blend with each other into a much less intense but broader peak as a result of the increased spread in particle size and the shear strain applied during the grinding. It has been documented, however, that in the case of the thermosalient materials this procedure oftentimes results in alleviation of the thermal or even complete suppression of the effect [41]. This property appears to correlate with the pronounced softness of the thermosalient crystals. The grinding might affect the thermal profile in at least two ways: there could be a low limit of the particle size (i.e. minimal number of unit cells) for the crystal to accumulate a sufficient amount of strain to be able to propel itself, or the pressure applied during grinding may cause defects that dissipate the mechanical strain that is required for motion. Further studies are required to investigate the relative contribution of these and possibly other factors in what appears as suppression of the effect by grinding.

The transition temperature depends on the chemical composition, and thus it may be possible to control the temperature at which the thermosalient effect occurs. On the example of solid solutions of a metal coordination complex, it has been demonstrated that indeed one can systematically modulate the TS transition temperature by replacing the central metal ion with another ion of similar size [57]. This elegant approach provides opportunity to control, within a limited temperature range, the temperature of the mechanical effect. The size dependence of the thermal behavior of thermosalient solids begs the question of whether there is a low size threshold at which the phenomenon can still be observed. Microcrystals obtained by fragmentation during a TS phase transition are known to remain active, but the effect is notably and visibly reduced. A recent attempt to prepare nanocrystals of TS compound was successful only with the preparation of such crystals [63]. An anodic aluminium oxide (AAO) template was used to prepare the nanocrystals of TBB with controllable size, similar to a method used by others [64, 65]. Most of the nanocrystals, however, were polycrystalline, were in the less stable (high-temperature) phase, and sublimed at higher temperature.

5.2.4 Directionality of Motion

The disintegrative nature of the salient effects and the significant amount of kinetic energy carried away by the motile crystals of their fragments poses a substantial challenge with imaging of these phenomena. This exercise requires a high-speed camera with sufficient spatial and temporal resolution coupled to a microscope, tracking software, and a significant number of samples to extract meaningful statistics. The usual sample size consists of >100 crystals. A detailed study of the photosalient effects that accompanies the photoinduced linkage isomerization of the complex $[Co(NH_3)_5(NO_2)]Cl(NO_3)$ to $[Co(NH_3)_5(ONO)]Cl(NO_3)$, where the coordination atom of the nitro group switches from nitrogen to oxygen, shows that the crystals of this compound move preferentially in the direction of their longest axis [66]. The motility of any crystal can be described as one of the five kinematic effects, and occasionally they combine to one of the seven combined kinematic patterns that describe complex motions. In addition to the types of motion and disintegration, this kinematic analysis provides preferred directions of motion of the center of mass, which can be correlated with the directionality of the strains in the crystal. It was found that the direction of motion coincides with the direction in which the highest strain is accumulated due to the directionality of the strong intermolecular interactions [66]. In another case, the dependence of the effect on the face on which the crystal was sitting was investigated, and this analysis was facilitated by the enhanced preservation of integrity of these crystals [62]. It was found that when they are placed against solid surface, the motion of the crystals occurs in the direction of the unit cell expansion and opposite of the solid support. This result confirms that, within the stochastic events that can sometimes determine the outcome, the motion of the photosalient and thermosalient crystals is generally directional, because it is induced by rapid and preferred expansion along certain directions (Figure 5.11).

5.2.5 Effect of Intermolecular Interactions

At the early stages of this research, the thermosalient solids were categorized in three groups based on the molecular shape, flexibility, and substitution [41]. This classification generally determines the molecular packing and the strength and directionality of the intermolecular interactions. Recent comparative studies on three related thermosalient *N*-salicylideneanilines confirm that the thermosalient effects depend on the crystal structure [67, 68]. A critically important factor for occurrence of the salient effects is the stage of rapid progression of the phase front throughout the crystal bulk. The efficacy of this process depends on the strength and direction of the intermolecular interactions, and thus it is expected that changes such as hydrogen bonds in these interactions will affect the thermosalient response. The compound *N*′-2-propylydene-4-hydroxybenzohydrazide, an IMACET appears as a good choice to examine the effect of the strength of intermolecular interactions, because it contains both acidic and nonacidic hydrogens that can be deuterated [49]. Indeed, it was found that the deuteration of the NH/OH and methyl protons results in opposite trend in shifting the phase transition temperature for crystals of similar size and habit; while crystals deuterated in the acidic protons move at lower temperature than that of the protiated sample, crystals with deuterated methyl protons move at higher temperatures [62]. This result can be explained by concomitant action of two factors that have opposite effects – strengthening of the hydrogen bonds in the former case and no significant effect on the hydrogen bonds but increased molecular mass in the latter case. This result shows that strong intermolecular interactions such as hydrogen bonds, if they are present, do not only have important effect on the integrity of the crystal, but they also affect the temperature of the phase transition. In a very direct elegant study of this compound, Centore et al. have recently shown that the relation between the molecular motion and the macroscopic motion depends on the dimensionality of the hydrogen bonds [69]. Specifically, when the crystal contains hydrogen bonds, the retention of the hydrogen bonds along two directions is likely to afford retention of single crystal integrity, while retention of the hydrogen bonds in only one or none of the directions will usually result in reconstructive transition (Figure 5.12).

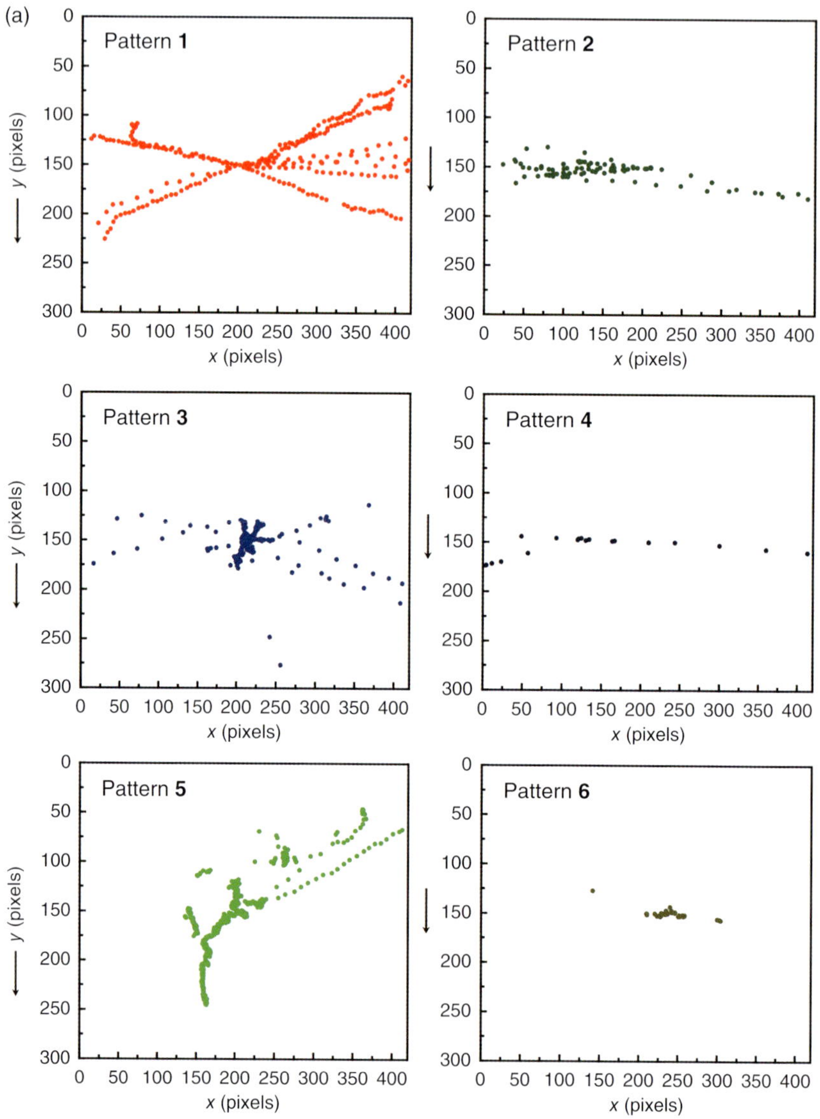

Figure 5.11 (a) Typical traces of the seven kinematic patterns observed with the photosalient crystals of 1 – *N* (the position on the *x* and *y* axes are expressed in pixels [px]). For tracking analysis, the crystals were approximated with a set of three quasicollinear points lying on the bisector of the longest crystal axis. All coordinates are normalized to the initial position of the crystal barycenter, which was set at (200150). The illustration on the top left typifies the motion of crystal 1 in 250 μs time-resolved snapshots. (b) Seven types of trajectories of the central tracking point simulating the barycenters of 1 – *N* crystals that perform photosalient motions, and a pie-type plot showing the frequency of occurrence of the effects. Source: Naumov et al. 2013 [66]. Reproduced with permission of Wiley-VCH Verlag GmbH & Co. KGaA, Weinheim.

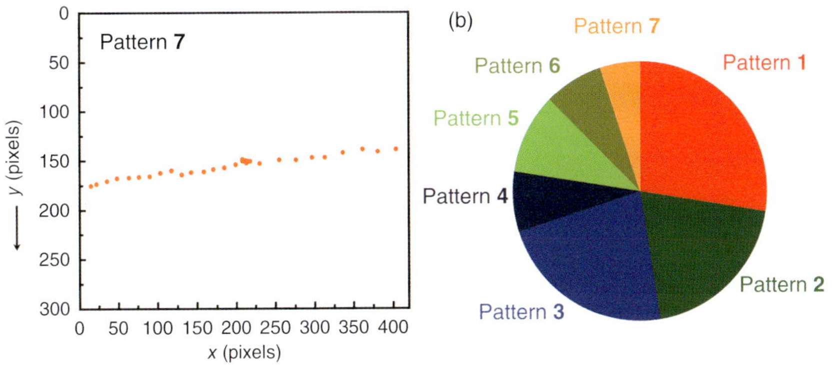

Figure 5.11 (*Continued*)

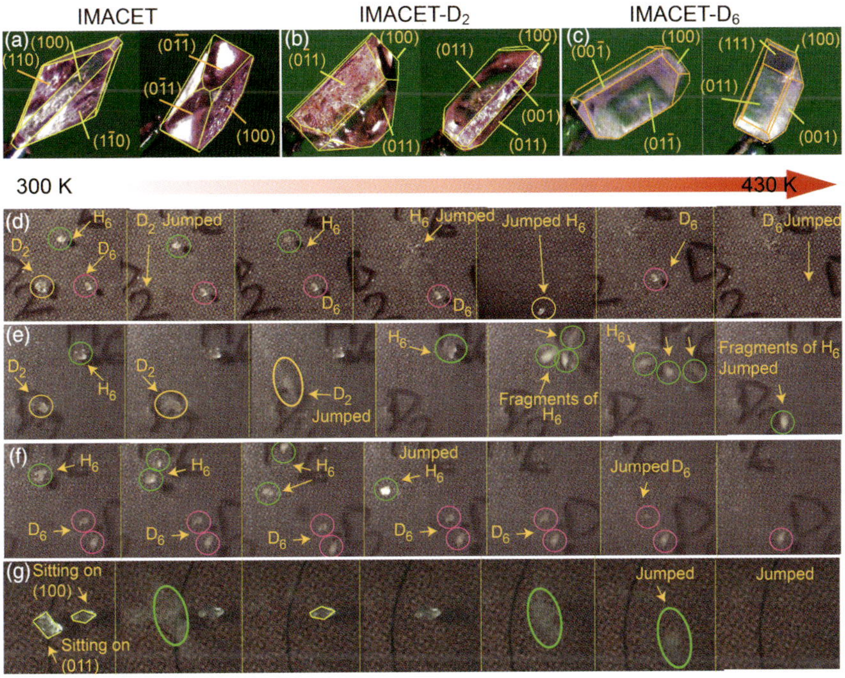

Figure 5.12 Face indices (a–c) and snapshots of the thermosalient effect in crystals of the three isotopic variants of IMACET (d–g). The images of protiated IMACET (denoted H_6 for clarity; green circles), IMACET-D_2 (D_2, yellow circles), and IMACET-D_6 (D_6, purple circles) in panels (d–g) were extracted from high-speed video recordings (recording rate: $2–5 \times 10^3\ s^{-1}$). Source: Naumov et al. 2013 [66]. Reproduced with permission of John Wiley & Sons.

5.2.6 Effect of Crystal Habit

The habit of the crystal, together with its structure, determine the distribution of the strain that develops inside the crystal, and thus it also affects the mechanical response. The correlation between the habit and the mechanical response has been thoroughly described in ref. [70], and it is also a subject of some of the other chapters in this book. It is empirically known that when exposed to light photoreactive needle-like crystals bend, thin plates tend to curl, and thick crystals neither bend nor curl. The correlation between the crystal aspect ratio and the type of the response in Figure 5.13 shows this relationship more clearly on a number of crystals of the same photoreactive compound having different habits expressed by the aspect ratio. As it can be concluded, indeed most of the long crystals simply bend, parts of thin crystals tend to curl, while the block-type crystals explode, that is they exhibit the photosalient effect.

5.2.7 Photosalient and Mechanosalient Effects

In addition to the thermosalient effect, which was reported first, this concept was later expanded to include crystal motion induced by light and pressure (force), and the so-called *photosalient* and *mechanosalient* effects were reported. In case of the photosalient effects, the motion or disintegration of the crystals is induced by light, which triggers a photochemical reaction (cases where light causes heating by photothermal effects are not included in this group). The chemical reaction should not release gaseous products that could also result in disintegration. Generally, in the photosalient crystals the mechanical effect is induced by the inability of the parent crystals lattice to accommodate the growing strain caused by the (nonvolatile) product. Although the exact mechanism remains unclear and requires further experiments to be confirmed, the mechanical effect can be a consequence of either the incompatibility of the growing product with the lattice of the reactant or a phase transition in the parent lattice induced by the growth of the product. In either case, the lattice of the reactant is not able to withstand the strain caused by the chemical reaction and it usually violently disintegrates. The kinetic profile of this reaction in case of dimerization provides experimental evidence that shows that this is a two-stage process [71]; however, further studies are needed to elucidate the nature of these two processes.

The notable examples of photosalient reactions are linkage isomerization in Co(III) complexes with small ligands [66] and photodimerization reactions, of either organic molecules or metal–organic compounds. The first example of a photosalient effect by dimerization of purely organic compound was provided by some of us with a GFP chromophore model [19], and the effect was investigated later with benzylidenedimethylimidazolinone [70]. The effect in metal–organic complexes has been demonstrated in a series of related compounds with chemically analogous ligands [71, 72]. The kinematics and the crystal structure were correlated in a set of nine structurally complex related materials AgL_2X_2 (L = 4-styrylpyridine, 2′-fluoro-4-styrylpyridine, and 3′-fluoro-4-styrylpyridine, X = BF_4, ClO_4, and NO_3). Crystals of six compounds were photosalient. It was concluded that the occurrence of the photosalient effect is determined by not only the nature of the ligand but also the crystal packing, which directs the magnitude, direction, and rate of volume expansion [72].

The mechanosalient effect is even a rarer phenomenon, and refers to motion of crystals when they are mechanically stimulated by applying a local pressure. The effect is a result of a delayed phase transition, where the crystal is taken

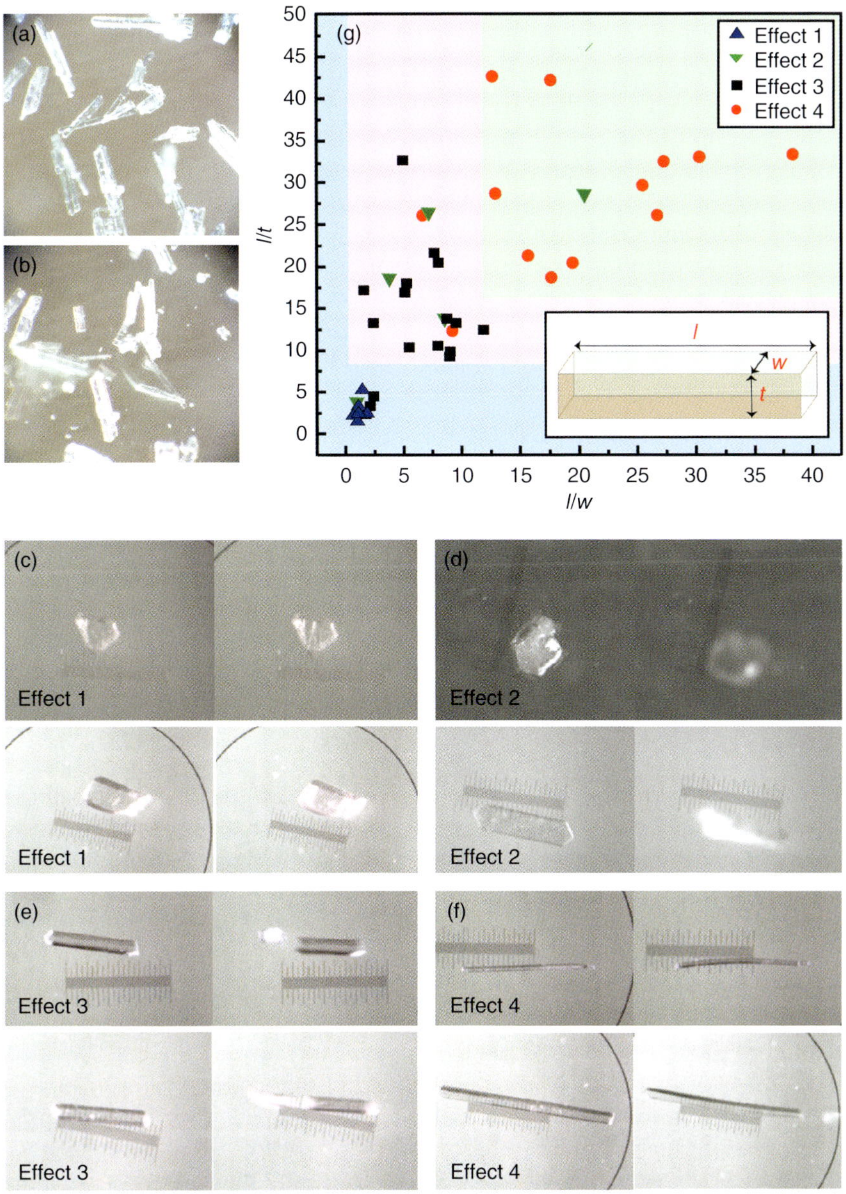

Figure 5.13 The effect of aspect ratio on the photomechanical effect in crystals of benzylidenedimethylimidazolinone (BDHF). (a, b) Photosalient effect in crystals of *E*-BDHF excited with UV light shown before (a) and after (b) short exposure to UV light. (c–f) Snapshots of the four kinematic effects observed during the photosalient effect of single crystals of *E*-BDHF (two examples are shown for each effect): cracking without motion (effect 1, c), hopping with preservation of macroscopic integrity (effect 2, d), splitting into fragments of nearly equal size that fly off (effect 3, e), and bending followed by breaking (effect 4, f). (g) Distribution of kinematic effects of *E*-BDHF over the crystal aspect ratio. The aspect ratio of crystals with size l (length) × w (width) × t (thickness) is presented as correlation between the ratios l/w (abscissa) and l/t (ordinate). The population of the upper half is not mirrored in the lower half due to the choice of the three dimensions ($l > w > t$), which implies $l/t > l/w$. Source: Nath et al. 2015 [70]. Reproduced with permission of American Chemical Society.

over the phase transition temperature, but does not transform or transforms only partially. When localized pressure is applied, domains of different phases are formed. If there are not many interfaces between phase domains, the strain that has accumulated results in elastic energy, which is released by mechanical reconfiguration of the crystal that manifests as motion. The most prominent example of this phenomenon is probably the terephthalic acid crystals [47]. When they are taken over the phase transition temperature and poked by a needle, they jump due to the fast phase transition. A striking example of stress-induced transition was reported by Tao and coworkers [73, 74], with the metastable form II of bis(8-hydroxyquinolinato)copper(II) tetracyanoquinodimethane (CuQ_2-TCNQ). A crystal of this compound when touched with a needle undergoes rapid phase transition to form I with expansion of nearly 100% in length and 50% reduction in the thickness.

5.2.8 Applications of the Salient Effects

Being the most rapid way to reshape a crystal, the salient phenomena appear as new and exciting platform for applications in devices, which can generate motion, that is as actuating elements. This has been already accomplished by other switching materials, such as spin-crossover compounds [75]. These are essential components that can be used in soft microrobots, organic electronics, or other applications. From the application viewpoint, the ability of the salient molecular crystals for very rapid transformation may be interesting to be considered as a way to rapidly convert energy from heat (kinetic energy) or light to mechanical motion. These are fundamentally important processes and favor the salient crystals as fast transducers of energy, much faster than other mechanically responsive molecular materials, such as photomechanically bending crystals. The rate by which the material responds is one of the most important criteria in the selection of materials for design of actuators and, in that respect, the salient phenomena should be considered as viable candidates. It should be mentioned, however, that ultimately the properties of the material should be selected to meet a particular application, so both fast response, such as that observed with salient crystals, or slow process, such as those observed with bending, twisting, and coiling crystals, can be considered depending on their intended application.

One of the main counter-arguments that is brought up in the considerations of the thermosalient effect and the related effects for applications is the propensity of these materials to disintegrate during the transition. Although this is generally the case, some of the TS compounds (particularly those that include hydrogen bonds in their structures) are more prone to retention of crystal integrity than others. Notable examples of such durable crystals are pyroglutamic acid [61] and terephthalic acid [47]. Moreover, for the crystals that do undergo disintegration, they can be considered to target applications that require a single actuation event. One of the many applications that require a single actuation event (single-stroke actuators), which requires a single act of rapid disintegration, is their use as electrical fuses [76]. Recently, it has been demonstrated that when TBB crystal is coated with a conductive (silver) layer, it can conduct electricity to the point when it is heated over the phase transition. At that point the crystal breaks and disconnects the electrical circuit. Although there is clearly a space for improvement, this property can be used to fabricate all-solid fuses that are devoid of arcing problems with the liquid (alloy) fuses. Other applications that make use of the propelling properties of the TS crystals can also be envisaged as precursors to high-energy materials.

Another way to preserve the integrity of the salient crystals for operations that require multiple cycles is to embed them in a flexible medium such as polymer, elastomer, or liquid crystal-based soft materials, which provides mechanical support. In a far analogy with biogenic actuators, on such *hybrid* salient-polymer material, the crystals play the role of a bone, while the polymer plays the role of the flesh. If there is a sufficient mechanical coupling between the crystals and the polymer, the momentum acquired by the crystals in these hybrid materials can be transferred to actuate the polymer, which in return provides the mechanical support. Although there is a challenge to obtain sufficient softness of the polymer so that there is sufficient transfer of momentum, yet the polymer is flexible to move, this has been accomplished by incorporating PHA crystals in sodium caseinate [77]. This hybrid material responds to both heat and light (due to the photoswitchable unit of the ligand) and the response is wavelength-specific. The magnitude of the photoinduced response reflects faithfully the absorption spectrum of the chromophore, which confirms that in the case of the photoinduced response the effects are due to excitation with light and are not a result of heating effects. This approach can be expanded to other mechanically responsive crystals and other polymers, as it has been demonstrated recently by using the so-called mixed matrix membrane strategy to incorporate crystals of anthracene derivatives that undergo [4+4] dimerization in polyvinylidene fluoride (PVDF) [78]. The photomechanical response required a high ratio of 1 : 1 (weight). Particles of about 11 μm were uniformly distributed in PVDF. The hybrid materials bend by UV light and straighten by heating, and were used to demonstrate grasping small objects.

5.3 Self-healing Crystals

Of all the dynamic motions and events organic crystals are capable of, one of the newest, and perhaps most unexpected, is self-healing. Self-healing materials are materials that are capable of recovering from external damage or deterioration. This property has been thoroughly explored in polymers, hydrogels, and other mesophasic materials for years [79–81]. The field of self-healing crystals was initiated in 2016 [82] and it is still in its infancy. The first self-healing crystals were discovered with dipyrazolethiuram disulfide [82]. The work utilized reversible disulfide bond shuffling of the disulfide group to form new bonds at a cracked interface and partially restore the integrity of the crystal. The degree of healing was quantified using a customized tensile tester. The crystal was adhered to a pair of arms on the tensile tester, fractured, mildly compressed, separated, and the recovery was determined to be 6.7% (Figure 5.14). The self-healing response was also measured using a variety of visualization methods such as

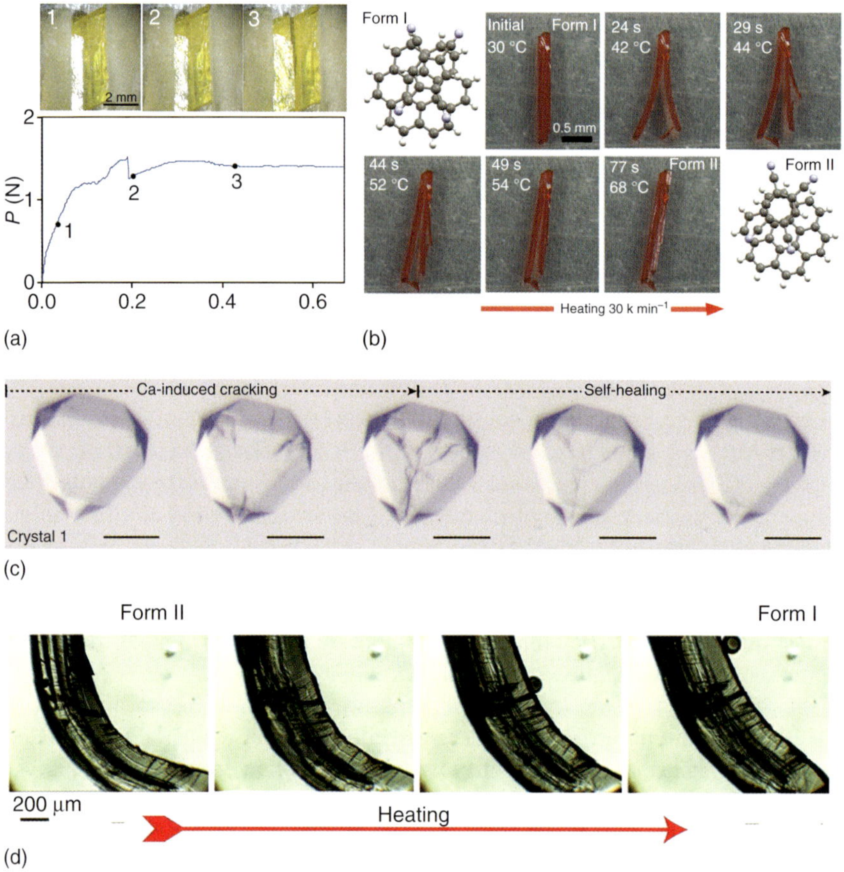

Figure 5.14 Self-healing crystals. (a) A tensile test on a healed crystal of dipyrazolethiuram disulfide. The drop in the load at 2 signified a 6.7% recovery of the tensile strength. Source: [82]. Reproduced with permission of John Wiley & Sons. KGaA, Weinheim. (b) The splitting and restoration of 1,2,4,5-tetracyanobenzene-coronene cocrystals upon heating. Source: [83]. Reproduced with permission of John Wiley & Sons.. (c) Ca-induced cracking and spontaneously self-healing of ferritin crystals integrated with a hydrogel polymer. Source: [84]. Reproduced with permission of John Wiley & Sons. (d) A bent crystal of terephthalic acid in form II, straightens out after being heated to 348 K (75 °C) and converting to form I. Source: Karothu et al. 2016 [47]. Reproduced with permission of American Chemical Society.

computerized tomography (CT) and scanning electron microscopy (SEM) to observe the internal and external structure of the recovered crystals. Both the SEM and CT-scanned images indicated regions of attachment and separation, which is expected for a compound that showed a healing percentage of 6.7%. The recovery percentage is quite modest, but indicated the effect was feasible and spurred other investigations into the self-healing phenomenon in crystals.

Since that time, a few other examples of self-healing crystals have been discovered [47, 83, 84]. The compounds are able to sustain damage and retain their macroscopic integrity but in a different fashion from other self-healing mesophasic materials. Typical self-healing materials are characterized by purposely damaging and then subsequently measuring the restoration of the material's mechanical properties using a tensile tester or dynamic mechanical analysis instrument. Crystals, however, tend to be more brittle and less ductile than mesophasic material and so they cannot be readily cast into traditional "dog bone" shapes for mechanical testing. However, the crystals are still clearly capable of sustaining external damage and recover their integrity. This is evidenced in the work by the Tezcan group where crystals of ferritin were integrated with a hydrogel polymer and the crystals were able to expand more than 500% of their original volume [84]. During the NaCl-induced expansions and Ca-induced contractions, the crystals occasionally cracked and over time they were capable of spontaneous self-healing (Figure 5.14c). This form of deformation and recovery is different from a typical tensile test that separates and reattaches pieces of the material, and as such the self-healing efficiency of the ferritin was not readily quantifiable by conventional mechanical tests, though the recovery is obvious.

Other self-healing effects have been observed during phase transitions from work by both the Tao and Naumov groups. The Tao group reported that cocrystals of coronene and 1,2,4,5-tetracyanobenzene began to undergo a single-crystal-to-single-crystal transformation and split when heated above 313 K (40 °C) and then rejoin after being heated up to 343 K (70 °C) (Figure 5.14b) [83]. The compounds maintained single crystallinity after undergoing the fracturing and mending process, but were found in the high-temperature phase (form II). The process was reversible and the macroscopic integrity was maintained after the transformation. Crystals of terephthalic acid were also able to recover from external deformation by undergoing a phase transition from its room temperature phase (form II) to its high-temperature phase (form I) (Figure 5.14d) [47]. A crystal purposefully bent on its (010) face partially straightens out when heated to 348 K (75 °C) and undergoes a phase transition to form I and then returns to its bent shape after cooling at 315 K (42 °C). The processes found in dipyrazolethiuram disulfide, ferritin, terephthalic acid, and 1,2,4,5-tetracyanobenzene-coronene cocrystals all display the same general effect; macroscopic damage of a crystal and its subsequent recovery. However, each deformation event varied between bifurcation, cracking, realignment, and splitting, and this shows there is a wealth of diverse effects to be discovered with self-healing crystals.

References

1 Abakumov, G.A. and Nevodchikov, V.I. (1982). Thermomechanical and photomechanical effects observed on crystals of a free-radical complex. *Dokl. Phys. Chem.* 266: 1407–1410.

2 Boldyreva, E.V., Sidelnikov, A.A., Chupakhin, A.P. et al. (1984). Deformation and mechanical fragmentation of the crystals $[Co(NH_3)_5NO_2]X_2$ (X = Cl, Br, NO_3) in the course of linkage photoisomerization. *Dokl. Phys. Chem.* 277: 893–896.

3 Ivanov, F.I. and Urban, N.A. (1986). Mechanism of photomechanical deformation of β-lead azide whisker crystals. *React. Solids* 1: 165–170.
4 Boldyreva, E. (1994). Intramolecular linkage isomerization in the crystals of some Co(III) – ammine complexes – a link between inorganic and organic solid state chemistry. *Mol. Cryst. Liq. Cryst. Sci. Technol., Sect. A* 242: 17–52.
5 Boldyreva, E.V. (2001). Crystal-structure aspects of solid-state innersphere isomerization in nitro(nitrito)pentaamminecobalt(III) complexes. *Russ. J. Coord. Chem.* 27: 297–323.
6 Irie, M. (2000). Diarylethenes for memories and switches. *Chem. Rev.* 100: 1685–1716.
7 Irie, M., Fukaminato, T., Sasaki, T. et al. (2002). A digital fluorescent molecular photoswitch. *Nature* 420: 759.
8 Kobatake, S., Takami, S., Muto, H. et al. (2007). Rapid and reversible shape changes of molecular crystals on photoirradiation. *Nature* 446: 778–781.
9 Uchida, K., Sukata, S.-i., Matsuzawa, Y. et al. (2008). Photoresponsive rolling and bending of thin crystals of chiral diarylethenes. *Chem. Commun.*: 326–328.
10 Zhu, L., Al-Kaysi, R.O., and Bardeen, C.J. Reversible photoinduced twisting of molecular crystal microribbons. *J. Am. Chem. Soc.* 2001 (133): 12569–12575.
11 Naumov, P., Chizhik, S., Panda, M.K. et al. (2015). Mechanically responsive molecular crystals. *Chem. Rev.* 115: 12440–12490.
12 Chizhik, S., Sidelnikov, A., Zakharov, B. et al. (2018). Quantification of photoinduced bending of dynamic molecular crystals: from macroscopic strain to kinetic constants and activation energies. *Chem. Sci.* 9: 2319–2335.
13 Kim, T., Zhu, L., Mueller, L.J., and Bardeen, C.J. (2014). Mechanism of photoinduced bending and twisting in crystalline microneedles and microribbons composed of 9-methylanthracene. *J. Am. Chem. Soc.* 136: 6617–6625.
14 Steed, J.W. (2018). 21st century developments in the understanding and control of molecular solids. *Chem. Commun.* 54: 13175–13182.
15 Nath, N.K., Panda, M.K., Sahoo, S.C., and Naumov, P. (2014). Thermally induced and photoinduced mechanical effects in molecular single crystals – a revival. *CrystEngComm* 16: 1850–1858.
16 Commins, P., Desta, I.T., Karothu, D.P. et al. (2016). Crystals on the move: mechanical effects in dynamic solids. *Chem. Commun.* 52: 13941–13954.
17 Etter, M.C. and Siedle, A.R. (1983). Solid-state rearrangement of (phenylazophenyl)palladium hexafluoroacetylacetonate. *J. Am. Chem. Soc.* 105: 641–643.
18 Gigg, J., Gigg, R., Payne, S., and Conant, R.J. (1987). The allyl group for protection in carbohydrate chemistry. Part 21. (±)-1,2:5.6- and (±)-1,2:3,4-di-*O*-isopropylidene-myo-inositol. The unusual behaviour of crystals of (±)-3,4-di-*O*-acetyl-1,2,5,6-tetra-*O*-benzyl-myo-inositol on heating and cooling: a “thermosalient solid”. *Chem. Soc., Perkin Trans. 1*: 2411.
19 Naumov, P., Kowalik, J., Solntsev, K.M. et al. (2010). Topochemistry and photomechanical effects in crystals of green fluorescent protein-like chromophores: effects of hydrogen bonding and crystal packing. *J. Am. Chem. Soc.* 132: 5845–5857.

20 Skoko, Ž., Zamir, S., Naumov, P., and Bernstein, J. (2010). The thermosalient phenomenon. "Jumping crystals" and crystal chemistry of the anticholinergic agent oxitropium bromide. *J. Am. Chem. Soc.* 132: 14191–14202.
21 Zamir, S. (1995). PhD thesis, Ben-Gurion University of the Negev.
22 Zamir, S., Bernstein, J., and Greenwood, D.J. (1994). A single crystal to single crystal reversible phase transition which exhibits the "Hopping effect". *Mol. Cryst. Liq. Cryst. Sci. Technol., Sect. A* 242: 193–200.
23 Alimi, L.O., van Heerden, D.P., Lama, P. et al. (2018). Reversible thermosalience of 4-aminobenzonitrile. *Chem. Commun.* 54: 6208–6211.
24 Martins, F.F., Joseph, A., Diogo, H.P. et al. (2018). Irreversible magnetic behaviour caused by the thermosalient phenomenon in an iron(III) spin crossover complex. *Eur. J. Inorg. Chem.* (25): 2976–2983.
25 Rawat, H., Samanta, R., Bhattacharya, B. et al. (2018). Thermosalient forms: carryover of thermosalient behavior of coformers from single component to multicomponent forms? *Cryst. Growth Des.* 18: 2918–2923.
26 So, H.-S., Minami, T., Jindo, T., and Matsumoto, S. (2018). Thermosalient effect of two polymorphs of a diketopyrrolopyrrole dye with different crystal systems and molecular arrangements. *CrystEngComm* 20: 5317–5320.
27 Tamboli, M.I., Karothu, D.P., Shashidhar, M.S. et al. (2018). Effect of crystal packing on the thermosalient effect of the pincer-type diester naphthalene-2,3-diyl-*bis*(4-fluorobenzoate): a new class II thermosalient solid. *Chem. Eur. J.* 24: 4133–4139.
28 Robertson, L., Penin, N., Blanco-Gutierrez, V. et al. (2015). $CuMo_{0.9}W_{0.1}O_4$ phase transition with thermochromic, piezochromic, and thermosalient effects. *J. Mater. Chem. C* 3: 2918–2924.
29 Ohtani, S., Gon, M., Tanaka, K., and Chujo, Y. (2017). A flexible, fused, azomethine–boron complex: thermochromic luminescence and thermosalient behavior in structural transitions between crystalline polymorphs. *Chem. Eur. J.* 23: 11827–11833.
30 Mittapalli, S., Perumalla, D.S., Nanubolu, J.B., and Nangia, A. (2017). Thermomechanical effect in molecular crystals: the role of halogen-bonding interactions. *IUCrJ* 4: 812–823.
31 Dharmarwardana, M., Arimilli, B.S., Luzuriaga, M.A. et al. (2018). The thermo-responsive behavior in molecular crystals of naphthalene diimides and their 3D printed thermochromic composites. *CrystEngComm* 20: 6054–6060.
32 Ghosh, S., Mishra, M.K., Ganguly, S., and Desiraju, G.R. (2015). Dual stress and thermally driven mechanical properties of the same organic crystal: 2,6-dichlorobenzylidene-4-fluoro-3-nitroaniline. *J. Am. Chem. Soc.* 137: 9912–9921.
33 Takeda, T. and Akutagawa, T. (2016). Anisotropic dissociation of $\pi-\pi$ stacking and flipping-motion-induced crystal jumping in alkylacridones and their dicyanomethylene derivatives. *Chem. Eur. J.* 22: 7763–7770.
34 Vicente, A.I., Joseph, A., Ferreira, L.P. et al. (2016). Dynamic spin interchange in a tridentate Fe(III) Schiff-base compound. *Chem. Sci.* 7: 4251–4258.
35 Singh, M. and Chopra, D. (2018). Diversity in mechanical response in donor–acceptor coupled cocrystal stoichiomorphs based on pyrene and 1,8-dinitroanthraquinone systems. *Cryst. Growth Des.* 18: 6670–6680.

36 Wang, H., Chen, P., Wu, Z. et al. (2017). Bending, curling, rolling, and salient behavior of molecular crystals driven by [2+2] cycloaddition of a styrylbenzoxazole derivative. *Angew. Chem. Int. Ed.* 56: 9463–9467.

37 Mulijanto, C.E., Quah, H.S., Tan, G.K. et al. (2017). Curved crystal morphology, photoreactivity and photosalient behaviour of mononuclear Zn(II) complexes. *IUCrJ* 44: 65–71.

38 Kitagawa, D., Kawasaki, K., Tanaka, R., and Kobatake, S. (2017). Mechanical behavior of molecular crystals induced by combination of photochromic reaction and reversible single-crystal-to-single-crystal phase transition. *Chem. Mater.* 29: 7524–7532.

39 Kitagawa, D., Okuyama, T., Tanaka, R., and Kobatake, S. (2016). Photoinduced rapid and explosive fragmentation of diarylethene crystals having urethane bonding. *Chem. Mater.* 28: 4889–4892.

40 Mishra, M.K., Mukherjee, A., Ramamurty, U., and Desiraju, G.R. (2015). Crystal chemistry and photomechanical behavior of 3,4-di-methoxycinnamic acid: correlation between maximum yield in the solid-state topochemical reaction and cooperative molecular motion. *IUCrJ* 2: 653–660.

41 Sahoo, S.C., Panda, M.K., Nath, N.K., and Naumov, P. (2013). Biomimetic crystalline actuators: structure-kinematic aspects of the self-actuation and motility of thermosalient crystals. *J. Am. Chem. Soc.* 135: 12241–12251.

42 Ahmed, E., Karothu, D.P., and Naumov, P. (2018). Crystal adaptronics: mechanically reconfigurable elastic and superelastic molecular crystals. *Angew. Chem. Int. Ed.* 57: 8837–8846.

43 Sato, O. (2016). Dynamic molecular crystals with switchable physical properties. *Nat. Chem.* 8: 644–656.

44 Chung, H., Dudenko, D., Zhang, F. et al. (2018). Rotator side chains trigger cooperative transition for shape and function memory effect in organic semiconductors. *Nat. Commun.* 9: 278.

45 Chung, H., Ruzié, C., Geerts, Y., and Diao, Y. (2018). Hybrid mechanism of nucleation and cooperative propogation in a single-crystal-to-single crystal transition of a molecular crystal. *Cryst. Growth Des.* 18: 4245–4251.

46 Anwar, J., Tuble, S.C., and Kendrick, J. (2007). Concerted molecular displacements in a thermally-induced solid-state tansformation in crystals of D,L-norleucine. *J. Am. Chem. Soc.* 129: 2542–2547.

47 Karothu, D.P., Weston, J., Desta, I.T., and Naumov, P. (2016). Shape-memory and self-healing effects in mechanosalient molecular crystals. *J. Am. Chem. Soc.* 138: 13298–13306.

48 Sahoo, S.C., Sinha, S.B., Kiran, M.S.R.N. et al. (2013). Kinematic and mechanical profile of the self-actuation of thermosalient crystal twins of 1,2,4,5-tetrabromobenzene: a molecular crystalline analogue of a bimetallic strip. *J. Am. Chem. Soc.* 135: 13843–13850.

49 Centore, R., Jazbinsek, M., Tuzi, A. et al. (2012). A series of compounds forming polar crystals and showing single-crystal-to-single-crystal transitions between polar phases. *CrystEngComm* 14: 2645–2653.

50 Klaser, T., Popović, J., Fernandes, J.A. et al. (2018). Does thermosalient effect have to concur with a polymorphic phase transition? The case of methscopolamine bromide. *Crystals* 8: 301.

51 Ko, J.-H., Lee, K.-S., Sahoo, S.C., and Naumov, P. (2013). Isomorphous phase transition of 1,2,4,5-tetrabromobenzene jumping crystals studied by Brillouin light scattering. *Solid State Commun.* 173: 46–50.
52 Zakharov, B.A., Michalchuk, A.A.L., Morrison, C.A., and Boldyreva, E.V. (2018). Anisotropic lattice softening near the structural phase transition in the thermosalient crystal 1,2,4,5-tetrabromobenzene. *Phys. Chem. Chem. Phys.* 20: 8523–8532.
53 Zaczek, A.J., Catalano, L., Naumov, P., and Korter, T.M. (2019). Mapping the polymorphic transformation gateway vibration in crystalline 1,2,4,5-tetrabromobenzene. *Chem. Sci.* https://doi.org/10.1039/C8SC03897J.
54 Lončarić, I., Popović, J., Despoja, V. et al. (2017). Reversible thermosalient effect of N'-2-propylidene-4-hydroxybenzohydrazide accompanied by an immense negative compressibility: structural and theoretical arguments aiming toward the elucidation of jumping phenomenon. *Cryst. Growth Des.* 17: 4445–4453.
55 Crottaz, O., Kubel, F., and Schmid, H. (1997). Jumping crystals of the spinels $NiCr_2O_4$ and $CuCr_2O_4$. *J. Mater. Chem.* 7: 143–146.
56 Tarantino, S.C., Giannini, M., Carpenter, M.A., and Zema, M. (2016). Cooperative Jahn–Teller effect and the role of strain in the tetragonal-to-cubic phase transition in $Mg_xCu_{1-x}Cr_2O_4$. *IUCrJ* 3: 354–366.
57 Nauha, E., Naumov, P., and Lusi, M. (2016). Fine-tuning of a thermosalient phase transition by solid solutions. *CrystEngComm* 18: 4699–4703.
58 Lusi, M. and Bernstein, J. (2013). On the propulsion mechanism of "jumping" crystals. *Chem. Commun.* 49: 9293–9295.
59 Panda, M.K., Runčevski, T., Sahoo, S.C. et al. (2014). Colossal positive and negative thermal expansion and thermosalient effect in a pentamorphic organometallic martensite. *Nat. Commun.* 5: 4811.
60 Panda, M.K., Etter, M., Dinnebier, R.E., and Naumov, P. (2017). Acoustic emission from organic martensites. *Angew. Chem. Int. Ed.* 56: 8104–8109.
61 Panda, M.K., Runčevski, T., Husain, A. et al. (2015). Perpetually self-propelling chiral single crystals. *J. Am. Chem. Soc.* 137: 1895–1902.
62 Panda, M.K., Centore, R., Causà, M. et al. (2016). Strong and anomalous thermal expansion precedes the thermosalient effect in dynamic molecular crystals. *Sci. Rep.* 6: 29610.
63 Khalil, A., Hu, C.T., and Naumov, P. (2018). Nanoscale crystallization and thermal behaviour of 1,2,4,5-tetrabromobenzene. *CrystEngComm* 20: 636–642.
64 Kwon, E., Oikawa, H., Kasai, H., and Nakanishi, H. (2007). A fabrication method of organic nanocrystals using stabilizer-free emulsion. *Cryst. Growth Des.* 7: 600–602.
65 Dong, X., Tong, F., Hanson, K.M. et al. (2019). Hybrid organic-inorganic photon-powered actuators based on aligned diarylethene nanocrystals. *Chem. Mater.* https://doi.org/10.1021/acs.chemmater.8b04568.
66 Naumov, P., Sahoo, S.C., Zakharov, B.A., and Boldyreva, E.V. (2013). Dynamic single crystals: kinematic analysis of photoinduced crystal jumping (the photosalient effect). *Angew. Chem. Int. Ed.* 52: 9990–9995.

67 Mittapalli, S., Perumalla, D.S., and Nangia, A. (2017). Mechanochemical synthesis of *N*-salycilideneaniline: thermosalient effect of polymorphic crystals. *IUCrJ* 4: 243–250.
68 Werny, M.J. and Vittal, J.J. (2017). Regulating thermosalient behaviour in three polymorphs. *IUCrJ* 4: 202–203.
69 Centore, R. and Causà, M. (2018). Translating microscopic molecular motion into macroscopic body motion: reversible self-reshaping in the solid state transition of an organic crystal. *Cryst. Growth Des.* 18: 3535–3543.
70 Nath, N.K., Runčevski, T., Lai, C.-Y. et al. (2015). Surface and bulk effects in photochemical reactions and photomechanical effects in dynamic molecular crystals. *J. Am. Chem. Soc.* 137: 13866–13875.
71 Medishetty, R., Husain, A., Bai, Z. et al. (2014). Single crystals popping under UV light: a photosalient effect triggered by a [2+2] cycloaddition reaction. *Angew. Chem. Int. Ed.* 53: 5907–5911.
72 Medishetty, R., Sahoo, S.C., Mulijanto, C.E. et al. (2015). Photosalient behavior of photoreactive crystals. *Chem. Mater.* 27: 1821–1829.
73 Liu, J., Liu, G., Liu, Y. et al. (2016). The role of weak interactions in the mechano-induced single-crystal-to-single-crystal phase transition of 8-hydroxyquinoline-based co-crystals. *Chem. Asian J.* 11: 1682–1687.
74 Liu, G., Liu, J., Liu, Y., and Tao, X. (2014). Oriented single-crystal-to-single-crystal phase transition with dramatic changes in the dimensions of crystals. *J. Am. Chem. Soc.* 136: 590–593.
75 Shepherd, H.J., Gural'skiy, I.A., Quintero, C.M. et al. (2013). Molecular actuators driven by cooperative spin-state switching. *Nat. Commun.* 4: 2607.
76 Khalil, A., Ahmed, E., and Naumov, P. (2017). Metal-coated thermosalient crystals as electrical fuses. *Chem. Commun.* 53: 8470–8473.
77 Sahoo, S.C., Nath, N.K., Zhang, L. et al. (2014). Actuation based on thermo/photosalient effect: a biogenic smart hybrid driven by light and heat. *RSC Adv.* 4: 7640–7647.
78 Yi, Q., Yang, X., Chen, Y. et al. (2018). Fabrication of light-triggered soft artificial muscles via a mixed-matrix membrane strategy. *Angew. Chem. Int. Ed.* 57: 10192–10196.
79 Li, G. and Meng, H. (2015). *Recent Advances in Smart Self-Healing Polymers and Composites.* Cambridge: Woodhead Publishing.
80 Yang, Y. and Urban, M.W. (2013). Self-healing polymeric materials. *Chem. Soc. Rev.* 42: 7446–7422.
81 Bekas, D.G., Tsirka, K., Baltzis, D., and Paipetis, A.S. (2016). Self-healing materials: a review of advances in materials, evaluation, characterization and monitoring techniques. *J. Compos., Part B* 87: 92–119.
82 Commins, P., Hara, H., and Naumov, P. (2016). Self-healing molecular crystals. *Angew. Chem. Int. Ed.* 55: 13028–13032.
83 Liu, G., Liu, J., Ye, X. et al. (2017). Self-healing behavior in a thermo-mechanically responsive cocrystal during a reversible phase transition. *Angew. Chem. Int. Ed.* 56: 198–202.
84 Zhang, L., Bailey, J.B., Subramanian, R.H. et al. (2018). Hyperexpandable, self-healing macromolecular crystals with integrated polymer networks. *Nature* 557: 86–91.

6

Shape Memory Molecular Crystals

Satoshi Takamizawa

Yokohama City University, Department of Materials System Science, Graduate School of Nanobioscience, 22-2 Seto, Kanazawa-ku, Yokohama, Kanagawa 236-0027, Japan

Introduction

Structural materials for robotics have to be endowed with mechanical strength, human friendliness, and environmental resistance [1, 2]. From the viewpoint of the last two points, organic materials have great advantages to conventional inorganic materials. For example, they have soft texture, are free from metal allergens, and are resistant to water and salt, which damage metallic materials, and so forth [2, 3]. On the other hand, mechanical properties have been of less interest in the research of organic chemistry because generally organic-based compounds are mechanically weak. In recent years, we and other researchers have conducted intense study on the mechanical properties of organic crystals [4–19], and in this chapter, some outstanding results of ours [8–19], including the first observation of superelasticity [8, 9] and superplasticity in organic compounds [12] as well as detailed study on ferroelasticity [14–19], will be presented. In common sense, organic crystals are fragile, and easily collapsed upon external force or distortion. However, organic crystals with those uncommon mechanical properties bear such external or internal loading, showing unique deformabilities. Thus, such materials are expected to be used as structural materials. Especially, superelastic materials can propel themselves while ferroelasticity and superplasticity are related to processability of materials.

Before showing our works, we start from brief explanation on deformability of common solid [20], as well as superelastic and ferroelastic materials [21, 22]. When external force (F) is applied to a solid, stress (σ) is generated and the solid deforms. This deformation, referred to as distortion (ε), is in proportion to F ($\sigma = E\varepsilon$, where E is Young's modulus) when F is small (Figure 6.1a). At this stage, the shape spontaneously recovers to the initial shape by removal of F. When F goes beyond a certain threshold, called elastic limit, the observed ε is larger than calculated ε. σ at which the observed ε is larger than calculated ε in 0.2% defines yielding point, which is an important parameter showing the strength of a material. The deformation caused by such large F is termed as plastic deformation, represented by irreversible deformation of fracture or permanent strain. Once a

Mechanically Responsive Materials for Soft Robotics, First Edition. Edited by Hideko Koshima.

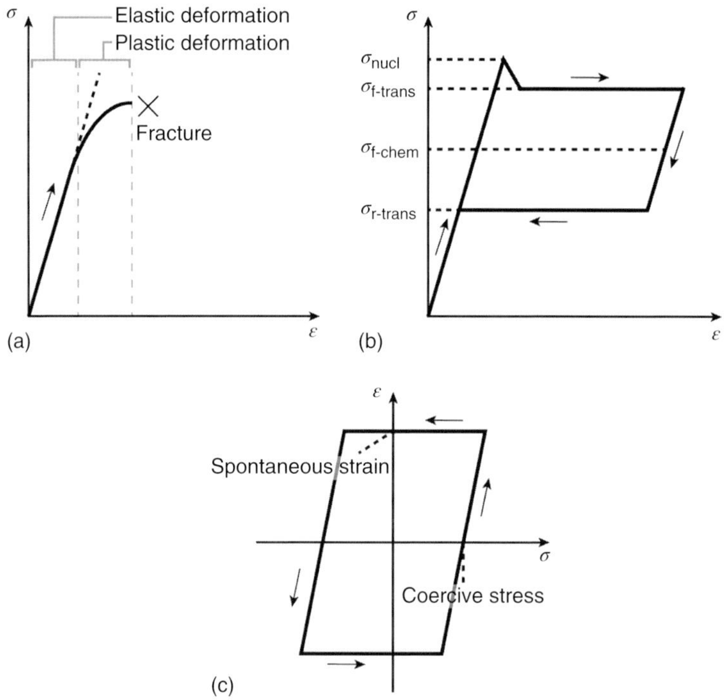

Figure 6.1 Typical forms of s–s curves for (a) common, (b) superelastic, and (c) ferroelastic materials.

common solid reaches such deformation, it does not recover the original shape by removal of F or application of inversed F.

Some solid materials reversibly recover their shapes from plastic deformation. "Superelasticity" or "ferroelasticity" are representative example of such anomalous behaviors, by which materials can reversibly recover their shape even from distortion of more than several decades percent. These deformabilities are both based on diffusionless transformation, which is a kind of phase transformation from the initial "mother phase" to the generated "daughter phase." The term "diffusionless" means that the component chemical entities keep one-to-one correspondence because the transformation is not accompanied by diffusion of chemical entities, like in gas ⇆ liquid transformation. Superelasticity and ferroelasticity are distinguished by how they recover their shapes; superelastic materials recover their initial shape by removal of F, while the structural distortion retains after the removal of F in ferroelastic materials and inverted F is required for shape recovery.

Figure 6.1b shows the typical strain–stress (s–s) curve of superelastic materials. When shear force was applied to the material, linear increase in stress is observed at the earliest stage of the deformation because of elastic deformation. As shearing proceeded, the stress abruptly decreases at a certain point, followed by a plateau. When the shear force is weakened, stress decreases in a certain degree, plateau is observed again, and the plot goes back to the origin. To summarize, superelasticity is featured by a hysteresis loop composed by a set of plateau and elastic linear σ–ε relationship with a single peak. The stresses around the loop are generally termed as σ_{nucl}, $\sigma_{f\text{-}trans}$, $\sigma_{r\text{-}trans}$, and σ_{chem}. σ_{nucl} (nucl: nucleation) is the stress for the shear-induced daughter phase to generate the nucleus, which is observed as a peak because the generation of the new crystalline domain causes some distortions in the structure owing to energetically unfavorable domain interfaces, and therefore the required stress is larger than that of domain growth. The stress of the following plateau is $\sigma_{f\text{-}trans}$ (f-trans: forward transformation), which is the stress for domain growth of the daughter phase. $\sigma_{r\text{-}trans}$ (r-trans: reverse transformation) is the stress generated when the daughter phase shows the spontaneous reverse transformation to the mother phase. In hypothetical perfect equilibrium condition (shear with infinitesimal small step with infinite wait time), hysteresis loop would not be observed ($\sigma_{f\text{-}trans} = \sigma_{r\text{-}trans}$). In other words, both $\sigma_{f\text{-}trans}$ and $\sigma_{r\text{-}trans}$ depend on the non-ideality in deformation, and it is difficult to treat in thermodynamics. Therefore, the true shear stress, i.e. $\sigma_{f\text{-}trans}$ (or $\sigma_{r\text{-}trans}$) in the perfect equilibrium condition, is assumed to be the average of observed $\sigma_{f\text{-}trans}$ and $\sigma_{r\text{-}trans}$, and termed as σ_{chem} (chem: chemical).

Figure 6.1c shows the s–s curve used to evaluate ferroelastic behaviors [22]. Note that vertical and horizontal axes are inverted from those of superelastic materials. The ferroelastic "retention of distortion after the removal of external force" appears in the graph as nonzero ε value at $\sigma = 0$, i.e. the intercept of vertical axis, which is termed as spontaneous strain. "The inverted force to cancel the strain" appears at $\varepsilon = 0$, i.e. the intercept of horizontal axis, and is termed as coercive stress.

The words "superelasticity" or "ferroelasticity" are not recognized quite widely, but actually, "shape memory effect" (SME) [21, 23, 24] based on these mechanical behaviors are used in every aspect of our lives [25–29]. For example, industrial products like temperature control bulb of a water tap [26], wearing apparels like underwear or frames of eyeglass [27], medical instruments like stent or orthodontic wire [28], and, moreover, toys and works of art [29]. SME is defined by two distinguished processes of "retention of deformation" and "shape recovery by warming." The first process is related to ferroelasticity, while the second is superelasticity. Until really recently, SME has been believed to be specific for metal alloys (shape memory alloys, SMAs) [21, 25, 30, 31], because superelasticity had been reported only in metal alloys, although ferroelasticity has been known in organic compound since 1970s [32–34]. However, 2014 witnessed a paradigm shift in this research field triggered by the surprising discovery of superelasticity in purely organic terephthalamide by the authors [8, 9], followed by other organic-based superelastic or ferroelastic compounds (Figure 6.2). In 2018, superplasticity was first observed in a metal complex crystal consisting of one-dimensional coordination polymers [13], making it highly promising to find more and more novel, useful, and attractive deformabilities.

6.1 Discovery of Organosuperelasticity

Well-formed single crystals of terephthalamide (**1**) (Figure 6.3a) were obtained by recrystallization of a reagent-grade terephthalamide from hot water and used

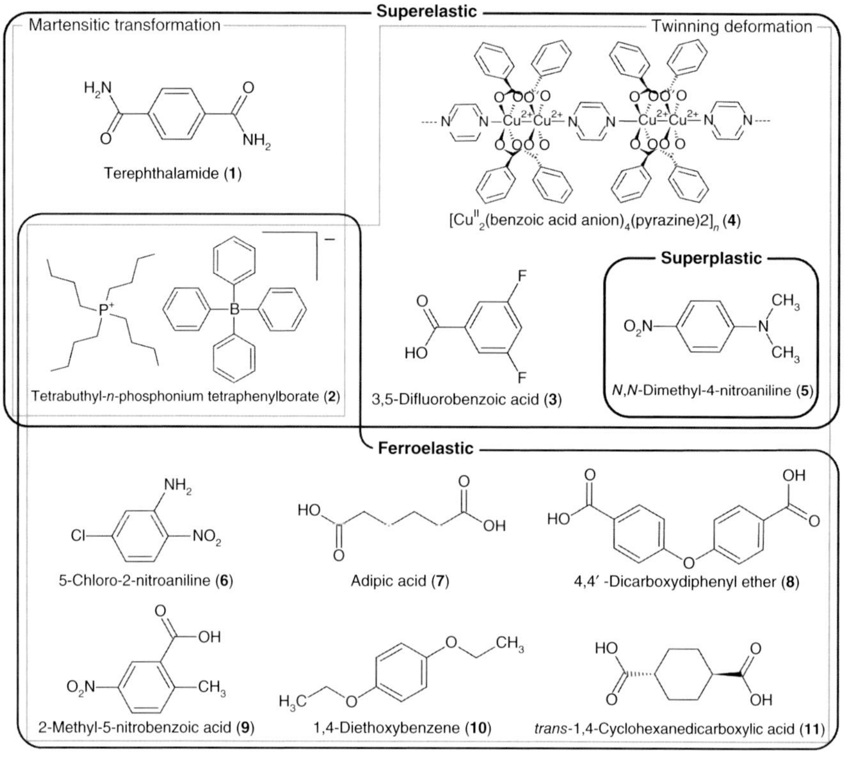

Figure 6.2 Molecular structures of the superelastic, superplastic, or ferroelastic molecules.

in the experiments after vacuum drying. X-ray structure analysis showed that molecules of **1** had column structure along [010] direction by N—H···O=C hydrogen bond network through end-to-end double hydrogen bonds (N···O distance: 2.932(4) Å) along the long axis of **1**. The columns are connected to each other by side-to-side double N—H···O=C hydrogen bonds (N···O distance: 2.912(3) Å) to form the uniform –A–A–A–A– sheet arrangement in {001}. We found that the shear stress (Figure 6.3b) on {010} surfaces of **1** bent its crystal shape generating another crystalline domain with a sharp interface across the crystal specimen. When the applied force was removed, the crystal recovered its original shape, and the interface propagated back in 3.33 mm s^{-1} at 298 K.

X-ray crystallographic study revealed that the shear-generated crystalline domain is in another crystal phase (β phase). The β phase had sheet type structure along the long axis of molecule of **1** as in α phase. The connection of columns were modified to form an alternative –A′–B–A′–B– arrangement. Also, the manner of offset stacking of phenyl rings remarkably changed from α (surface separation: 3.500 Å) to β (3.209 Å in the column A′ and 3.900 Å in the column B).

X-ray study for the crystal in the bending form also help clarifying how each crystal domain is connected at the interface under the α/β coexistent state. The lattice mismatching between α and β phases caused some distortion. A certain

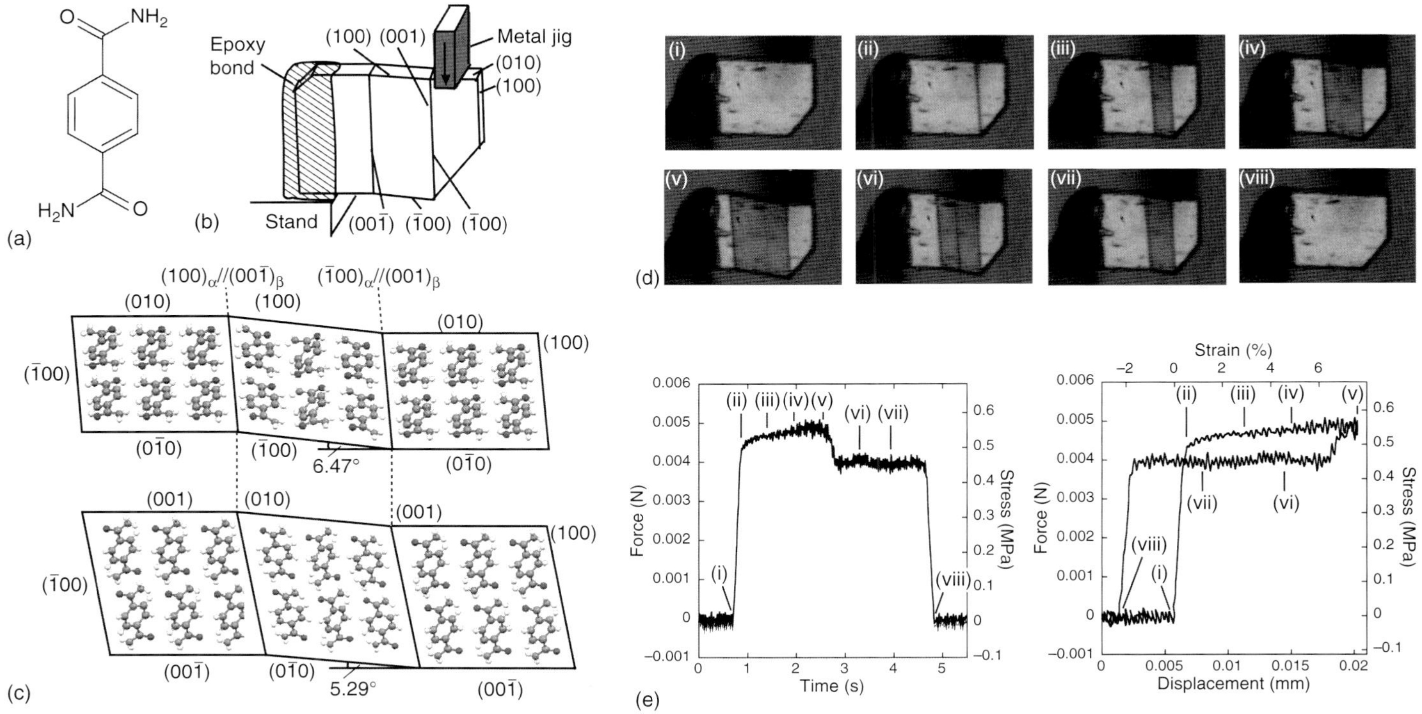

Figure 6.3 (a) Molecular structure of **1**. (b) Experimental setting for shearing. (c) Elucidated phase connection under a shear-induced transition state. (d) Side view photographs taken with a polarization microscope during the shear-strain test with the letters and numbers showing the position in panel (e) for the time course and the displacement course, respectively. Source: Takamizawa and Takasaki (2014) [8]. Reproduced with permission of John Wiley & Sons.

difference in the length of *c*-axis (α) and *b*-axis (β) (*c*(α) (7.1853 Å)//*b*(β) (7.268 Å)) indicates the mainly unidirectional distortion along these axes. The deduced connection view at the interface (Figure 6.3c) clearly revealed the martensitic mechanism assisted by the change of molecular packing. The directions in [010] and [00$\bar{1}$] in α crystal can be effective axes for shear-induced transformation. With a microscope, the bending angles were observed to be in the range of 5–8° and 3–5° judged from the projection directions of [001] and [010], respectively, which agree with the expected angles from the X-ray data in 6.47° and 5.29°, respectively. The agreement to the structural description demonstrates the harmonious support of the macro- and microscopic structural changes in **1**. The relationship between shear stress and strain was investigated by loading on the portion of crystal surface {010}, which should be nearly parallel to the shear-transition easy axis.

For the quantitative discussion on the observed superelastic behavior, an s–s curve was collected to have the hysteresis loop characteristic for superelastic materials (Figure 6.3d,e). With one end of a crystal fixed, a metal blade was pushed across the (010) surface of the crystal specimen. Stress was detected when the blade reached the crystal, and increased linearly in response to the increasing loaded force (photographs (i–v) in Figure 6.3d). Then, the stress was constant at where the thin β phase domain was generated. This plateau continued during the growth of the β phase, i.e. the continuous proceeding of α → β transformation. At unloading process (pulling the blade back), a slight decrease in the stress was observed followed by a plateau during the transformation from the β to the α phase until the β domain narrowed to the boundary line. Then, the stress linearly decreased down to zero along the vanishing line boundary (photographs (vi–viii) in Figure 6.3d) until the blade was removed from the crystal surface.

The shear force was normalized by the cross-sectional area of domain interface to have the intrinsic shear parameters for the observed superelastic behavior at room temperature. Normalized shear force for the transformation was $4.96 \times 10^5\ \mathrm{N\,m^{-2}}$ (0.496 MPa) and for the reverse transformation was $4.59 \times 10^5\ \mathrm{N\,m^{-2}}$ (0.459 MPa). σ_M is 1000 times smaller than that of the typical Ti–Ni alloy (558 MPa) [31]. The energy storage density (*E*) and energy storage efficiency (η) of **1** were estimated at $0.062\ \mathrm{MJ\,m^{-3}}$ ($0.042\ \mathrm{J\,g^{-1}}$, $6.96\ \mathrm{J\,mol^{-1}}$) and 0.925, respectively. The expected maximum strain is up to 11.34%, based on the crystallographic data. *E* is 226 times less than that of typical Ti–Ni alloy ($14\ \mathrm{MJ\,m^{-3}}$, $2.15\ \mathrm{J\,g^{-1}}$, $114.76\ \mathrm{J\,mol^{-1}}$). The energy storage ability is related to the lattice energy, and because of the components bulkiness, the volumetric energy density becomes smaller by the extent of the lattice. Consequently, superelastic molecular material can produce a large transformation from a small energy input, and, vice versa, can generate a uniform level of small power with a large reverse transformation.

In superelasticity, spontaneous reversed transformation, which is attributed to the shape recovery, is often explained in terms of thermodynamics [35]. Gibbs energy (*G*) linearly changes to the temperature in the first approximation ($G = H - TS$, where *H* is enthalpy, *T* is temperature, and *S* is entropy). In a material with several phases, because the phases have different *S*, *G* of different phases cross at a finite temperature, and stable phase and metastable phase are switched. However, phase transformation is not triggered immediately here. Phase transformation in increased or decreased temperature occurs when difference in G is enough to compensate energetic loss due to nucleation of daughter phase. The temperatures at which phase transition actually occurs are denoted by A_s (warming process) and M_s (cooling process) in the study of metal alloys. Likewise, the temperatures at which the phase transition finishes are denoted by A_f (warming process) and M_f (cooling process). Basically, these temperatures are in a relation of $M_f < M_s < A_s < A_f$, dividing the temperature into five regions, in which different deformabilities are observed. For example, superelastic hysteresis loop in s–s curve is observed above A_f, because the effect is phase transformation to metastable state by shearing, and consecutive reverse transformation to the truly stable state accompanied by the shape recovery by removal of the shear.

This temperature dependence should be observed also in organosuperelastic materials. Because SME is a combination of ferroelasticity under A_s and superelasticity above A_f, organic compounds can exhibit SME if they show both superelasticity and ferroelasticity. Actually, ferroelasticity has been known in organic materials [32–34], and therefore, the discovery of organosuperelasticity of **1** opened up the possibility of SME in organic-based compounds. In 2016, we have reported SME by confirming ferroelastic deformation by twinning deformation and superelastic behavior of an organic salt, tetrabutyl-n-phosphonium tetraphenylborate ($(P^nBu_4)^+(BPh_4)^-$, **2**) by collecting the temperature dependence of s–s curves [10].

Single crystals of **2** were prepared by recrystallization from acetone solution of crude powder obtained from mixed aqueous solution of PBu_4Cl and $NaBPh_4$. Through application of shear force at room temperature, a colorless rod-shaped crystal (α phase) underwent a shear-induced ferroelastic twinning in growing a twin (α^+/α^- coexistence state) with mobile planar domain interface across the crystal rod keeping an invariant bending angle of 12°. Sequential displacement of the shear position on the crystal rod formed zigzag crystals. Upon warming, the deformed crystal regained its straight shape (thermal shape recovery) (Figure 6.4a). On the other hand, homogeneous straight shape crystal in α phase converges into high temperature phase (β phase) by warming (Figure 6.4b), resulting in the bending in 6°. When the crystal is cooled down to room temperature, the initial straight shape is recovered. This indicates that the successive "coexisting $\alpha^\pm \rightarrow$ homogeneous $\beta \rightarrow$ homogeneous α'' process is attributed to the practical SME of $(P^nBu_4)^+(BPh_4)^-$. This process can be induced in a spatio-sequential manner, showing the clear manifestation of SME (Figure 6.4c).

Single-crystal X-ray diffraction analysis gave the crystal connections under the shear-induced bending state. By warming the crystal of **2**, the crystal system changed from triclinic P-1 in the α phase (394 K) into monoclinic $P2_1/n$ in the β phase (403 K). Crystal face indexing showed the interfaces for shear-induced twinning of α^+/α^- ($(103)_{\alpha^+}//(103)_{\alpha^-}$ or $(\bar{1}0\bar{3})_{\alpha^+}//(\bar{1}0\bar{3})_{\alpha^-}$), and for the heat-induced crystal of $\alpha^\pm/\beta$ ($(205)_{\alpha^+}//(\bar{3}\bar{1}3)_\beta$ or $(205)_{\alpha^-}//(3\bar{1}\bar{3})_\beta$), respectively (Figure 6.5a).

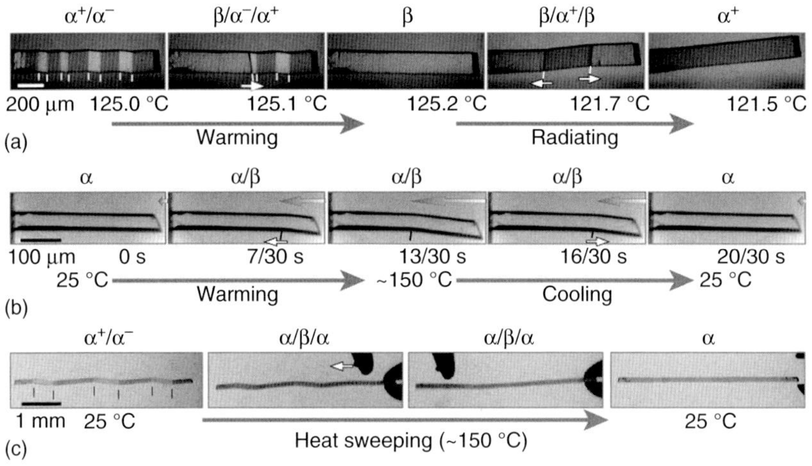

Figure 6.4 (a) Thermal shape recovery from the deformed zigzag α^+/α^- form prepared at room temperature into the "remembered" straight rod β form around 125.0 °C by warming, (b) thermal actuation for cyclic $\alpha \leftrightarrows \beta/\alpha$ transformation via heat fluctuation with an iron at 150 °C, and (c) spatio-sequential shape recovery via a heat sweep over the deformed zigzag pattern of α^+/α^-. Source: Takamizawa and Takasaki (2016) [10]. Reproduced with permission of The Royal Society of Chemistry.

The crystal bending angles for the α^+/α^- and $\alpha^{\pm}/\beta$ ($=\alpha^+/\beta$ or α^-/β) connections are estimated based on X-ray crystallographic study to be 12.2° and 5.3°, respectively, which agree with the ones observed under a microscope (12° and 6°). This demonstrated how to preserve one-to-one molecular correspondence during transformation to satisfy the martensitic manner (Figure 6.5b). The α^+/α^- connection takes a rotary reflection plane regarded as a type II twin (Figure 6.5c) [36], which only requires the rotation of the molecules in the conversion. On the other hand, in the appearance of the $\alpha^{\pm}/\beta$ interface, the method of deviation of the molecular density determines the interface of either α^+/β or α^-/β (Figure 6.5d). Since molecular components have certain complexities in structure unlike the atomic components in SMAs, the molecular correspondence at the interface can be regulated via a change of orientation and conformation as well as their relative position. A minimal molecular rotation with a harmonic conformational change can achieve molecular correspondence at the interface without breaking the crystal.

Thus, deformability of **2** is closely related to temperature. Here, we investigated the thermodynamic aspect of the deformability via differential scanning calorimetry (DSC) and temperature dependence of s–s curves. First, DSC curve was collected to obtain four pertinent transition temperatures of M_f (118.56 °C), M_s (120.46 °C), A_s (123.00 °C), and A_f (124.97 °C). From the thermal perspective, the crystal phase should be α below M_s (L region) and β over A_s (H region) (Figure 6.6a). Either phase is allowed between M_s and A_s (L/H region) depending on the temperature profile. Thermal hysteresis was defined by a separation between A_s and M_s ($A_s - M_s$) in the current crystal, and had a width of 2.5 K. Then, temperature dependence of the s–s curves was collected by shearing on

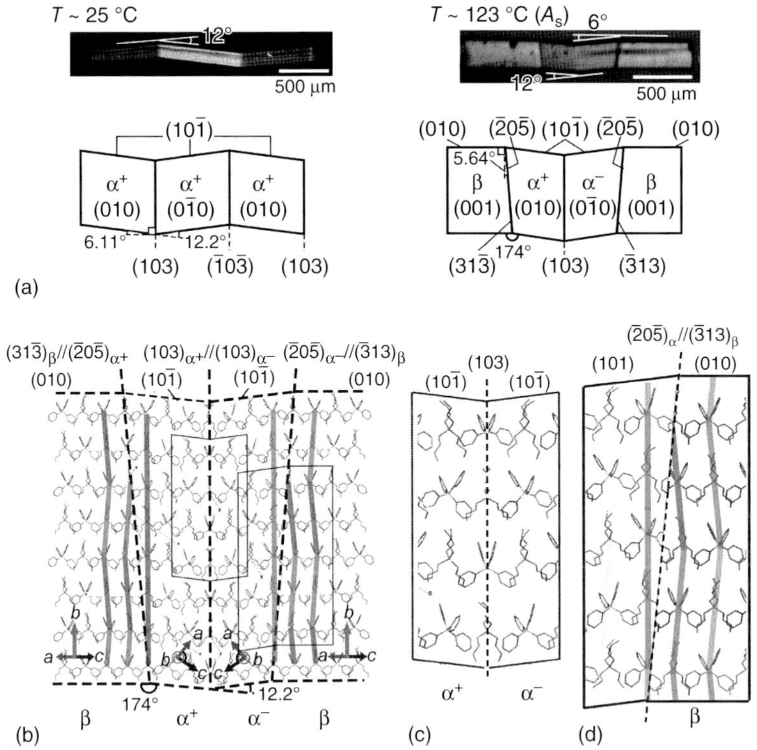

Figure 6.5 (a) Photographs taken with a polarized microscope with accompanying crystal face. The molecular correspondence at the boundary for (b) $\beta/\alpha^+/\alpha^-/\beta$, and magnified figures for (c) α^+/α^- and (d) α^-/β. Source: Takamizawa and Takasaki (2016) [10]. Reproduced with permission of The Royal Society of Chemistry.

$(10\bar{1})_\alpha$ and $(010)_\beta$ at various temperatures between −50 and 130 °C. The kinetic properties can be divided into the martensitic twinning (T) and superelastic (S) behaviors (Figure 6.6b). T in L region occurs below A_s (123 °C), in which typical ferroelastic s–s curves for $\alpha^+ \rightarrow \alpha^-$ conversion were observed. On the other hand, S occurs above A_s in H region. The crystal was spontaneously restored above the A_s.

Figure 6.6c summarizes the mechanical responses of **2** by categorizing temperature into three parts: α^+/α^- below $T_{\text{chem}-\delta}$, $\alpha^+/\alpha^-/\beta$ between $T_{\text{chem}-\delta}$ and T_{chem}, and β/α between T_{chem} and $T_{\text{chem}+\delta}$. Thus, the shape memory region (mechanically deformed state) can take α^+/α^- below A_s (T_{chem}). The $\alpha^+/\alpha^-/\beta$ deformation can be prepared between $T_{\text{chem}-\delta}$ and T_{chem} and superelastically regain the β solid shape if crossing over $T_{\text{chem}+\delta}$ with elevating temperature. The superelastic recovery of the deformed crystal needs a δ rise in temperature against the thermal transition temperature of A_s (T_{chem}). From a general view, the observed thermomechanical diagram for **2** is similar to the thermoelastic martensitic transformation for SMAs proposed by Otsuka and Shimizu in 1986 [24].

The shape recovery velocity ($v_{\text{r-trans}}$) in the spontaneous superelastic reverse transformation increases in proportion to the square of the temperature as

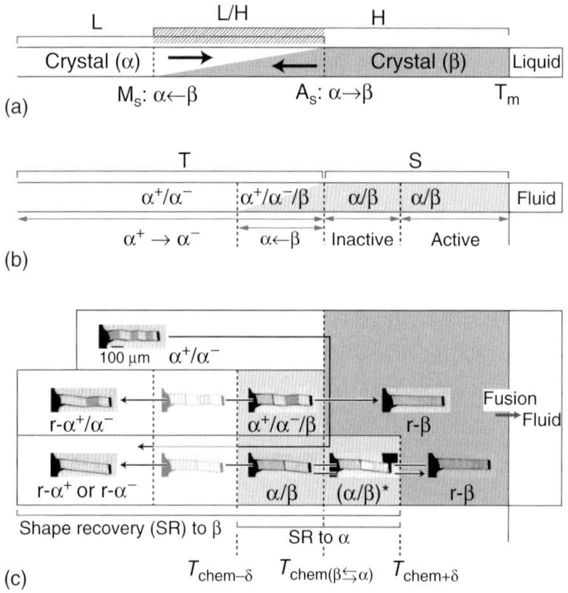

Figure 6.6 Phase diagram for **2**: (a) thermal transition, (b) kinetic phase diagram for force-induced mechanical phase, and (c) shape recovery (SR) diagram for thermomechanical transformation (SR, r-α, or r-β mean shape recovery and crystal phase restored into the original shape. *Shear is needed to conserve the deformation). Source: Takamizawa and Takasaki (2016) [10]. Reproduced with permission of The Royal Society of Chemistry.

the strength of the induced shear increases in relation to $v_{\text{r-trans}} \propto \sigma_{\text{r-trans}}^{\ 2}$ ($\because \sigma_{\text{r-trans}} \propto (T - T_{\text{chem}+\delta})$), which is suggestive of the conversion of the potential energy to kinetic energy. In the reversible thermal transformation, the crystal of **2** transforms across the transition temperatures of M_s ($\beta \rightarrow \alpha$) and A_s ($\alpha \rightarrow \beta$) in the manner of an isothermal martensitic transformation. Due to the ease in the interface shuttle in **2**, a mere thermal transition can recover the shape upon generating a slight shear of 0.03 MPa at M_s and less than 0.01 MPa at A_s.

In this work, finally, we demonstrated the actuation of **2** based on SME. A single crystal of **2** with a 1 cm^2 cross-sectional area is expected to lift a 100 kg weight horizontally with 10 MPa recovery shear generation around 200 °C. In order to visually observe the lifting work, we prepared a large single crystal and cleaved it into a 3.8 cm long cross-sectional area (Figure 6.7a), which had a 1.8 kg lift capability with a **2** specimen of 0.6 g in self-weight with a 0.18 cm^2 (0.6 × 0.3 cm^2) MPa recovery shear around 130 °C. Considering the current experimental configuration, we hung stainless steel (SUS) weights of 1, 10, and 100 g (more than 10^2 times heavier than the specimen's own weight) with a SUS filament on the single-crystal specimen after the specimen was manually deformed below A_s. After the experimental system was warmed in an oven, the specimen regained its default straight shape via sequential lifting as the recovery shear increased during the rise in temperature, demonstrating the practical and feasible work capability of the SME of **2** (Figure 6.7b), although the lifting started at 131.0 (1 g), 131.8 (10 g), and 133.4 °C (100 g), which are higher by 5.9 (131.0 − 125.1) − 8.2 °C (133.4 − 125.2) (Figure 6.7c) than the expected starting temperatures from the microcrystal experiments depicted in Figure 6.7a, probably due to the nonideal strain in a large crystal specimen hanging heavy weights. This demonstrates the desired characteristics of shape memory materials for miniaturization and conversion of heat into a practical work output within a

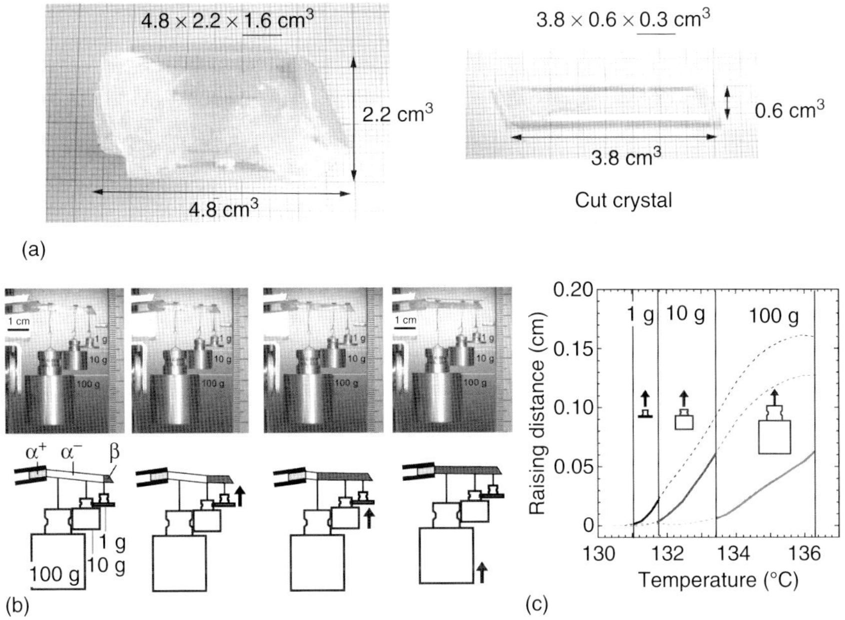

Figure 6.7 (a) Photographs of a large single crystal and a cut crystal for the lifting experiment. (b) Snapshots during the lifting work of **2** hung by 1, 10, 100 g SUS weights in an oven with a schematic for the explanation and (c) a sequential lifting course for each weight with temperature rise. Source: Takamizawa and Takasaki (2016) [10]. Reproduced with permission of The Royal Society of Chemistry.

narrow temperature cycle. It should be noted that **2** exhibits a high A_f point of 124.97 °C, which is higher than 110 °C known as the highest limit for binary Ni–Ti alloys without any additive.

6.2 Twinning Organosuperelasticity

The organosuperelastic effects shown above are based on the phase transformation between different phases (α and β phases). Superelasticity in metal alloys is known to also occur due to twinning deformation, in which mother and daughter phases have the same crystal phase with different orientation [36, 37]. We have reported several organosuperelastic materials based on twinning deformation [11, 12].

Single crystals of 3,5-difluorobenzoic acid (**3**) were prepared by sublimation of as-purchased sample (TCI Co., Ltd.) at about 1 kPa and 50 °C, which gave colorless rod-like single crystals with typical lengths of 0.3–1.0 mm [11]. The shear force on the crystal surface of $(0\bar{1}\bar{1})$ bent the shapes of single crystals of **3** accompanying the generation of the daughter crystal domain (α_1) from the original mother crystal domain (α_0) with a sharp interface propagating upon twinning deformation. The bending angle of the crystal specimen measured under an optical microscope had significantly large value of 27° (Figure 6.8a). The bent crystal spontaneously restored to its original straight shape when the shear force was

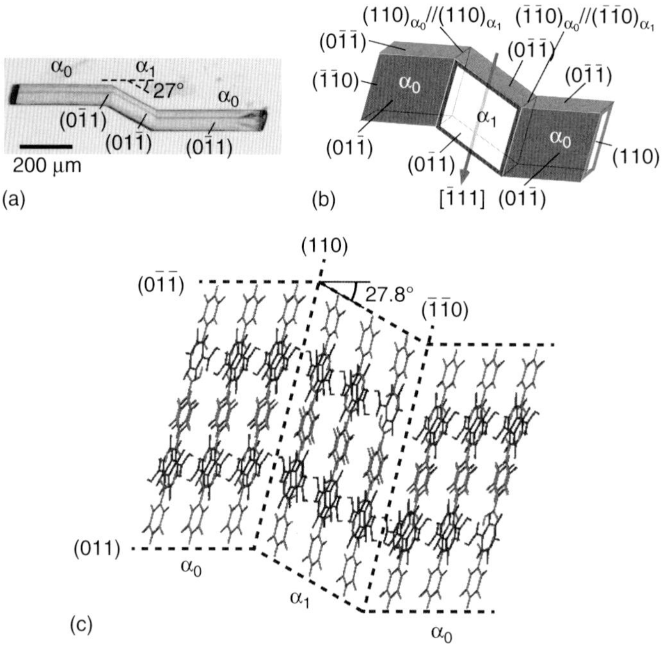

Figure 6.8 (a) Photograph of a crystal and (b) crystal face indices of **3** bent along $[\bar{1}11]$. (c) Connection between crystal packing on twinning interfaces. Source: Takamizawa, S. and Takasaki, Y. (2015) [11]. Reproduced with permission of John Wiley & Sons.

removed and the interface propagated back by 2.5 mm s^{-1} at 298 K. The interface is necessary to shape recovery motion and one interface transfers to the crystal end by knocking out of another interface from one end of the crystal.

X-ray study on the bent crystal of **3** revealed the mechanical twinning state with an interface of $(110)_{\alpha_0}//(110)_{\alpha_1}$ (or $(\bar{1}\,\bar{1}0)_{\alpha^1}//(\bar{1}\,\bar{1}0)_{\alpha_0}$), which is formed by connecting the identical lattices at the interface, which is relatively rotated 180° around the direction of shear force $[\bar{1}11]$ (type II twinning [36]) (Figure 6.8b). The crystal structures of both crystal domains are the same, which does not produce the Bain strain on the interface. In the crystals, molecules of **3** were dimerized by double O—H· · ·O=C hydrogen bonds of carboxylic acid (O· · ·O distance: 2.625 Å) along the long axis of each molecule with a dihedral angle of 9.48° between the carboxylic acid and the phenyl plane. The dimers associated to form a polymer sheet bound by the weak hydrogen bond as C—H· · ·O=C (C· · ·O distance: 3.378 Å) between the carbon atom of the phenyl ring and the oxygen atom of the adjacent molecule. The dimers stacked upon each other by π–π interaction with distances of 3.397 Å (C–C) and 3.218 Å (C–O). Considering the crystal connection in the twinning, the estimated bending angle is 27.8° from the projection directions of $[0\bar{1}1]_{\alpha_0}$ and $[01\bar{1}]_{\alpha_1}$ (Figure 6.8c), which agrees with the observed bending angle of 27° measured under a microscope. The proportion in the displacement of the shear position on the $(0\bar{1}\,\bar{1})_{\alpha_0}$ surface and the moving (far) side of the α_1/α_0 interface was 1 : 2.02, which also agrees with the expected value of

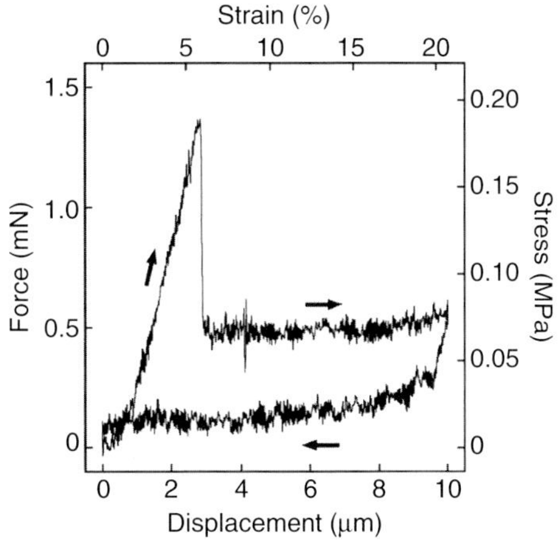

Figure 6.9 The s–s curve of **3** under shear stress at 298 K. Source: Takamizawa, S. and Takasaki, Y. (2015) [11]. Reproduced with permission of John Wiley & Sons.

1 : 1.96 based on the crystallographic data. Therefore, the harmonious support of the macro- and microscopic structural changes demonstrates deformation twinning of **3**.

As expected from the macroscopic crystal transformation, a superelastic loop in the s–s curve was obtained by loading shear force on $(0\bar{1}\,\bar{1})_{\alpha_0}$ face (Figure 6.9) by a metal blade (25 mm wide) with one end of a single crystal fixed with epoxy bond. σ_{nucl}, $\sigma_{f\text{-}trans}$, $\sigma_{r\text{-}trans}$ observed following the linear increase in the stress in elastic deformation were 0.189 MPa (1.37 mN), 0.07 MPa, 0.01–0.03 MPa, respectively. The energy storage density (E) and energy storage efficiency (η) were calculated as 11.9 kJ m^{-3} and 0.178, respectively. The values of $\sigma_{f\text{-}trans}$, E, and η in **3** were one-seventh to one-fifth of the values of **1** ($\sigma_{f\text{-}trans}$: 0.496 MPa, E: 62 kJ m^{-3}, and η: 0.925).

Here, we consider the molecular shapes and orientation at the mismatched twinning interface, which are expected to be deviated from the ones in virgin crystals because they are regulated by crystal packing. On the other hand, some intermolecular interactions such as hydrogen bond should be maintained to prevent fracture of the crystal. For the in-plane direction of the interface, slight shifts of molecules (Figure 6.10a) can preserve π–π interactions on the interface because their phenyl planes are nearly parallel to the interface. On the other hand, for the out-of-plane direction of the interface, a conformational adjustment of molecules (Figure 6.10b) is required because the C_2 symmetry should cleave the hydrogen bond at the center of the dimerized **3**. By considering the dissymmetric molecular distortion, the pseudorotation axis orthogonal to the C(phenyl)—C(carbonyl) bond axis of the molecules avoids the cleavage of the double hydrogen bonds across the interface (Figure 6.10c). The slight molecular conformational change on the interface is a reasonable hypothesis for the origin of the recovery stress. The distortion energy (1.69×10^{-4} J m^{-2}, 4.9×10^{5} J m^{-3}, and 210 J mol^{-1} (the case for mere benzoic acid)) gained from the dihedral twist between the carbonyl and phenyl planes from 8.3° to 12.7° is large enough to explain the energy required for the recovery stress in the transfer of the interface (6.5×10^{-6} J m^{-2}, 1.9×10^{4} J m^{-3}, and 8.0 J mol^{-1}), which was estimated by the s–s curve.

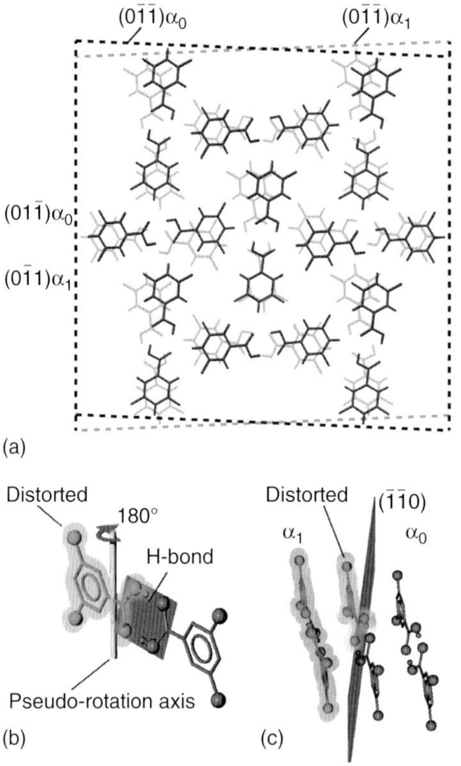

Figure 6.10 Origin of the recovery stress. (a) Mismatches of molecular orientations and positions between α_0 and α_1 domains on the twinning interface. Molecular distortion around (b) the symmetry axis and (c) the interface. Source: Takamizawa, S. and Takasaki, Y. (2015) [11]. Reproduced with permission of John Wiley & Sons.

To summarize, detailed structural study using X-ray revealed that the interface propagates to ease the distortion at the interface, and spontaneously recovers the initial crystal shape.

Superelasticity based on twinning deformation is also observed in $[Cu^{II}_2(\text{benzoic acid anion})_4(\text{pyrazine})]_n$ (**4**), microporous single-crystal host [12], which can be a microfluidic via open pores along [100] direction for gaseous fluid (Figure 6.11a,b). By shearing (00$\bar{1}$) surface with a glass needle at room temperature, a thin crystal domain in a daughter (α′) domain appeared from {1$\bar{1}\bar{1}$}, and grew along the [010] direction with a velocity of 133 μm s^{-1} keeping a domain width c. 5 μm. After going across the crystal, the thin domain broadened in 0.5 μm s^{-1} to separate the interfaces. The α′ domain spontaneously contracted and disappeared by removal of the stress through the organosuperelastic reverse transition to mother domain (α).

Crystal phase indexing under the coexisting state of α and α′ domains, which is bent at the domain interface by 14.6° along the projected direction of $[010]_\alpha$ and $[010]_{\alpha_0}$, revealed a type II twin, i.e. connected two crystalline domains in the same lattice with different orientation. This twinning skewed the direction of 0.8-nm-width channel of α′ domains (Figure 6.11c). Thus, the generation of the α′ domain alters the direction of gas permeation by mechanical twinning and the width or number of channels in the generated α′ domains is precisely regulated by the shear range of the original α crystal. Figure 6.11d,e shows the snapshots

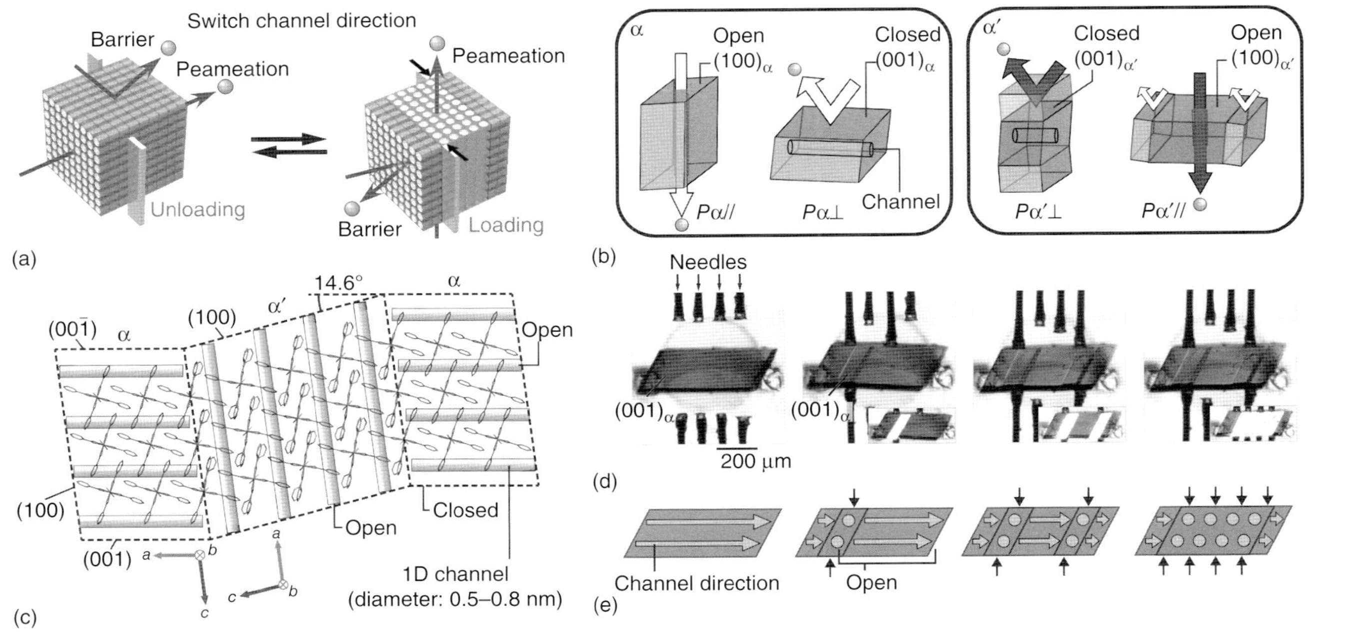

Figure 6.11 (a) Expected gas flow switch in shear-induced transition in direction. (b) Schematic explanation of the gas permeation of the embedded crystals. Gas permeability (*P*) in α (left) and α′ phases (right). (c) Connection of α and α′ phases accompanied by rotation of channel direction (bands). (d) Active generation/degeneration of α′ domains by shearing the crystal of **4** with movable microneedles and (e) schematics showing directions of the penetrating channels. Source: Takasaki and Takamizawa (2015) [12]. Reproduced with permission of Springer Nature.

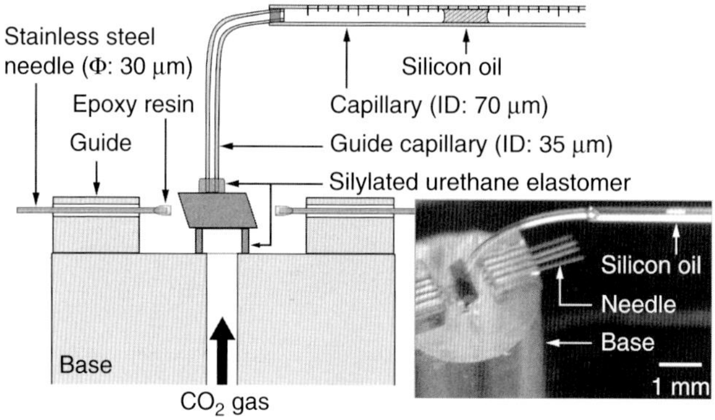

Figure 6.12 Experimental system of spatiotemporal gas permeation measurement with a shearing procedure at room temperature in continuous introduction of CO_2 gas from the base. Source: Takasaki and Takamizawa (2015) [12]. Reproduced with permission of Springer Nature.

and corresponding schematic illustration of spatio-resolved induction of twinning from α to α′, accompanied by the alternation of channel direction.

To demonstrate the dynamic switch of the permeated gas flow at the designed time and positions on the crystal **4**, we measured spatiotemporal gas permeation, which was traced by the movement of silicone oil inside capillaries guided through the capillaries attached to the crystal surfaces (Figure 6.12).

In this experiment, the permeation of CO_2 gas to horizontal (H) or vertical (V) directions was investigated. In the original α crystal, gas flowed out from the open surface in the H direction, whereas it was blocked on the closed surface in the V direction (white region in Figure 6.13a). By applying shear force on the crystal, the permeation/barrier directions were dynamically interchanged in generating α′ domain through the rotation of the channel direction (gray region in Figure 6.13a). The gas permeation starts immediately after α′ phase appears. The interchange was spontaneously switched back by removal of the shear force and could be reproduced repeatedly. Also, applying shear force at multiple positions dynamically generated the gas flow positions in the crystal (Figure 6.13b), which was demonstrated by alternate gas permeations at positions V_1 and V_2, depending on the switch of the sheared positions (white and gray regions in Figure 6.13b). Therefore, a single crystal of **4** provides spatiotemporal controllability of molecular flow by mechanical twinning.

With respect to the controllability of molecular flow, a single-crystal host of **4** can be regarded as an assemblage of pore units, which can be called "porons (pore + on(s))" as they are quasi-particles, with transformability assisted by flexible host skeletons. Dynamic rearrangement of "porons" can alter the pore connections, which produces the spatiotemporal controllability of flow directivity within a single solid. In this study, organosuperelasticity is an optimal property for the rearrangement of "porons." This theory raises a novel strategy for constructing microfluidic devices. We have demonstrated the active switchability of gas flow in directions and positions through a microporous molecular crystal of **4** at a designed time caused by shear-induced martensitic transition. The generated α′ domain in mechanical twining involved rotation of

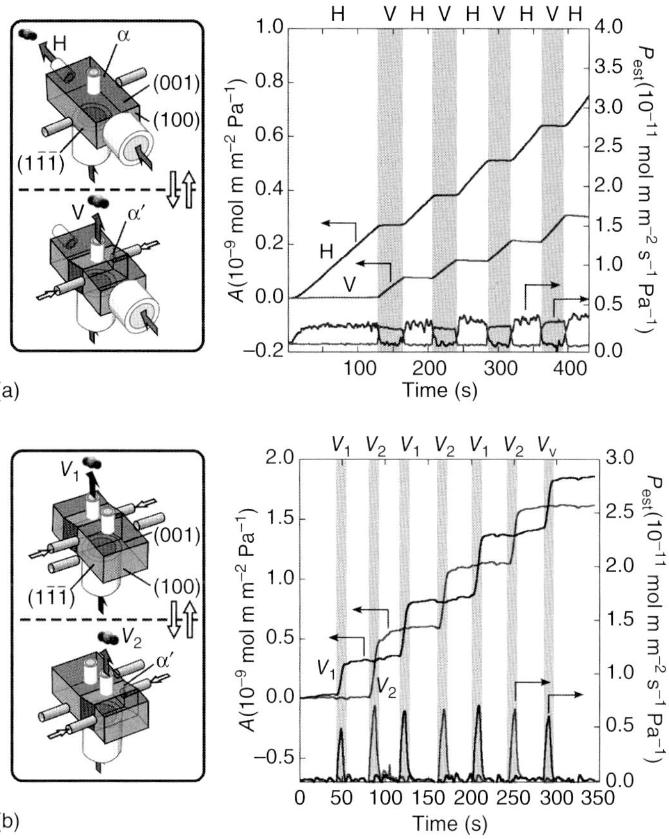

Figure 6.13 Schematic explanation for gas flow switch and permeated amount of CO_2 gas (A) and estimated permeability (P_{est}) (a) through the crystal in H and V directions and (b) through two positions (V_1 and V_2) in the V direction. Source: Takasaki and Takamizawa (2015) [12]. Reproduced with permission of Springer Nature.

the channel directions. The domain spontaneously contracted and disappeared by removal of the force as a superelasticity. In fact, a single crystal of **4** provides spatiotemporal controllability of molecular flow by mechanical twinning. This microporous single crystal would be used as a device with precise flux and/or directional controllability of a molecular flow manipulated by mechanical force. Furthermore, by using a shear-induced mechanism in a microdevice, flammable fluid such as high-pressure hydrogen gas would be safely controlled due to the lack of need for electric actuation. Consequently, the introduction of shear-induced transition phenomena into channel solids enables dynamic control of molecular flow in the solids.

Through this study on the crystal of **4**, two significantly important results were obtained. First, the crystal is a metal complex, in which superelasticity have never been reported, that is the novel class of potentially superelastic materials was discovered following the crystal of **1**. Second, the crystals of **4** are porous, and the twinning deformation causes the switching of the permeability. This, the control of the functionality, is not achievable in superelastic metal alloys, and is totally development of science of superelasticity based on non-alloy materials.

6.3 Organosuperplasticity Through Multilayered Sliding

In organic materials, molding processing methods for stiff and hard-to-work crystalline polymers have been developed according to the properties of individual polymer. Common crystalline materials, particularly single crystals, least deserve useful plasticity, in the common view. For this reason, less attention has been paid to the plasticity in organic crystalline solids for a long time. We found the superplasticity (organosuperplasticity) in organic single crystals of *N*,*N*-dimethyl-4-nitroaniline (**5**), which shows the unique deformability of superplasticity as well as superelasticity based on twinning deformation [13]. Although superplasticity had been believed to be manifested only in metal and alloys and traditionally studied in metallurgy [38], the discovery of superplasticity in purely organic compound **5** showed that the limitation in the materials type was merely a superstition.

5 (0.5–5.0 × 0.2–0.8 × 0.03–0.10 mm, $P2_1$ Sohncke group) were recrystallized from an acetone solution as well-formed yellow platelet crystal, of which face index are shown in Figure 6.14a. X-ray crystallographic study revealed that the molecules tilted in 75.4° to {001} surface, and formed closed-packed plane by aligning in ⟨100⟩ direction via π–π stacking. The sheet structures are connected by hydrogen bonding between N=O of nitro group and H—C of dimethylamino group (c. 3.6 Å from the oxygen to carbon atoms) to stack in ⟨001⟩ direction.

The crystal shows the superelastic behavior based on twinning deformation by shearing to ⟨201⟩ direction on the $\{20\bar{1}\}$ (Figure 6.14d, top). The interface was $(\bar{2}01)_{\alpha_0}//(20\bar{1})_{\alpha_1}$ (or $(20\bar{1})_{\alpha_0}//(\bar{2}01)_{\alpha_1}$) and the twinning deformation was caused by a 180° rotation around the axis perpendicular to the interface. During the twinning, the major axes of the components in each domain tilted 4.9° against the interface (Figure 6.14d, middle). The spontaneous shape recovery was therefore attributable to frustration of molecular assembly, which was caused by the tilt, at the twinning interface (Figure 6.14d, bottom).

s–s curves collected at room temperature showed that crystal of **5** exhibits either superelastic or superplastic behavior depending on the direction of loaded forces (Figure 6.15) as described below.

Shear force was applied along the normal vector of the $(00\bar{1})$ face (Figure 6.16) with the (001) face fixed on a base. The bending angle between the $(001)_{\alpha_0}$ and $(001)_{\alpha_1}$ faces was 23°, which was close to the calculated angle of 20.3° based on X-ray crystallographic study. Typical superelastic s–s cycles was observed, i.e. successive observation of appearance of twinning domain at the critical shear stress of 0.028 MPa, followed by deformation under a constant stress (0.030–0.032 MPa) along with the growth of the α_1 domain, and the spontaneous shape recovery with a constant recovery stress (0.01 MPa).

The shearing on (001) and $(00\bar{1})$ faces to ⟨100⟩ direction of the crystal caused the surprisingly large distortion of 500% and deformation into parallelogram

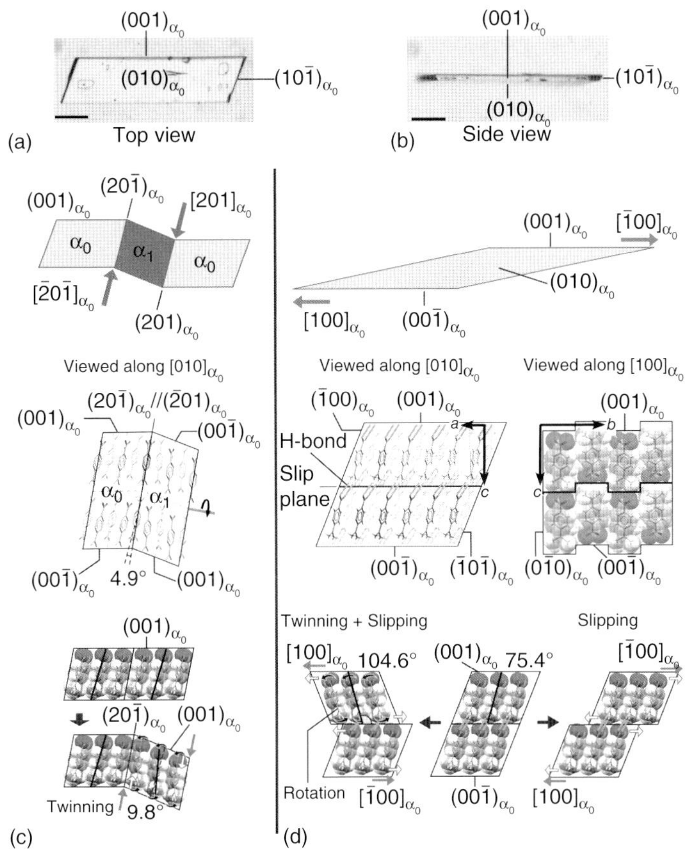

Figure 6.14 (a, b) Crystal photos of **5** with face indices. (c) Bending shape (top), microstructure around an interface between twins (middle), and orientation change (bottom) in superelastic deformation. (d) Deformed shape of the crystal in superplastic deformation, with 500% strain (top), and the microstructures (middle) and slip mechanism (bottom). Source: Takamizawa et al. 2018 [13]. Reproduced with permission of Springer Nature.

shape. Considering the crystal structure, this deformation is supposed to be originated in multiplane slip by the simple slip between closely packed plane (shear to $[\bar{1}00]$ direction) or a combination of slip and twinning (shear to [100] direction). To have an s–s curve, (001) and $(00\bar{1})$ surfaces of the crystal were attached to glass jigs separately at room temperature, and then, the jig on the (001) face was moved along the [100] direction at a constant speed of $50\,\mu m\,min^{-1}$ (Figure 6.17). Shear-strain rate ε was calculated by $x/h \times 100\%$, where x is displacement and h is crystal height (338 μm). As a result, the crystal started to deform plastically by slipping at the yield point of 0.34 MPa (0.8%). In the region of $\varepsilon = 0$–325%, the slip deformation proceeded in the whole crystal (shear stress $\sigma = 0.25$–0.34 MPa).

The remarkable merit in superplasticity of **5** is that the deformation does not suffer crystallinity, and therefore the properties derived from crystallinity is retained. Here, we present a superelastic deformation as an example (Figure 6.18).

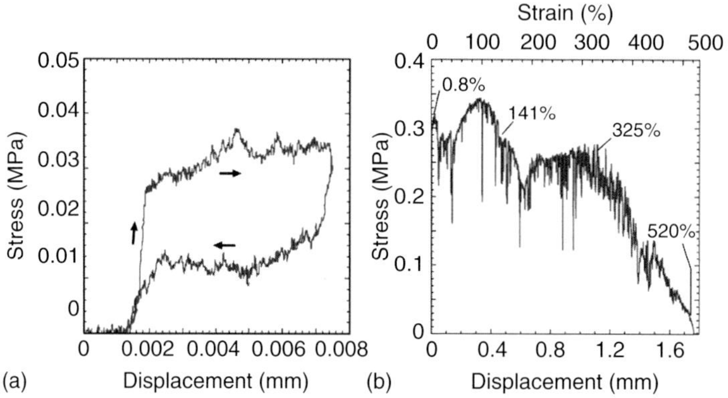

Figure 6.15 s–s curves of **5** collected by loading forces to (a) normal direction to (00$\bar{1}$) face and (b) ⟨100⟩ direction. Source: Takamizawa et al. 2018 [13]. Reproduced with permission of the Nature Publishing Group (open access).

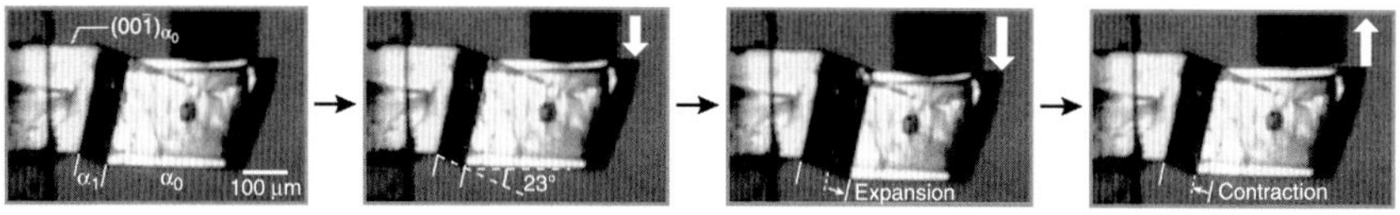

Figure 6.16 Snapshots of a crystal of **5** during superelastic deformation. Source: Takamizawa et al. 2018 [13]. Reproduced with permission of Springer Nature.

We conducted shear tests and single-crystal X-ray structure analysis on a superplastically deformed crystal of **5** at room temperature. Superelastic behavior was observed in superplastically deformed moieties of the crystal with a shear-strain rate of c. 400%. The retention of single crystallinity of the crystals (deformation rate: 100–150% and 330%) was confirmed by single-crystal X-ray diffraction measurements.

In a common sense, organic crystals have less opportunities for industrial applications owing to their brittle and less processable nature. However, superplastic materials can be processed while their single crystallinity is retained as well as the functions derived from the single crystallinity, making it possible to prepare highly processable functional crystals.

6.4 Twinning Ferroelasticity

As shown above, we have reported several superelastic materials based on twinning deformation. Actually, it seems that the deformation of organic crystals based on twinning deformation tends to cause the ferroelasticity, as several ferroelastic materials have been reported since 1970s [32–34]. Although these previous researches have different background from the researches of superelasticity and ferroelasticity are no doubt closely related phenomena because the twinning also potentially causes superelasticity from our findings described above. It is promising that these deformabilities can be systematically

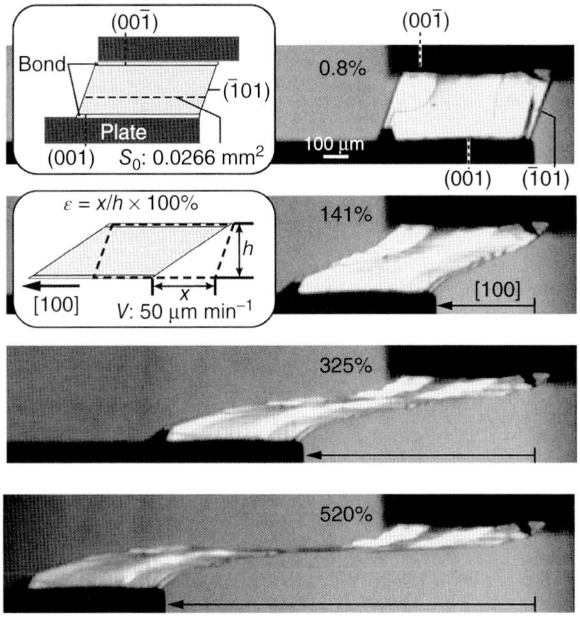

Figure 6.17 Snapshots during superplastic deformation of **5**. Source: Takamizawa et al. 2018 [13]. Reproduced with permission of Springer Nature.

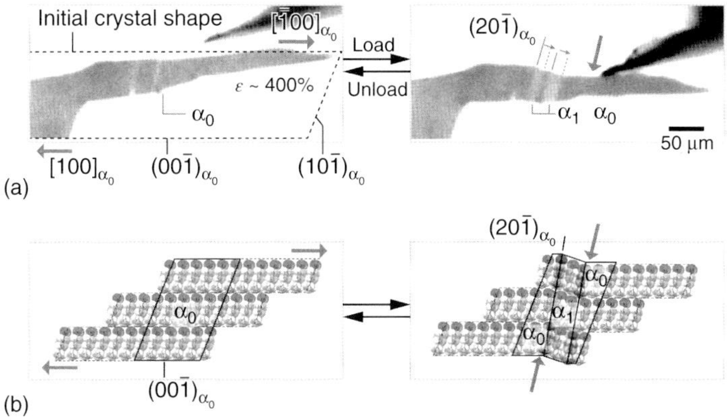

Figure 6.18 Coupling of superplasticity and superelasticity. (a) Superelastic behavior of crystals of **5** after superplastic deformation. (b) The schematic model represents the superelastic deformation in single crystalline regions. Source: Takamizawa et al. 2018 [13]. Reproduced with permission of Springer Nature.

understood by the detailed investigation on the macroscopic crystal deformation, microscopic molecular deformation, and stress. Then the condition for the manifestation of the deformabilities will be unveiled.

Based on this idea, ferroelasticity of 5-chloro-2-nitroaniline (**6**) (Figure 6.19a) was investigated [14]. Single crystals of **6** were recrystallized from acetone at room temperature. Well-formed yellow plates were obtained with typical lengths

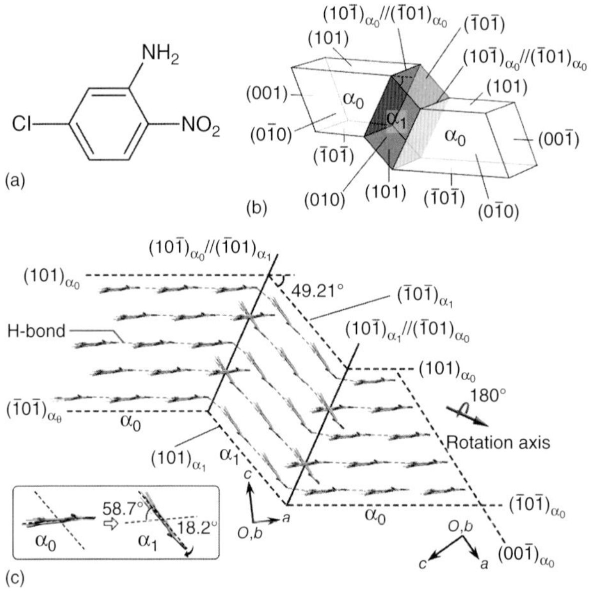

Figure 6.19 (a) Molecular structure of **6**. (b) Crystal face indices of the twinned crystal. (c) Crystal packing comparison of the mother phase (α_0) and the daughter phase (α_1) of **6** viewed parallel to $(101)_{\alpha_0}$ .Crossed molecules on each interface are illustrated by superimposing the packing of α_0 and α_1 based on an assumption that either orientation is possible. Source: Mir et al. 2017 [14]. Reproduced with permission of John Wiley & Sons.

of 0.5–1.5 mm and thicknesses of 0.05–0.1 mm. A crystal of **6** was mechanically twinned at 298 K by shearing along the normal vector of (101) using tweezers.

Single-crystal X-ray structure analysis of the bent crystal at 298 K confirmed that the α_0 and α_1 domain have the same crystal structure with $P2_1/n$ symmetry, and that lattice correspondence for rotational twinning exists at the twinning interface, $(10\bar{1})_{\alpha_0}//(\bar{1}01)_{\alpha_1}$ (or $(10\bar{1})_{\alpha_1}//(\bar{1}01)_{\alpha_0}$) (Figure 6.19b). This lattice connection will give the bending angle of 49.21° for the projection on $(101)_{\alpha_0}$ and $(\bar{1}0\bar{1})_{\alpha_1}$, which agrees well with the value of 49° obtained by measurement under an optical microscope. Because the twinning deformation manner is strictly restricted by the crystallographic symmetry, only the observed rotational twinning axis, the twofold screw axis along ⟨010⟩, is attributed to the twinning caused by the shear parallel to ⟨101⟩ in the current crystal system. Also, the deformation involves $\alpha_0 \rightarrow \alpha_1$ conversion by a 180° rotation of the α_0 domain about the rotational axis perpendicular to the twinning interface, which requires a shift in molecular orientation (Figure 6.19c).

To avoid fracture at the interface, the connection between the two domain should not be severely unstable. Here, lattice matching and intermolecular interactions are considered. First, the unit cells fit perfectly at the α_0/α_1 interface because the interface of each phase is crystallographically equivalent. Also, three different intermolecular interaction in the current system, i.e. hydrogen bonding between the amino and nitro groups (N—H· · ·O=N, N· · ·O distance: 2.923 Å) (intermolecular dotted lines in Figure 6.19c), a weak close-contact distance of

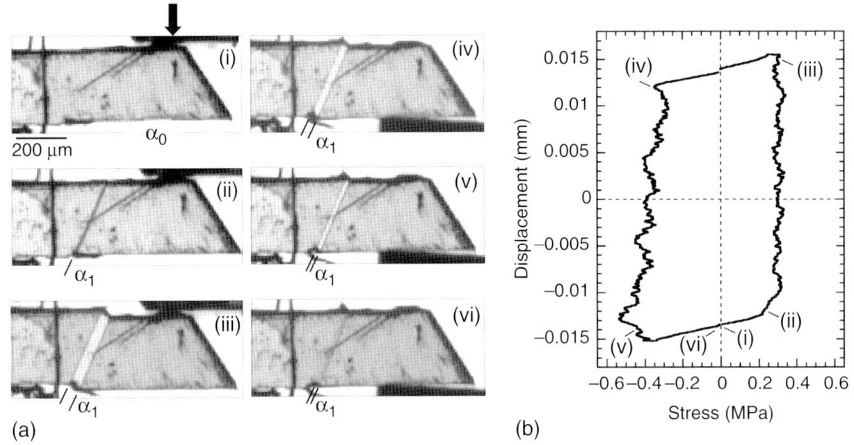

Figure 6.20 (a) Snapshots of twinning deformation upon application of shear stress. The shear force was applied to forward direction in (i)–(iii) and reverse direction in (iv)–(vi). (b) Ferroelastic hysteresis loop. Source: Mir et al. 2017 [14]. Reproduced with permission of John Wiley & Sons.

Cl with 3.408 Å distance, aromatic π–π interactions, with a distance of 3.873 Å (C· · ·C) between adjacent phenyl rings are in part maintained by adjustments of the molecular orientation.

The mechanical property of ferroelasticity in **6** was confirmed by recording a s–s curve of a single-crystal specimen. For the measurement, a glass jig was pushed against the crystal surface of $(101)_{\alpha_0}$ at a constant speed of 50 μm min^{-1} to load shear force (Figure 6.20) with one end of the crystal fixed to a glass stage with epoxy. As shown in Figure 6.20, shear stress (σ) was detected when the glass jig reached the crystal surface at (i) and a linear increase in the loading force began in (i)–(ii). The stress reached 0.327 MPa at which point the twinning interface was generated at (ii). The α_1 domain was subsequently observed in a midst of the crystal specimen (α_0). The applied shear stress then remained roughly constant as a function of displacement during growth of the α_1 domain in (ii)–(iii). On holding the displacement of the jig, the α_1 domain remained present and a spontaneous strain was recorded by removing the force in (iii)–(iv). Soon after applying force across the crystal $(\bar{1}0\bar{1})_{\alpha_0}$, using the jig to induce the reverse stress at (iv), a reverse deformation, contraction of the α_1 domain, proceeded. Simultaneously, a similar critical stress of 0.39 MPa was shown, followed by a gradual increase of the stress to 0.57 MPa (iv–v). At the point of force unloading, shortly after (v), the α_1 domain disappeared and the right twin interface was completely reached to the left α_0 domain.

Thus, we have confirmed the ferroelastic behavior of **6** by shear experiment, and X-ray study revealed a good lattice conformity at the α_0/α_1 interface without breaking the intermolecular hydrogen-bonding interactions. This indicates that some structural trick to cancel distortion at the interface may lead to ferroelasticity. To realize this, flexible molecular structure may be suitable. Here, we present an adipic acid (**7**, Figure 6.21a) crystal case for example, of which the

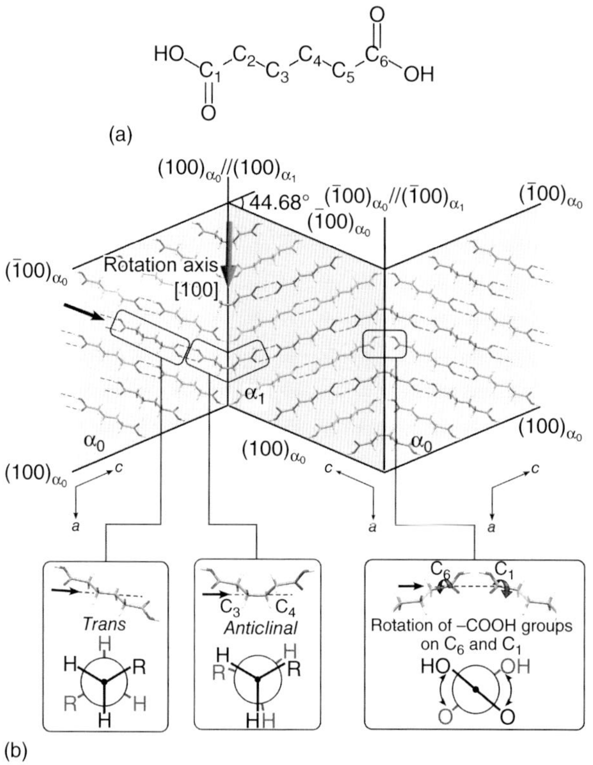

Figure 6.21 (a) Molecular structure of **7**. (b) Packing diagrams of overlapping α_0 and α_1 of **7** viewed along $[010]_{\alpha_0}$. Left-side inset b indicates the trans and anticlinal conformations in an alkyl chain. Each "R" is $-CH_2-COOH$. Right-side inset (b) indicates an estimated rotation of –COOH groups on C6 and C1 atoms at the interface to keep the hydrogen bond linkage. Source: Mir et al. 2018 [15]. Reproduced with permission of the Royal Society of Chemistry.

crystal deformation is geometrically related to the deformation of the molecules themselves [15].

Single crystals of 7 were recrystallized from a 1 : 1 mixture of tetrahydrofuran and toluene at room temperature to obtain well-formed colorless plates or rods with typical lengths of 0.2–1.0 mm and thickness 0.05–0.15 mm. Upon the shear force application perpendicular to $(\bar{1}00)_{\alpha_0}$ (α_0: mother domain) at 298 K, a crystal of 7 showed the typical ferroelastic bending, i.e. generation of the daughter domain (α_1), retention of deformed shape after removal of the applied force, and shape recovery by applying the inverted force, which is perpendicular to $(100)_{\alpha_0}$. The bending angle of the crystal specimen was 42°. The s–s curves of 7 was recorded by loading force on $(\bar{1}00)_{\alpha_0}$ by a glass jig moving with a constant speed of 50 μm min^{-1} to obtain a ferroelastic hysteresis loop with coercive stress (σ_c) of 0.50 MPa and a certain spontaneous strain.

Single-crystal X-ray structure analysis confirmed the twinning deformation under shear force (Figure 6.21b). The α_0 and α_1 domains have an identical crystal structure with the same symmetry, i.e. monoclinic crystal system and $P2_1/c$

space group. According to the crystal face indexing, the twinning mode has been confirmed as a 180° rotation about an axis along $[100]_{\alpha_0}$ (illustrated as bold arrows in Figure 6.21b), which is parallel to the $\{001\}_{\alpha_0}$ and $\{010\}_{\alpha_0}$. According to the crystallographic data, molecules of **7** are in all trans conformation in the solid state. The predicted microscopic bending angle between α_0 and α_1 lattices is 44.68°, which well agreed with the experimentally measured angle of 42°.

Figure 6.21b also shows the crystal packing diagrams of the adjacent α_0 and α_1 domains. Molecules form an infinite 1D hydrogen bonding along $[101]_{\alpha_0}$ (thin arrow on the left side of Figure 6.21b) by O—H···O=C, which is, actually, preserved at the conversion from α_0 to α_1 because of the flexibility of **7**. Changes in the molecular orientations, which is usually required in the twinning deformation, is realized by sequential bending of alkyl chain of **7** by rotation around C—C bond. For example, the trans to anticlinal conformational change about each C3—C4 bond is considered when the interface exists on the center of the molecule (left-side inset Figure 6.21b). In addition, at the neighboring 1D chain (illustrated with a low contrast color in Figure 6.21b), either of facing –COOH groups should rotate about the C5(alkyl)—C6(carboxylic) bond (and/or C1—C2 bond) at the interface for keeping the continuity of the hydrogen bond linkage (right-side inset Figure 6.21b). The crystal integrity during the twinning deformation is also supported by the nearly no-plane strain on the interface because S_{α_1} and S_{α_0} are identical, where S is cross-sectional area parallel to the interface calculated by multiplying the lengths of the a and b axes. Here, the flexibility of alkyl chain would reduce the structural distortion of facing planes at the interface.

4,4′-Dicarboxydiphenyl ether (**8**) is another example of conformation change in the twinning deformation. Flipping of the phenyl group helps avoiding the distortion at the domain interface, leading to the ferroelastic deformation [16].

We obtained single crystals of **8** as a mixture of two polymorphs with dissimilar crystal shapes of thick needles and prism-shaped crystals. The prism-shaped crystals readily undergo twinning deformation when subjected to shear force (Figure 6.22a,b), and are picked for experiments. An s–s curve confirmed the ferroelastic behavior of **8** with nucleation stress of c. 1.05 MPa, coercive stress c. 0.21 or 0.14 MPa (Figure 6.22c,d). The nucleation stress is relatively high compared with the coercive stress, indicating a large initial energy barrier.

Single-crystal X-ray diffraction data were collected for a mechanically twinned crystal of **8**. The sample was deformed manually using tweezers, while the separation of crystalline domain was confirmed via observation under an optical microscope. The crystal faces of the mother domain (α_0), the daughter domain (α_1), and the twinning interface were determined. It is evident from face indexing that the deformation occurs by 180° rotational twinning about the axis perpendicular to the interfacial plane $(\bar{1}\,\bar{1}\,\bar{1})_{\alpha_0}//(111)_{\alpha_1}$. This interface was unambiguously confirmed by measuring the bending angles related to the twinning deformation.

The lattice matching at the interface was also investigated. The molecular packing diagrams show that the molecular orientation of **8** is mostly preserved during the deformation from α_0 to α_1. On the other hand, at the twining interface, phenyl rings are expected to flip in c. 30.2° as shown in Figure 6.23. Although rotation or flipping of the entire molecule is avoided in the conversion from α_0 to α_1, this conformational change adjusts molecule in the chemical environment of α_1. The

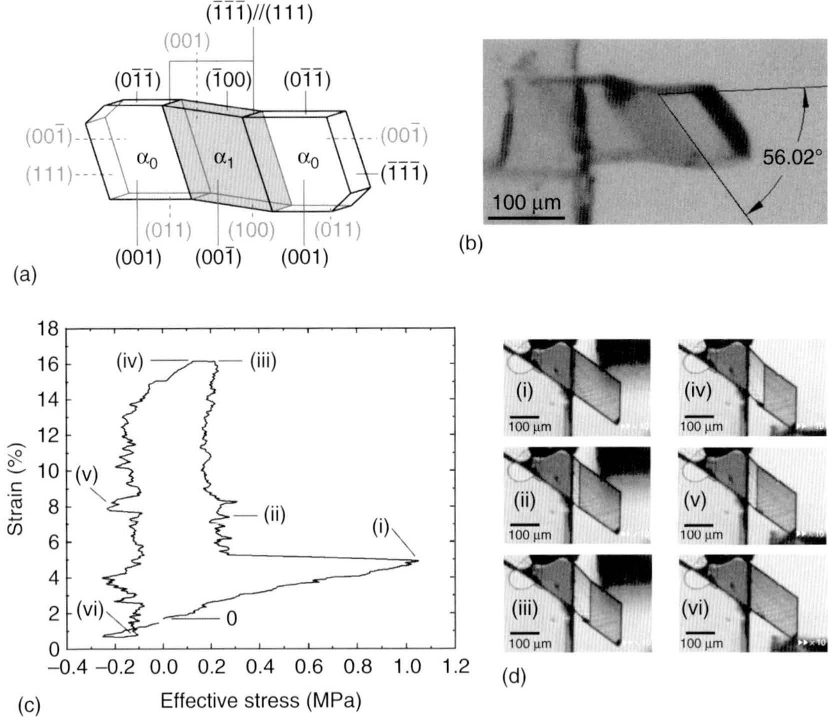

Figure 6.22 (a) Face indices for the twinning deformation in **8** with α_0 and α_1. (b) Snapshots of mechanical twinning. (c) s–s curve for a single ferroelastic cycle of **8** and (d) snapshots of mechanical twinning of **8** during the s–s experiment. Source: Engel et al. 2018 [16]. Reproduced with permission of The Royal Society of Chemistry.

relatively high nucleation stress is due to the steric hindrance of phenyl group, which is still large even though the flipping is easier than the rotation of entire molecule.

The hydrogen bonding of the carboxylic acid group in the crystal of **8** might collapse in the deformation because of the flipping of phenyl rings. Actually, free rotation about the carboxylic acid C—C bond helps preserving the hydrogen bond and it is likely that some rotation (c. 0–20°) of the –COOH group occurs about this C—C bond and the planarity of the hydrogen-bonded dimers is mostly preserved. This would relieve molecular distortion at the interface and is necessary given the experimental evidence that spontaneous recovery does not occur, suggesting that the overall strain at the twinning interface itself is relatively low.

Here, let us consider dissipated energy (E_d) based on the shear experiment for the ferroelastic compounds **6**, **7**, and **8** because the large molecular deformation may cause large E_d. As a result, **6**, **7**, and **8** had large E_d values of 216, 420.98, and 53.77216 kJ m^{-3}. On the other hand, the stress to induce ferroelastic deformation in organic compound is smaller than that of general inorganic compound [31]. This is expected to be good for some applications such as vibration absorber. Therefore, large E_d is one important goal of designing organic ferroelastic materials. We have reported that 2-methyl-5-nitrobenzoic acid (**9**)

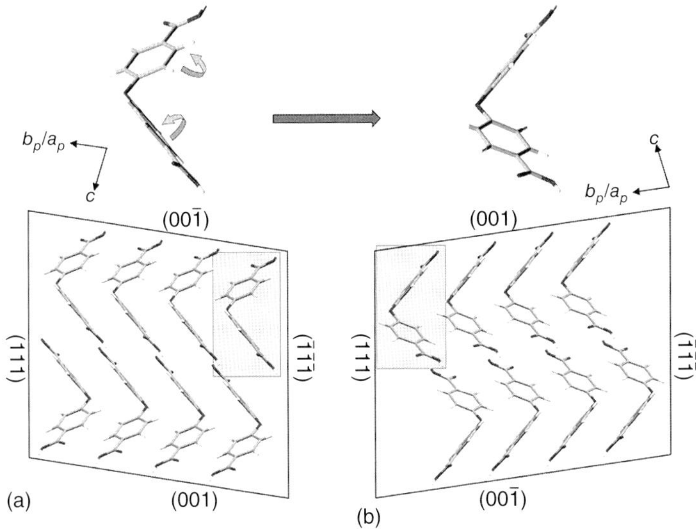

Figure 6.23 Proposed molecular conformational changes occurring during the deformation from (a) α_0 viewed along [$\bar{1}$10] to (b) α_1 viewed along [1$\bar{1}$0]. Source: Engel et al. 2018 [16]. Reproduced with permission of The Royal Society of Chemistry.

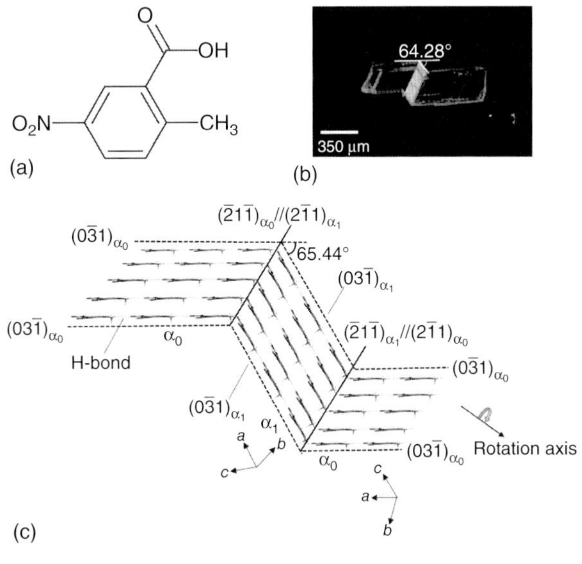

Figure 6.24 Molecular structure of **9**. (b) Optical image of the twinned crystal, prepared by compression on crystal surface $(110)_{\alpha_0}$, along [013] direction. Crystal packing diagram projected along $(0\bar{3}1)_{\alpha_0}$. Source: Mir et al. 2018 [17]. Reproduced with permission of The Royal Society of Chemistry.

(Figure 6.24a) possesses the largest E_d as a molecular-based compound [17]. **9** is a beige-colored aromatic compound with a high melting point of 184 °C. Well-shaped, pale yellow-colored rods with typical lengths of 0.2–1.0 mm and thicknesses of 0.1–0.3 mm were obtained by recrystallization from acetone at 298 K.

Twinning deformation was induced by compression stress perpendicular to $(110)_{\alpha_0}$ and $(\bar{1}\,\bar{1}0)_{\alpha_0}$ (α_0: mother domain) crystal faces at 298 K. The twinning caused crystal bending in surprisingly large angle of 64.28° (Figure 6.24b), which agreed with the calculated bending angle of 65.44° between $(0\bar{3}1)_{\alpha_0}$ and $(03\bar{1})_{\alpha_1}$ (α_1: twinned domain) based on the crystal domain connection at $(\bar{2}1\bar{1})_{\alpha_0}//(2\bar{1}1)_{\alpha_1}$ interface confirmed by single-crystal X-ray structure analysis (Figure 6.24c).

Both the α_0 and α_1 crystal structures have an identical crystal structure with the same symmetry, i.e. triclinic crystal system and P-1 space group. The α_1 domain is related to α_0 domain by a 180° rotation about the axis along $[013]_{\alpha_0}$. Although the deformation requires drastic molecular rearrangement, crystal cleavage is prevented because of strong hydrogen network by closely packed molecules surrounded by up to four neighboring molecules in the structure. In the $(013)_{\alpha_0}$ plane, 2D molecular arrangements are formed by two different hydrogen bonding. Typical O—H···O hydrogen bonding between carboxylic acid groups occurs along [013] with the C–O distance of 2.639 Å, as well as C—H···O interactions along $[0\bar{3}1]$ between the aromatic hydrogen H and O on the nitro group with C–O distance of 2.673 Å. The hydrogen bonding between carboxylic acid groups is the dominant force over weak van der Waals interactions between nitro and aromatic hydrogen atoms.

The maximum strain of **9** is expected to be 218.81%, which is higher than that for previous reports on organic crystals that have comparatively smaller bending angles than **9**. For example, the strain values for tetra-methyl-tetra-selena-fulvalene $(TMTSF)_2ClO_4$, $(TMTSF)_2PF_6$, 3,5-difluorobenzoic acid, and 5-chloro-2-nitroaniline are 33.3%, 36%, 52.7%, and 115.9%, respectively. Moreover, the strain value is higher than for certain twinned metals. For examples, the values for gold nanopillars and copper nanowires are 5% and 7.2%, respectively.

We conducted compression measurements of a single-crystal specimen to have an s–s curve (Figure 6.25a), which derives the effective force along the angle 57.28° between twinning interface and effective shear. A glass jig was pushed against the $(110)_{\alpha_0}$ crystal surface at a constant speed of 30 μm min^{-1} with one end of a single-crystal fixed to a glass stage. As shown in Figure 6.25b,c, a ferroelastic hysteresis was observed with the appearance of α_1 domain at 6.891 MPa, and with the largest coercive stress among all reports. The α_1 domain showed the unsmooth growth from both sides of the domain, represented by spikes in the s–s curve (Figure 6.25c(ii, iii)). This could be ascribed to the multiple generation of α_1 domains.

The E_d was calculated as 1000 kJ m^{-3} (0.6796 kJ kg^{-1}, and 123.109 J mol^{-1}) based on the equation ($E_d = \sigma_c \tan\theta$). The value of **9** is exceptionally large as an organic crystal. For example, 65.24 times larger than the corresponding value for 3,5-difluorobenzoic acid (15.47 kJ m^{-3}, 0.010 kJ kg^{-1}, 1.642 J mol^{-1}), and 4.67 times larger than in the case of 5-chloro-2-nitroaniline (216 kJ m^{-3}, 0.136 kJ kg^{-1}, 23.46 J mol^{-1}). This large E_d value is even comparable to that of alloys. Since E_d is a derivative of the applied shear stress (σ) and deformation angle (θ) of the crystal, such dissipated energy density is expected to be increased by enlarging the bending angle of the crystal, which in the case of **9** is 64.28°, and/or the required shear stress for the deformation. The high E_d will be effective for the damping of mechanical vibrations.

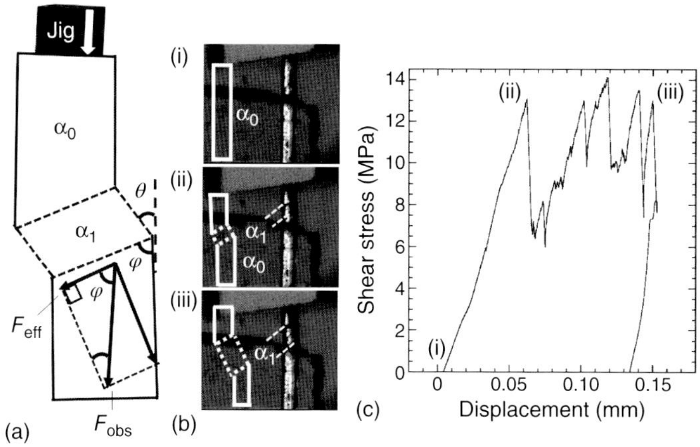

Figure 6.25 Measurement of an s–s curve: (a) schematic of crystal deformation pattern and force components, (b) snapshots of the twinning deformation of shear stress (i–iii) with inset sketches of the deformation pattern, and (c) s–s curve at 298 K. Source: Mir et al. 2018 [17]. Reproduced with permission of The Royal Society of Chemistry.

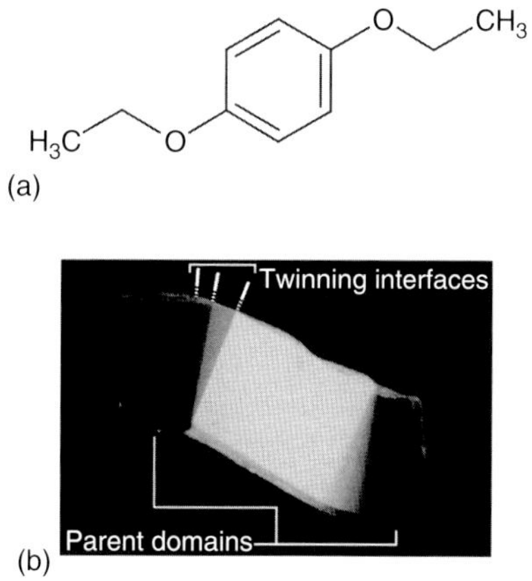

Figure 6.26 (a) Molecular structure of **10** and (b) mechanically deformed crystal of **10** with parent domains and observable twinning interfaces highlighted. Source: Engel and Takamizawa 2018 [18]. Reproduced with permission of John Wiley & Sons, Inc.

The organic ferroelastic materials presented above can deform in a specific angle. This is, actually, quite common because twinning deformation is essentially strictly limited in terms of the structure. Surprisingly, however, 1,4-diethoxybenzene (**10**) (Figure 6.26a) can bend in free angle in a certain plane because twinning deformation can occur in any of a set of crystal zone (Figure 6.26b) [18].

Single crystals of **10** were prepared by concentrating ethanol solution and cutting to show (010) surface. Under an optical microscope, the crystal was manipulated by tweezers to find out that the most readily induced and well-defined twinning interface is associated by bending angle of c. 10° (Figure 6.27a). X-ray diffraction data collected for one such mechanically twinned crystal at 298 K showed that the compound **10** crystallizes in the Sohncke monoclinic space group $P2_1$, and that the parent (α_0) and daughter (α_1) domains are related by a 180° rotation about [1 10 0], which is parallel to

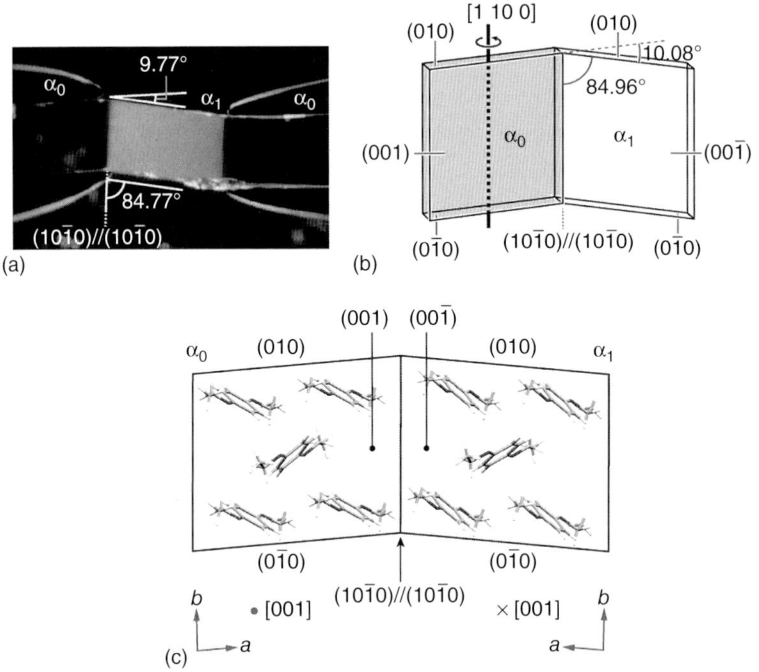

Figure 6.27 (a) Snapshot of bending crystal. (b) Face indices, interface angle, and bending angle determined by single-crystal X-ray diffraction. (c) Crystal packing of α_0 and α_1 projected onto $(001)_{\alpha_0}$ (and $(00\bar{1})_{\alpha_1}$). Source: Engel and Takamizawa 2018 [18]. Reproduced with permission of John Wiley & Sons, Inc.

the twinning interface $(10\,\bar{1}\,0)$ (Figure 6.27b). For twinning interface $(10\bar{1}0)$, the observed interface angle 84.7° – for (010) against $(10\,\bar{1}\,0)$ – and crystal bending angle 9.7°, are in good agreement with the calculated corresponding angles 85.0° and 10.1°, respectively. Partial packing diagrams of the adjacent α_0 and α_1 domains are presented in Figure 6.27c. Although hydrogen-bonding network do not exist in the crystal of **10**, intermolecular electrostatic interactions maintain the connectivity of molecules across the twinning interface.

Remarkably, a shear force across the (010) or $(0\bar{1}0)$ face of **10** induced multiple discrete modes of twinning deformation with sequential twin interfaces of $(5\bar{1}0)$, $(4\bar{1}0)$, $(3\bar{1}0)$, and $(2\bar{1}0)$, beyond the mode with the interface $(10\,\bar{1}\,0)$. Simultaneous compression and shearing induced the twinning with interface $(1\bar{1}0)$. In all bending modes, a shear force opposite to the original direction induces recovery in a manner consistent with ferroelasticity.

Figure 6.28 represents the proposed general mechanism for multiple modes of mechanical twinning in **10**. The angle θ is the interface angle, φ is the molecular-level bending angle, which is equivalent to the required angle of molecular rotation, and Φ is the macroscopic bending angle (Figure 6.28a). The crystals of **10** are twinned with interfaces of nth twinning mode of $(n\,\bar{1}0)$ with 180° rotation axes of $[1n0]$, i.e. the crystallographic zone, of which zone axis is

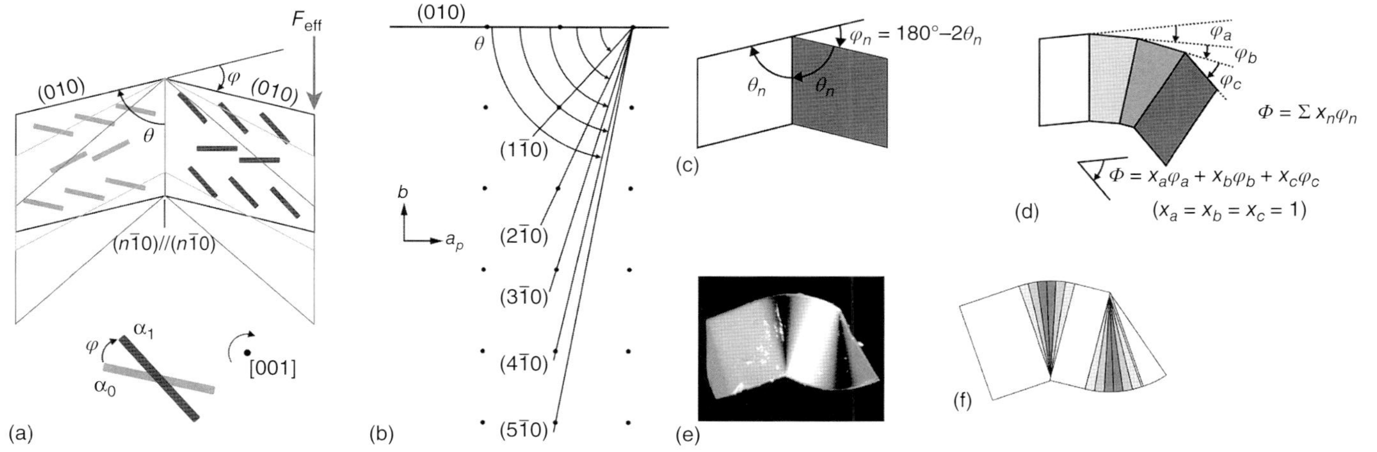

Figure 6.28 Mechanistic aspects: (a) general mechanism with molecules represented as rectangles, φ the molecular-level bending angle and θ the interface angle; (b) calculated θ for several different twin planes in **10**; (c) relationship between φ and θ; (d) general formula for Φ, the macroscopic bending angle; (e) curvature of **10** where multiple small domains of different modes have been activated; and (f) schematic representation of curvature. Source: Engel and Takamizawa 2018 [18]. Reproduced with permission of John Wiley & Sons, Inc.

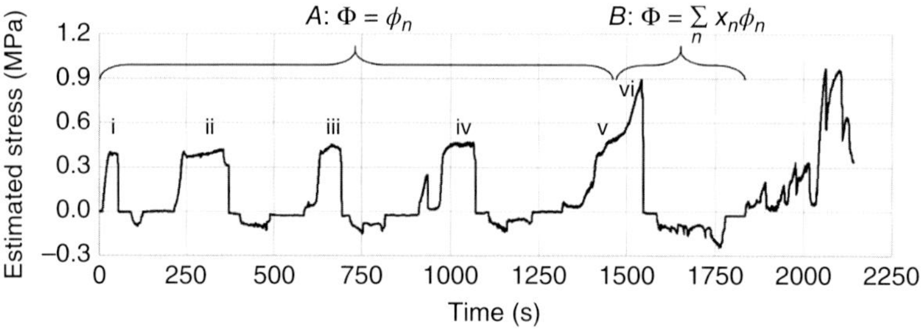

Figure 6.29 Stress–strain experiment where the displacement was increased stepwise by applying the shear force first across (010) and second across (0$\bar{1}$0) in multiple cycles. Source: Engel and Takamizawa 2018 [18]. Reproduced with permission of John Wiley & Sons, Inc.

[001], is attributed to the twinning (Figure 6.28b). Because some twinning planes can be activated simultaneously, a crystal bends in various Φ, which is calculated by $\Phi = \sum x_i \varphi_n$, where $x_i = 0, 1, 2, \ldots, \infty$ is the number of times the ith mode is activated, and φ_n is molecular-level bending angle calculated by $\varphi_n = 180° - 2\,\theta_n$ (θ_n: the interface angle) (Figure 6.28c,d). Actually, a sharp domain interface is unobservable when α_1 domain is first generated because simultaneous induction of twinning at the ∞th, …, 12th, 11th modes, of which difference in consecutive bending angles $\Delta\varphi = \varphi_n - \varphi_{n-1}$ are all less than 1°, provides practically smooth bending (Figure 6.28e,f). On the other hand, $\Delta\varphi$ can be distinguished when n is 10 or smaller, diving a discrete step.

An s–s curve was collected for four cycles of twinning deformation and recovery in two opposing directions for a single crystal of **10** (Figure 6.29). The first direction involved a force applied across (010) while the second direction involved application of a force across (0$\bar{1}$0). The crystal was mounted with (010) parallel to the horizontal dimension and the total shear force was applied perpendicular to the (010) face. The observed σ_{nucl} of the first deformation was 0.46–0.49 MPa. For cycle 1, the critical stress (coercive stress) for propagation of the daughter domain is approximately 0.40 MPa, while the stress associated with recovery is in the range 0.05–0.11 MPa. After cycle 1, the twinning interface remained, thus large σ_{nucl} is not detected in subsequent cycles. Cycles 2–4 for the force applied across (010) show a critical shear stress of 0.050–0.084 MPa for propagation of α_1 and 0.029–0.047 MPa for recovery of α_0. For cycles 2–4 where the force was applied across (0$\bar{1}$0) the critical shear stress is 0.042–0.063 MPa for propagation and 0.47–0.67 MPa for recovery.

A second s–s experiment involved cyclic application of the shear force first across (010) and second across (0$\bar{1}$0). With each cycle the total displacement was increased. Despite a progressively increasing bending angle by accessing of lower n-valued twinning modes, the estimated coercive stress again remained relatively constant at 0.35–0.46 MPa. In the range 0–1490 seconds the n-values are determined based on $\Phi = \varphi_n$. Beyond 1490 seconds or 0.2285 mm of displacement, we assume $\Phi = \sum x_i \varphi_n$, speculating which modes are active, and beyond 1830 seconds the model tends to break down.

Evidently the initial cycle of twinning deformation in **10** involves similar nucleation and coercive stresses, approximately in the range 0.3–0.5 MPa, regardless of

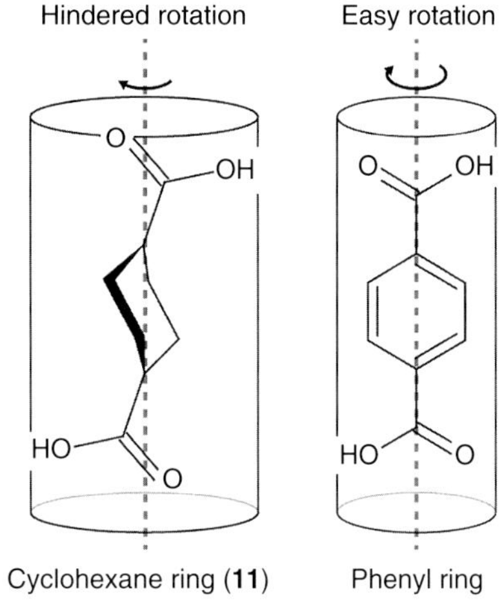

Figure 6.30 A schematic illustrating the molecular rotation of cyclohexane and phenyl rings. Source: Mir et al. 2018 [19]. Reproduced with permission of the Royal Society of Chemistry.

the direction of the applied force or which twinning modes are accessed. Since the total observed force remains approximately constant, the effective shear force is proportional to cos $\Phi_n/2$ according to $F_{eff} = F_{obs} \cos \Phi_n/2$. Recall that Φ_n represents both the molecular-level bending angle and the required angle of molecular rotation. As the molecular rotation angle increases, which involves progression to twinning modes with smaller interface angles, the shear force component of F_{obs} decreases.

trans-1,4-Cyclohexanedicarboxylic acid (**11**) (Figure 6.30) is another example of a control of ferroelastic behavior, i.e. the control of spontaneous strain [19]. **11** has a nonplanar C_6 ring in a chair conformation, due to which the rotatable single C—C bonds between each –COOH group and the cyclohexane ring are parallel, but are not on the same axis. This makes ring flipping difficult in dense crystal packing. In addition, bulkiness of molecular structure prevents hydrogen bonding via –COOH group.

Colorless crystals of **11** with typical lengths of 0.25–0.40 mm and thicknesses of 0.04–0.1 mm were obtained from a 1 : 1 mixture of methanol and acetone at 298 K. Upon the stress application perpendicular to the crystal surface $(001)_{\alpha_0}$ (mother domain: α_0) at 298 K, atypical ferroelastic behavior was observed. Shear stress generated the multiple thin daughter crystal domains (α_1) from the mother domain (α_0). Also, the α_1 domain disappeared spontaneously if the shear stress is removed before α_0 domain is converged into an α_1 domain completely.

Single-crystal X-ray structure analysis on the twinned crystal revealed that the crystal is in a monoclinic crystal system and $P2_1/c$ space group, and the twinning interface is $(\bar{1}00)_{\alpha_0}//(\bar{1}00)_{\alpha_1}$. The calculated bending angle was 33.28°, which agrees with the observed value of 31.42° under an optical microscope. In the deformation, the α_1 domain is formed by the 180° rotation of the α_0 domain about the rotational axis along $[001]_{\alpha_0}$ (Figure 6.31), which is parallel to $\{100\}_{\alpha_0}$ and $\{010\}_{\alpha_0}$. This conversion requires ring flipping of the bulky C_6 rings of molecule **11**. Adjacent molecules along [001] form hydrogen bonds with 3.369 Å long O—H···O=C between carboxylic group (intermolecular dotted lines in

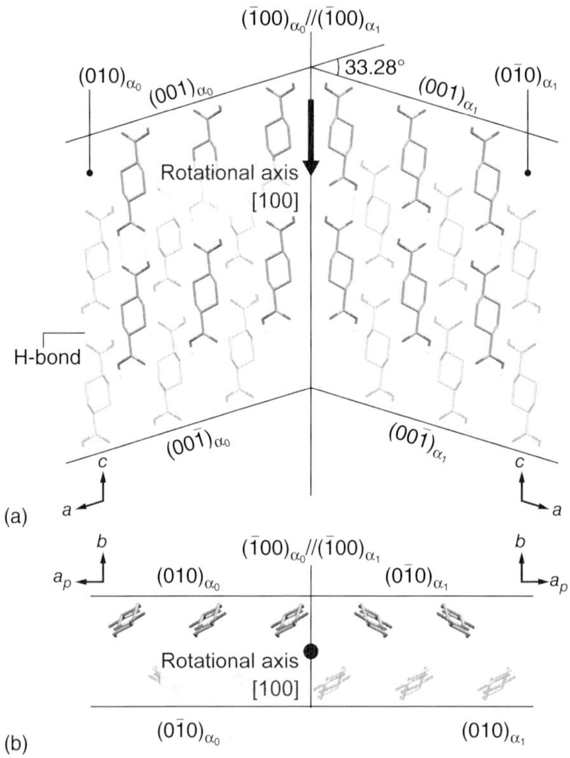

Figure 6.31 Crystal packing diagram projected along (a) $(001)_{\alpha_0}$ and (b) $(010)_{\alpha_0}$. Source: Mir et al. 2018 [19]. Reproduced with permission of the Royal Society of Chemistry.

Figure 6.31), forming 1D molecular chains, while C_6 rings have relatively weak van der Waals interactions.

The s–s curve had atypical shape with many spikes toward horizontal direction, instead of typical two sets of parallel plateaus (Figure 6.32). The spikes represent the generation of fresh domain interface along with the multiple generation of α_1 domains. This unique deformation is due to the bulky C_6 rings, of which hindrance in the rotation during twinning causes the nonuniform deformation. The bulkiness is also attributed to relatively large coercive stress of 0.7 MPa and the σ_{nucl} of 2.0 MPa compared to the values of other ferroelastic crystals described above. This shows the possibility to control the value of σ through the incorporation of a bulky moiety into the molecular structure. The shear to the opposite direction simultaneously triggered contraction of the α_1 domain (expansion of the original α_0 domain), and the nucleation of an α_0 domain from the α_1 domain (v in Figure 6.32a), which generates fresh interfaces, resulting in a spike in the s–s curve. The E_d (=$\sigma_c \tan\theta$) in this deformation had the relatively large value of 142.450 kJ m^{-3} (0.1039 kJ kg^{-1}, and 17.903 J mol^{-1}), in which the large shear stress (σ) is attributed.

Because of the large steric hindrance, the crystal of **11** has large spontaneous strain as well as the large E_d, and the unique mechanical behavior of spontaneous shape recovery by removing the shear before the completion of shear-induced phase transformation. This showed that ferroelastic behavior can be designed by molecular structure.

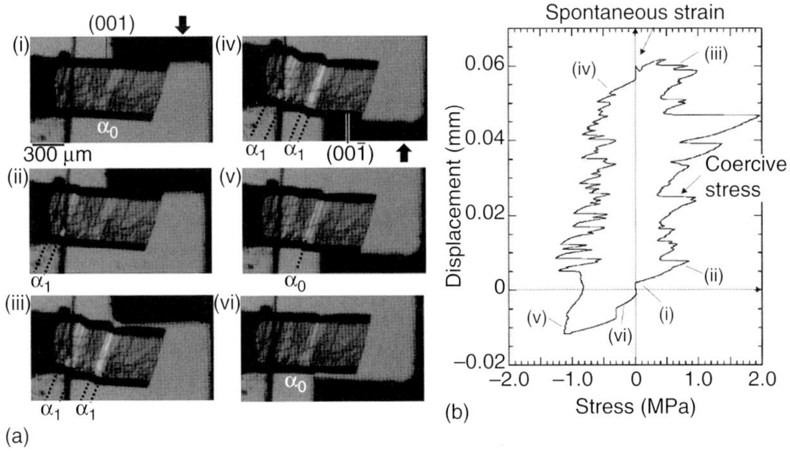

Figure 6.32 Measurement of the s–s curve. (a) Snapshots of the twinning deformation upon the application of shear stress: (i–iii) forward direction; (iv–vi) reverse direction. (b) The ferroelastic hysteresis loop under shear stress at 298 K. Source: Mir et al. 2018 [19]. Reproduced with permission of the Royal Society of Chemistry.

6.5 Summary

As presented in this chapter, unique deformability of solid materials are consecutively found. It is noteworthy that anisotropy of the molecule and the crystal structure are attributed to the manifestation of the observed unique mechanical behavior, and that chemical entities composing the organic crystal are molecules with certain structures, unlike metal or alloys composed from atoms, which are approximated by simple spheres from the viewpoint of structure. Therefore, the molecular structure can lead to the design of the deformability of the whole crystal. Throughout our research, we have investigated the deformation by reliable experimental techniques, connecting macroscopic and microscopic deformation, based on the fact that the deformations do not suffer the crystallinity of samples. Via this approach, scientific foundation is surely being established.

References

1 Ashby, M.F. (1999). *Materials Selection in Mechanical Design*. Oxford: Butterworth-Heinemann.

2 Wang, L., Nurzaman, S.G., and Iida, F. (2017). Soft-material robotics. *Found. Trends Robot.* 5 (3): 191–259.

3 Ilievski, F., Mazzeo, A.D., Shepherd, R.F. et al. (2011). Soft robotics for chemists. *Angew. Chem. Int. Ed.* 50 (8): 1890–1895.

4 Commins, P., Desta, I.T., Karothu, D.P. et al. (2016). Crystals on the move: mechanical effects in dynamic solids. *Chem. Commun.* 52 (97): 13941–13954.

5 Kobatake, S., Takami, S., Muto, H. et al. (2007). Rapid and reversible shape changes of molecular crystals on photoirradiation. *Nature* 446: 778–781.
6 Panda, M.K., Ghosh, S., Yasuda, N. et al. (2015). Spatially resolved analysis of short-range structure perturbations in a plastically bent molecular crystal. *Nat. Chem.* 7: 65–72.
7 Krishna, G.R., Devarapalli, R., Lal, G., and Reddy, C.M. (2016). Mechanically flexible organic crystals achieved by introducing weak interactions in structure: supramolecular shape synthons. *J. Am. Chem. Soc.* 138 (41): 13561–13567.
8 Takamizawa, S. and Miyamoto, Y. (2014). Superelastic organic crystals. *Angew. Chem. Int. Ed.* 53 (27): 6970–6973.
9 Ikeda, T. and Ube, T. (2014). A superelastic organic crystal. *Nature* 511: 300–301.
10 Takamizawa, S. and Takasaki, Y. (2016). Shape-memory effect in an organosuperelastic crystal. *Chem. Sci.* 7 (2): 1527–1534.
11 Takamizawa, S. and Takasaki, Y. (2015). Superelastic shape recovery of mechanically twinned 3,5-difluorobenzoic acid crystals. *Angew. Chem. Int. Ed.* 54 (16): 4815–4817.
12 Takasaki, Y. and Takamizawa, S. (2015). Active porous transition towards spatiotemporal control of molecular flow in a crystal membrane. *Nat. Commun.* 6: 8934.
13 Takamizawa, S., Takasaki, Y., Sasaki, T., and Ozaki, N. (2018). Superplasticity in an organic crystal. *Nat. Commun.* 9: 3984.
14 Mir, S.H., Takasaki, Y., Engel, E.R., and Takamizawa, S. (2017). Ferroelasticity in an organic crystal: a macroscopic and molecular level study. *Angew. Chem. Int. Ed.* 56 (50): 15882–15885.
15 Mir, S.H., Takasaki, Y., and Takamizawa, S. (2018). An organoferroelasticity driven by molecular conformational change. *Phys. Chem. Chem. Phys.* 20 (7): 4631–4635.
16 Engel, E.R., Takasaki, Y., Mir, S.H., and Takamizawa, S. (2018). Twinning ferroelasticity facilitated by the partial flipping of phenyl rings in single crystals of 4,4′-dicarboxydiphenyl ether. *R. Soc. Open Sci.* 5 (1): 171146.
17 Mir, S.H., Takasaki, Y., Engel, E.R., and Takamizawa, S. (2018). Enhancement of dissipated energy by large bending of an organic single crystal undergoing twinning deformation. *RSC Adv.* 8 (39): 21933–21936.
18 Engel, E.R. and Takamizawa, S. (2018). Versatile ferroelastic deformability in an organic single crystal by twinning about a molecular zone axis. *Angew. Chem. Int. Ed.* 57 (37): 11888–11892.
19 Mir, S.H., Takasaki, Y., Engel, E.R., and Takamizawa, S. (2018). Controllability of coercive stress in an organoferroelasticity by incorporation of bulky flipping moiety in molecular crystal. *CrystEngComm* 20 (27): 3807–3811.
20 White, M.A. (1999). *Properties of Materials*. New York, NY: Oxford University Press, Inc.
21 Otsuka, K. and Ren, X. (2005). Physical metallurgy of Ti–Ni-based shape memory alloys. *Prog. Mater. Sci.* 50 (5): 511–678.
22 Salje, E.K.H. (2012). Ferroelastic materials. *Annu. Rev. Mater. Res.* 42: 265–283.

23 Chang, L.-C. and Read, T.A. (1951). Plastic deformation and diffusionless phase changes in metals – the gold-cadmium beta phase. *Trans. AIME* 191: 47–52.
24 Otsuka, K. and Shimizu, K. (1986). Pseudoelasticity and shape memory effects in alloys. *Int. Met. Rev.* 31 (1): 93–114.
25 Otsuka, K. and Kakeshita, T. (2002). Science and technology of shape-memory alloys: new developments. *MRS Bull.* 27 (2): 91–100.
26 Schetky, L.M. (1993). A review of: "Engineering aspects of shape memory alloys" by T.W. Duerig, K.N. Melton, D. Stockel, and CM.Waymanb Butterworth-Heinemann 499 pages, hardcover, 1990. *Mater. Manuf. Process.* 8 (4–5): 567–569.
27 Mattila, H.R. (2006). *Intelligent Texitiles and Clothing*. Cambridge: Woodhead Publishing Limited.
28 Fischer, H., Vogel, B., and Welle, A. (2004). Applications of shape memory alloys in medical instruments. *Minim. Invasive Ther. Allied Technol.* 13 (4): 248–253.
29 Philippe, J.-M. (1989). Art and shape memory alloys. *Leonardo* 22 (1): 117–120.
30 Ölander, A. (1932). An electrochemical investigation of solid cadmium-gold alloys. *J. Am. Chem. Soc.* 54 (10): 3819–3833.
31 Pieczyska, E., Gadaj, S., Nowacki, W.K. et al. (2005). Characteristics of energy storage and dissipation in TiNi shape memory alloy. *Sci. Technol. Adv. Mater.* 6 (8): 889–894.
32 Suzuki, I. and Okada, K. (1979). Direct observation of ferroelasticity in squaric acid $C_4H_2O_4$. *Solid State Commun.* 29 (11): 759–762.
33 Wadhawan, V.K. (1978). Ferroelastic effect in orthoboric acid. *Mater. Res. Bull.* 13 (1): 1–8.
34 Kroupa, J. and Fuith, A. (1993). Optical study of the successive phase transitions of *n*-alkyl ammonium dihydrogen phosphate crystals. *Phys. Rev. B* 48 (6): 4119–4121.
35 Kaufman, L. and Cohen, M. (1958). Thermodynamics and kinetics of martensitic transformation. *Prog. Met. Phys.* 7: 165–246.
36 Bilby, B.A. and Crocker, A.G. (1965). The theory of the crystallography of deformation twinning. *Proc. R. Soc., A: Math. Phys. Eng. Sci.* 288 (1413): 240–255.
37 Cahn, R.W. (1954). Twinned crystals. *Adv. Phys.* 3 (12): 363–445.
38 Langdon, T.G. (2009). Seventy-five years of superplasticity: historic developments and new opportunities. *J. Mater. Sci.* 44 (22): 5998–6010.

Part II

Mechanically Responsive Polymers and Composites

7

Mechanical Polymeric Materials Based on Cyclodextrins as Artificial Muscles

Akira Harada[1], Yoshinori Takashima[2], Akihito Hashidzume[2], and Hiroyasu Yamaguchi[2]

[1] *Osaka University, The Institute of Scientific and Industrial Research, Ibaraki, Osaka 567-0047, Japan*
[2] *Osaka University, Department of Macromolecular Science, Graduate School of Science, Toyonaka, Osaka, 560-0043, Japan*

7.1 Introduction

In recent years much attention has been focused on a mechanism that can move automatically: robot. Robots have been conventionally made from metals, ceramics, and inorganic materials. These are hard, heavy, and hazardous. Recently, soft, flexible, and lightweight materials have attracted much attention for the construction of new robots: soft robots. In order to design and produce such soft robots, organic compounds, especially polymeric materials have been used for the bodies and attachments. In robotics, actuators play a central role in their functions. Molecular motors, molecular devices, and molecular machines have attracted much attention from chemists, physicists, and biologists. The 2016 Nobel Prize in Chemistry was awarded jointly to Professor Sauvage [1], Professor Stoddart [2], and Professor Feringa [3] for the design and synthesis of molecular machines. The word "molecular machine" is found 20 000 times in SciFinder Scholar. Most of the work has been devoted to the design and synthesis of molecular-level machines. There are only a few on the development of macroscopic devices using such molecular machines.

The authors have been working on integration of such molecular-level machines to macroscopic devices. This is reminiscent of muscle composed of actin and myosin in sarcomere. In this chapter, design and synthesis of polymeric materials based on cyclodextrins (CDs) [4–8] as artificial muscles are described. CDs are cyclic molecules consisting of six to eight glucose units. They are known to form inclusion complexes with various small molecules and simple ions. Previously we found that CDs form inclusion complexes not only with low-molecular-weight compounds but also with large molecules such as macromolecules (polymers). Polyrotaxanes in which many CD rings are entrapped along a polymer chain [9] and various interlocked molecules consisting of CDs and axle molecules are also prepared. CDs are able to move along the axle molecule. We synthesized molecular shuttle [10], electric trap

Mechanically Responsive Materials for Soft Robotics, First Edition. Edited by Hideko Koshima.

[11], and various daisy chains [12] and stimuli-responsive supramolecules by combining the function of azobenzene (Azo), stilbene (Sti), or ferrocene (Fc).

We describe molecular machines containing CD [13, 14], and their integration for the macroscopic devices [15–47] such as muscle [33, 42, 45–47]. We mainly focus on main chain and side chain recognition of CD in forming supramolecular polymer networks exhibiting stimuli-responsive movement on a macroscopic scale. In particular, applying external stimuli to supramolecular materials with spatiotemporal control can lead to macroscopic movement of materials. Stimuli-responsive guest molecules can be used to design materials having movable or transportable molecular systems through molecular recognition.

7.2 Artificial Muscle Regulated by Cross-Linking Density

If a gel cross-linked by inclusion complexes of CDs and stimuli-responsive guest residues is obtained, external stimuli should cause the gel to expand and shrink due to change in cross-linking density (Figure 7.1). Accordingly, the gel can move linearly or bend responding to external stimuli. A mixture of CD polymer and Azo polymer formed a supramolecular hydrogel exhibiting gel to sol transition on ultraviolet (UV) irradiation [22]. However, gel containing only a small number of chemical cross-links did not change to a sol by UV irradiation. Instead, it expanded because cross-linking density changed inside the chemically cross-linked gel. This mechanism can be applied to produce a photo-responsive soft actuator that bends away from the light source [33]. When a chemically cross-linked gel containing βCD and Fc was treated with an oxidizing agent, the gel expanded because of dissociation of βCD-Fc cross-links. However, when the gel was treated with reducing reagents, the gel shrank because Fc^+ was reduced to Fc. The gel could lift up a weight heavier than the gel itself [34]. Combinations of two kinds of host–guest pairs gave multifunctional hydrogels. When inclusion complexes of βCD-Fc and βCD-Ad were introduced into the gel network, the gel expanded on oxidation. This gel exhibited not only expansion–contraction but also self-healing ability and redox-responsive shape memory properties [40].

7.2.1 A Host–Guest Gel with αCD and Azo

The authors selected Azo compounds as guest molecules because the affinity of αCD for *trans*-Azo is higher than that for *cis*-Azo [22, 48]. Azo affects the photo-induced deformation and remote controllability. Using supramolecular hydrogels, which exhibit an expansion–contraction behavior responsive to the photo-stimulus, the authors successfully prepared a photo-responsive supramolecular actuator reminiscent of a natural muscle.

A host–guest gel with αCD and Azo (αCD-Azo gel) [33] was prepared by radical copolymerization of a mixture of αCD-modified acrylamide (αCD-AAm), azobenzene acrylamide (Azo-AAm), methylene bis-acrylamide (MBAAm), and acrylamide (AAm) in dimethyl sulfoxide (DMSO). The contents (x) of αCD-AAm and Azo-AAm units are 1–3 mol%. The polymer chains in αCD-Azo gel were cross-linked with MBAAm (the mole percentage content of MBAAm is 2 or 4 mol%). Figure 7.2a shows the chemical structures of αCD-Azo gel(x, y), αCD

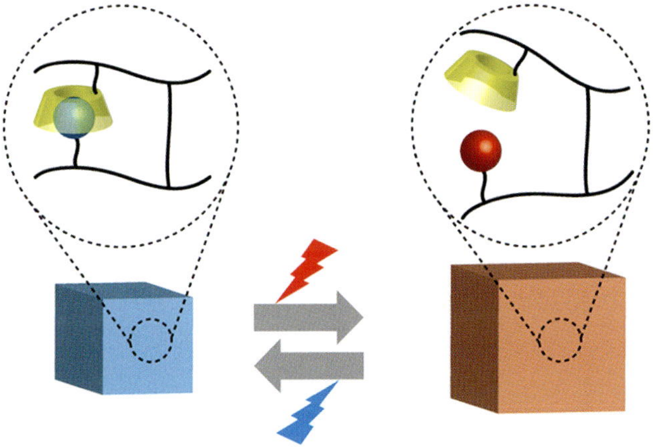

Figure 7.1 Schematic illustration of artificial muscle regulated by cross-linking density. Source: Nakahata et al. 2017 [45]. Reproduced with permission of the Pharmaceutical Society of Japan.

gel(1, 2) (without the Azo-AAm unit), Azo gel(1, 2) (without the αCD-AAm unit), and AAm gel(0, 2) (without αCD-AAm and Azo-AAm units). αCD-Azo gels feature three types of gels with host–guest units (x = 1, 2, and 3 mol%) and cross-linking units (y = 2 and 4 mol%). After gelation in DMSO, rinsing with water replaces the DMSO contained in the αCD-Azo gel(x, y).

Figure 7.2b shows the weight ratio of the gels upon substituting DMSO with water. The weight ratio of gel absorbed with DMSO is defined as 100%. Removing the DMSO contained in the αCD-Azo gel(x, y) significantly decreased the weight ratio with an increase in the mole percentage of the αCD and Azo unit. As shown in Figure 7.2c, substituting DMSO with water caused αCD-Azo gel(x, 4) to contract. In addition, the weights of αCD-Azo gel(2, 2) and (3, 2) decreased. On the other hand, the weight ratio of αCD gel(1, 2), Azo gel(1, 2), and AAm gel(0, 2) increased upon solvent manipulation. αCD-Azo gel(x, y) contracted upon substituting DMSO with water because the host–guest complexation forms cross-links, which were confirmed using creep rupture measurements. These results indicated that the formation of an inclusion complex between αCD and the Azo units caused αCD-Azo gels to shrink owing to the increase in cross-links.

7.2.2 Photo-Responsive Volume Change of αCD-Azo Gels

The effects of photo-stimuli on the expansion–contraction behavior of αCD-Azo gels were investigated by irradiating flat plates of αCD-Azo gels immersed in water for an hour. Photo-irradiation with UV light (λ = 365 nm) isomerizes the *trans*-Azo group into the *cis*-Azo group, whereas the reverse occurs with visible (Vis) light (λ = 430 nm). Figure 7.3a shows the weight change of the αCD-Azo gel upon UV and Vis light irradiation. UV irradiation of the αCD-Azo gel increased the weight of the hydrogels, whereas continuous irradiation of Vis light to the αCD-Azo gels restored the initial weight and volume. These volume changes of αCD-Azo gels are correlated with the inclusion complex formation between αCD

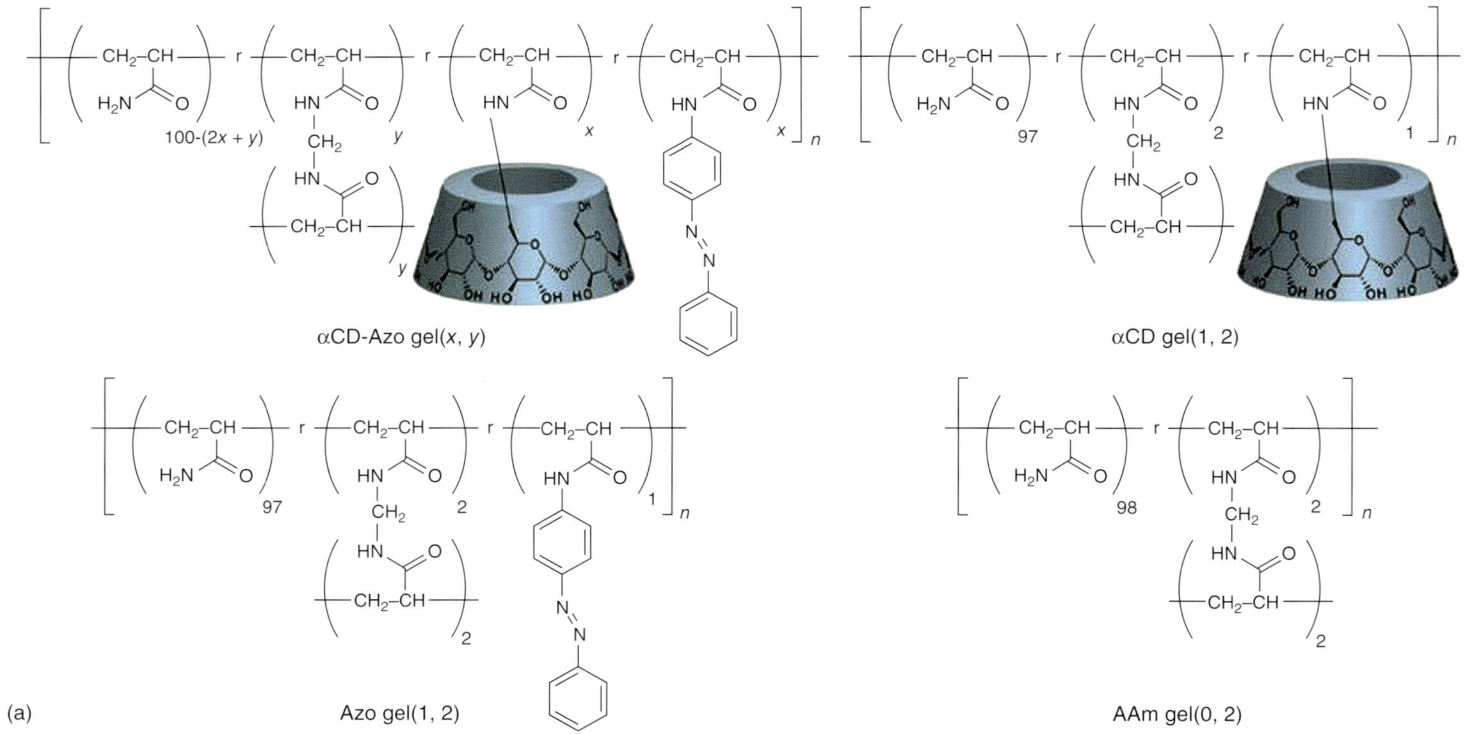

Figure 7.2 Polymer gels used to prepare photo-responsive supramolecular actuators and their expansion–contraction behavior. Chemical structures of αCD-Azo gel, αCD gel (without the Azo-AAm unit), Azo gel (without the αCD-AAm unit), and AAm gel (without αCD-AAm and Azo-AAm units) (a). *x* is the mole percentage of the host and guest units. *y* is the mole percentage of the cross-linking unit (MBAAm). Weight ratio change in the gels upon replacing DMSO with water (b). Gels initially absorb DMSO. The weight ratio of gel absorbed with DMSO is defined as 100%. Error bars, standard deviation for five measurements. Photographs of the volume change of αCD-Azo gel (*x*, 4) upon replacing DMSO with water (c). Source: Takashima et al. 2012 [33]. Reproduced with permission of Springer Nature.

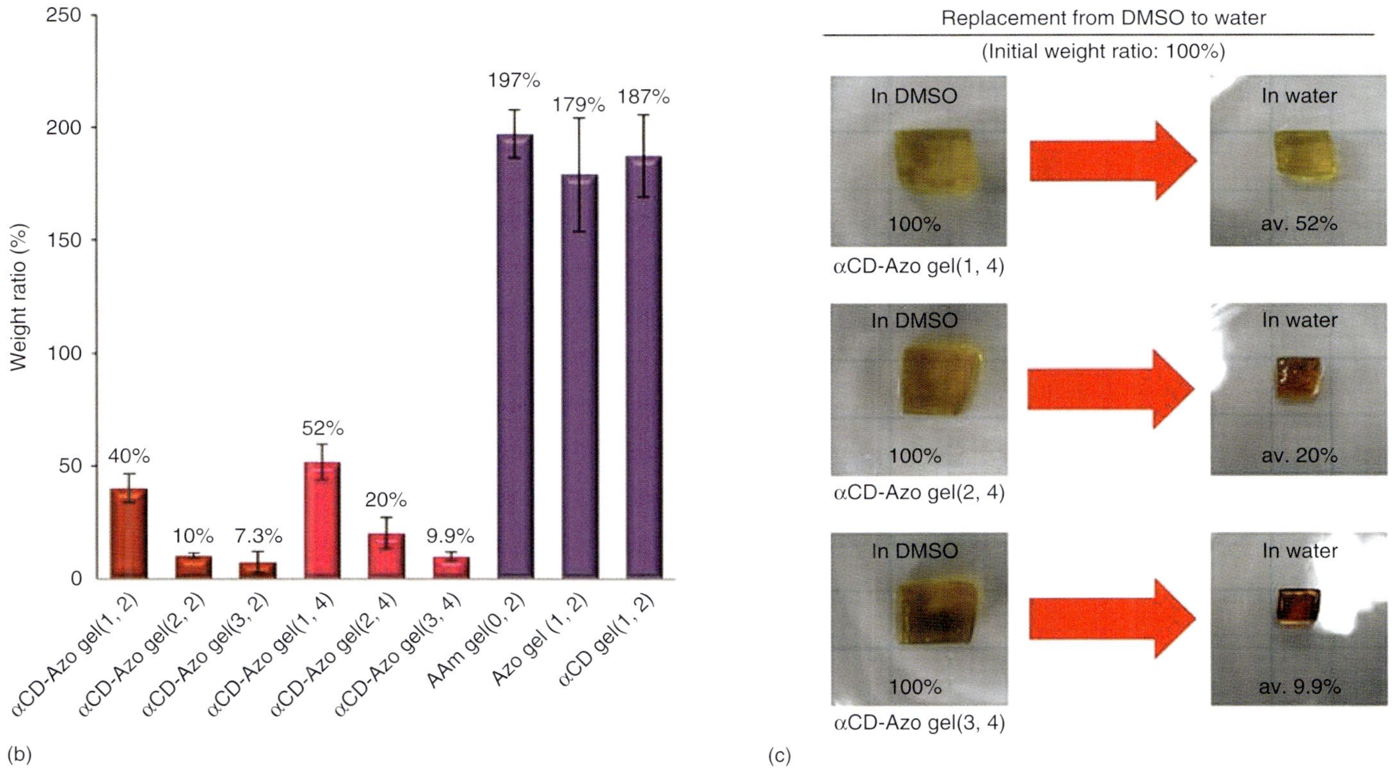

Figure 7.2 (*Continued*)

and Azo units (Figure 7.3b). The difference in the association constants of αCD for the Azo isomers creates the expansion–contraction behavior of αCD-Azo gels upon UV and Vis light irradiation. The weight ratio of αCD-Azo gel(x, 2) with 2 mol% of MBAAm was larger than that of αCD-Azo gel(x, 4) with 4 mol% of MBAAm, indicating that a smaller cross-linking ratio induces a larger volume change of the gel. Similarly, the weight ratio of αCD-Azo gel(2, 2) was larger than that of αCD-Azo gel(3, 2). The inside of the Azo unit of αCD-Azo gel(3, 2) did not isomerize from the trans- to the cis-form because the concentration of the Azo group was too high to optically transmit through the opposite side, meaning UV light was absorbed on the surface of αCD-Azo gel(3, 2). UV irradiation causes the *trans*-Azo group of the αCD-Azo gel to isomerize into the cis-form, whereas Vis irradiation causes the *cis*-Azo group to isomerize into the trans-form. Before UV irradiation, the isomer contents of αCD-Azo gel was trans:cis = 70 : 30, whereas afterward, the ratio changed to trans:cis = 5 : 95. However, the isomer contents recovered to trans:cis = 69 : 31 upon Vis light irradiation. Consequently, the photo-isomerization of the Azo unit was reversible even in the αCD-Azo gel. Figure 7.3c shows the proposed scheme for the expansion–contraction behavior of αCD-Azo gels by photo-irradiation. Before photo-irradiation, αCD-Azo gels contracted forcefully to form supramolecular noncovalent cross-links between αCD and *trans*-Azo units through host–guest interactions. After UV irradiation (365 nm), the trans-form isomerized into the cis-form, decreasing the number of noncovalent cross-links as the inclusion complexes between αCD and the Azo units were dissociated, causing the αCD-Azo gels to expand. However, after subsequent Vis irradiation the trans-form recovered, increasing the number of noncovalent cross-links and forming inclusion complexes, which caused αCD-Azo gels to contract.

7.2.3 Photo-Responsive Property of αCD-Azo Gels

Irradiating the *α*CD-Azo gel(2, 2) with UV light from the left side bent the gel to the right as shown in Figure 7.4a, whereas irradiating the bent gel with Vis light from the same side for an hour restored the initial state (Figure 7.4b). Similarly, irradiating the plate gel from the right side caused the gel to bend to the left side, while irradiating with Vis light restored the initial state (Figure 7.4c). This bending behavior can be repeated for at least five cycles. Figure 7.4e shows the flexion angle of behavior *α*CD-Azo gel(2, 2) irradiated with UV and Vis lights for an hour. The flexion angle (θ) defined in Figure 7.4d became saturated after UV irradiation for about an hour, and did not significantly decrease upon standing under its own weight for an hour in the dark. Conversely, irradiation with Vis light for an hour restored the bent gel to the initial state and the flexion angle decreases. The Vis irradiation time required for the bent gel to return to a flat gel was similar to the UV irradiation time. Figure 7.4f shows the repeated experiments of αCD-Azo gel(2, 2) irradiated with UV and Vis lights for five minutes. The gel plate clearly showed back-and-forth motion depending on the wavelength without irradiation history. Moreover, to observe the deformation of the gel with irradiation, a ribbon-shaped αCD-Azo gel(1, 2) was prepared. The ribbon-shaped gel turned to a coil by the irradiation of UV light from the left side (Figure 7.4g).

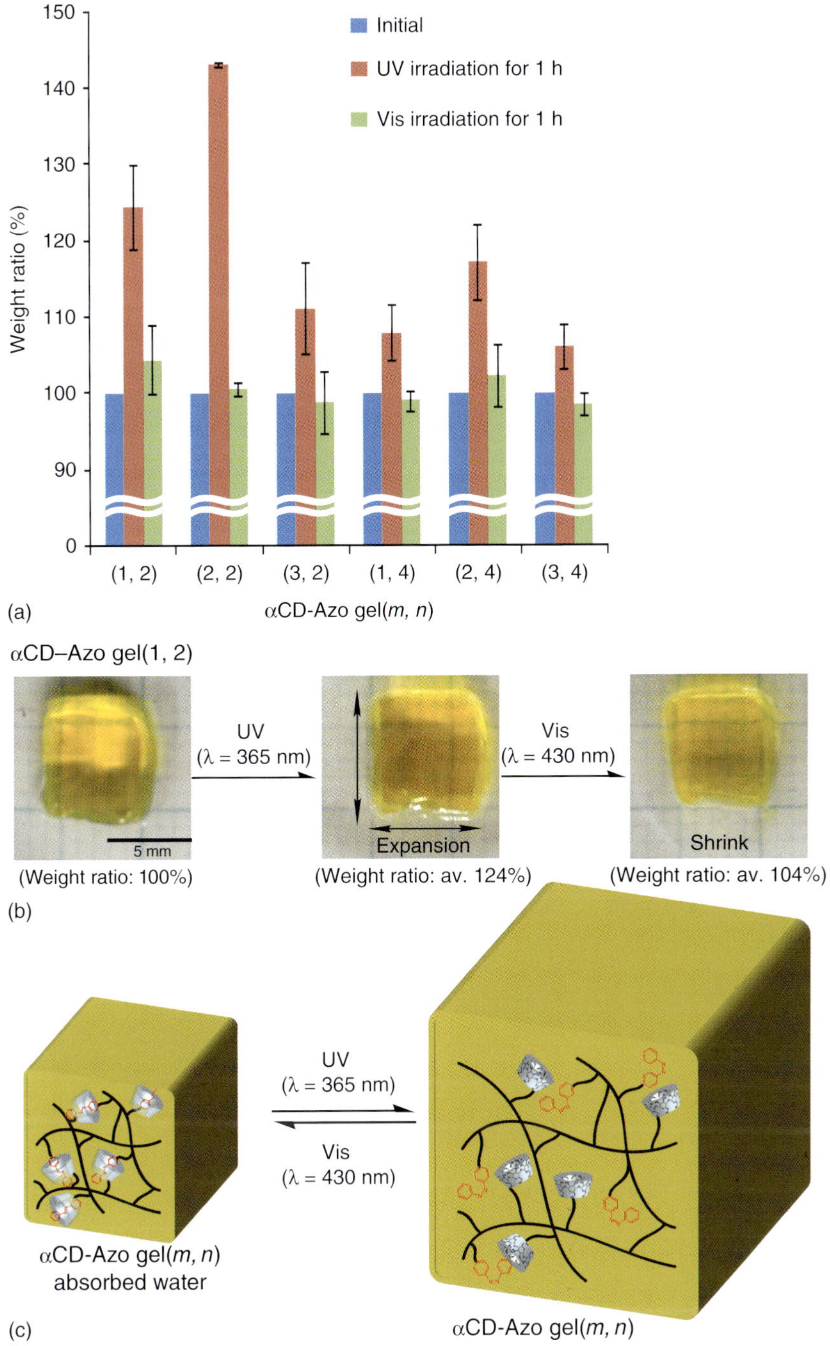

Figure 7.3 Photo-responsive weight change of αCD-Azo gels in water. Weight change of αCD-Azo gels before and after photo-irradiation with UV light (365 nm) and Vis light (430 nm) (a). Photographs of the volume change of αCD-Azo gel(1, 2) irradiated by UV and Vis light. Scale bar, 5 mm (b). Schematic illustration of the expansion–contraction of αCD-Azo gel irradiated by UV and Vis light (c). Source: Takashima et al. 2012 [33]. Reproduced with permission of Springer Nature.

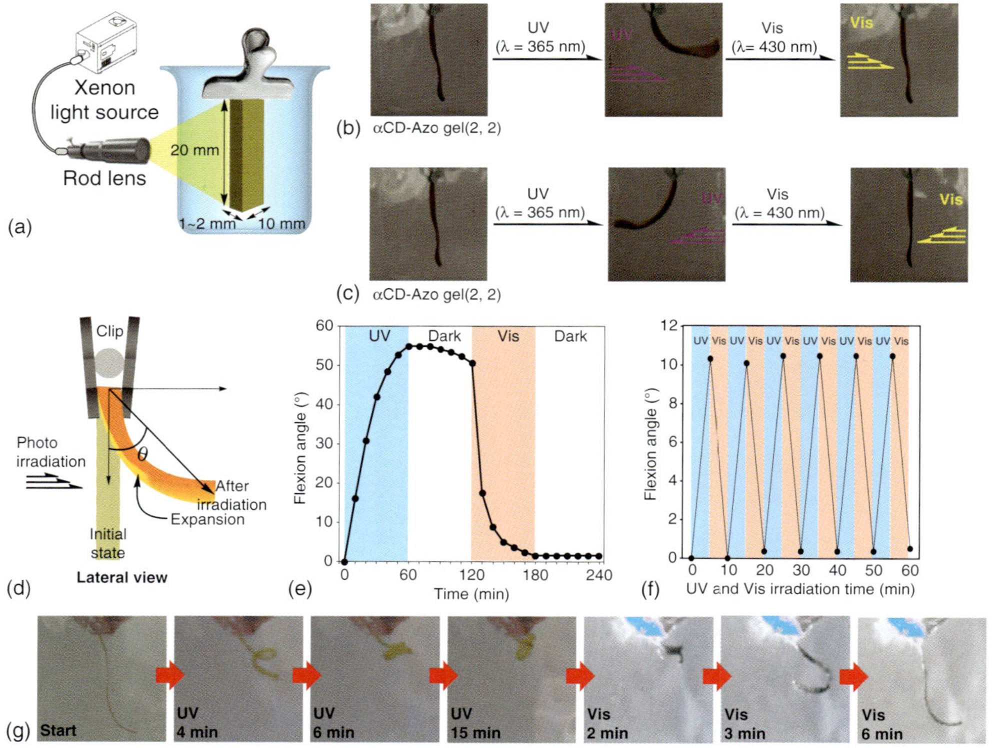

Figure 7.4 Photo-responsive actuator of αCD-Azo gel in water. Experimental devices and the size of αCD-Azo gel(2, 2) in water (a). Light irradiation from the left side of αCD-Azo gel(2, 2) for an hour. After UV irradiation, αCD-Azo gel(2, 2) bends to the right side. Subsequent irradiation with Vis light restores the initial state (b). Light irradiation from the right side of αCD-Azo gel(2, 2) for an hour. After UV irradiation, αCD-Azo gel(2, 2) bends to the left side. Subsequent irradiation with Vis light restores the initial state (c). Lateral view of αCD-Azo gel(2, 2) hung with a clip. Flexion angle (θ) is defined here (d). Plots of irradiation time versus θ in αCD-Azo gel(2, 2) (e). The repeated experiment of αCD-Azo gel(2, 2) irradiated with UV and Vis lights for 5 minutes (f). Plots show the correlation between irradiation time and θ. Light irradiation from the left side of the ribbon-shaped αCD-Azo gel(1, 2) for 15 minutes (g). After UV irradiation, αCD-Azo gel(1, 2) forms a coil. Subsequent irradiation with Vis light restores the initial state. Source: Takashima et al. 2012 [33]. Reproduced with permission of

The coil-shaped gel returned to the ribbon-shaped gel by Vis light irradiation. The ribbon–coil transition can be repeated for at least five cycles. These results indicate that photo-isomerization of the Azo group is correlated to the flex behavior of the αCD-Azo gel plates. The plate or ribbon gels bent in the opposite direction of the incident light because the surface of the gel plate exposed to UV light expands in water, but the volume of the surface not exposed to UV light remained constant, suggesting that the strain deformation between the exposed and unexposed areas creates the flex behavior of αCD-Azo gels.

7.3 Artificial Muscle Regulated by Sliding Motion

Photo-responsive supramolecular materials as shown in Figure 7.5 also have a potential as an artificial muscle. The introduction of rotaxane structure into polymeric materials is an effective method in the molecular design of entities with both mechanically functional and responsive properties. Movable (topological) cross-linkers in polymeric materials indirectly connect polymers with mechanically interlocked rotaxane architectures. The movable cross-linking points along the polymer chain enable unique mechanical properties of materials, such as high flexibility and entropic elasticity, by the stress-dispersive deformation process. The authors designed a new type of photo-responsive polymeric actuator

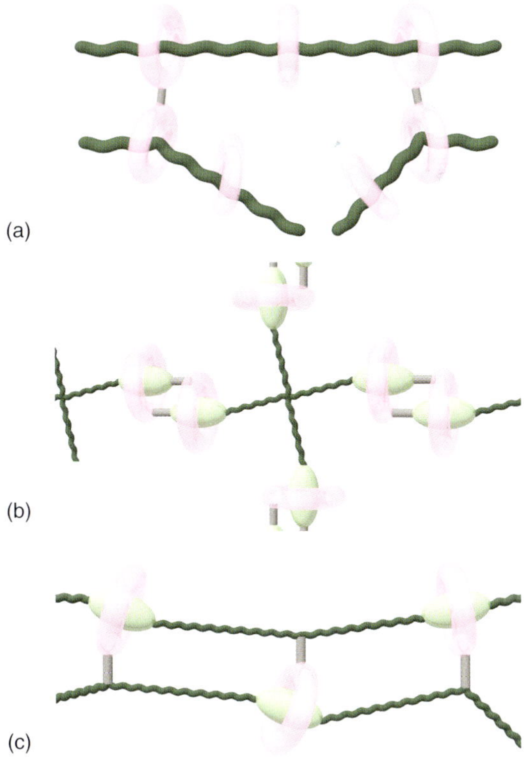

Figure 7.5 Schematic illustration of artificial muscle regulated by sliding motion. Polyrotaxane materials with cross-links between cyclic molecules [49] (a), [c2]daisy chain materials with polymer chains cross-linked by [c2]daisy chain molecules (b), and [2]rotaxane materials with cross-links between cyclic molecules and axis molecules (c). Source: Takashima et al. 2018 [46]. Reproduced with permission of American Chemical Society.

based on a [2]rotaxane, which can be obtained by polycondensation reaction between multifunctional pseudo-rotaxane and poly(ethylene glycol) (PEG) [46]. The topologically cross-linked αCD-Azo gels showed photo-responsive deformation behavior even in the dry state.

7.3.1 Preparation of a Topological Hydrogel (αCD-Azo Hydrogel)

Figure 7.6a shows αCD-Azo hydrogel prepared via a polycondensation reaction between *pseudo*[2]rotaxanes of Lys-αCD with Azo and PEG derivatives. PEG derivatives are highly water-soluble and lack photo-absorption bands in the UV and Vis light regions. The interaction between the PEG and Azo units is negligible. In addition, because the glass transition temperature (T_g) of PEG derivatives is around −20 °C, the flexible PEG-based materials are thought to exhibit photo-responsive properties in the dry state. After the polycondensation reaction, the reaction solution became a self-standing hydrogel (αCD-Azo hydrogel). To elucidate the effect of the topological structure of the αCD-Azo hydrogel, a covalently cross-linked Azo hydrogel was prepared as a reference by the polycondensation reaction among a bis-lysine derivative (BisLys), Am_2Azo, and Su_2PEG (Figure 7.6b). Although the reaction of the Lys-αCD/Am_2Azo complex and Su_2PEG in DMSO did not yield the αCD-Azo organogel, the reaction of BisLys, Am_2Azo, and Su_2PEG in DMSO gave the Azo organogel. The Lys-αCD/Am_2Azo complex dissociated in DMSO. These results indicate that the Lys-αCD/Am_2Azo complex in water plays an important role as a cross-linking point to form αCD-Azo gel.

7.3.2 Mechanical and Photo-Responsive Properties of the αCD-Azo Hydrogel

The mechanical properties of the αCD-Azo hydrogels and the Azo hydrogel were evaluated by tensile measurements (Figure 7.7a). Although the rupture strain of the Azo hydrogel(3.4k) cross-linked with Su_2PEG molecular weight (Mw = 3.4 k g mol^{-1}) was 350%, that of the αCD-Azo hydrogel(3.4k) cross-linked with Su_2PEG(3.4k) (Mw = 3.4 k g mol^{-1}) was four times higher (1350%). The rupture strain of the αCD-Azo hydrogel(10k) cross-linked with Su_2PEG(10k) increased to 2800%. Similarly, the rupture energy of the αCD-Azo hydrogels also increased with an increase in the molecular weight of PEG, indicating that sliding motion of Lys-αCD along the PEG chains resulted in stress distribution within the polymer network, allowing the gel to endure the external stress (Figure 7.7b). Figure 7.7d shows the change in the cross-link density in the αCD-Azo hydrogel(3.4k), measured as shown in Figure 7.7c, depending on UV and Vis light irradiations. After UV light irradiation for one minute, the Young's modulus decreased. In addition, continuous irradiation with Vis light resulted in recovery of the initial Young's modulus. After UV light irradiation, the *cis*-Azo unit dethread from the αCD cavity led to an increase of the sliding mobility of the Lys-αCD unit and a decreased Young's modulus.

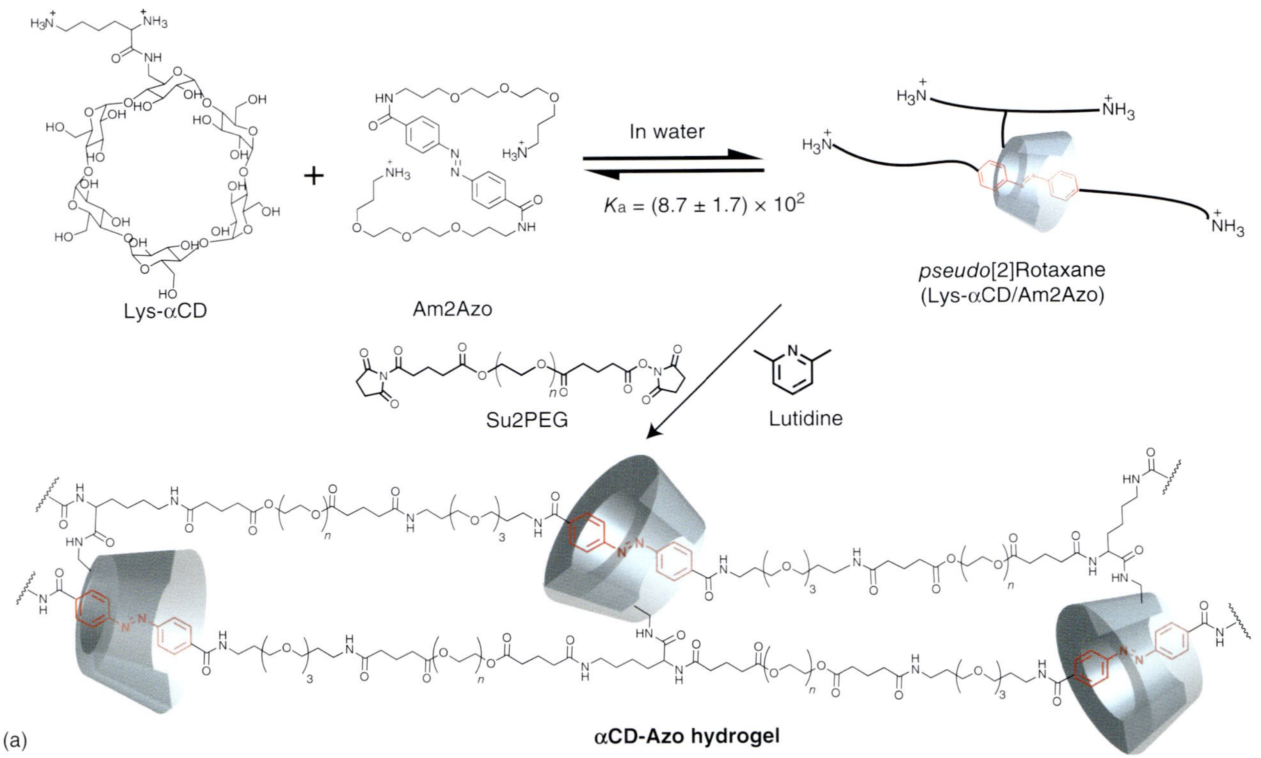

(a)

Figure 7.6 Schematic illustration of the preparation of the αCD-Azo hydrogel as a topological material (a) and the Azo hydrogel as a chemically cross-linked material (b). Source: Takashima et al. 2018 [46]. Reproduced with permission of American Chemical Society.

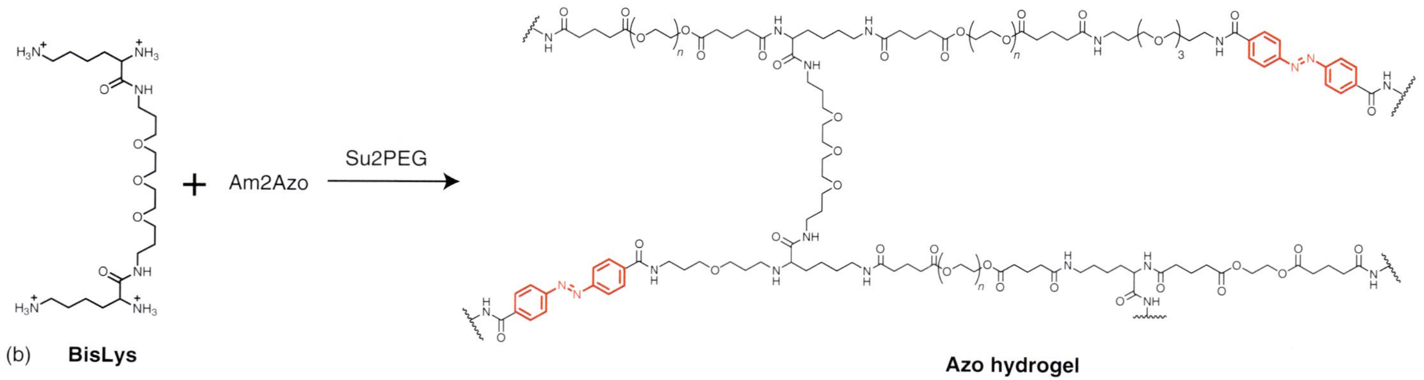

Figure 7.6 (*Continued*)

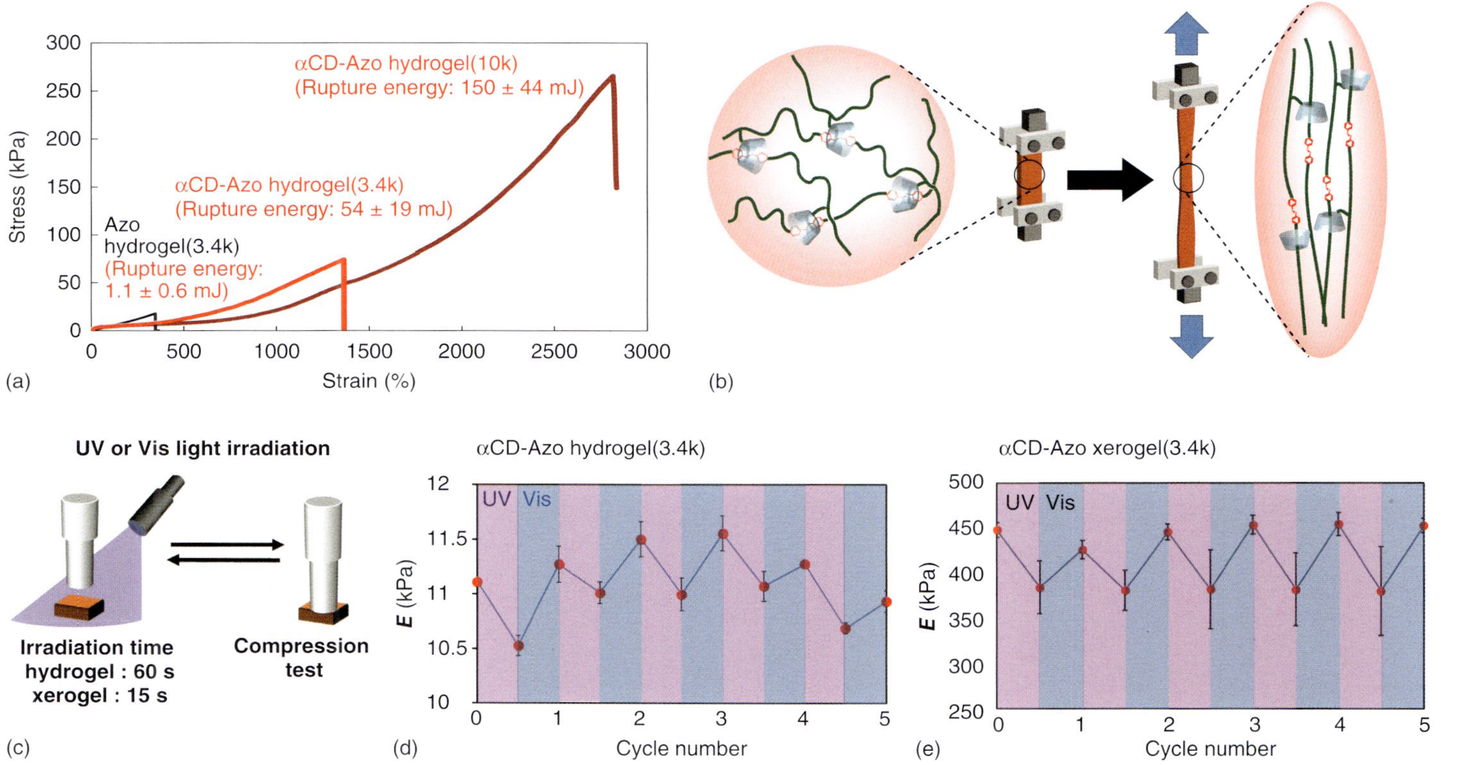

Figure 7.7 Stress–strain curves for the αCD-Azo hydrogel(3.4k), the αCD-Azo hydrogel(10k), and the Azo hydrogel(3.4k) (a). Proposed scheme of the stretching property for the αCD-Azo hydrogels during the tensile tests (b). Compression experimental setup of the αCD-Azo hydrogel(3.4k) and the αCD-Azo xerogel(3.4k) irradiated with UV and Vis light (c). The repeated testing of the Young's modulus of the αCD-Azo hydrogel(3.4k) with exposure to UV and Vis light for 60 seconds (d). The repeated testing of the Young's modulus of the αCD-Azo xerogel (3.4k) with exposure to UV and Vis light for 15 seconds (e). Source: Takashima et al. 2018 [46]. Reproduced with the permission of American Chemical Society.

7.3.3 UV and Vis Light-Responsive Actuation of the αCD-Azo Xerogel

To align the Lys-αCD/Am_2Azo unit, the αCD-Azo hydrogel(3.4k) was uniaxially elongated and dried under vacuum to obtain a fiber-like αCD-Azo xerogel(3.4k). The stretched αCD-Azo xerogel(3.4k) showed anisotropic X-ray diffraction patterns, which are characteristic of an anisotropic structure. On the other hand, the unstretched αCD-Azo xerogel(3.4k) (Figure 7.7e) showed a Debye–Scherrer ring, which is characteristic of an isotropic structure. Figure 7.8 shows the results of deformation testing of the stretched αCD-Azo xerogel in response to light irradiation. The stretched αCD-Azo xerogels(3.4k and 10k) quickly bent toward the light source side within three seconds upon UV light irradiation (Figure 7.8c). Here, the flexion angle θ is defined as shown in Figure 7.8b. The flexion angle of the covalently cross-linked Azo xerogel did not change by photo-irradiation. The αCD-Azo xerogel(3.4k) showed slightly faster deformation and a larger amount of displacement. The αCD-Azo hydrogel(3.4k) has higher density of the cross-linking point than the αCD-Azo hydrogel(10k), and the higher cross-linking density is supposed to result in the relatively faster deformation. In addition, the change in flexion angle per second was found to depend on the strain applied to the αCD-Azo hydrogel(3.4k) before drying (Figure 7.8e), suggesting that the applied strain contributed to the efficient response and that anisotropic rotaxane network of the actuator showed an efficient response. Figure 7.8f–h shows energy conversion from light to mechanical work. When an αCD-Azo xerogel(3.4k) sample on which a weight of 174 mg was hung was irradiated with UV light, it contracted and lifted the weight vertically upward by 3.3 mm within 10 seconds. The mechanical work (W) produced by the αCD-Azo xerogel(3.4k) sample was 5.6 μJ.

7.4 An Artificial Molecular Actuator with a [c2]Daisy Chain ([c2]$AzoCD_2$)

A four-arm poly(ethylene glycol) (tetraPEG) was cross-linked with photo-responsive [c2]daisy chains (Figure 7.9a). Two types of polymeric gel (hydrogel and xerogel) containing CD-based [c2]daisy chains as cross-linkers (Figure 7.9b) were prepared [42].

The volume changes of the [c2]$AzoCD_2$ hydrogel and the Azo hydrogel upon UV and Vis light irradiations were investigated (Figure 7.10). Figures 7.10b,c show photographs of the volume changes of the [c2]$AzoCD_2$ hydrogel and the Azo hydrogel following UV and Vis light irradiations, respectively. Irradiation of the cuboid [c2]$AzoCD_2$ hydrogel with UV light for three hours as shown in Figure 7.10a decreased the volume, and continuous irradiation of the shrunken hydrogel with Vis light for three hours resulted in recovery of the initial volume. In contrast, irradiation of the Azo hydrogel with UV light increased the volume, and continuous irradiation with Vis light restored the initial volume (Figure 7.10c). Figure 7.10d shows the correlation between the weight ratio and irradiation wavelength. The [c2]$AzoCD_2$ hydrogel and Azo hydrogel exhibited opposite contraction and expansion behavior. The contraction–expansion

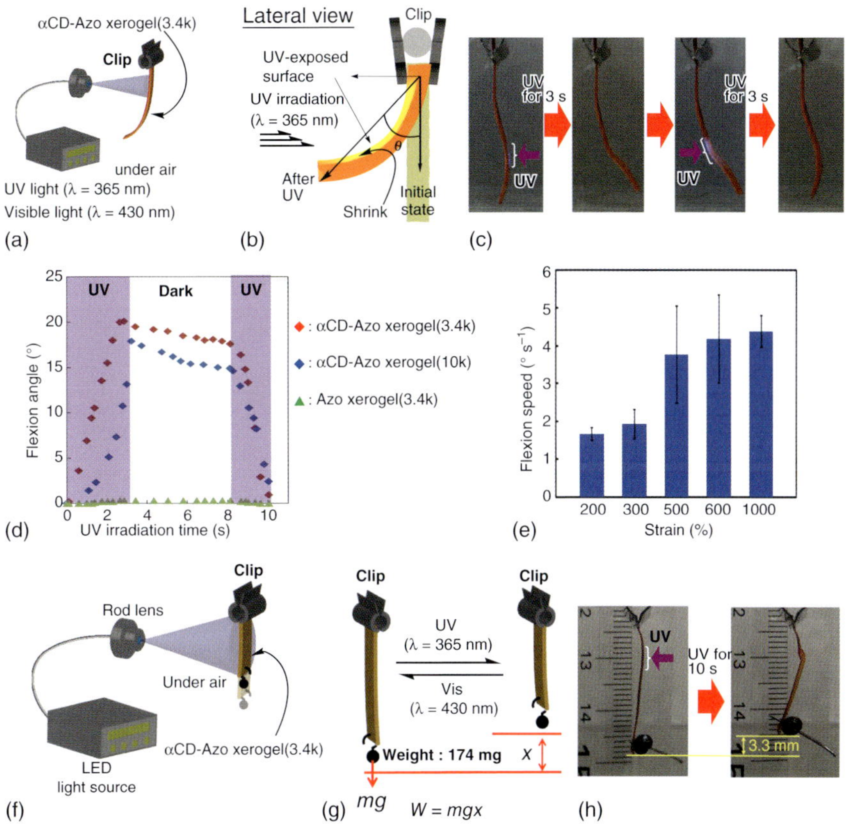

Figure 7.8 Experimental equipment and irradiation experimental setup of the αCD-Azo xerogel(3.4k) in air (a). Lateral views of the αCD-Azo xerogel(3.4k) hung with a clip. Flexion angle (θ) is defined here (b). Light irradiation from the right side of the αCD-Azo xerogel(3.4k) for three seconds. After UV irradiation, the αCD-Azo xerogel(3.4k) bends to the right side. Subsequent irradiation with UV light from the left side restores the initial form (c). Plots of the flexion angle (θ) as a function of the irradiation time of the αCD-Azo xerogel (3.4k) (d). The dependency of flexion speed on the strains applied to the αCD-Azo hydrogel(3.4k) before drying process (e). The equipment of energy conversion test of αCD-Azo xerogel(3.4k) (f). The energy conversion from light to mechanical work is evaluated by using the values and equation shown in (g). Lateral views of the αCD-Azo xerogel(3.4k) lifting up the weight (h). Source: Takashima et al. 2018 [46]. Reproduced with permission of American Chemical Society.

motion of [c2]AzoCD$_2$ induced a volume change in the [c2]AzoCD$_2$ hydrogel. The conformational change induced by UV light irradiation discharges the water contained in the [c2]AzoCD$_2$ hydrogel, shrinking the gel (Figure 7.10e). The contracted state of [c2]AzoCD$_2$ was restored to the initial expanded state. Recovery of the conformation resulted in uptake of external water under Vis irradiation due to the swelling of the hydrogel. In contrast, before UV irradiation, the Azo hydrogel formed a contracted state because the hydrophobic *trans*-Azo groups assembled to function as noncovalent cross-linkers (Figure 7.10f). After UV irradiation, the assembly dissociated, decreasing the cross-linking density

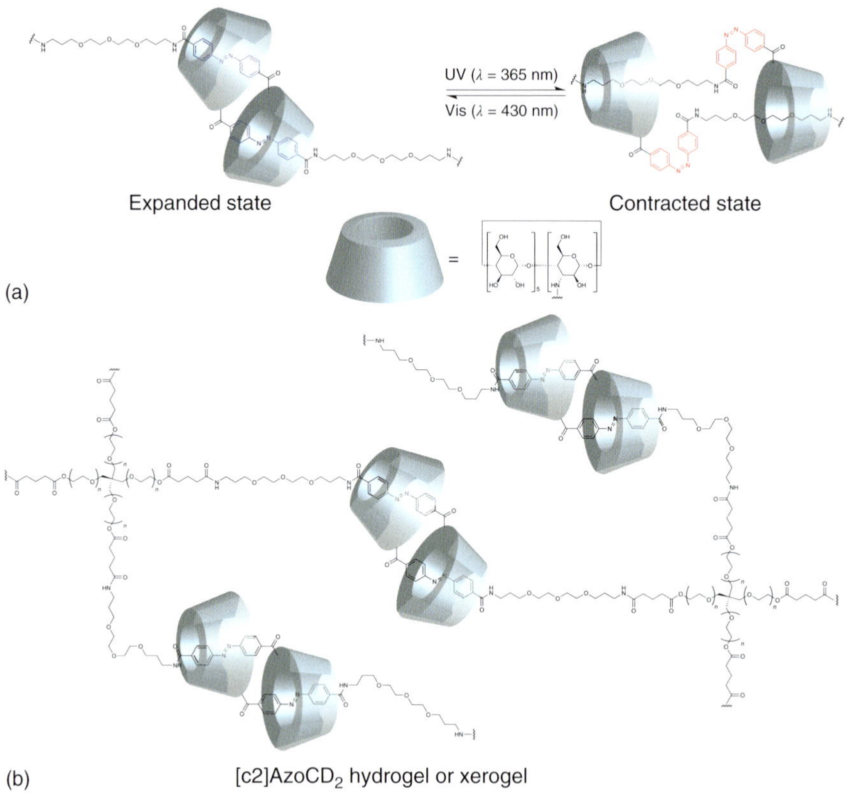

Figure 7.9 Photo-isomerization scheme of an αCD-based [c2]daisy chain with an Azo derivative as the axis (a). Chemical structure of a [c2]$AzoCD_2$ hydrogel or xerogel consisting of four-arm poly(ethylene glycol) (tetraPEG) (b). Source: Iwaso et al. 2016 [42]. Reproduced with permission of Springer Nature.

with photo-isomerization from *trans*- to *cis*-Azo. However, the Azo organogel and the [c2]$AzoCD_2$ organogel, which were swollen with DMSO, did not exhibit any volume changes because hydrophobic interaction of the Azo moieties does not work in DMSO. The αCD unit of the [c2]$AzoCD_2$ organogel containing DMSO lacks molecular recognition.

7.4.1 Photo-Responsive Actuation of the [c2]$AzoCD_2$ Hydrogel

The correlation between the directions of the incident light and deformation with UV and Vis light was investigated. The [c2]$AzoCD_2$ hydrogel or the Azo gel actuator was immersed in water. The [c2]$AzoCD_2$ hydrogel bent to the right following UV light irradiation using a rod lens and a xenon lamp from the right side of the hydrogel, and irradiation with Vis light from the right restored the initial condition (Figure 7.11a). Similarly, irradiation of the [c2]$AzoCD_2$ hydrogel from the left side with UV light caused the hydrogel to bend to the left, and the initial state was restored following irradiation of the bent hydrogel with Vis light. In contrast,

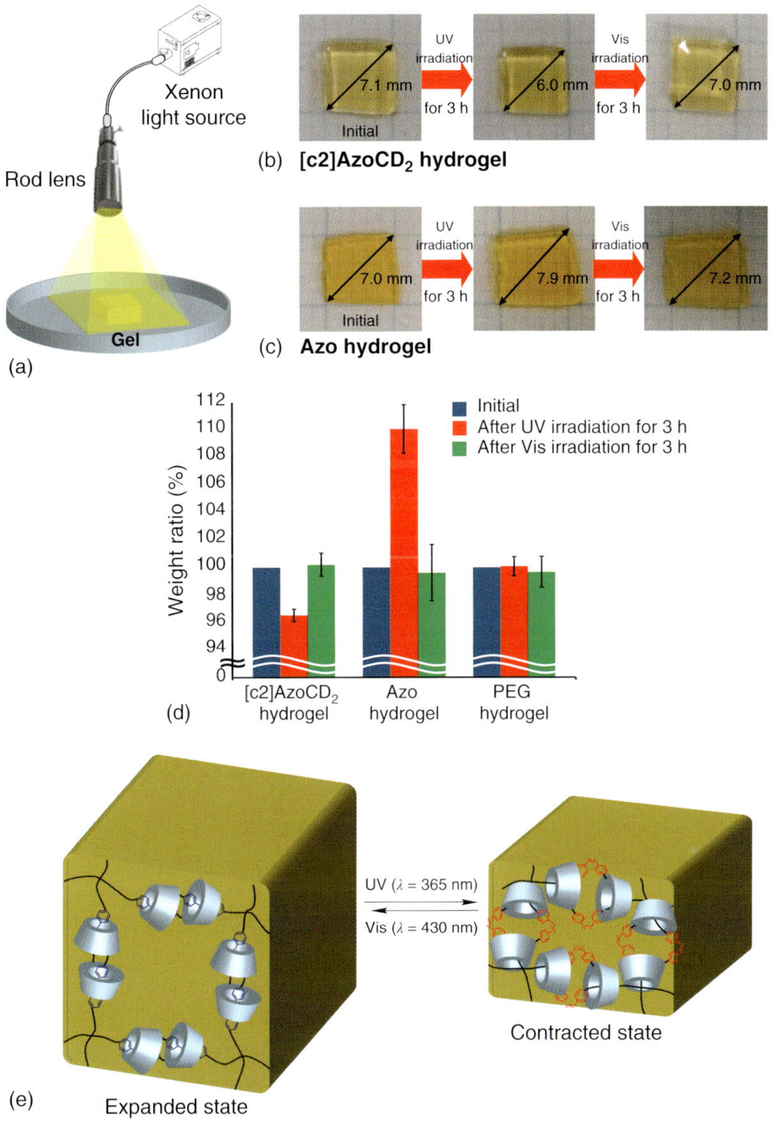

Figure 7.10 Photo-responsive properties of the [c2]AzoCD$_2$ hydrogel and the Azo hydrogel. Irradiation experimental setup of the cuboid gels ($5 \times 5 \times 1$ mm^3) in water (a). Photographs of the volume changes of the [c2]AzoCD$_2$ hydrogel (b) and the Azo hydrogel (c) following irradiation with UV and Vis light for three hours, respectively. Weight change of the [c2]AzoCD$_2$ hydrogel, Azo hydrogel, and PEG hydrogel before and after photo-irradiation with UV and Vis light (d). The weight ratio (%) is measured with an electric balance. Schematic illustration of the expansion–contraction of the [c2]AzoCD$_2$ hydrogel upon photo-irradiation (e) and the Azo hydrogel upon photo-irradiation (f). Source: Iwaso et al. 2016 [42]. Reproduced with permission of Springer Nature.

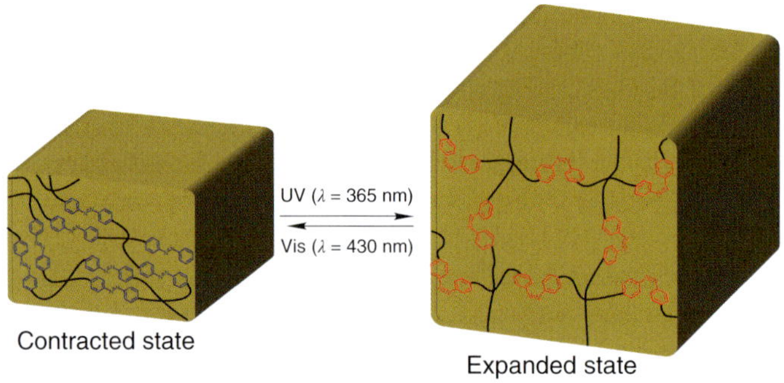

Figure 7.10 (*Continued*)

irradiation of the Azo hydrogel with UV light from the right side bent the hydrogel to the left and irradiation of the bent Azo hydrogel with Vis light restored the initial state (Figure 7.11c). These results demonstrate that the flexion properties of the [c2]AzoCD$_2$ hydrogel and Azo hydrogel have an opposite motion dependence on the light direction. Figure 7.11b,d shows the flexion angle θ of the hydrogels irradiated with UV and Vis light, respectively, for three hours. The θ value becomes saturated after approximately three hours of UV irradiation and does not significantly decrease on standing under its own weight for one hour in the dark. In contrast, irradiation with Vis light immediately restores the bent gel to the initial state. Figure 7.11e presents the results of repetition experiments for the [c2]AzoCD$_2$ hydrogel under alternating periods of irradiation (one hour) with UV and Vis light. The bending behavior of the [c2]AzoCD$_2$ gel can be repeated for at least 10 cycles without hysteresis. These observations indicate that the flex behavior of the [c2]AzoCD$_2$ hydrogel is the opposite of that of the Azo hydrogel. Angle θ of the [c2]AzoCD$_2$ hydrogel is positive, but θ for the Azo hydrogel is negative. The [c2]AzoCD$_2$ hydrogel bent toward the light source because the exposed surface of the hydrogel preferentially absorbs UV light energy. The conformational change of the [c2]AzoCD$_2$ unit induced by UV light irradiation discharges the absorbed water in the [c2]AzoCD$_2$ gel, shrinking the gel. UV light irradiation decreased the volume of the exposed surface and the nonexposed side remained constant (Figure 7.11f). The Azo hydrogel bent to the opposite direction of the light source because, before UV irradiation, the Azo gel is in the contracted state due to the formation of the hydrophobic *trans*-Azo assembly. After UV irradiation, the assembly dissociates, which decreases the cross-linking density, and the volume of the exposed surface then increases upon photo-isomerization from *trans*- to *cis*-Azo. The [c2]AzoCD$_2$ hydrogel and the Azo hydrogel have different contraction–expansion mechanisms.

7.4.2 Photo-Responsive Actuation of the [c2]AzoCD$_2$ Xerogel

The photo-responsiveness of the [c2]AzoCD$_2$ xerogel and Azo xerogel were compared. First, the [c2]AzoCD$_2$ hydrogel or the Azo hydrogel was slowly frozen on an ice plate. The icy hydrogels were then slowly lyophilized to afford the [c2]AzoCD$_2$ xerogel or the Azo xerogel. The [c2]AzoCD$_2$ xerogel immediately

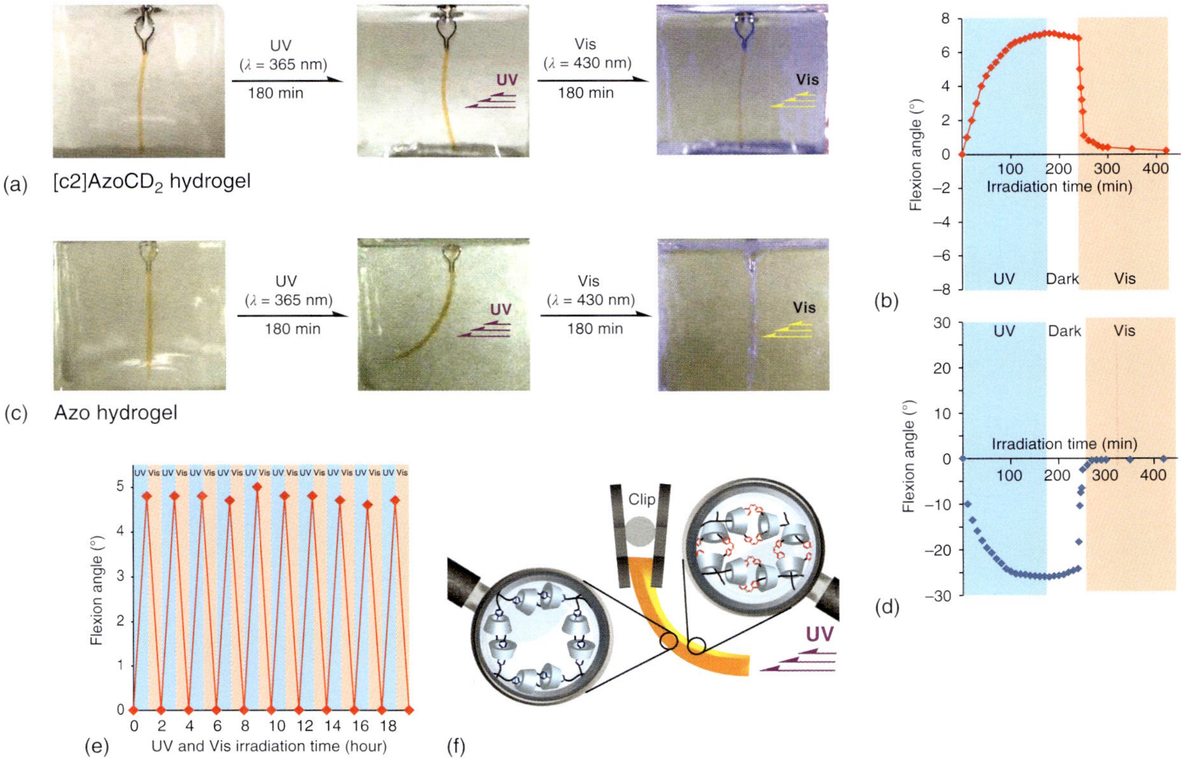

Figure 7.11 Photo-responsive hydrogel actuators in water. Photographs of the [c2]AzoCD$_2$ hydrogel irradiated with UV and Vis light (a). Plot of flexion angle θ versus irradiation time of the [c2]AzoCD$_2$ hydrogel (b). Photographs of the Azo hydrogel irradiated by UV and Vis light (c). Plot of flexion angle θ as a function of irradiation time for the Azo hydrogel (d). Repetition experiment for the [c2]AzoCD$_2$ hydrogel irradiated with UV and Vis light for one hour (e). Schematic illustration of the bending mechanism of the [c2]AzoCD$_2$ hydrogel upon photo-irradiation (f). Source: Iwaso et al. 2016 [42]. Reproduced with permission of Springer Nature.

responded to the UV light irradiation in the dry state and bent toward the light source. The bending angle reached 52° in ca. 3.6 seconds, which is approximately 30 times faster than that of the [c2]AzoCD_2 hydrogel (Figure 7.12a). More importantly, the response speed of the [c2]AzoCD_2 xerogel is very fast. The [c2]AzoCD_2 xerogel exhibits a speed of 7.1 deg s^{-1}. Although the [c2]AzoCD_2 hydrogel requires approximately three hours to bend 7.2°, the [c2]AzoCD_2 xerogel bent in about one second resulting in a response speed that is more than 10 800 times faster than that of the [c2]AzoCD_2 hydrogel. Continuous irradiation with Vis light from the same side of the [c2]AzoCD_2 xerogel did not restore the initial form because the hydrophobic interaction (the CD unit/the *trans*-Azo unit) and the swelling pressure fail to function in the dry state. However, irradiation of the opposite side with UV light restored the initial position. The size was reduced by the sequence of UV light irradiation from the right and left sides. The [c2]AzoCD_2 xerogel therefore exhibits *pseudo*-reversible deformation using a sequence of UV light irradiation. Figure 7.12b shows the zigzag conformational change of the [c2]AzoCD_2 xerogel following irradiation with UV light. After

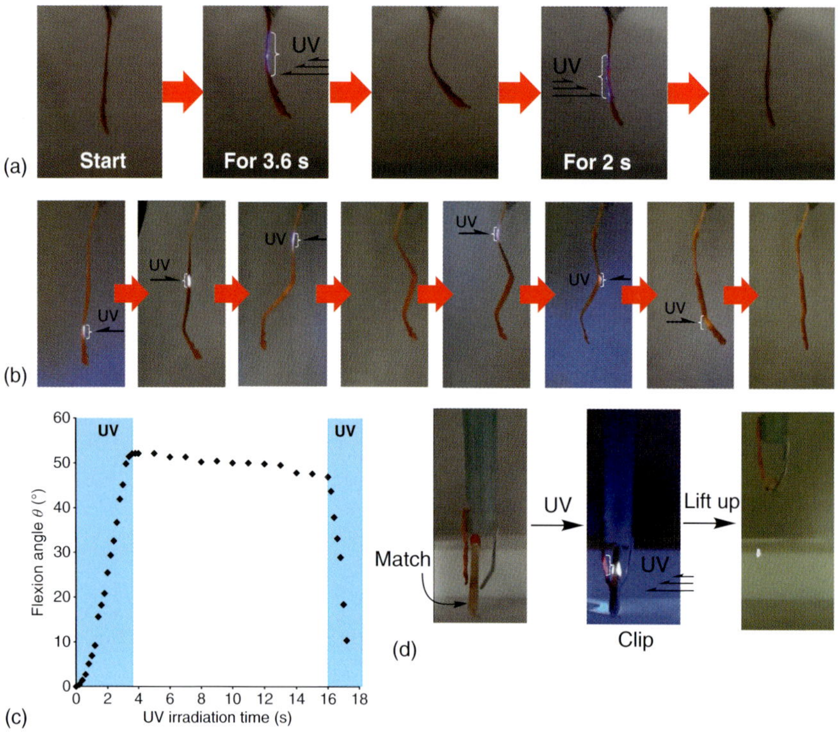

Figure 7.12 Photo-responsive xerogel actuators under dry conditions. Photographs of the [c2]AzoCD_2 xerogel irradiated with UV light from the right side (a). Photographs of the [c2]AzoCD_2 xerogel, which was alternately irradiated with UV light from the right and left sides, resulting in the formation of a zigzag conformation (b). Plot of flexion angle θ as a function of irradiation time for the [c2]AzoCD_2 xerogel (c). Photograph of a hand with stimuli-responsive fingers (the [c2]AzoCD_2 xerogel) (d). Source: Iwaso et al. 2016 [42]. Reproduced with permission of Springer Nature.

formation of the zigzag conformation, the initial form is restored by irradiation of the opposite side with UV light. Figure 7.12c shows the flexion angle θ of the xerogel irradiated with UV light. The bending mechanism of the [c2]AzoCD$_2$ xerogel involves a sliding motion of the [c2]AzoCD$_2$ unit, which shrinks the end-to-end distance of the [c2]daisy chain and reduces the xerogel. A lifting experiment using a 107.8 mg match and the [c2]AzoCD$_2$ xerogel was conducted. When reaching the hand position, the xerogel was irradiated to grip the match. The hand successfully pinches and lifts up the object, confirming that the fingers of the [c2]AzoCD$_2$ xerogel have sufficient stiffness to provide weight resistance for the match (Figure 7.12d).

7.5 Supramolecular Materials Consisting of CD and Sti

The authors focused on a [c2]daisy chain composed of αCD and Sti as a molecular machine to perform polymeric material actuation [47]. Figure 7.13 shows the proposed structure of the Sti-based [c2]daisy chain complex (αCD-Sti)$_2$, which is a precursor for [c2]daisy chain cross-linking between PEG chains. The association constant (K_a) of αCD with a *trans*-Sti derivative is high ($K_a = 1200\,M^{-1}$), but that of αCD with a *cis*-Sti derivative is low ($K_a = 150\,M^{-1}$) enough that the [c2]daisy chain complex easily dissociates in an aqueous solution. Therefore, the association/dissociation behavior of the (αCD-Sti)$_2$ complex can be controlled by isomerization of Sti, which is triggered by photo-irradiation (Figure 7.13a). A UV/Vis spectral analysis revealed the photo-isomerization of the (αCD-Sti)$_2$ complex. The irradiation of an aqueous solution of the (αCD-Sti)$_2$ complex with UV-A (UV light by a xenon lamp with a bandpass filter, $\lambda = 350\,nm$, radiation intensity: $1.7\,mW\,cm^{-2}$) resulted in isomerization from the trans-isomer to the cis-isomer within 20 seconds. Irradiation with UV-C ($\lambda = 280\,nm$) resulted in cis to trans isomerization of the (αCD-Sti)$_2$ complex, indicating that the photo-isomerization is reversible. The isomerization rates were $0.11\,s^{-1}$ (trans to cis) and $0.12\,s^{-1}$ (cis to trans). These values are 60 times faster than those of the Azo-based [c2]daisy chain complex (αCD-Azo)$_2$, which is our former precursor. UV/Vis spectroscopies also revealed that the trans-isomer of (αCD-Sti)$_2$ was converted to the cis-isomers in 83% yield, whereas (αCD-Azo)$_2$ gave only 39% of the cis-isomer under the same conditions. Moreover, the quantum yields of the photo-isomerization of Sti were 250 times higher than those of Azo. Sti showing high reaction rate, conversion, and quantum yield in photo-isomerization is suitable for the component of molecular machine driven by light irradiation.

7.5.1 (αCD-Sti)$_2$ Hydrogel

The (αCD-Sti)$_2$ cross-linker upon photo-irradiation changes the volume of the αCD-Sti hydrogel(10k) (Figure 7.13b) via a unique mechanism that differs from that of the Sti hydrogel(10k) (Figure 7.13c and Figure 7.14a-f). When the αCD-Sti hydrogel(10k) was irradiated by UV-A in water, the αCD-Sti hydrogel(10k) bent toward the light source and increased the flexion angle to 12.0° within seven seconds. UV–Vis spectroscopies revealed that the αCD-Sti hydrogel(10k) showed

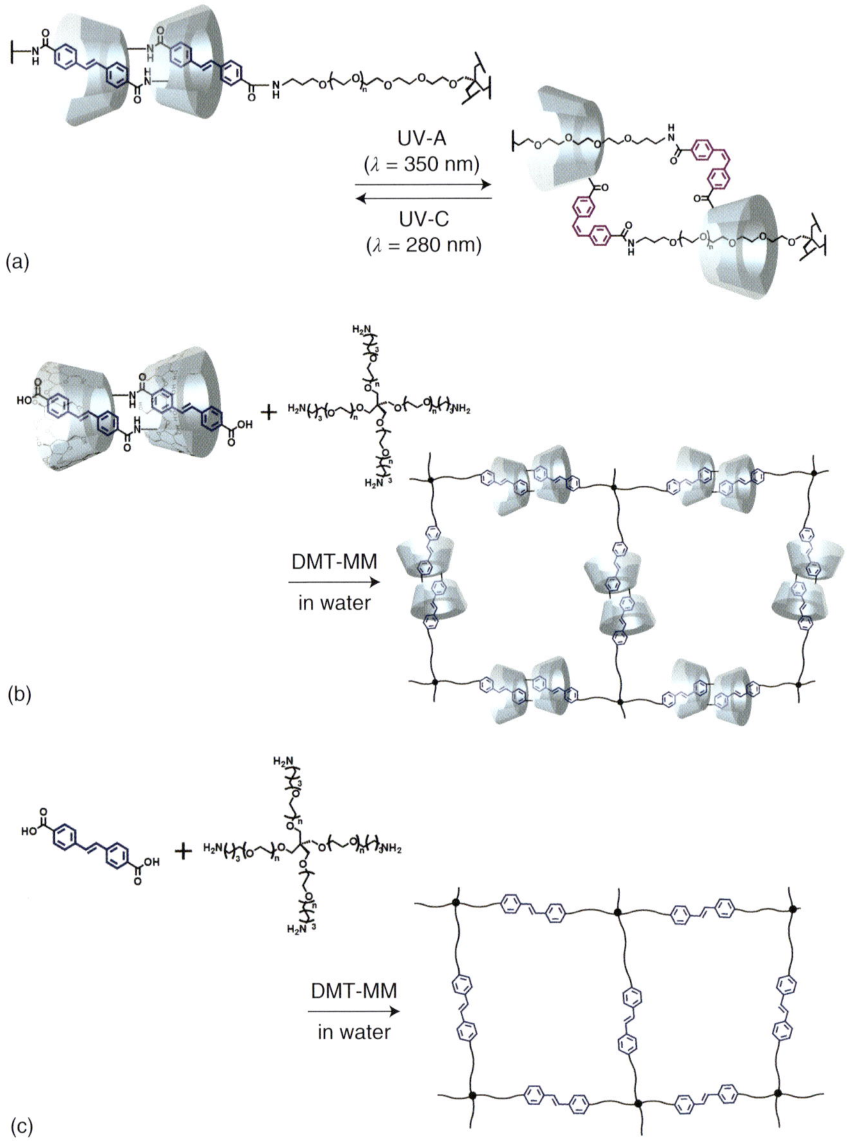

Figure 7.13 Structure of the [c2]daisy chain prepared from αCD and Sti and actuators with [c2]daisy chain cross-linking (αCD-Sti hydrogel) and covalent bond cross-linking (Sti hydrogel). (a) Isomerization of [c2]daisy chain crosslinks, (b) αCD-Sti hydrogel: [c2]daisy chain crosslinks, and (c) Sti hydrogel: crosslinks with covalent bonds. Source: Ikejiri et al. 2018 [47]. Reproduced with permission of American Chemical Society.

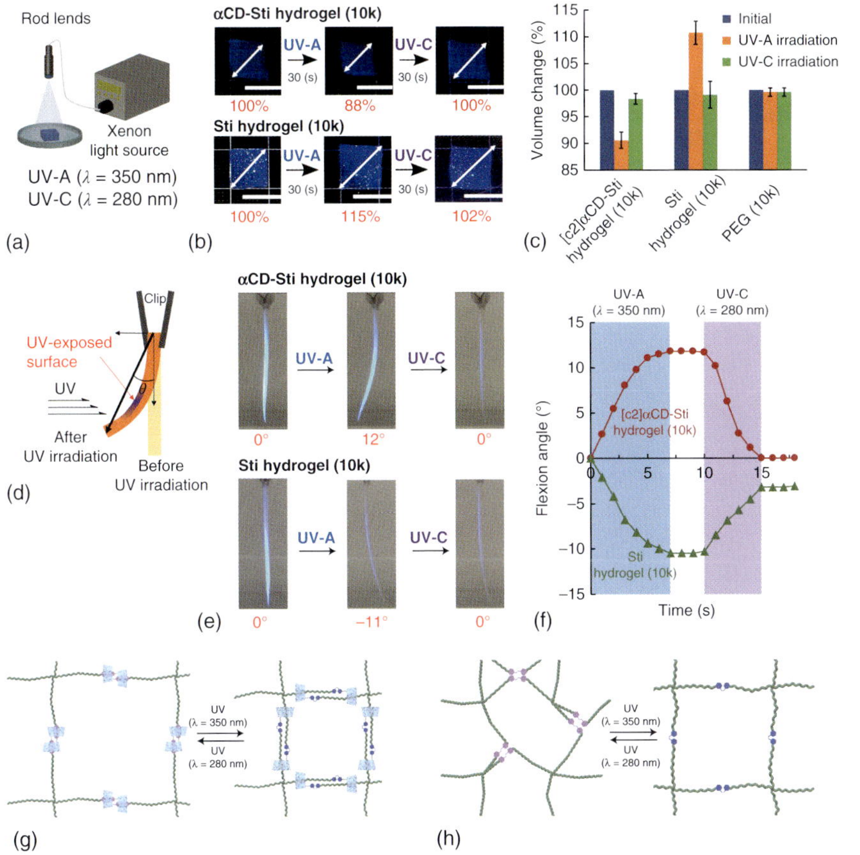

Figure 7.14 Experimental equipment and irradiation experimental setup of the αCD-Sti hydrogel(10k) (a). Contraction and expansion of the αCD-Sti hydrogel(10k) and the Sti hydrogel(10k) irradiated by UV-A ($\lambda = 350$ nm, 30 seconds) and UV-C ($\lambda = 280$ nm, 30 seconds) (b). Volume change of the αCD-Sti hydrogel(10k), the Sti hydrogel(10k), and the PEG hydrogel (10k) upon UV light irradiations (c). Lateral views of the αCD-Sti hydrogel(10k) hung with a clip. Flexion angle (θ) is defined here (d). Light irradiation from the left side of the αCD-Sti hydrogel(10k) for seven seconds. After irradiation of UV-A, the αCD-Sti hydrogel(10k) bends to the left side. Subsequent irradiation with UV-C from the left side restores the initial form. The Sti hydrogel(10k) bent to the opposite side by the photo-irradiations (e). Plots of the flexion angle (θ) as a function of the irradiation time of the αCD-Sti hydrogel(10k) and Sti hydrogel(10k) (f). Schematic illustrations of the mechanisms for the deformation of the αCD-Sti hydrogel(10k) (g) and the Sti hydrogel(10k) (h) upon irradiation with UV-A and UV-C. Source: Ikejiri et al. 2018 [47]. Reproduced with permission of American Chemical Society.

the photo-isomerization in 55% yield ([*trans*-Sti]/[*cis*-Sti] = 45/55) within 40 seconds. In contrast, the flexion angle of the Sti hydrogel(10k) indicated smaller bending to the other side. These results indicate that the contraction and expansion of the hydrogels were induced on the side near the light source and resulted in the bending behavior. The bending directions agree with the expanding/contracting properties of the αCD-Sti hydrogel(10k) and Sti hydrogel(10k). The initial speed of the bending motion was fast, i.e. 2.6° s^{-1} in αCD-Sti hydrogel(10k). Although the actuation of the αCD-Azo hydrogel(10k) took three hours to reach to 12.0°, the αCD-Sti hydrogel(10k) accomplished within seven seconds, which is 1600 times faster than that observed in our former work. The αCD-Sti hydrogel(10k) returned to its original state (flexion angle $\theta = 0°$) upon UV-C irradiation, indicating that isomerization of the Sti unit from the cis-isomer to the trans-isomer leads to reverse deformation (Figure 7.14g). The Sti hydrogel(10k) also showed reversible deformation upon UV-A and UV-C irradiation, indicating that the deformation is triggered by the isomerization of the Sti unit. Before UV-A irradiation, the *trans*-Sti moiety in the [c2]daisy chain is included by αCD to form the $(\alpha CD\text{-}Sti)_2$ complex. Therefore, the [c2]daisy chain cross-linker takes its longest form. When the αCD-Sti hydrogel(10k) was irradiated by UV-C irradiation, the *trans*-Sti unit isomerized to the cis-form, and the αCD unit allowed the *cis*-Sti moiety to slide onto the PEG chain, shrinking the length of the [c2]daisy chain. This molecular motion is supposed to result in the macroscopic contraction of the αCD-Sti hydrogel(10k). The behavior of Sti hydrogel(10k) was studied by fluorescence spectra of model Sti dye (stilbene dicarboxylic acid) in water, which revealed that excimer's peaks were observed at 420–600 nm, indicative of dimerization of Sti residues. Through the photo-irradiation ($\lambda = 350$ nm), intensity of the excimer's peaks decreased whereas that of monomer ($\lambda = 408$ nm) increased. This result indicates that *trans*-Sti forms dimer in water and that the aggregates can be dissociated by the photo-isomerization (Figure 7.14h). The *trans*-Sti moieties are supposed to form aggregations as cross-linking points in the Sti hydrogel(10k). When Sti was isomerized by UV-A irradiation, the cis-isomers lost their planar symmetry and their aggregates in the polymer network disassociated. Thus, the Sti hydrogel can be expanded by UV-A irradiation.

7.5.2 $(\alpha CD\text{-}Sti)_2$ Dry Gel

The αCD-Sti hydrogels were dried via lyophilization to obtain a dry gel. The αCD-Sti dry gels were transparent and self-standing materials. The response of the αCD-Sti dry gel to photo-stimuli was investigated (Figure 7.15a–c). When the αCD-Sti dry gel(10k) was subjected to UV-A irradiation, the αCD-Sti dry gel(10k) bent toward the light source, and the flexion angle of the dry gel increased to 6.0° within three seconds, showing that the αCD-Sti dry gel(10k) can work as a dry polymeric actuator. That is the αCD-Sti unit is a versatile material that can work in both wet and dry states. The αCD-Sti dry gel(20k) quickly bent to 26.0° within three seconds, indicating that the degree of deformation increases as the Mw of PEG increases. Moreover, when, the αCD-Sti dry gels were irradiated with UV-C, they returned to their original state, leading to reverse actuation. At least 15 bending cycles were performed with the same dry gel using UV-A and UV-C irradiations. In contrast, the covalently cross-linked Sti dry gels showed slight bending to the opposite side of the light source. These results suggest that the reversible, large, and rapid deformations in both the wet and dry states are due to the sliding motion of the $(\alpha CD\text{-}Sti)_2$ unit. Bending actuation of the αCD-Azo dry gels to arbitrary directions was performed by changing sides of the UV irradiation. On the contrary, the αCD-Sti dry gel realized

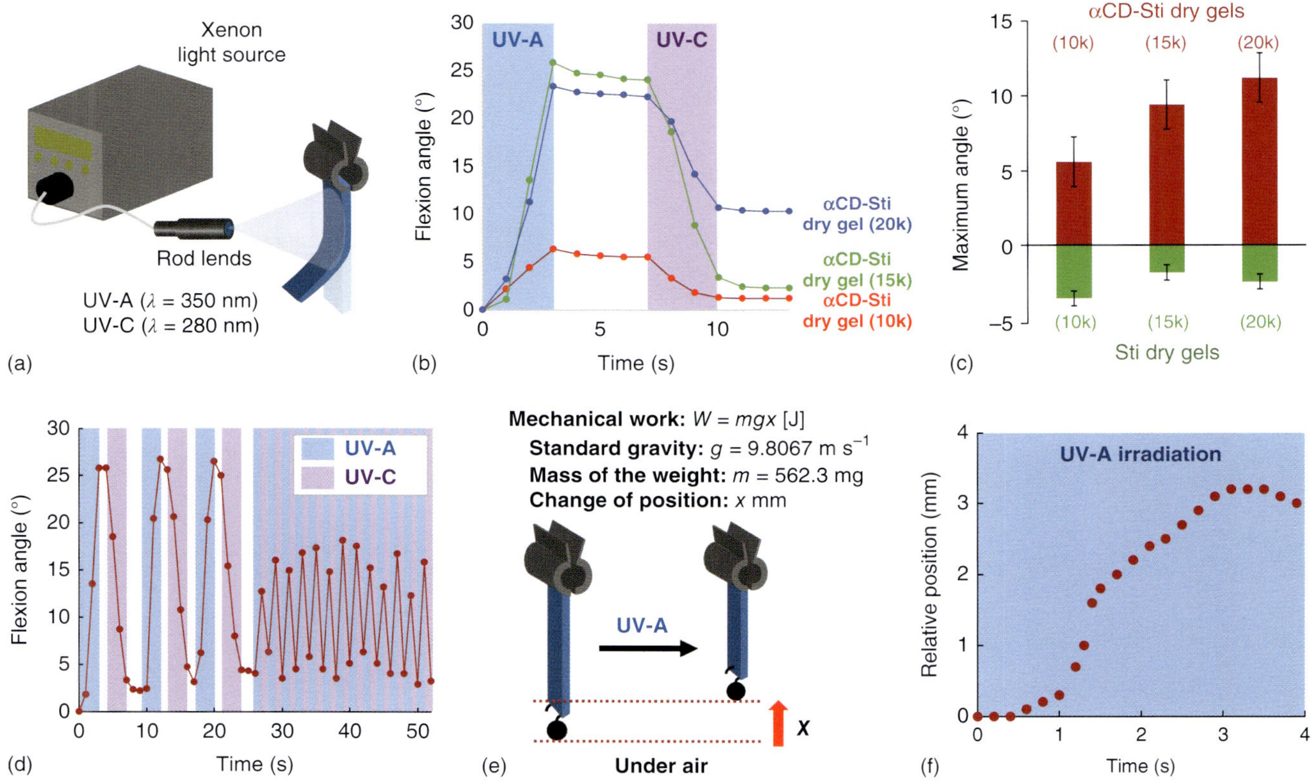

Figure 7.15 Experimental equipment and irradiation experimental setup of the αCD-Sti dry gel (10k) in air (a). Plots of the flexion angle (θ) as a function of the irradiation time of the αCD-Sti dry gels (10k), (15k), and (20k) (b). The maximum flexion angles of the αCD-Sti dry gels and Sti dry gels upon UV-A irradiation (c). Repeated tests of the αCD-Azo hydrogel(10k) bending upon exposure to UV-A and UV-C irradiation (d). The energy conversion test of the αCD-Sti dry gel (20k). The energy conversion from light to mechanical work was evaluated by using the values and equation shown in (e). Plot of the position of the weight as a function of the irradiation time during the energy conversion testing of the αCD-Sti dry gel (20k) (f). Source: Ikejiri et al. 2018 [47]. Reproduced with permission of American Chemical Society.

reversible actuation in air. Switching the wavelength of light ($\lambda = 350/280$ nm) can repeatedly let the αCD-Sti dry gel bend and return to the original shape as shown in Figure 7.15d. The hydrophobic Sti residue can easily return to the αCD cavity. That is the hydrophobicity of Sti supports reversible motion of the [c2]daisy chain. The dependence on PEG length during photo-actuation is summarized in Figure 7.15c. The maximum flexion angle of the αCD-Sti dry gels increased to 11.5° when the αCD-Sti dry gel(20k) was employed. The longer PEG chains resulted in the larger flexion angle of the αCD-Sti dry gel. The length of the PEG chain, which the αCD unit can slide on, is supposed to affect the degree of contraction. The hydrophobic property of Sti should fully enhance the ability of the [c2]daisy chain with higher Mw to show the fast and reversible actuation depending on the Mw of PEGs. These dry state results agree with those of the αCD-Sti hydrogels and Sti hydrogels, respectively. The mechanisms shown for the hydrogels are also reasonable for the dry gels. As shown in Figure 7.15e,f, an energy conversion from light to mechanical work was performed. We used 36.5 mg of the αCD-Sti dry gel(20k) for an energy conversion test as the αCD-Sti dry gel(20k) showed the fastest and largest deformation. When the αCD-Sti dry gel(20k) hung with a weight of 562.3 mg was irradiated with UV-A irradiation, the αCD-Sti dry gel (20k) contracted and lifted the weight vertically by 3.2 mm within three seconds. The mechanical work (W) produced by the αCD-Sti dry gel(20k) was 17.8 μJ. This value is 150 times higher than that obtained in the αCD-Azo dry gel system. Additionally, the power ($P = W/t$, t: time) generated by the αCD-Sti dry gel(20k) was 5.9 μW, which is 10 times higher than that of the [2]rotaxane-type actuator using Azo. The αCD-Sti dry gel(20k) performed the rapid and large mechanical work upon photo-irradiation.

7.6 Concluding Remarks

Reversible expansion–contraction supramolecular hydrogels and a supramolecular actuator-like artificial molecular muscle system consisting of αCD-Azo gel were successfully constructed. Supramolecular hydrogels are thought to be important to realize "soft machines" like biological systems. These stimulus-responsive expansion–contraction properties are similar to that of muscle fibrils, such as sarcomere consisting of actin filaments. Moreover, photo-responsive materials have many general applications, including remotely controlled materials and medical devices.

A tough and photo-responsive polymeric material was also prepared with [2]rotaxane structures as topological cross-links in the polymer network. The stress distribution property based on the sliding motion of the topological cross-links contributed to this excellent extensibility. In addition, the material can be reversibly deformed by repeated irradiation with UV or Vis light in aqueous media. Importantly, the dried material uniaxially extended during the drying process showed a faster response than the hydrogel. The stretched xerogel quickly bent toward the light source side upon UV light irradiation. The orientation of the rotaxane network within the material might enable this efficient response.

The [c2]AzoCD$_2$ hydrogel exhibited photo-stimuli-induced back–and-forth motion, photo-responsive actuation in the [c2]AzoCD$_2$ xerogel was thought to be difficult. The [c2]AzoCD$_2$ xerogel achieves *pseudo*-reversible deformation by a sequence of UV light irradiation. The ability to slide was found to be important for the photo-driven movement of a xerogel on a macroscopic scale in the dry state.

Photo-responsive actuators as both hydrogels and dry gels consisting of PEG molecules cross-linked by [c2]daisy chains (αCD-Sti)$_2$ were constructed. The obtained gels underwent fast and large deformations triggered by UV irradiation in both wet and dry states. The actuation was driven by photo-isomerization of the Sti moiety in the [c2]daisy chain with high quantum yields. Repeatable actuation under dry conditions through reversible and rapid photo-isomerization of the Sti moiety were accomplished by using UV light with different wavelengths. The authors are now developing a uniaxially oriented molecular actuator with a [c2]daisy chain that selectively responds to polarized light and amplifies the uniaxial deformation.

References

1 Sauvage, J.-P. (2017). From chemical topology to molecular machines (nobel lecture). *Angew. Chem. Int. Ed.* 56: 11080–11093.

2 Stoddart, J. F. (2017). Mechanically interlocked molecules (MIMs) – molecular shuttles, switches, and machines (nobel lecture). *Angew. Chem. Int. Ed.* 56: 11094–11125.

3 Feringa, B.L. (2017). The art of building small: from molecular switches to motors (nobel lecture). *Angew. Chem. Int. Ed.* 56: 11060–11078.

4 Harada, A., Hashidzume, A., Yamaguchi, H., and Takashima, Y. (2009). Polymeric rotaxanes. *Chem. Rev.* 109: 5974–6023.

5 Harada, A., Takashima, Y., and Yamaguchi, H. (2009). Cyclodextrin-based supramolecular polymers. *Chem. Soc. Rev.* 38: 875–882.

6 Harada, A. and Takashima, Y. (2013). Macromolecular recognition and macroscopic interactions by cyclodextrins. *Chem. Rec.* 13: 420–431.

7 Harada, A., Takashima, Y., and Nakahata, M. (2014). Supramolecular polymeric materials via cyclodextrin–guest interactions. *Acc. Chem. Res.* 47: 2128–2140.

8 Hashidzume, A., Takashima, Y., Yamaguchi, H., and Harada, A. (2017). Cyclodextrin. In: *Comprehensive Supramolecular Chemistry II* (ed. J. Atwood), 269–316. Oxford: Elsevier.

9 Harada, A., Li, J., and Kamachi, M. (1992). The molecular necklace: a rotaxane containing many threaded α-cyclodextrins. *Nature* 356: 325–327.

10 Kawaguchi, Y. and Harada, A. (2000). A cyclodextrin-based molecular shuttle containing energetically favored and disfavored portions in its dumbbell component. *Org. Lett.* 2: 1353–1356.

11 Kawaguchi, Y. and Harada, A. (2000). An electric trap: a new method for entrapping cyclodextrin in a rotaxane structure. *J. Am. Chem. Soc.* 122: 3797–3798.

12 Hoshino, T., Miyauchi, M., Kawaguchi, Y. et al. (2000). Daisy chain necklace: tri[2]rotaxane containing cyclodextrins. *J. Am. Chem. Soc.* 122: 9876–9877.
13 Harada, A. (2001). Cyclodextrin-based molecular machines. *Acc. Chem. Res.* 34: 456–464.
14 Hashidzume, A., Yamaguchi, H., and Harada, A. (2014). Cyclodextrin-based molecular machines. *Top. Curr. Chem.* 354: 71–110.
15 Tomatsu, I., Hashidzume, A., and Harada, A. (2005). Photoresponsive hydrogel system utilizing molecular recognition of alpha-cyclodextrin. *Macromolecules* 38: 5223–5227.
16 Kuad, P., Miyawaki, A., Takashima, Y. et al. (2007). External stimulus-responsive supramolecular structures formed by a stilbene cyclodextrin dimer. *J. Am. Chem. Soc.* 129: 12630–12631.
17 Deng, W., Yamaguchi, H., Takashima, Y., and Harada, A. (2007). A chemical-responsive supramolecular hydrogel from modified cyclodextrins. *Angew. Chem. Int. Ed.* 46: 5144–5147.
18 Inoue, Y., Kuad, P., Okumura, Y. et al. (2007). Thermal and photochemical switching of conformation of poly(ethylene glycol)-substituted cyclodextrin with an azobenzene group at the chain end. *J. Am. Chem. Soc.* 129: 6396–6397.
19 Ogoshi, T., Takashima, Y., Yamaguchi, H., and Harada, A. (2007). Chemically-responsive sol-gel transition of supramolecular single-walled carbon nanotubes (SWNTs) hydrogel made by hybrids of SWNTs and cyclodextrins. *J. Am. Chem. Soc.* 129: 4878–4879.
20 Yamauchi, K., Takashima, Y., Hashidzume, A. et al. (2008). Switching between supramolecular dimer and non-threaded supramolecular self-assembly of stilbene amide alpha-cyclodextrin by photoirradiation. *J. Am. Chem. Soc.* 130: 5024–5025.
21 Deng, W., Yamaguchi, H., Takashima, Y., and Harada, A. (2008). Construction of chemical-responsive supramolecular hydrogels from guest-modified cyclodextrins. *Chem. Asian J.* 3: 687–695.
22 Tamesue, S., Takashima, Y., Yamaguchi, H. et al. (2010). Photoswitchable supramolecular hydrogels formed by cyclodextrin and azobenzene polymers. *Angew. Chem. Int. Ed.* 49: 7461–7464.
23 Li, S., Taura, D., Hashidzume, A. et al. (2010). Photocontrolled size changes of doubly-threaded dimer based on an α-CD derivative with two recognition sites. *Chem. Lett.* 39: 242–243.
24 Li, S., Taura, D., Hashidzume, A., and Harada, A. (2010). Light-switchable Janus [2]rotaxanes based on α-cyclodextrin derivatives bearing two recognition sites linked with oligo(ethylene glycol). *Chem. Asian J.* 5: 2281–2289.
25 Harada, A., Kobayashi, R., Takashima, Y. et al. (2011). Macroscopic self-assembly through molecular recognition. *Nat. Chem.* 3: 34–37.
26 Nakahata, M., Takashima, Y., Yamaguchi, H., and Harada, A. (2011). Redox-responsive self-healing materials formed from host-guest polymers. *Nat. Commun.* 2: 511.
27 Wang, Z., Takashima, Y., Yamaguchi, H., and Harada, A. (2011). Photoresponsive formation of pseudo[2]rotaxane with cyclodextrin derivatives. *Org. Lett.* 13: 4356–4359.

28 Tamesue, S., Takashima, Y., Yamaguchi, H. et al. (2011). Photochemically-controlled supramolecular curdlan/single-walled carbon nanotube composite gel: preparation of molecular distaff by cyclodextrin modified curdlan and phase transition control. *Eur. J. Org. Chem.* 2011: 2801–2806.

29 Hu, J., Hashidzume, A., and Harada, A. (2011). Photoregulated switching of the recognition site of α-cyclodextrin in a side chain polyrotaxane bearing two recognition sites linked with oligo(ethylene glycol). *Macromol. Chem. Phys.* 212: 1032–1038.

30 Omori, K., Takashima, Y., Yamaguchi, H., and Harada, A. (2011). pH responsive [2]rotaxanes with 6-modified-α-cyclodextrins. *Chem. Lett.* 40: 758–759.

31 Yamaguchi, H., Kobayashi, Y., Kobayashi, R. et al. (2012). Photoswitchable gel assembly based on molecular recognition. *Nat. Commun.* 3: 603.

32 Zheng, Y., Hashidzume, A., Takashima, Y. et al. (2012). Switching of macroscopic molecular recognition selectivity using a mixed solvent system. *Nat. Commun.* 3: 831.

33 Takashima, Y., Hatanaka, S., Otsubo, M. et al. (2012). Expansion-contraction of photoresponsive artificial muscle regulated by host–guest interactions. *Nat. Commun.* 3: 1270.

34 Nakahata, M., Takashima, Y., Hashidzume, A., and Harada, A. (2013). Redox-generated mechanical motion of a supramolecular polymeric actuator based on host–guest interactions. *Angew. Chem. Int. Ed.* 52: 5731–5735.

35 Kobayashi, Y., Takashima, Y., Hashidzume, A. et al. (2013). Reversible self-assembly of gels through metal–ligand interactions. *Sci. Rep.* 3: 1243.

36 Zheng, Y., Hashidzume, A., and Harada, A. (2013). pH-responsive self-assembly by molecular recognition on a macroscopic scale. *Macromol. Rapid Commun.* 34: 1062–1066.

37 Nakamura, T., Takashima, Y., Hashidzume, A. et al. (2014). A metal-ion-responsive adhesive material via switching of molecular recognition properties. *Nat. Commun.* 5: 4622.

38 Nakahata, M., Takashima, Y., and Harada, A. (2014). Redox-responsive macroscopic gel assembly based on discrete dual interactions. *Angew. Chem. Int. Ed.* 53: 3617–3621.

39 Nakahata, M., Mori, S., Takashima, Y. et al. (2014). pH- and sugar-responsive gel assemblies based on boronate–catechol interactions. *ACS Macro Lett.* 3: 337–340.

40 Miyamae, K., Nakahata, M., Takashima, Y., and Harada, A. (2015). Self-healing, expansion-contraction, and shape-memory properties of preorganized supramolecular hydrogel through host–guest interactions. *Angew. Chem. Int. Ed.* 54: 8984–8987.

41 Harada, A. and Hashidzume, A. (2015). Cyclodextrin-based chemo- and pH-responsive polymer systems for pharmaceutical and biomedical applications. In: *Chemoresponsive Materials: Stimulation by Chemical and Biological Signals*, vol. 7 (ed. H.-J. Schneider), 167–207. Royal Society of Chemistry.

42 Iwaso, K., Takashima, Y., and Harada, A. (2016). Fast response dry-type artificial molecular muscles with [c2]daisy chains. *Nat. Chem.* 8: 625–632.

43 Takashima, Y., Yonekura, K., Koyanagi, K. et al. (2017). Multifunctional stimuli-responsive supramolecular materials with stretching, coloring, and self-healing properties functionalized via host–guest interactions. *Macromolecules* 50: 4144–4150.
44 Takashima, Y. and Harada, A. (2017). Functioning via host-guest interactions. *J. Incl. Phenom. Macrocycl. Chem.* 87: 313–330.
45 Nakahata, M., Takashima, Y., and Harada, A. (2017). Supramolecular polymeric materials containing cyclodextrins. *Chem. Pharm. Bull.* 65: 330–335.
46 Takashima, Y., Hayashi, Y., Osaki, M. et al. (2018). A photoresponsive polymeric actuator topologically cross-linked by movable units based on a [2]rotaxane. *Macromolecules* 51: 4688–4693.
47 Ikejiri, S., Takashima, Y., Osaki, M. et al. (2018). Solvent-free photoresponsive artificial muscles rapidly driven by molecular machines. *J. Am. Chem. Soc.* 140: 17308–17315.
48 Tomatsu, I., Hashidzume, A., and Harada, A. (2006). Contrast viscosity changes upon photoirradiation for mixtures of poly(acrylic acid)-based a-cyclodextrin and azobenzene polymers. *J. Am. Chem. Soc.* 128: 2226–2227.
49 Okumura, Y. and Ito, K. (2001). The polyrotaxane gel: a topological gel by figure-of-eight cross-links. *Adv. Mater.* 13: 485–487.

8

Cross-Linked Liquid-Crystalline Polymers as Photomobile Materials

Toru Ube and Tomiki Ikeda

Research and Development Initiative Chuo University 1-13-27 Kasuga, Bunkyo-ku, Tokyo 112-8551, Japan

Introduction

Stimuli-responsive polymers that exhibit macroscopic deformation under heat, electricity, humidity, and light have been extensively studied due to their potential as soft actuators. Polymer materials possess tremendous variation in structures ranging from subnano- to macroscale, enabling flexible design of materials for various applications. Through adequate design of polymer materials, structural changes at nanoscales can be effectively transferred to macroscopic structures and properties. Among the various stimuli, light is advantageous in spatial selectivity, remote controllability, and variation in wavelength and polarization. Wireless conversion of light into mechanical work is a fascinating approach to develop untethered soft robots with minimum additional components [1, 2].

Pioneering studies focusing on photomechanical effects were undertaken in the 1960s [3]. Polymer systems doped or functionalized with dyes have been investigated since initial studies. Merian first reported the photoinduced shrinking of about 0.1% for polymer fibers containing azobenzene in 1966 [4]. Large deformation of gels such as expansion and bending was accomplished in the 1980s [5–7]. On the other hand, photoinduced deformation in dry polymer systems was typically <1% before 2000 [8, 9].

Development of cross-linked liquid-crystalline polymers (CLCPs) opened a new field of stimuli-responsive polymers [10, 11]. Self-assembling properties of liquid crystals (LCs) allow homogeneous molecular alignment at large scales, which is readily switched by external field as seen in LC displays. Incorporation of LC moieties (mesogens) into polymer chains induces strong correlation between the alignment of mesogens and conformation of polymer chains. Furthermore, deformation of polymer chains can be directly related with macroscopic deformation through formation of cross-linked structure. Thus, the alignment of mesogens is altered when CLCPs are macroscopically deformed under external force [12, 13]. Moreover, deformation of CLCPs can be induced by alignment change of mesogens, which is triggered by external stimuli such as heat, electricity, and light. This correlation was theoretically predicted by de Gennes in 1975 [14, 15]. Finkelmann and coworkers experimentally developed

Mechanically Responsive Materials for Soft Robotics, First Edition. Edited by Hideko Koshima.

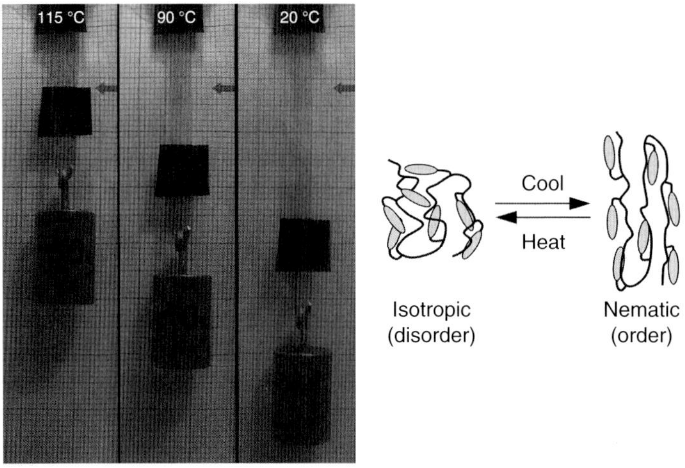

Figure 8.1 Deformation of a CLCP film induced by the change in temperature. Source: Ohm et al. 2010 [11]. Reproduced with permission of John Wiley & Sons.

CLCPs with polysiloxane backbone in 1981 [16]. CLCPs with soft nature can also be referred to as LC elastomers. They demonstrated that mesogens in a polydomain film (i.e. mesogens align in the same direction only in local domains and the director of each domain is random) can be homogeneously aligned under external force [12, 13]. Furthermore, they developed a two-step cross-linking method to fabricate monodomain CLCPs (i.e. mesogens align in the same direction within an entire sample) [17]. They succeeded in inducing reversible macroscopic deformation of a monodomain film through nematic–isotropic phase transition triggered by the change in surrounding temperature (Figure 8.1).

The alignment change of mesogens can be induced by light with the aid of photochromic moieties [18]. Isothermal phase transition was found in LC systems containing azobenzene [19]. When irradiated with UV light in nematic phase, LC–isotropic phase transition occurs along with trans–cis isomerization of azobenzene moieties (Figure 8.2). The LC phase is restored by irradiation with visible light or heating. This phenomenon is based on the LC nature of azobenzene moieties. Rod-like *trans*-azobenzene stabilizes LC phases, whereas bent *cis*-azobenzene disturbs them. Photoinduced phase transition has been observed both in low-molecular-weight and polymer systems. In liquid-crystalline polymers (LCPs), disturbed alignment of mesogens can be preserved under glass transition temperature, enabling application for photomemory materials [20–22].

Photomechanical effects in dry polymer systems have been significantly enhanced by incorporation of photochromic moieties into CLCPs. Finkelmann et al. introduced azobenzene moieties in their monodomain LC elastomers, enabling photoinduced contraction–expansion as large as 20% [23]. Furthermore, Ikeda and coworkers developed CLCPs with polyacrylate backbone, bearing high concentration of azobenzene moieties in side chains and cross-linking points [24, 25]. Upon UV and visible light irradiations, the film showed

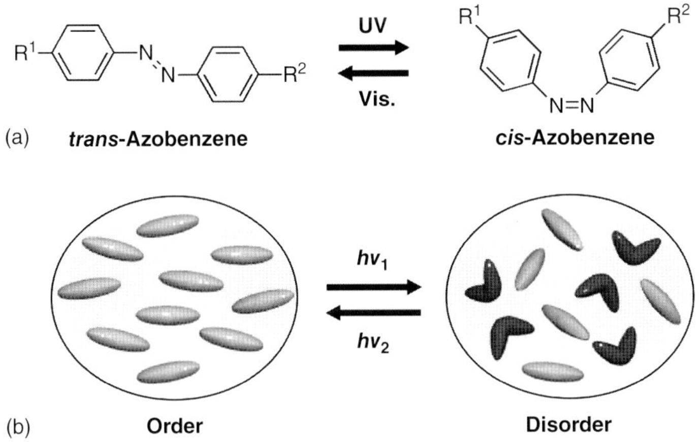

Figure 8.2 (a) Photoisomerization of azobenzene. (b) Order–disorder phase transition of LCs induced by photoisomerization.

bending–unbending motion. These systems are detailed in Sections 8.1.1 and 8.1.2.

Since the initial development of CLCPs, photomobile polymer materials with various 3D motions have been developed [26–30]. In this chapter, structural variation of photomobile LCPs is overviewed, providing the relationship among structures, mechanical properties, and photoresponsive properties.

8.1 Structures and Functions of Photomobile Materials Based on LCPs

8.1.1 Polysiloxanes

Photoinduced deformation of CLCPs was first reported for polysiloxanes containing photochromic moieties and mesogens in side chains [23]. These systems show large strain due to their soft nature of polysiloxane (Young's modulus: 10^5–10^6 Pa at room temperature). CLCPs with polysiloxane backbones are prepared through hydrosilylation reaction of poly(hydrogenmethylsiloxane) (PHMS) and LC vinyl compounds (Figure 8.3). Monodomain films can be fabricated by a two-step cross-linking method. In the first step, linear polymers are partially cross-linked to obtain a polydomain film. Then, the film is stretched to induce uniaxial alignment, and the second cross-linking reaction is performed in the stretched state to fix the alignment.

In the pioneering work by Finkelmann et al., a monodomain film with azobenzene moieties showed contraction of 20% upon UV light irradiation (Figure 8.4a) [23]. This deformation is triggered by trans–cis isomerization of azobenzene moieties, which leads to disordering of mesogens, and then contraction of polymer chains along the initial alignment direction. After the UV irradiation was switched off, the film gradually reverted to the initial shape along with thermal cis–trans isomerization. Terentjev and coworkers

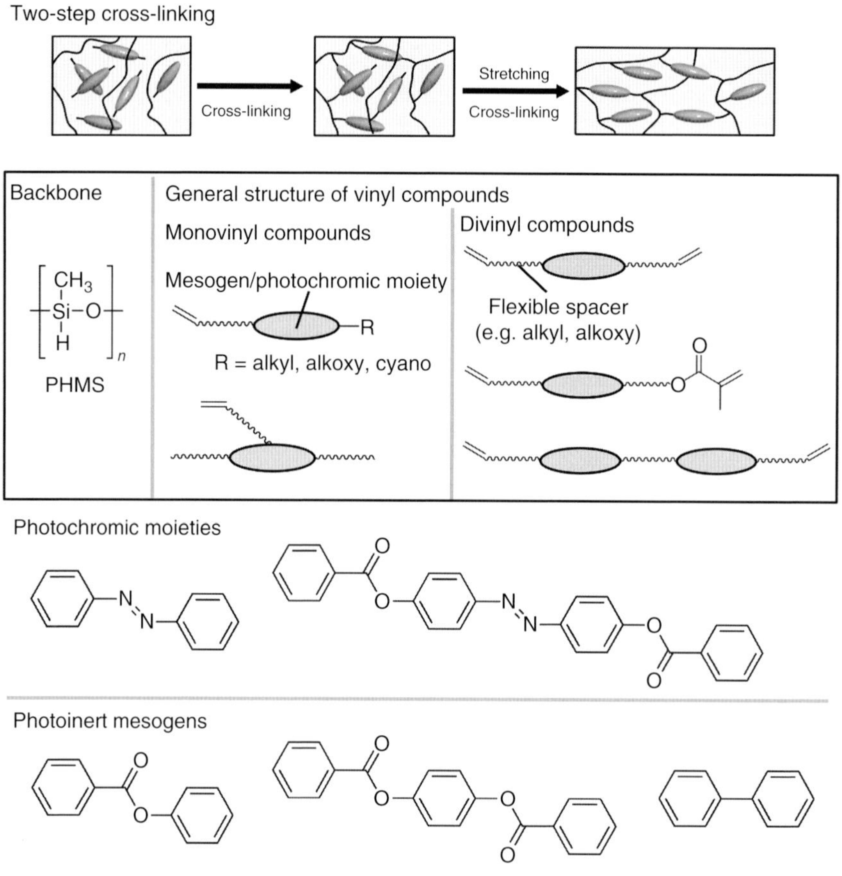

Figure 8.3 Chemical structures of CLCPs with polysiloxane backbones.

investigated the photogenerated force of CLCP films with polysiloxane backbone [31, 32]. The films showed photoinduced stress of 1–3 kPa, which was measured using mechanical analyzers under constant strain upon UV light irradiation (Figure 8.4b). Actually, photogenerated force depends on various factors such as chemical structure, alignment of mesogens, wavelength and intensity of actinic light, temperature, and sample size. With optimized condition, photoinduced stress of 40 kPa has been reported for these systems [33].

Polysiloxanes containing mesogens and azobenzene moieties in main chain were also investigated for photomechanical effects [34]. Main chain CLCPs are generally prepared by a two-step reaction: step-growth reaction between difunctional mesogens and chain extenders, followed by endlinking with multifunctional cross-linkers, i.e. hydrosilylation reaction of divinyl mesogens with siloxane chain extenders and multifunctional cross-linkers [35, 36]. Under control of cross-linking density in main chain CLCPs with polysiloxane backbone, a large strain of 400% has been achieved for thermal actuation [37]. These systems exhibit higher photoinduced stress (70 kPa) than side chain types due to their

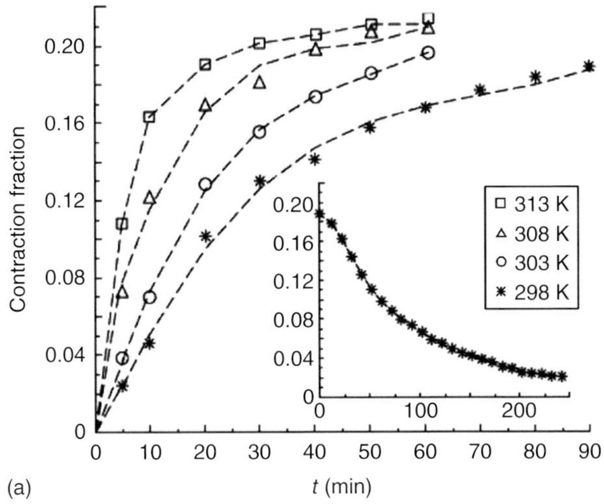

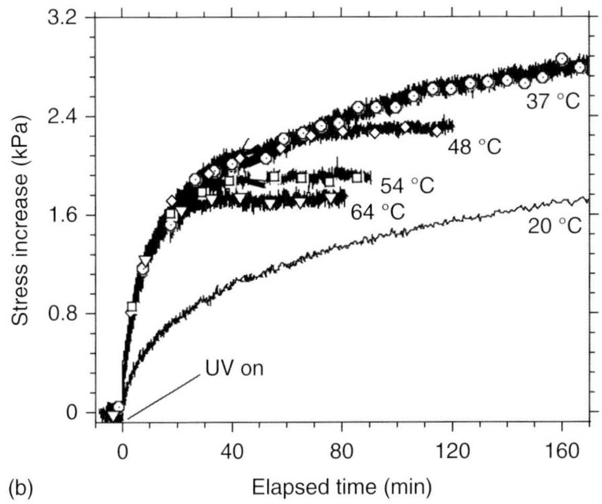

Figure 8.4 Photomechanical effects of CLCP films with polysiloxane backbone. (a) Photoinduced contraction of a film upon UV light irradiation. Source: Finkelmann et al. 2001 [23]. Reproduced with permission of The American Physical Society. (b) Photogenerated stress under constant strain upon UV light irradiation (365 nm, 6 mW cm^{-2}). Source: Cviklinski et al. 2002 [32]. Reproduced with permission of Springer-Verlag.

strong correlation between the alignment of mesogens and conformation of polymer chains [34].

8.1.2 Polyacrylates

Polyacrylates bearing azobenzene moieties in side chains are mostly studied systems as photoresponsive CLCPs. Chemical structures of monomers and cross-linkers are shown in Figure 8.5. Methacrylate monomers and cross-linkers have also been used in the same way. Monodomain samples are usually prepared

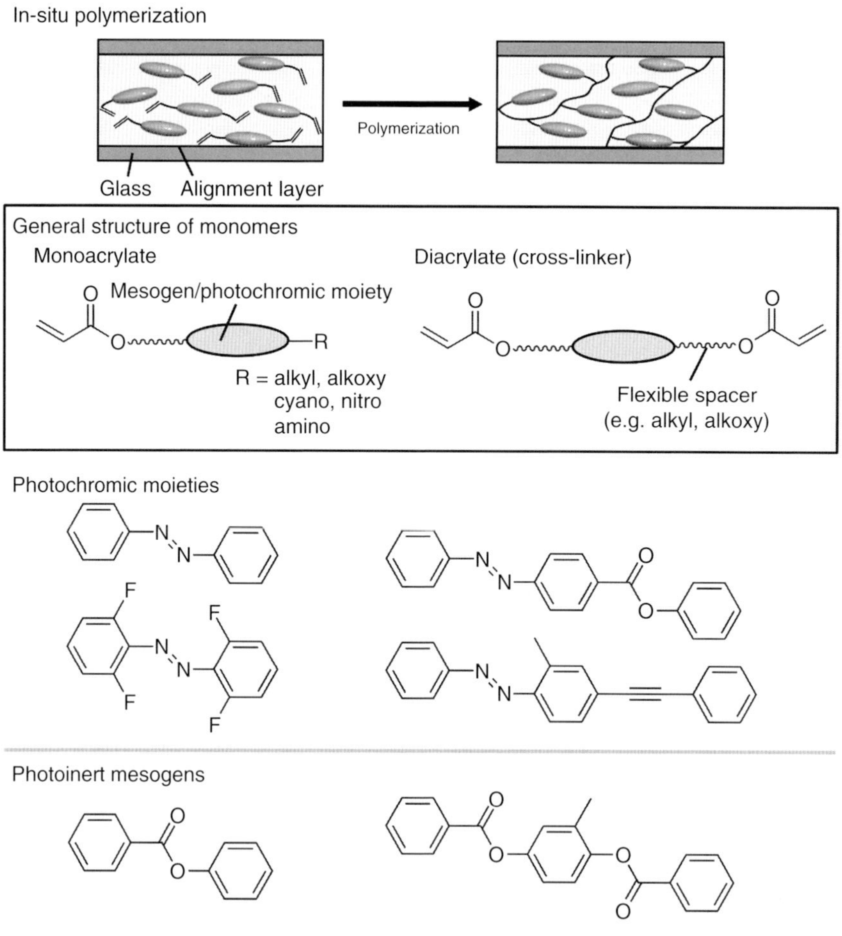

Figure 8.5 Chemical structures of CLCPs with polyacrylate backbones.

from LC mono- and diacrylates through in situ polymerization [38]. A mixture of monomer, cross-linker, and initiator is injected into a glass cell coated with alignment layers such as rubbed polyimides. Then the photopolymerization is carried out at a LC phase to form polymer networks, which memorize the alignment of mesogens during polymerization. These CLCPs typically show glassy nature (Young's modulus: 10^8–10^9 Pa at room temperature) in contrast to polysiloxane systems.

CLCPs with high concentration of photochromic moieties can be prepared through copolymerization of photochromic LC monomers and cross-linkers [24, 25]. When a monodomain film is irradiated with UV light, the film bends toward the light source [25]. The film reverts to the initial flat shape when irradiated with visible light. In systems containing photochromic moieties with high concentration, the incident light is mainly absorbed near the surface of the film in accordance with Beer–Lambert law. Therefore, photoisomerization, alignment change of mesogens, and deformation of polymer chains are induced only in the

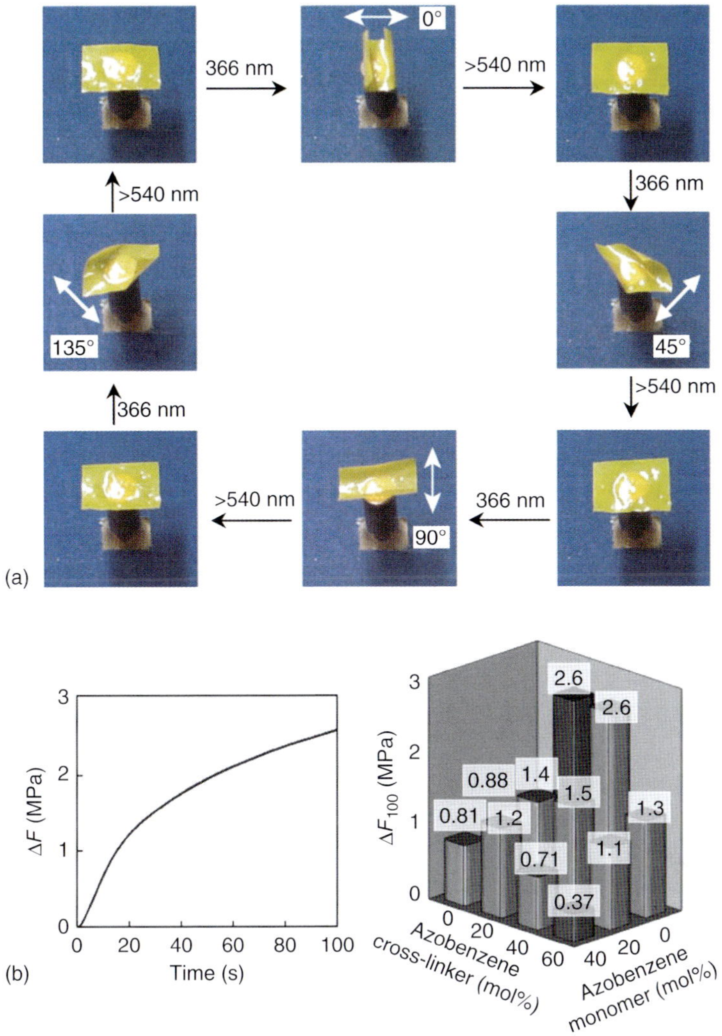

Figure 8.6 Photomechanical effects of CLCP films with polyacrylate backbone. (a) Photoinduced deformation of a polydomain CLCP film upon irradiation with linearly polarized UV light. Source: Yu et al. 2003 [24]. Reproduced with permission of Macmillan Publishers. (b) Photoinduced stress of monodomain CLCP films upon UV light irradiation (366 nm, 20 mW cm^{-2}), which were prepared with various concentration of azobenzene monomers and cross-linkers. Film size: 5 mm × 7 mm × 20 μm. Source: Kondo et al. 2010 [48]. Reproduced with permission of The Royal Society of Chemistry.

surface region of the film, which results in bending of the whole sample. Moreover, photoactuation of polydomain films has also been realized by the use of linearly polarized UV light (Figure 8.6a) [24]. In this case, azobenzene moieties oriented parallel with respect to the polarized direction of incident light efficiently absorb light. Thus, the direction of surface contraction, and consequently the bending axis, can be selected through the polarization state of the incident

UV light. Photoinduced bending behavior has been reproduced by theoretical models [39–44]. Finite element method is a powerful approach to predict and visualize 3D deformation [45, 46].

Concentration of azobenzene moieties in CLCPs can be controlled by introducing photoinert mesogens such as phenyl benzoate [47, 48]. The penetration depth of the incident light is an essential factor in determining photomechanical properties. In addition, the position of azobenzene moieties also affects the photoresponsive properties of CLCPs [48]. It has been found that azobenzene moieties at cross-links largely contribute to deformation and force, whereas those at noncross-linked side chains show photoisomerization with high efficiency. In optimized composition, photoinduced stress of 2.6 MPa has been reported (Figure 8.6b) [48].

Various substituents have been introduced into azobenzene to control the absorption wavelength and the rate of thermal relaxation. Azotolane shows absorption peak of π–π^* transition at longer wavelength compared with azobenzene due to the larger π-conjugation length, which enables to effectively induce trans–cis isomerization with visible light [49, 50]. Photoactuation of CLCP films with azotolane moieties has been demonstrated with sunlight through color filters. Azobenzene moieties substituted with fluorine atoms at *ortho*-position show separation of n–π^* absorption bands of trans- and cis-isomers [51, 52]. Continuous chaotic oscillation has been observed for CLCP films with fluorinated azobenzene moieties upon exposure to ambient sunlight due to continuous forward and backward switching [53]. Furthermore, fluorinated azobenzenes show prolonged half-life of cis-isomers, which enables to preserve the photodeformed state for more than eight days [54]. On the contrary, azobenzene moieties with push–pull substitution patterns with strong electron-donating and -withdrawing substituents, such as amino and nitro groups, exhibit short thermal half-life of cis-isomers [55]. Formation of tautomerizable azohydrozone also reduces the stability of cis-isomers [110]. CLCPs with these azobenzene moieties immediately revert to initial shapes after switching off the actinic light. Applying this strategy, continuous wave-like motions have recently been realized [56].

Initial alignment of mesogens can be controlled upon polymerization through the alignment layers coated on the inner surfaces of cells. Even CLCP films with the same chemical composition can exhibit different deformation depending on the initial alignment of mesogens. A film with homogeneous alignment (parallel to the long axis of the film) shows bending toward the light source as described previously. On the other hand, a film with homeotropic alignment (normal to the film surface) shows bending away from the light source when irradiated with visible light, which is attributed to the expansion of the film surface (Figure 8.7a) [57]. Katsonis and coworkers extensively studied the photoinduced deformation of CLCP films with twist-nematic alignment, enabling winding and unwinding of spring, and helix inversion, mimicking the motions of plant tendrils [58, 59].

Photoalignment technique enabled the preparation of CLCP films with complex alignment of mesogens, which can be controlled through the alignment layers that are patterned through the irradiation with linearly polarized light (Figure 8.7b) [60]. Broer and coworkers succeeded in preparation of CLCP films with radial and azimuthal alignment of mesogens, which deformed into

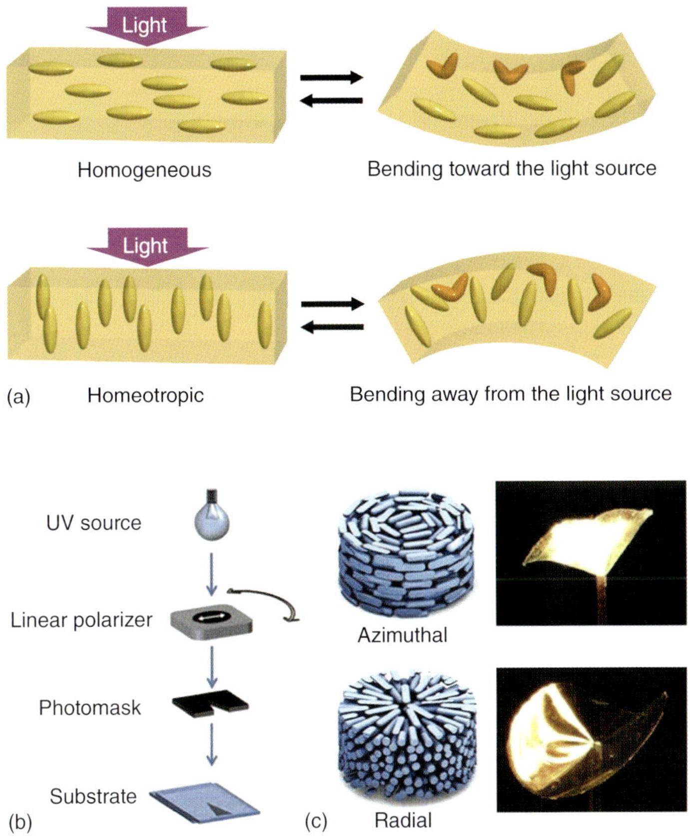

Figure 8.7 (a) Schematic illustration of photoinduced bending of CLCP films with homogeneous and homeotropic alignment. (b) Setup for the preparation of patterned alignment cells. (c) Deformation of CLCP films with azimuthal and radial alignments upon heating with an IR lamp. (b,c) Source: de Haan et al. 2012 [61]. Reproduced with permission of John Wiley & Sons.

cone and saddle shapes upon heating with an infrared (IR) lamp (Figure 8.7c) [61]. They also fabricated films with strips and checkerboard patterns, showing accordion-like folding and periodic buckling [62].

Various fabrication methods of CLCPs have been developed instead of in situ polymerization. As CLCPs are insoluble and infusible due to the presence of chemical cross-links, the shape of samples obtained by in situ polymerization using cells has been restricted to films. Post-cross-linking of linear acrylates or methacrylates has been demonstrated under alignment control by external force or alignment layers, using polymer reaction [63–65], photocross-linking [66], and electron beam irradiation [67]. Melt drawing of linear polymers results in fibers with uniaxial alignment of mesogens [63–66]. The bending direction of fibers can be selected through the incident direction of light due to cylindrical symmetry. Moreover, fabrication of a CLCP film with large size (14 cm × 14 cm) has been demonstrated using electron beam irradiation [67]. Inkjet printing has also been applied for preparation of micrometer-sized CLCPs, mimicking the motion of natural cilia [68].

8.1.3 Polyacrylate Elastomers Prepared from LC Macromers

Recently, novel synthetic procedures for polyacrylate networks have been developed to obtain CLCPs showing both large actuation strain (similar to conventional polysiloxanes) and versatile alignment (similar to conventional polyacrylates). White and coworkers applied a two-step reaction procedure: chain extension reaction through Michael addition to form LC macromers, and cross-linking with radical polymerization (Figure 8.8) [69, 70]. They used nematic diacrylates and a primary amine as starting materials. Importantly, these reactions can be conducted under control with alignment layer, enabling complicated patterning in combination with photoalignment technique. Topological defects were imprinted in these LC elastomers, which exhibited large out-of-plane deformation (100 times taller than the initial film thickness of 50 μm) along with the change in temperature. Thiol-ene reaction has also been applied for chain extension reaction instead of Michael addition [71]. Stacking of these films into layers enabled the thermal generation of large force. The lifting of 56 g of load over 0.4 mm was demonstrated with 26 mg of the stack of LC elastomer films [72]. The out-of-plane deformation has also been realized by photoactuation for LC elastomers containing azobenzene moieties (Figure 8.8) [73]. Conical domes within a film can be selectively actuated with the aid of spatial selectivity of light.

These synthetic routes for preparing CLCPs from LC oligomers have also been applied to various fabrication processes, such as replica molding [74], fiber drawing [75], and 3D printing [76–80]. Flow-induced alignment can be effectively utilized by controlling viscosity through the molecular weight of LC oligomers. In fabrication with 3D printing, a mixture containing LC oligomer is extruded, followed by photoisomerization to form network structure and fix the alignment of mesogens. Although the main focus of 3D printing for CLCPs has been thermoactuation systems, this technology is also promising for photoresponsive systems with complicated architecture.

8.1.4 Systems with Multiple Polymer Components

In CLCPs containing photochromic moieties with high concentration, light is mainly absorbed near the surface of the film. Consequently, chromophores located deeply within a film remain inactive. Therefore, a CLCP film can be replaced with a bilayer structure of a CLCP layer and a photoinert substrate. As CLCPs with high cross-linking density tend to be brittle, application of flexible substrates such as polyethylene improves mechanical properties of photoactuators. A light-driven plastic motor has been realized with a bilayer film of a photodeformable CLCP and polyethylene (Figure 8.9) [81]. Improvement in mechanical properties allows photogeneration of large stress without fracture, which is especially advantageous in photoactuation with high intensity (Figure 8.9b). Simultaneous irradiation with UV and visible light led to rotation of the belt and the pulleys (Figure 8.9c). The bending of the UV irradiated part

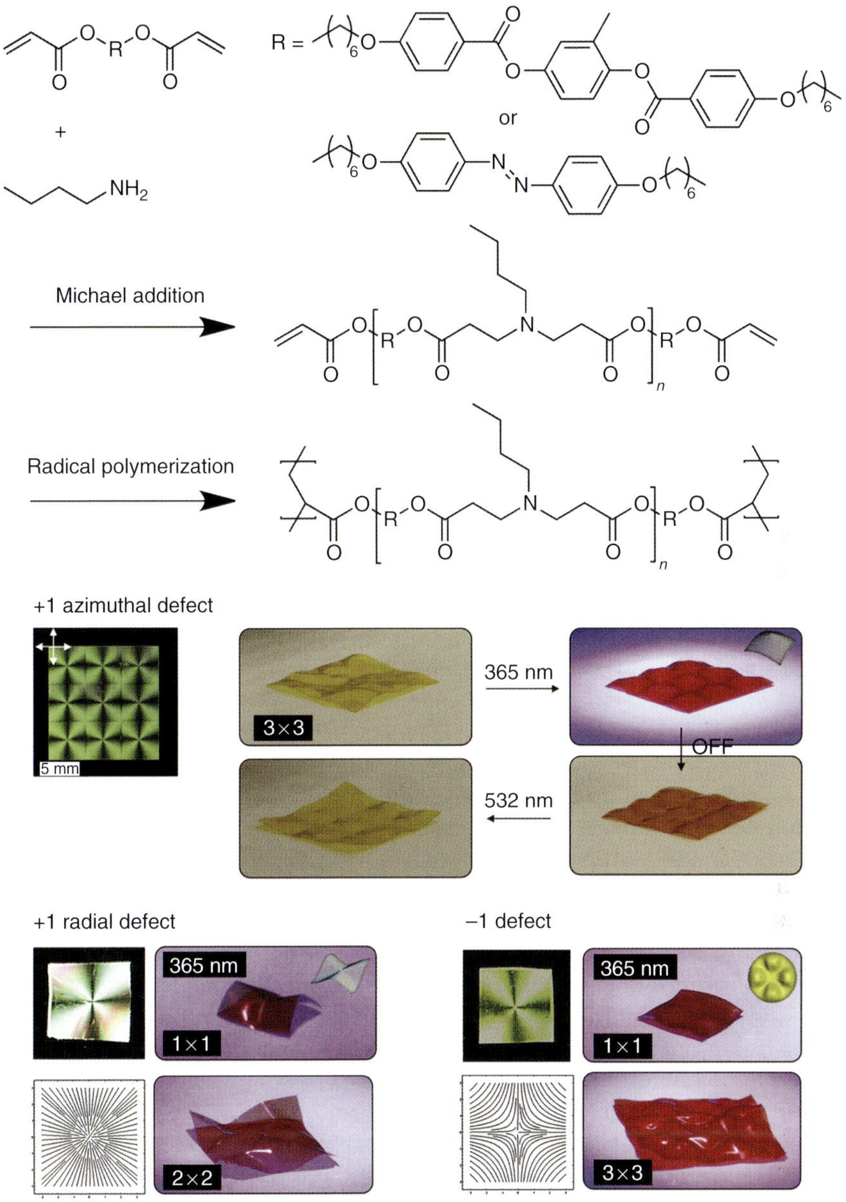

Figure 8.8 Synthetic procedure and photoinduced deformation of CLCPs prepared via LC oligomers under patterned alignment. Source: Ahn et al. 2016 [73]. Reproduced with permission of John Wiley & Sons.

generates torque on the small pulley, which leads to rotation. In addition, CLCP films can be laminated at multiple parts of a polyethylene film to enable motions like a robot arm with a high degree of freedom (Figure 8.9d) [82–84]. Photomanipulation of objects using these artificial robot arms has been demonstrated [83, 84]. Combination of CLCP films with layers containing upconversion agents allows photoactuation with red light, which is advantageous in biological applications due to high penetration ability into tissues and a negligible thermal effect [85].

Mechanical properties of polymer materials can be improved by the use of multiple polymer components. Interpenetrating polymer network (IPN) is a combination of two or more polymer networks synthesized in juxtaposition [86]. The properties of IPN films are controlled through the selection of constituent polymers and their composition. Photomobile polymer materials with IPN structure were developed by introducing amorphous polymers into CLCP networks [87, 88]. Sequential formation of CLCPs and amorphous polymer networks yields IPNs without disturbing the alignment of mesogens in CLCPs. Mechanical and photoresponsive properties were investigated for IPNs containing various poly(alkyl methacrylate)s (Figure 8.10) [87]. Young's modulus of IPN films could be controlled by changing the amorphous polymer components. IPN films containing poly(dodecyl methacrylate) (PDDMA) showed fast bending behavior upon UV light irradiation, which is attributed to low elastic modulus of PDDMA. Furthermore, IPN films containing poly(dimethylsiloxane) were found to exhibit much faster bending behavior than pristine CLCP films [88].

8.1.5 Composites

Composite systems of CLCPs and carbon nanotubes (CNTs) have been extensively studied. CNTs absorb light over a wide range of wavelength and convert it directly into local heat [89, 90]. This photothermal effect has been applied to induce thermal phase transition and deformation of CLCPs (Figure 8.11a). CNTs can be homogeneously dispersed in CLCP matrices with the aid of adequate chemical treatments. CNT/CLCP composites show deformation when irradiated with visible to IR light with an increase in temperature of the samples. When the actinic light is switched off, the samples revert to the initial shapes along with heat transfer to the surroundings. A bilayer sample of CLCP/silicone exhibited fast bending motion with large strain (Figure 8.11b), which could be applied to induce sophisticated motions resembling robots [91]. Composite systems with gold nanoparticles were also applied for actuation of CLCP microparticles through photothermal effects (Figure 8.11c) [92]. Irradiation with a focused IR laser beam on a submicrometer region within the microparticle induced deformation into various shapes. Furthermore, photoactuation of CLCPs with near-IR light has been demonstrated for CLCPs containing upconversion nanophosphors, which consist of lanthanide dopants embedded in a nanocrystalline host lattice [93].

The combination of azobenzene CLCPs and CNT sheets was reported to enhance mechanical toughness [94, 95]. In this system, the CNT sheet serves as an alignment layer to induce homogeneous alignment of mesogens. The

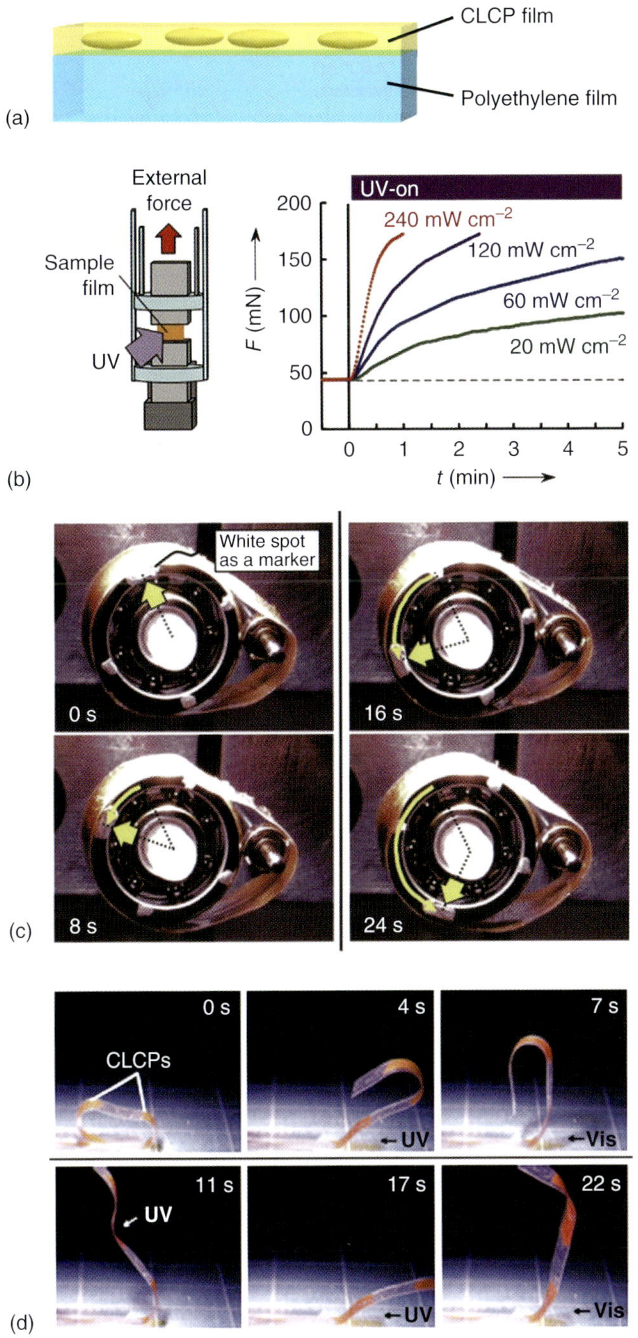

Figure 8.9 Bilayer films composed of CLCP and polyethylene layers. (a) Schematic illustration. Source: Yamada et al. 2008 [81]. Reproduced with permission of John Wiley & Sons. (b) Photoinduced force upon UV light irradiation (366 nm). Film size: 2.5 mm × 5 mm; thickness of CLCP layer: 15 μm; thickness of polyethylene layer: 50 μm. (c) Light-driven plastic motor. (b,c) Source: Yamada et al. 2008 [81]. Reproduced with permission of Wiley-VCH. (d) Light-driven robotic arm. Source: Yamada et al. 2009 [82]. Reproduced with permission of The Royal Society of Chemistry.

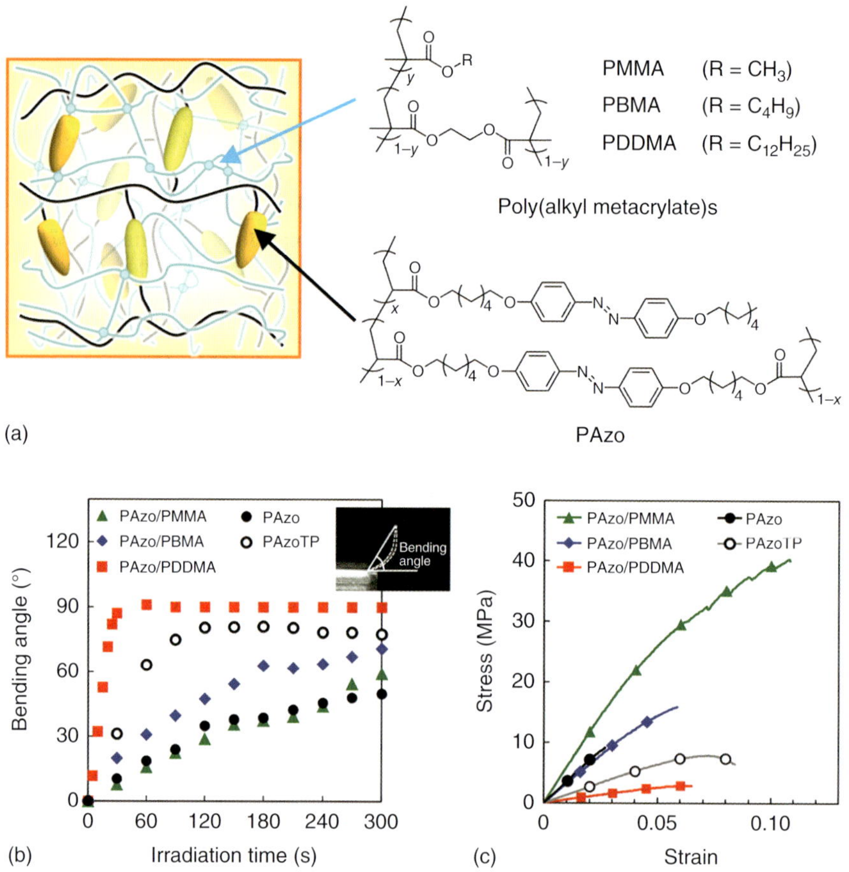

Figure 8.10 Photomobile polymer materials with IPN structure. (a) Schematic illustration. (b) Photoinduced bending behavior of films upon UV light irradiation (366 nm, 10 mW cm^{-2}). Film size: 3 mm × 1 mm × 16 μm. (c) Stress–strain curves of films upon stretching along the director axis of azobenzene moieties. Source: Ube et al. 2017 [88]. Reproduced with permission of The Royal Society of Chemistry.

resultant composite film showed bending and unbending behavior upon alternate irradiation with UV and visible light. The introduction of aligned CNTs remarkably increased the tensile strength up to 30 MPa along the CNT-aligned direction and provided electrical conductivity for the CLCP film.

8.1.6 Linear Polymers

Network structures of CLCPs have been conventionally formed by covalent bonds, which leads to insolubility. Cross-linking with noncovalent bonds allows better processing of LCPs (Figure 8.12). Hydrogen-bonded networks were fabricated from linear LCPs containing carboxyl group (hydrogen bond donors) at side chains and azobenzene derivatives functionalized with two pyridyl rings (hydrogen bond acceptors) [96]. Photoinduced bending of a film was successfully observed even in the absence of covalently bonded cross-links. The film was soluble in tetrahydrofuran (THF), enabling the recycle of linear polymers and cross-linkers. Covalently bonded cross-links can also be replaced to physical cross-links [97]. Yu and coworkers applied olefin metathesis polymerization to develop linear LCPs with mechanical robustness to enable manipulation of fluid slugs in LCP microtubes [98], which is detailed in Chapter 9.

Cross-links with noncovalent bonds have been also applied to linear main chain LCPs. Photoinduced deformation was reported for linear LCPs with polyurethane [99, 100] and polyester [101] backbones, which form cross-links through hydrogen bonding and π–π interaction, respectively.

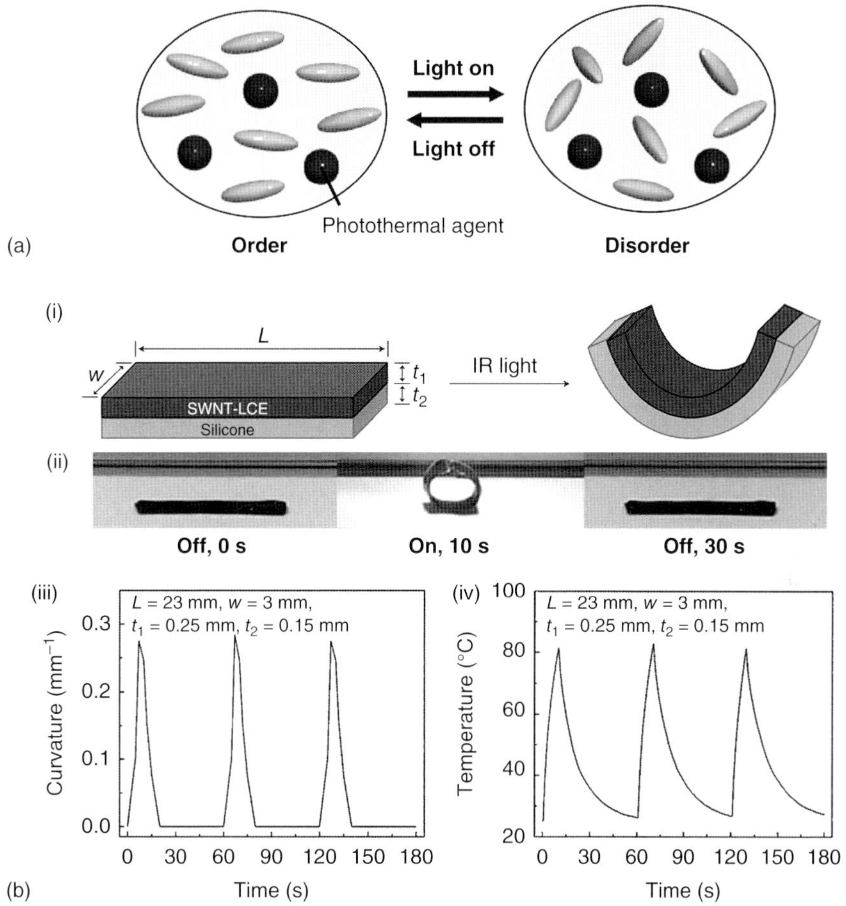

Figure 8.11 Composite films of CLCP and photothermal agents. (a) Schematic illustration of phase transition induced by photothermal process. (b) Bilayer films composed of CLCP with dispersed CNTs and silicone. (i) Schematic illustration. (ii) Bending behavior upon exposure to NIR light (11 mW cm^{-2}). Time evolutions of (iii) curvature and (iv) temperature of the elastomer. Source: Kohlmeyer and Chen [91]. Reproduced with permission of John Wiley & Sons. (c) Deformation of CLCP microparticles doped with gold nanocrystals upon scanning a laser beam along blue arrows shown in the insets. Scale bar: 20 μm. Source: Sun et al. 2012 [92]. Reproduced with permission of The American Institute of Physics.

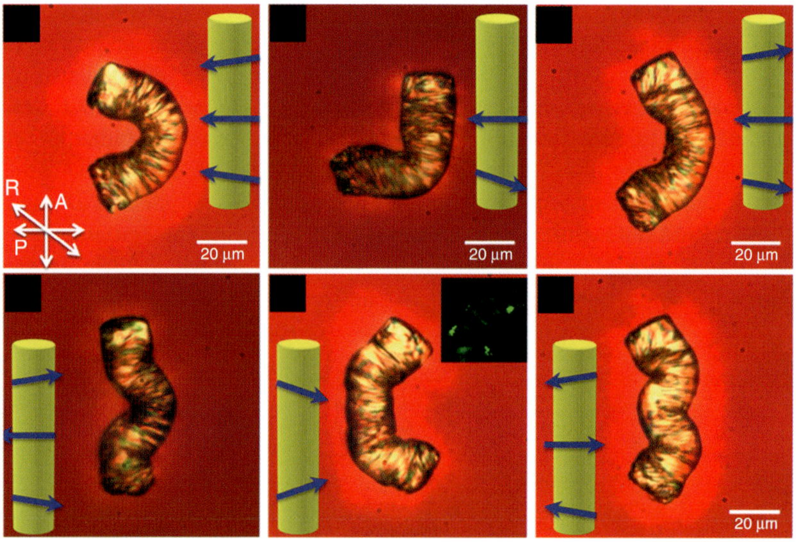

(c)

Figure 8.11 (*Continued*)

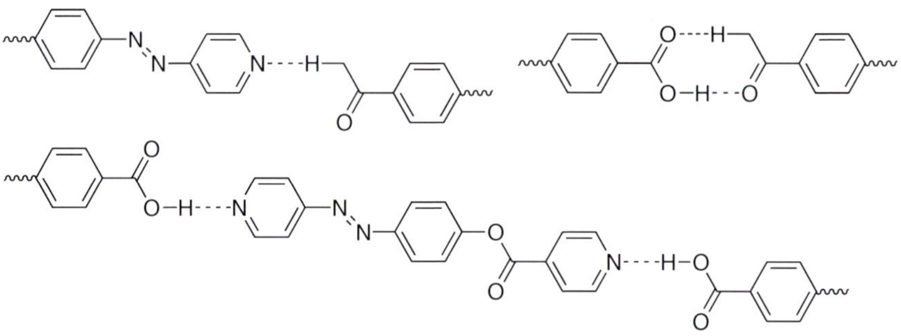

Figure 8.12 Typical examples of cross-links formed by hydrogen bonds.

8.1.7 Rearrangeable Network with Dynamic Covalent Bonds

Incorporation of dynamic covalent bonds (covalent bonds capable of exchange reactions) provides another way to enhance processability of CLCPs. Leibler and coworkers enabled to reshape and remold cross-linked polymers with the aid of rearrangeable networks [102, 103]. They prepared covalently bonded epoxy networks, which contain ester and hydroxy groups. Upon heating in the presence of catalyst, the network topology is rearranged through transesterification reaction (Figure 8.13a). These materials can be regarded as a new class of polymer materials in addition to conventional thermoplastic polymer and thermoset polymer.

The strategy of network rearrangement has been applied to reshaping of CLCPs. Ji and coworkers incorporated biphenyl groups as mesogens in an epoxy network to fabricate thermoresponsive CLCPs [104, 105]. A polydomain film is stretched under heating to induce uniaxial alignment of mesogens, and further heating allows to rememorize the alignment of mesogens to yield a

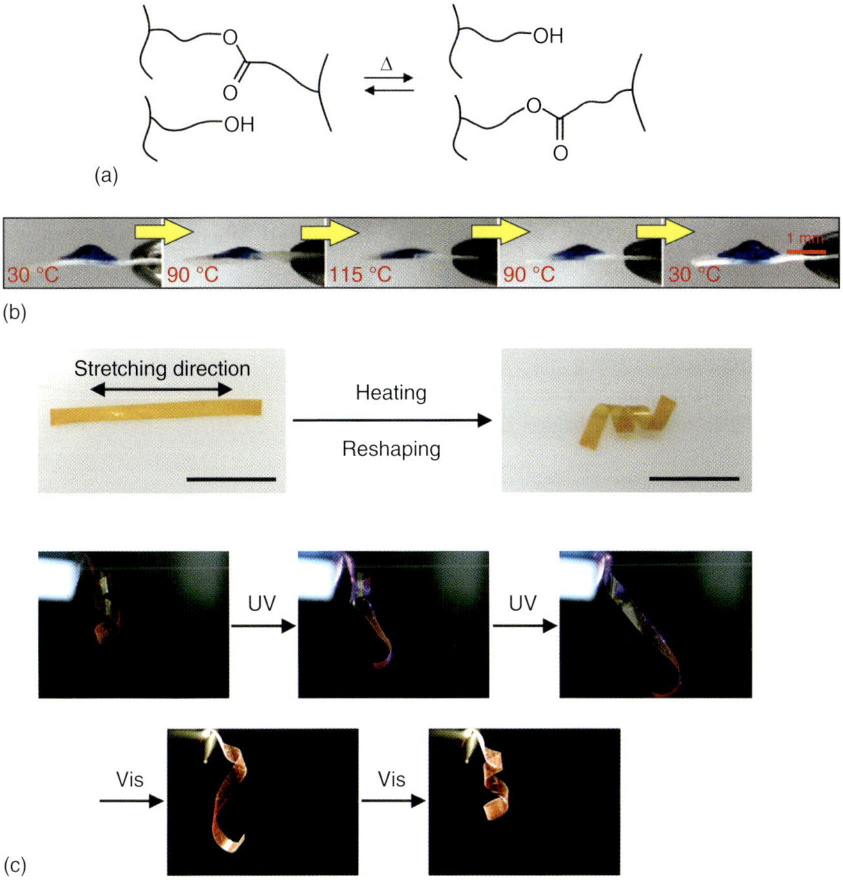

Figure 8.13 Rearrangeable networks with dynamic covalent bonds. (a) Schematic illustration of the rearrangement of network topology through transesterification. (b) Deformation of a dome-shaped sample upon heating. Source: Pei et al. 2014 [104]. Reproduced with permission of Macmillan Publishers. (c) Reshaping and photoactuation of a CLCP film with polysiloxane backbone containing exchangeable links. Source: Ube et al. 2016 [106]. Reproduced with permission of John Wiley & Sons.

monodomain film through the rearrangement of network structure triggered by transesterification. The monodomain film showed contraction and expansion with change in temperature. Furthermore, films were remolded into various 3D shapes. A dome-shaped sample could be reversibly switched to flat shape upon heating (Figure 8.13b).

Reshaping through the network rearrangement has also been demonstrated for polysiloxane CLCPs containing azobenzene moieties (Figure 8.13c) [106]. Introducing phenyl-OH groups into polysiloxane CLCPs with phenyl benzoate cross-linkers allowed reshaping and realignment under heating without catalyst. A monodomain film showed reversible bending behavior upon irradiation with UV and visible light. The monodomain film could be reshaped into spiral and helicoid under heating at LC temperature. The spiral sample showed unwinding motion upon UV light irradiation along with contraction of the outer surface (Figure 8.13c). The helicoidal sample could be bent toward the UV light source at various directions due to the identical flexural rigidity regardless of azimuthal angles. Photoactuation of epoxy-based rearrangeable networks has also been realized by introducing azobenzene moieties [107–109]. Zhao and coworkers investigated the photoresponsive properties of an epoxy-based system, on which was imposed high uniaxial strain of 100–200% for realignment of mesogens [108, 109]. They found that the photoinduced deformation is triggered not only by photoisomerization of azobenzene moieties but also by photothermal relaxation of polymer chains. Synergetic use of these effects improved photoinduced motion of reshaped samples such as wheels and spring-like motors.

8.2 Summary

Variation in the structure of photomobile polymer materials has been significantly enhanced in addition to conventional polysiloxanes and polyacrylates. Programming of 3D motions through the control of initial molecular alignment is a great advantage of these systems. Advances in fabrication processes of LCPs allow flexible design of soft actuators with sufficient photoresponsive and mechanical properties, and demanded architectures. Further breakthrough would be enabled by adequate design of photomechanical systems based on multidisciplinary studies between material science and robotics.

References

1 Rich, S.I., Wood, R.J., and Majidi, C. (2018). Untethered soft robotics. *Nat. Electron.* 1: 102–112.

2 Palagi, S. and Fischer, P. (2018). Bioinspired microrobots. *Nat. Rev. Mater.* 3: 113–124.

3 White, T.J. (2017). *Photomechanical Materials, Composites, and Systems: Wireless Transduction of Light into Work*. Hoboken, NJ: Wiley.

4 Merian, E. (1966). Steric factors influencing the dyeing of hydrophobic fibers. *Text. Res. J.* 36: 612–618.

5 Irie, M. and Kungwatchakun, D. (1984). Photoresponsive polymers. Mechanochemistry of polyacrylamide gels having triphenylmethane leuco derivatives. *Macromol. Chem. Rapid Commun.* 5: 829–832.

6 Irie, M. (1986). Photoresponsive polymers – Reversible bending of rod-shaped acrylamide gels in an electric-field. *Macromolecules* 19: 2890–2892.

7 Suzuki, A. and Tanaka, T. (1990). Phase-transition in polymer gels induced by visible-light. *Nature* 346: 345–347.

8 Eisenbach, C.D. (1980). Isomerization of aromatic azo chromophores in poly(ethyl acrylate) networks and photomechanical effect. *Polymer* 21: 1175–1179.

9 Zimmermann, E. and Stille, J. (1985). Photoresponsive polyquinolines. *Macromolecules* 18: 321–327.

10 Warner, M. and Terentjev, E.M. (2003). *Liquid Crystal Elastomers*. Oxford: Oxford University Press.
11 Ohm, C., Brehmer, M., and Zentel, R. (2010). Liquid crystalline elastomers as actuators and sensors. *Adv. Mater.* 22: 3366–3387.
12 Finkelmann, H., Kock, H.-J., Gleim, W., and Rehage, G. (1984). Investigations on liquid crystalline polysiloxanes, 5. Orientation of LC-elastomers by mechanical forces. *Macromol. Chem. Rapid Commun.* 5: 287–293.
13 Zentel, R. (1989). Liquid crystalline elastomers. *Adv. Mater.* 1: 321–329.
14 de Gennes, P.G. (1975). Reflexions sur un type de polymeres nematiques. *C. R. Acad. Sci. B* 281: 101.
15 de Gennes, P.G., Hébert, M., and Kant, R. (1997). Artificial muscles based on nematic gels. *Macromol. Symp.* 113: 39–49.
16 Finkelmann, H., Kock, H., and Rehage, G. (1981). Investigations on liquid-crystalline polysiloxanes. 3. Liquid-crystalline elastomers – a new type of liquid-crystalline material. *Macromol. Chem. Rapid Commun.* 2: 317–322.
17 Küpfer, J. and Finkelmann, H. (1991). Nematic liquid single-crystal elastomers. *Macromol. Chem. Rapid Commun.* 12: 717–726.
18 Ikeda, T. (2003). Photomodulation of liquid crystal orientations for photonic applications. *J. Mater. Chem.* 13: 2037–2057.
19 Tazuke, S., Kurihara, S., and Ikeda, T. (1987). Amplified image recording in liquid-crystal media by means of photochemically triggered phase-transition. *Chem. Lett.* 16: 911–914.
20 Ikeda, T., Horiuchi, S., Karanjit, D. et al. (1990). Photochemically induced isothermal phase-transition in polymer liquid-crystals with mesogenic phenyl benzoate side-chains. 2. photochemically induced isothermal phase-transition behaviors. *Macromolecules* 23: 42–48.
21 Ikeda, T., Kurihara, S., Karanjit, D., and Tazuke, S. (1990). Photochemically induced isothermal phase-transition in polymer liquid-crystals with mesogenic cyanobiphenyl side-chains. *Macromolecules* 23: 3938–3943.
22 Ikeda, T. and Tsutsumi, O. (1995). Optical switching and image storage by means of azobenzene liquid-crystal films. *Science* 268: 1873–1875.
23 Finkelmann, H., Nishikawa, E., Pereira, G.G., and Warner, M. (2001). A new opto-mechanical effect in solids. *Phys. Rev. Lett.* 87: 015501.
24 Yu, Y., Nakano, M., and Ikeda, T. (2003). Directed bending of a polymer film by light. *Nature* 425: 145.
25 Ikeda, T., Nakano, M., Yu, Y. et al. (2003). Anisotropic bending and unbending behavior of azobenzene liquid-crystalline gels by light exposure. *Adv. Mater.* 15: 201–205.
26 Ikeda, T., Mamiya, J., and Yu, Y. (2007). Photomechanics of liquid-crystalline elastomers and other polymers. *Angew. Chem. Int. Ed.* 46: 506–528.
27 Ikeda, T. and Ube, T. (2011). Photomobile polymer materials: from nano to macro. *Mater. Today* 14: 480–487.
28 Palffy-Muhoray, P. (2012). Liquid crystal elastomers and light. In: *Liquid Crystal Elastomers: Materials and Applications*, vol. 250 (ed. W.H. DeJeu), 95–118. Berlin: Springer-Verlag.

29 Ube, T. and Ikeda, T. (2014). Photomobile polymer materials with crosslinked liquid- crystalline structures: molecular design, fabrication, and functions. *Angew. Chem. Int. Ed.* 53: 10290–10299.
30 White, T.J. and Broer, D.J. (2015). Programmable and adaptive mechanics with liquid crystal polymer networks and elastomers. *Nat. Mater.* 14: 1087–1098.
31 Hogan, P.M., Tajbakhsh, A.R., and Terentjev, E.M. (2002). UV manipulation of order and macroscopic shape in nematic elastomers. *Phys. Rev. E* 65: 041720.
32 Cviklinski, J., Tajbakhsh, A.R., and Terentjev, E.M. (2002). UV isomerisation in nematic elastomers as a route to photo-mechanical transducer. *Eur. Phys. J. E Soft Matter* 9: 427–434.
33 Sánchez-Ferrer, A., Merekalov, A., and Finkelmann, H. (2011). Opto-mechanical effect in photoactive nematic side-chain liquid-crystalline elastomers. *Macromol. Rapid Commun.* 32: 671–678.
34 Sánchez-Ferrer, A. and Finkelmann, H. (2013). Opto-mechanical effect in photoactive nematic main-chain liquid-crystalline elastomers. *Soft Matter* 9: 4621–4627.
35 Bergmann, G.H.F., Finkelmann, H., Percec, V., and Zhao, M.Y. (1997). Liquid-crystalline main-chain elastomers. *Macromol. Rapid Commun.* 18: 353–360.
36 Donnio, B., Wermter, H., and Finkelmann, H. (2000). Simple and versatile synthetic route for the preparation of main-chain, liquid-crystalline elastomers. *Macromolecules* 33: 7724–7729.
37 Wermter, H. and Finkelmann, H. (2001). Liquid crystalline elastomers as artificial muscles. *E-Polymers* 1: 111–123.
38 Broer, D., Finkelmann, H., and Kondo, K. (1988). In-situ photopolymerization of an oriented liquid-crystalline acrylate. *Macromol. Chem. Phys.* 189: 185–194.
39 Warner, M. and Mahadevan, L. (2004). Photoinduced deformations of beams, plates, and films. *Phys. Rev. Lett.* 92: 134302.
40 Corbett, D. and Warner, M. (2007). Linear and nonlinear photoinduced deformations of cantilevers. *Phys. Rev. Lett.* 99: 174302.
41 Corbett, D., van Oosten, C.L., and Warner, M. (2008). Nonlinear dynamics of optical absorption of intense beams. *Phys. Rev. A* 78: 013823.
42 van Oosten, C.L., Corbett, D., Davies, D. et al. (2008). Bending dynamics and directionality reversal in liquid crystal network photoactuators. *Macromolecules* 41: 8592–8596.
43 Corbett, D. and Warner, M. (2008). Polarization dependence of optically driven polydomain elastomer mechanics. *Phys. Rev. E* 78: 061701.
44 Modes, C.D., Warner, M., van Oosten, C.L., and Corbett, D. (2010). Anisotropic response of glassy splay-bend and twist nematic cantilevers to light and heat. *Phys. Rev. E* 82: 041111.
45 Dunn, M.L. (2007). Photomechanics of mono- and polydomain liquid crystal elastomer films. *J. Appl. Phys.* 102: 013506.
46 Zhu, W., Shelley, M., and Palffy-Muhoray, P. (2011). Modeling and simulation of liquid–crystal elastomers. *Phys. Rev. E* 83: 051703.

47 Harris, K.D., Cuypers, R., Scheibe, P. et al. (2005). Large amplitude light-induced motion in high elastic modulus polymer actuators. *J. Mater. Chem.* 15: 5043–5048.
48 Kondo, M., Sugimoto, M., Yamada, M. et al. (2010). Effect of concentration of photoactive chromophores on photomechanical properties of crosslinked azobenzene liquid-crystalline polymers. *J. Mater. Chem.* 20: 117–122.
49 Yin, R., Xu, W., Kondo, M. et al. (2009). Can sunlight drive the photoinduced bending of polymer films? *J. Mater. Chem.* 19: 3141–3143.
50 Cheng, F., Zhang, Y., Yin, R., and Yu, Y. (2010). Visible light induced bending and unbending behavior of crosslinked liquid-crystalline polymer films containing azotolane moieties. *J. Mater. Chem.* 20: 4888–4896.
51 Bleger, D., Schwarz, J., Brouwer, A.M., and Hecht, S. (2012). *o*-Fluoroazobenzenes as readily synthesized photoswitches offering nearly quantitative two-way isomerization with visible light. *J. Am. Chem. Soc.* 134: 20597–20600.
52 Knie, C., Utecht, M., Zhao, F. et al. (2014). *ortho*-Fluoroazobenzenes: visible light switches with very long-lived *Z* isomers. *Chem. Eur. J.* 20: 16492–16501.
53 Kumar, K., Knie, C., Bléger, D. et al. (2016). A chaotic self-oscillating sunlight-driven polymer actuator. *Nat. Commun.* 7: 11975.
54 Iamsaard, S., Anger, E., Aßhoff, S.J. et al. (2016). Fluorinated azobenzenes for shape-persistent liquid crystal polymer networks. *Angew. Chem. Int. Ed.* 55: 9908–9912.
55 Barrett, C.J., Mamiya, J., Yager, K.G., and Ikeda, T. (2007). Photo-mechanical effects in azobenzene-containing soft materials. *Soft Matter* 3: 1249–1261.
56 Gelebart, A.H., Mulder, D.J., Varga, M. et al. (2017). Making waves in a photoactive polymer film. *Nature* 546: 632–636.
57 Kondo, M., Yu, Y., and Ikeda, T. (2006). How does the initial alignment of mesogens affect the photoinduced bending behavior of liquid-crystalline elastomers? *Angew. Chem. Int. Ed.* 45: 1378–1382.
58 Iamsaard, S., Aßhoff, S.J., Matt, B. et al. (2014). Conversion of light into macroscopic helical motion. *Nat. Chem.* 6: 229–235.
59 Aßhoff, S.J., Lancia, F., Iamsaard, S. et al. (2017). High-power actuation from molecular photoswitches in enantiomerically paired soft springs. *Angew. Chem. Int. Ed.* 56: 3261–3265.
60 Yaroshchuk, O. and Reznikov, Y. (2012). Photoalignment of liquid crystals: basics and current trends. *J. Mater. Chem.* 22: 286–300.
61 de Haan, L.T., Sánchez-Somolinos, C., Bastiaansen, C.M.W. et al. (2012). Engineering of complex order and the macroscopic deformation of liquid crystal polymer networks. *Angew. Chem. Int. Ed.* 51: 12469–12472.
62 de Haan, L.T., Gimenez-Pinto, V., Konya, A. et al. (2014). Accordion-like actuators of multiple 3D patterned liquid crystal polymer films. *Adv. Funct. Mater.* 24: 1251–1258.
63 Deng, W., Li, M.-H., Wang, X., and Keller, P. (2009). Light-responsive wires from side-on liquid crystalline azo polymers. *Liq. Cryst.* 36: 1023–1029.

64 Yoshino, T., Kondo, M., Mamiya, J. et al. (2010). Three- dimensional photomobility of crosslinked azobenzene liquid-crystalline polymer fibers. *Adv. Mater.* 22: 1361–1363.
65 Lv, J., Wang, W., Wu, W., and Yu, Y. (2015). A reactive azobenzene liquid-crystalline block copolymer as a promising material for practical application of light-driven soft actuators. *J. Mater. Chem. C* 3: 6621–6626.
66 Liu, Y., Wu, W., Wei, J., and Yu, Y. (2017). Visible light responsive liquid crystal polymers containing reactive moieties with good processability. *ACS Appl. Mater. Interfaces* 9: 782–789.
67 Naka, Y., Mamiya, J., Shishido, A. et al. (2011). Direct fabrication of photomobile polymer materials with an adhesive-free bilayer structure by electron-beam irradiation. *J. Mater. Chem.* 21: 1681–1683.
68 van Oosten, C.L., Bastiaansen, C.W.M., and Broer, D.J. (2009). Printed artificial cilia from liquid-crystal network actuators modularly driven by light. *Nat. Mater.* 8: 677–682.
69 Ware, T.H., McConney, M.E., Wie, J.J. et al. (2015). Voxelated liquid crystal elastomers. *Science* 347: 982–984.
70 Fuchi, K., Ware, T.H., Buskohl, P.R. et al. (2015). Topology optimization for the design of folding liquid crystal elastomer actuators. *Soft Matter* 11: 7288–7295.
71 Ware, T.H., Perry, Z.P., Middleton, C.M. et al. (2015). Programmable liquid crystal elastomers prepared by thiol-ene photopolymerization. *ACS Macro Lett.* 4: 942–946.
72 Guin, T., Settle, M.J., Kowalski, B.A. et al. (2018). Layered liquid crystal elastomer actuators. *Nat. Commun.* 9: 2531.
73 Ahn, S.-K., Ware, T.H., Lee, K.M. et al. (2016). Photoinduced topographical feature development in blueprinted azobenzene-functionalized liquid crystalline elastomers. *Adv. Funct. Mater.* 26: 5819–5826.
74 Yang, H., Buguin, A., Taulemesse, J.-M. et al. (2009). Micron-sized main-chain liquid crystalline elastomer actuators with ultralarge amplitude contractions. *J. Am. Chem. Soc.* 131: 15000–15004.
75 Gelebart, A.H., Bride, M.M., Schenning, A.P.H.J. et al. (2016). Photoresponsive fiber array: toward mimicking the collective motion of cilia for transport applications. *Adv. Funct. Mater.* 26: 5322–5327.
76 Ambulo, C.P., Burroughs, J.J., Boothby, J.M. et al. (2017). Four-dimensional printing of liquid crystal elastomers. *ACS Appl. Mater. Interfaces* 9: 37332–37339.
77 Kotikian, A., Truby, R.L., Boley, J.W. et al. (2018). 3D printing of liquid crystal elastomeric actuators with spatially programed nematic order. *Adv. Mater.* 30: 1706164.
78 López-Valdeolivas, M., Liu, D., Broer, D.J., and Sánchez-Somolinos, C. (2018). 4D printed actuators with soft-robotic functions. *Macromol. Rapid Commun.* 39: 1700710.
79 Hagaman, D.E., Leist, S., Zhou, J., and Ji, H.-F. (2018). Photoactivated polymeric bilayer actuators fabricated via 3D printing. *ACS Appl. Mater. Interfaces* 10: 27308–27315.

80 Kuang, X., Roach, D.J., Wu, J. et al. (2019). Advances in 4D printing: materials and applications. *Adv. Funct. Mater.* 29: 1805290.

81 Yamada, M., Kondo, M., Mamiya, J. et al. (2008). Photomobile polymer materials: towards light-driven plastic motors. *Angew. Chem. Int. Ed.* 47: 4986–4988.

82 Yamada, M., Kondo, M., Miyasato, R. et al. (2009). Photomobile polymer materials – various three-dimensional movements. *J. Mater. Chem.* 19: 60–62.

83 Cheng, F., Yin, R., Zhang, Y. et al. (2010). Fully plastic microrobots which manipulate objects using only visible light. *Soft Matter* 6: 3447–3449.

84 Wei, J. and Yu, Y. (2012). Photodeformable polymer gels and crosslinked liquid-crystalline polymers. *Soft Matter* 8: 8050–8059.

85 Jiang, Z., Xu, M., Li, F., and Yu, Y. (2013). Red-light-controllable liquid-crystal soft actuators via low-power excited upconversion based on triplet–triplet annihilation. *J. Am. Chem. Soc.* 135: 16446–16453.

86 Sperling, L. (1994). Interpenetrating polymer networks – an overview. In: *Interpenetrating Polymer Networks*, vol. 239 (eds. D. Klempner, L.H. Sperling and L.A. Utracki), 3–38. Washington, DC: American Chemical Society.

87 Ube, T., Takado, K., and Ikeda, T. (2015). Photomobile materials with interpenetrating polymer networks composed of liquid-crystalline and amorphous polymers. *J. Mater. Chem. C* 3: 8006–8009.

88 Ube, T., Minagawa, K., and Ikeda, T. (2017). Interpenetrating polymer networks of liquid- crystalline azobenzene polymers and poly(dimethylsiloxane) as photomobile materials. *Soft Matter* 13: 5820–5823.

89 Koerner, H., Price, G., Pearce, N.A. et al. (2004). Remotely actuated polymer nanocomposites – stress-recovery of carbon-nanotube-filled thermoplastic elastomers. *Nat. Mater.* 3: 115–120.

90 Ahir, S.V. and Terentjev, E.M. (2005). Photomechanical actuation in polymer-nanotube composites. *Nat. Mater.* 4: 491–495.

91 Kohlmeyer, R.R. and Chen, J. (2013). Wavelength-selective, IR light-driven hinges based on liquid crystalline elastomer composites. *Angew. Chem. Int. Ed.* 52: 9234–9237.

92 Sun, Y., Evans, J.S., Lee, T. et al. (2012). Optical manipulation of shape-morphing elastomeric liquid crystal microparticles doped with gold nanocrystals. *Appl. Phys. Lett.* 100: 241901.

93 Wu, W., Yao, L., Yang, T. et al. (2011). NIR-light-induced deformation of cross-linked liquid-crystal polymers using upconversion nanophosphors. *J. Am. Chem. Soc.* 133: 15810–15813.

94 Wang, W., Sun, X., Wu, W. et al. (2012). Photoinduced deformation of crosslinked liquid-crystalline polymer film oriented by a highly aligned carbon nanotube sheet. *Angew. Chem. Int. Ed.* 51: 4644–4647.

95 Sun, X., Wang, W., Qiu, L. et al. (2012). Unusual reversible photomechanical actuation in polymer/nanotube composites. *Angew. Chem. Int. Ed.* 51: 8520–8524.

96 Mamiya, J., Yoshitake, A., Kondo, M. et al. (2008). Is chemical crosslinking necessary for the photoinduced bending of polymer films? *J. Mater. Chem.* 18: 63–65.

97 Petr, M., Katzman, B.-a., DiNatale, W., and Hammond, P.T. (2013). Synthesis of a new, low-T_g siloxane thermoplastic elastomer with a functionalizable backbone and its use as a rapid, room temperature photoactuator. *Macromolecules* 46: 2823–2832.
98 Lv, J.-a., Liu, Y., Wei, J. et al. (2016). Photocontrol of fluid slugs in liquid crystal polymer microactuators. *Nature* 537: 179–184.
99 Chen, S., Yuan, H., Zhuo, H. et al. (2014). Development of liquid-crystalline shape-memory polyurethane composites based on polyurethane with semi-crystalline reversible phase and hexadecyloxybenzoic acid for self-healing applications. *J. Mater. Chem. C* 2: 4203–4212.
100 Ban, J., Mu, L., Yang, J. et al. (2017). New stimulus-responsive shape- memory polyurethanes capable of UV light-triggered deformation, hydrogen bond-mediated fixation, and thermal-induced recovery. *J. Mater. Chem. A* 5: 14514–14518.
101 Zhong, H.-Y., Chen, L., Yang, R. et al. (2017). Azobenzene-containing liquid crystalline polyester with π–π interactions: diverse thermo-and photo-responsive behaviours. *J. Mater. Chem. C* 5: 3306–3314.
102 Montarnal, D., Capelot, M., Tournilhac, F., and Leibler, L. (2011). Silica-like malleable materials from permanent organic networks. *Science* 334: 965–968.
103 Capelot, M., Montarnal, D., Tournilhac, F., and Leibler, L. (2012). Metal-catalyzed transesterification for healing and assembling of thermosets. *J. Am. Chem. Soc.* 134: 7664–7667.
104 Pei, Z., Yang, Y., Chen, Q. et al. (2014). Mouldable liquid-crystalline elastomer actuators with exchangeable covalent bonds. *Nat. Mater.* 13: 36–41.
105 Yang, Y., Pei, Z., Li, Z. et al. (2016). Making and remaking dynamic 3D structures by shining light on flat liquid crystalline vitrimer films without a mold. *J. Am. Chem. Soc.* 138: 2118–2121.
106 Ube, T., Kawasaki, K., and Ikeda, T. (2016). Photomobile liquid-crystalline elastomers with rearrangeable networks. *Adv. Mater.* 28: 8212–8217.
107 Li, Y., Rios, O., Keum, J.K. et al. (2016). Photoresponsive liquid crystalline epoxy networks with shape memory behavior and dynamic ester bonds. *ACS Appl. Mater. Interfaces* 8: 15750–15757.
108 Lu, X., Guo, S., Tong, X. et al. (2017). Tunable photocontrolled motions using stored strain energy in malleable azobenzene liquid crystalline polymer actuators. *Adv. Mater.* 29: 1606467.
109 Lu, X., Zhang, H., Fei, G. et al. (2018). Liquid-crystalline dynamic networks doped with gold nanorods showing enhanced photocontrol of actuation. *Adv. Mater.* 30: 1706597.
110 Brode, W. R., Gould, J. H., and Wyman, G. M. (1952). The relation between the absorption spectra and the chemical constitution of dyes. XXV. Phototropism and cis-trans isomerism in aromatic azo compounds. *J. Am. Chem. Soc.***74**: 4641–4646.

9

Photomechanical Liquid Crystal Polymers and Bioinspired Soft Actuators

Chongyu Zhu, Lang Qin, Yao Lu, Jiahao Sun, and Yanlei Yu

Fudan University, Department of Materials Science and State Key Laboratory of Molecular Engineering of Polymers, 220 Handan Road, Shanghai, 200433, China

9.1 Background

Actuators in a traditional process control system perform specific actions in cooperation with motors, gears, and connecting rods. Although these actuators play an important part in most machinery, a further weight and size reduction of these structures are desirable to improve the integration and minimization of current metal-based facilities. Due to the low-cost, lightweight, and easy-to-process polymers, fabrication of new generation actuators has been simplified, especially bioinspired ones used for flexible motions. With the development of new polymer materials, responsive polymers, also known as "smart" polymers materials, have demonstrated the potentials in soft actuators thanks to their unique stimuli-responsive property. These polymers can change physical and/or chemical properties in response to the external stimuli including light [1], moisture [2], magnetism [3, 4], heat [5, 6], and pH [7, 8], providing alternative energy inputs for actuations.

Among these smart polymer materials, photoresponsive polymers are of particular interest for constructing soft actuators as light is a clean source that exerts contactless and precise control over materials, allowing the delicate motions of photoresponsive actuators for diverse functions. Particularly, a complex of polymers and liquid crystals (LCs), liquid crystal polymers (LCPs), exhibit unique properties such as elasticity, anisotropy, stimuli-responsiveness, and molecular cooperation effect. In response to external stimuli, the mesogens in the LCP realigned, coupled with the conformation of polymer backbones, to produce controllable and reversible macroscopic shape change of the whole sample. Incorporating photo-chromophores or photothermal particles endows the LCP with photoresponsiveness. With different photoresponsive mechanisms (i.e. photochemical phase transition, Weigert effect, and photothermal effect), various photomechanical actuators including motors, pumps, and valves have been fabricated using LCPs to act in similar functions as the traditional ones. Moreover, the photomechanical LCPs have been used to design more complex structures to mimic the motions from nature. Artificial worm, fish, iris, arm,

Mechanically Responsive Materials for Soft Robotics, First Edition. Edited by Hideko Koshima.

and other flexible actuators have successfully performed complex actions such as movement, high frequency vibration, gripping of specific objects, automatic control of luminous flux, precise control of small amounts of liquid, and so on. In this chapter, the photochemical mechanism of bioinspired actuators and some specific examples will be described in detail.

9.2 Actuation Principles

De Gennes proposed the possibility of using cross-linked liquid crystal polymers (CLCPs) as artificial muscles, by taking advantage of their significant contraction in the direction of the director axis (Figure 9.1a) [9]. The basic principle behind the reversible shape change lies in the conformational change of the polymer backbone upon the LC–isotropic phase transition [10–13]. The polymer chains in an anisotropic LC environment are different from those in the isotropic conformation, which leads to the distinct coil dimensions parallel and perpendicular to the LC director. If CLCPs lose anisotropic alignment order of LC, an isotropic chain conformation will be adopted and whole sample has to change its shape [14, 15]. What follows is a discussion of three classical mechanisms of shape change driven by light and different modes of complex deformation.

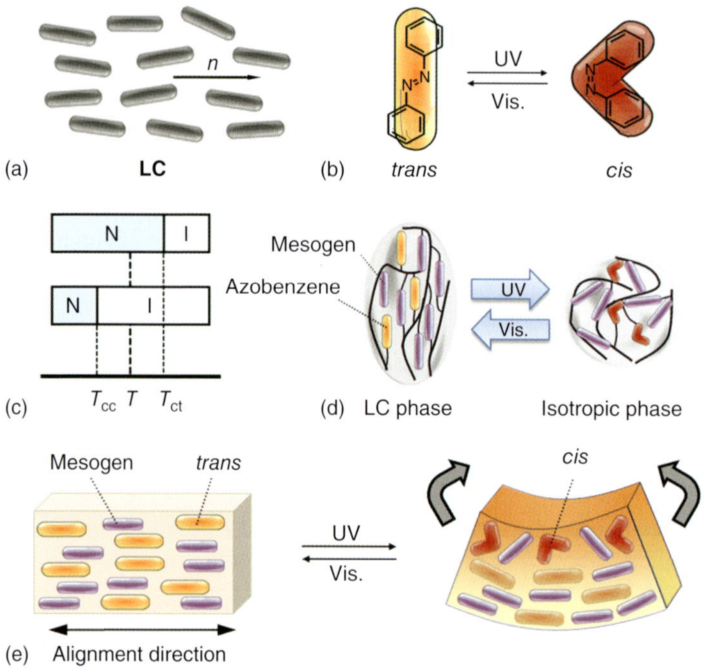

Figure 9.1 (a) Schematic illustration showing the direction of the director axis (n). (b) Schematic illustration showing photoisomerization of azobenzene driven by UV and visible light. (c) Phase diagrams of the photochemical phase transition of azobenzene/LC systems (N, nematic; I, isotropic). (d) Schematic illustration of reversible LC–isotropic photochemical phase transition and corresponding shape change in LC polymers. (e) Plausible mechanism of light-driven bending of a CLCP film based on photochenmical phase transition.

9.2.1 Photochemical Phase Transition

Cooperative motion of molecules in LC phases may be most advantageous in changing the molecular alignment. When photochromic mesogens, such as azobenzene, are incorporated into the polymer network and undergoes shape change upon light irradiation, the alignment of whole LC system will change because of the cooperative motion. This phenomenon illustrates that LC molecules only require a small amount of energy to change the alignment, i.e. the energy required for an alignment change of only 1 mol% LC molecules is enough to bring about the LC order change of the whole system.

Azobenzene is a well-known chromophore that can serve as a mesogen and a photoresponsive group in LCPs. It undergoes trans–cis and cis–trans isomerization upon exposure to ultraviolet (UV) and visible light (Figure 9.1b). The rodlike trans form of the azobenzene stabilizes the phase structure of the LC, whereas its bent cis form tends to disturb the LC order. As a consequence of two different configurations, the LC–isotropic phase transition temperature (T_c) of the cis form (T_{cc}) is much lower than that of trans form (T_{ct}). Therefore, when the sample is exposed to UV light, T_c of the sample gradually decreases as the cis form accumulates in the sample. If the sample temperature (T) is between T_{ct} and T_{cc}, the LC–isotropic phase transition of the sample will be induced once T_c is below T, giving rise to a general contraction along the alignment direction of the mesogens (Figure 9.1c). Through a similar process, the sample expands back to the initial size via the photochemical isotropic–LC phase transition induced by the cis–trans photoisomerizaton of azobenzene (Figure 9.1d) [16]. In addition to contraction and expansion, a three dimensional (3D) motion such as bending behavior of CLCP films can also be realized. Due to the large extinction coefficient of azobenzene, the surface region facing the incident light with a thickness <1 μm will absorb more than 99% incident photons when a high concentration of azobenzene molecules is present in the CLCP film. Thus, only the azobenzene on the film surface undergoes trans–cis photoisomerizaton, lowering the LC order, and leads to an anisotropic bending deformation of the film (Figure 9.1e). Depending on the initial alignment of mesogens in the CLCP films, exposure to UV light will lead to a different bending mode of the film, i.e. parallel aligned films will bend toward the light while homeotropic aligned films will bend away from the direction of the light irradiation [17].

Various monodomain CLCP films have been prepared by incorporating azobenzene chromophores in different forms (e.g. cross-linkers and monomers, Figure 9.2) [17–22] and exhibited photomechanical performances upon exposure to UV light with the aid of conventional LC cells. For example, Finkelmann et al. prepared an polysiloxane-based CLCP and demonstrated a 20% contraction upon UV irradiation [21]. In this case, the subtle variation in nematic LC order upon trans–cis photoisomerization causes a significant uniaxial deformation of the LCs along the director axis when the LC molecules are covalently cross-linked to form a polymer network.

The polydomain CLCP films can be considered as an integration of many microscale monodomains aligned in each direction. By means of selective absorption of linear polarized UV light, Ikeda and coworkers succeeded in

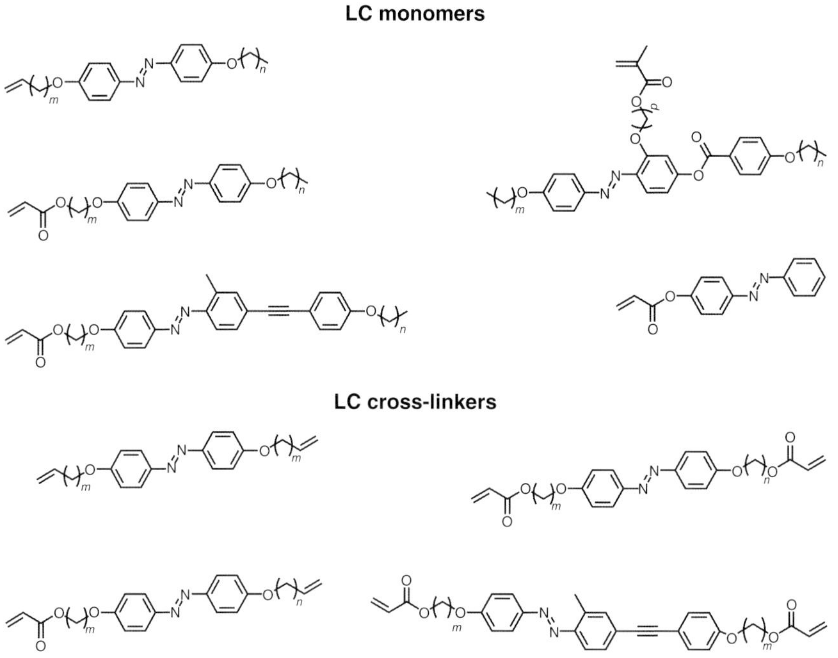

Figure 9.2 Examples of chemical structures of LC monomers and cross-linkers often used to prepare CLCP [17–22].

manipulating the bending direction in a polydomain CLCP film [23]. Although the direction of LC alignment in the film is random in macroscale, the selective absorption of linear polarized light leads to a contraction in specific domains where the azobenzene moieties are aligned along the direction of light polarization (Figure 9.3). Furthermore, Broer and coworkers prepared CLCP films with a densely cross-linked, twisted alignment of the mesogen units and realized the coiling actuation of the CLCP films in response to light [24]. Based on the 90° twisted LC alignment, the films showed large amplitude helical coiling motions as well as bending motions upon exposure to UV light.

Ikeda and coworkers prepared a kind of ferroelectric CLCP film in smectic C phase under an applied electric field [25]. In the obtained ferroelectric CLCP films, the long axis of each mesogen is tilted along the same direction in smectic layers oriented roughly perpendicular to the surface. These films bent toward the light source with a tilt to the rubbing direction of the alignment layer upon exposure to UV light, and reverted to initial flat state upon exposure to visible light (Figure 9.4a). Owing to the higher degree of order, these ferroelectric films possessed greater deformation amplitude and faster response speed. Iamsaard et al. reported photoresponsive polymer films with a twisted geometry. Embedded with different chiral dopants, the photoresponsive LC arranged in a left-handed or right-handed twist–nematic molecular organization in the films. Resulting from different cutting direction, a series of photoresponsive LCP springs was obtained, displaying complex motions such as winding, unwinding, or helix inversion irradiated by UV light (Figure 9.4b) [26].

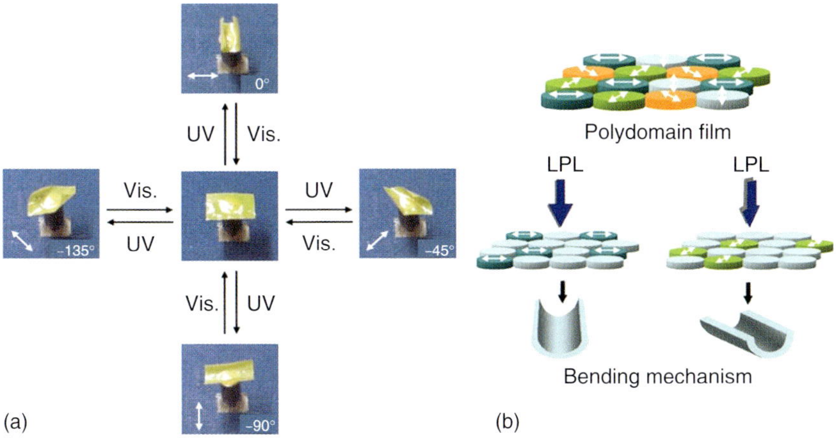

Figure 9.3 (a) Precise control of the bending direction of the polydomain CLCP film by linear polarized UV light at different angles of polarization (white arrows) at 366 nm, and being flattened again by visible light longer than 540 nm. Source: Yu et al. 2003 [23]. Reproduced with permission of Springer Nature. (b) Schematic illustration of the plausible bending mechanism.

9.2.2 Weigert Effect

Excited states of molecules are generated after light absorption when the electric field vector of light is consistent with the direction of transition moment of the molecule (photoselection). For example trans isomer of azobenzene exhibits $\pi-\pi^*$ transition moments approximately parallel to long molecular axis, thereby the trans isomer shows angular-dependent absorption of linear polarized light. Trans isomers with transition moment parallel to the polarization of light are activated to excited states, which is followed by trans–cis isomerization. In contrast, isomers with transition moments perpendicular to the polarization of the light are inactive for configurational change. With the repetition of trans–cis–trans isomerization cycles, trans isomers will be eventually located perpendicular to the polarization direction of the light and become inactive toward the irradiation (Figure 9.5a). This phenomenon indicates that the majority of azobenzene molecules will be aligned perpendicular to the light polarization upon continuous light irradiation, which is well-known as the Weigert effect [16].

In the case of unpolarized light, only the propagation direction is generally perpendicular to the electric vector of the light. The azobenzene molecules are inactive for photoisomerization when aligned with the long molecular axis along the propagation direction of the actinic light, because the propagation direction of light is always perpendicular to its electric field vector (Figure 9.5b). Therefore, when unpolarized light is employed, it is expected that the azobenzene moieties tend to be aligned only in the propagation direction of the actinic light [27].

Tabiryan et al. reported a CLCP film with extraordinarily strong and fast mechanical response to the influence of laser beam [28]. The direction of the bending or twisting can be reversed by altering the polarization of the laser beam. The CLCP film bent away from the laser when the alignment of the

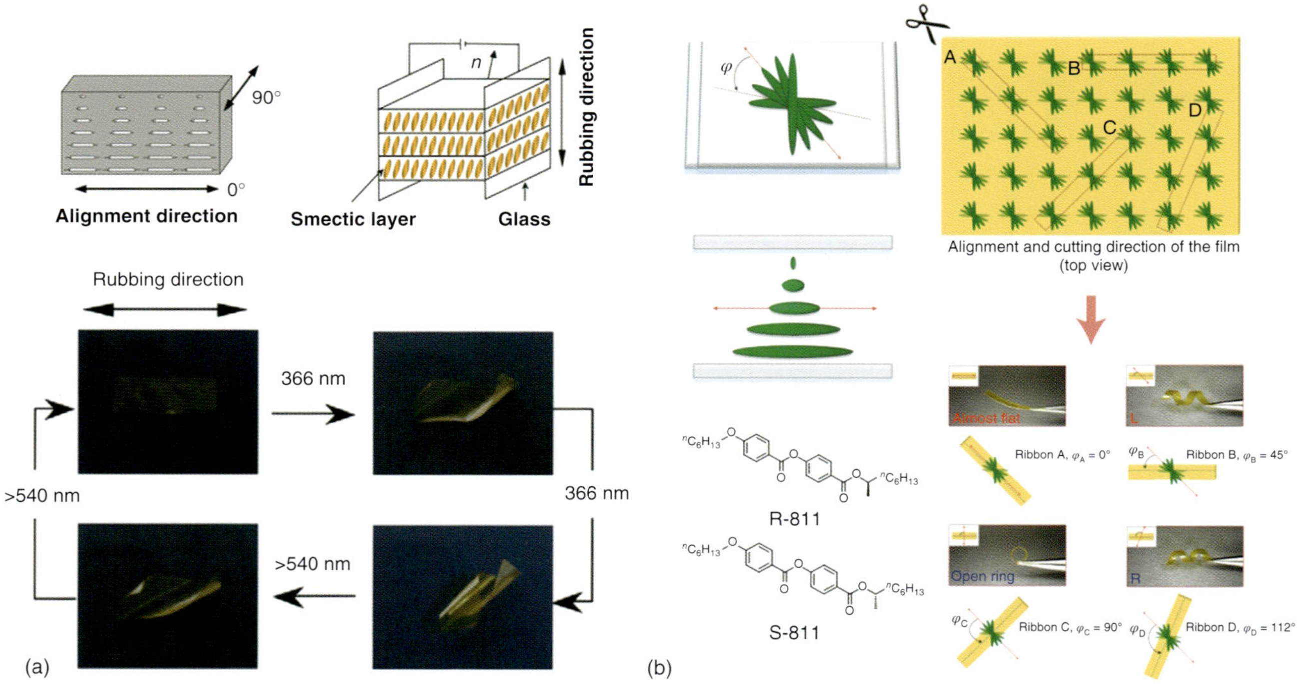

Figure 9.4 (a) Bending and unbending of the ferroelectric CLCP film upon alternate irradiation with UV and visible light at room temperature. Source: Yu et al. 2007 [25]. Reproduced with permission of John Wiley & Sons. (b) A photoresponsive liquid crystal in a twist–nematic molecular organization. The ribbons display a variety of shapes that depend on the direction in which they are cut. Source: Iamsaard et al. 2014 [26]. Reproduced with the permission of Springer Nature.

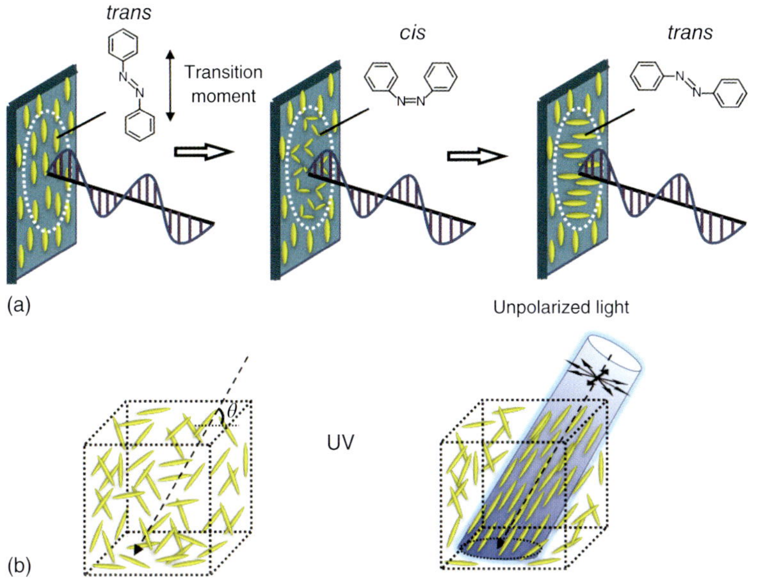

Figure 9.5 (a) Schematic illustration showing the procedure of photoalignment of azobenzene molecules upon linear polarized light in the dashed white circle. (b) Schematics showing reorientation of mesogens in azobenzene-containing LC systems with nonpolarized light that is incident at angle θ. Double arrows show the polarization direction of the light.

mesogens is orthogonal to the beam polarization (E ⊥ n), and it bent toward the laser when the alignment of the mesogens is parallel to the beam polarization (E || n). The film orientation could be varied within ±70 °C, resulting from the reorientation of azobenzene mesogens in the CLCP film driven by polarized light (Figure 9.6).

9.2.3 Photothermal Effect

Once heated above the LC–isotropic transition temperature, the mesogens lose their orientation and become randomly aligned, which brings polymer chains back to the random coil conformation. Accompanying the LC–isotropic phase transition, the CLCP contracts along the orientation alignment but expands in the perpendicular directions (Figure 9.7a) [10, 11, 29]. When the sample is cooled into the LC phase, below T_c, the mesogens recover spontaneously the initial orientation and the shape of the sample goes back to the original state. This is a brief statement for thermal-induced shape variation. Thermal-responsive CLCPs can be readily transformed into photoresponsive actuators by taking advantage of photothermal effect. To realize efficient photothermally driven actuation, in principle, CLCPs need to be loaded with a variety of photothermal agents. They can absorb light and convert optical energy into thermal energy by releasing heat into the polymer matrix, which triggers the LC–isotropic phase transition and shape change at the same time.

Broer and coworkers prepared circularly patterned CLCP films using a photoalignment technique, which exhibited reversible and exotic shape

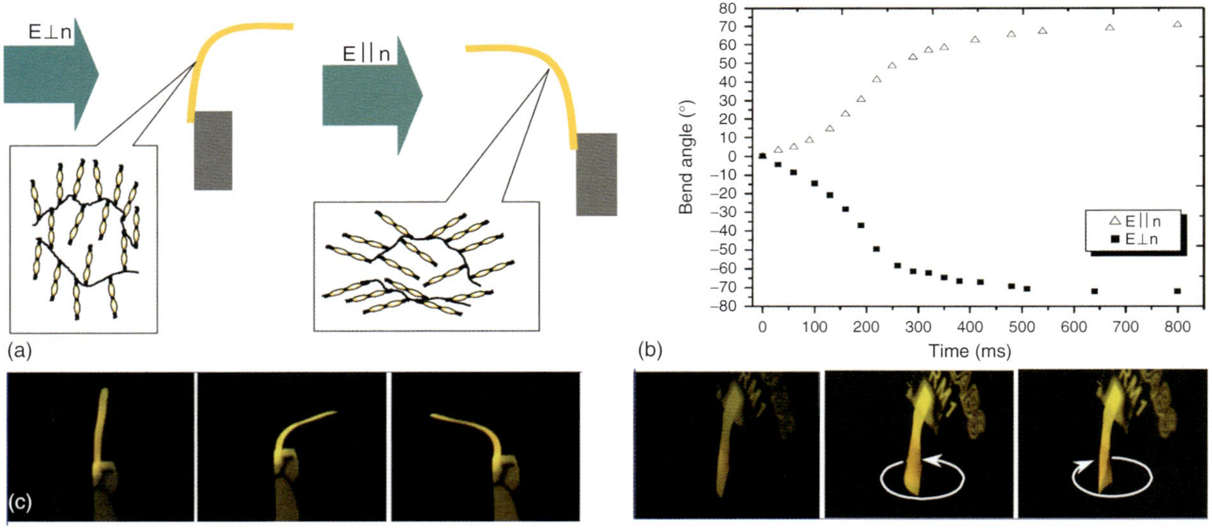

Figure 9.6 (a) Schematic of experimental geometry and the effect of laser-induced LC alignment. (b) Photos of optically induced polymer deformation obtained in the experiment. (c) Deformation dynamics of the polymer film (1 mm × 7 mm × 20 μm) at the power density of the laser beam equal to 0.25 W cm^{-2}. Source: Tabiryan et al. 2005 [28]. Reproduced with permission of OSA Publishing. https://creativecommons.org/licenses/by/4.0/.

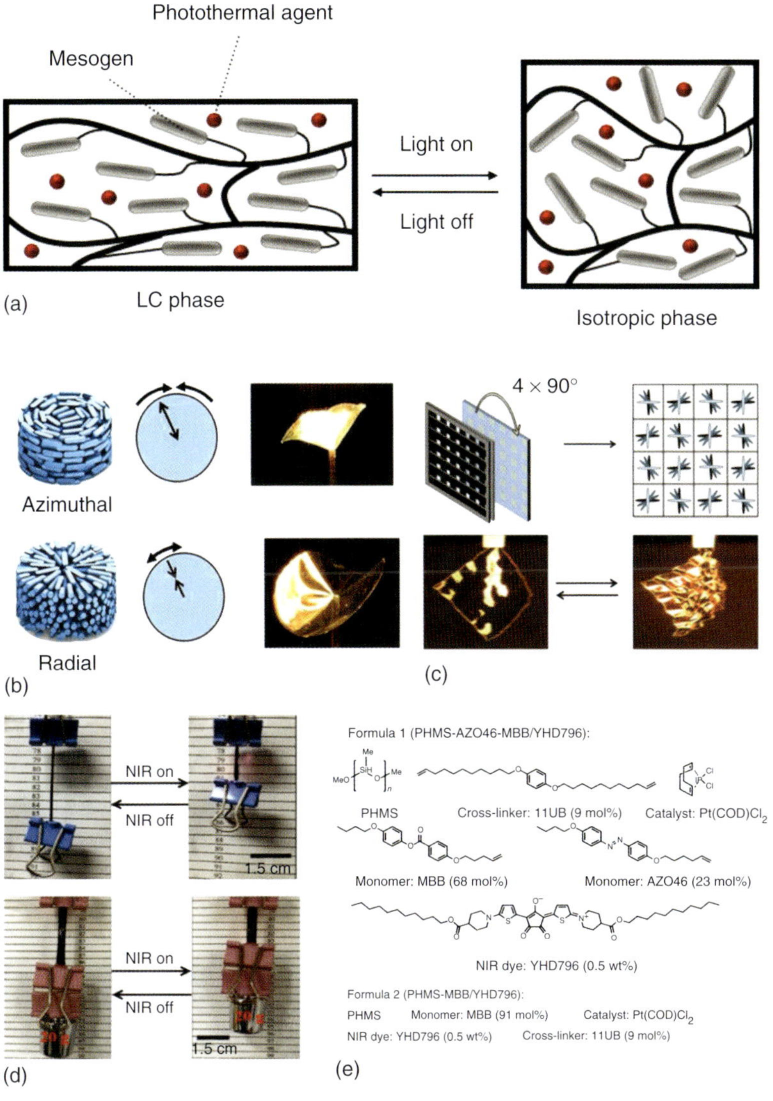

Figure 9.7 (a) Schematic illustration of shape change driven by photothermal effect. (b) IR light-driven deformation of the CLCP films with azimuthal and radial alignment. The arrows along the radius and the azimuth indicate the deformation directions. Source: de Haan et al. 2012 [30]. Reproduced with permission of John Wiley & Sons. (c) Deformation of a CLCP film with a checkerboard pattern upon IR irradiation. Source: de Haan et al. 2014 [31]. Reproduced with permission of John Wiley & Sons. (d) Photos of a CLCP film lifting up a binder clip load under NIR light irradiation. (e) The chemical compositions of PMHS-AZO46-MBB/YHD796 composite (Formula 1) and PMHS-MBB/YHD796 composite (Formula 2). Source: Wang et al. 2016 [33]. Reproduced with permission of Springer Nature. https://creativecommons.org/licenses/by/4.0/.

deformations in response to infrared (IR) light due to the photothermal effect of the dye [30]. As shown in Figure 9.7b, the azimuthal film deforms into a conical shape with the cone apex located at the center, while the radial film deforms into an anti-cone shape. In the case of an azimuthal alignment pattern, a reduction of the LC order upon temperature increase leads to compression along the azimuthal direction and an expansion along the radial direction. These stresses cannot be accommodated within the sheet plane, causing deformation of the flat sheet out of the plane into a cone. In the case of a radial alignment pattern, the opposite deformations take place. This results in an anti-cone, which is recognizable as a saddle shape. In order to achieve even more complex reversible deformation under IR light irradiation, they used the same technique to prepare accordion-like actuators [31]. In this case, the director profile was striped or checkerboard in the plane with a 90° twist across the film thickness. When the actuator is exposed to IR light, the accordion-like folds can be observed in the CLCP film with a striped director profile, whereas the film with a checkerboard director profile forms a periodic square pattern of peaks, depressions, and saddle points (Figure 9.7c).

Yang and coworkers designed and synthesized a CLCP actuator bearing chemically bound near-infrared (NIR) dye with a concentration up to 17.06 wt% [32]. Owing to the high content of the photothermal agent, the surface temperature of their CLCP actuator could jump over the LC–isotropic transition temperature (116 °C) in two seconds and reach 260 °C within eight seconds, exhibiting ultrafast photoresponsive speed. This actuator also shows superior mechanical properties by lifting up 5680 times its own weight upon NIR light-induced contraction. Furthermore, they prepared bilayer polysiloxane-based CLCP actuators by exhibiting two different 3D transformations (bending and chiral twisting) by tuning the wavelength band of light stimuli [33]. This bilayer actuator was fabricated by gluing two different uniaxially aligned CLCP samples together with a crossed angle of either 45° or −45°. Upon NIR irradiation, due to the photothermal effect of the incorporated NIR dye, the two layers of the CLCP actuator contract along their own alignment directions with a crossed angle of 45° or −45° (Figure 9.7d,e). Such a nonuniform contraction leads to different contraction ratios, resulting in two bending trends along the two alignment directions. The vector sum of the two deformations creates an inclined angle with the long axis of the CLCP ribbon and thus drives the dual-layer ribbon to twist or curl in either a right-handed or a left-handed helix.

9.3 Bioinspired Actuators and Their Applications

In contrast to conventional mechanically constructed robots, soft actuators can achieve similar functions with simpler structures, omitting additional components such as motors, gears, and connecting rods, thanks to the response behavior of photoresponsive polymers. Herein, we highlight the recent progress in bioinspired soft actuators based on photomechanical LCP and focus on their macro or micro behaviors mimicking natural creatures.

9.3.1 Soft Actuators Driven by Photothermal Effect

Combining light-inactive soft actuators with photothermal agents will endow them photoresponsive performances. For example Cai and coworkers fabricated a prototype of water–air interface robotic swimmer that combined the photothermal effect of polydopamine and the thermal responsiveness of CLCP. Controlled by a NIR laser, the CLCP fish swam at the highest speed of 1 mm s^{-1} through performing "swimming strokes" induced by reversible bending and unbending motions [34].

Single-walled carbon nanotube (SWNT) is another common photothermal material that is responsive to IR. Chen and coworker reported an IR light-driven inchworm walker that could crawl up a hill at 50° incline. The inchworm walker was comprised of a bilayer film, composed of one active CLCP layer with SWNT and one passive silicone layer, and two polycarbonate films with different shapes. Upon IR irradiation, the SWNT–CLCP layer in the bilayer film underwent a significant in-plane negative strain and further induced a large, fast, and reversible bending of the film allowing the inchworm walker to crawl (Figure 9.8a) [35]. Heliotropism is a phenomenon found in many plants such as sunflower. After the sunflower is exposed to sunlight, the tubular florets in the faceplate will heat up and the fibers at the base will shrink, causing the faceplate to change direction to receive sunlight. Inspired by this, Jiang et al. reported an artificial heliotropic

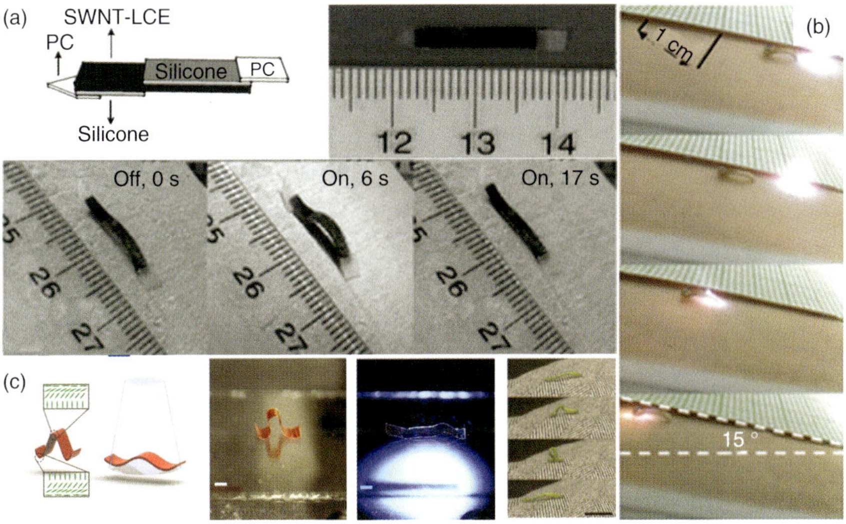

Figure 9.8 (a) Scheme of an inchworm walker device consisting of an asymmetric SWNT–CLCP/silicone bilayer film. The inchworm walker crawling up the wood substrate at a 508 incline in response to on and off cycles of continuous-wave (CW) NIR light (28.2 mW mm^{-2}). Source: Kohlmeyer et al. 2013 [35]. Reproduced with permission of John Wiley & Sons. (b) Photographs showing climbing at a speed of 12 mm min^{-1} on an inclined surface at 15°, laser intensity being 1.4 W cm^{-2}. Source: Dong et al. 2018 [37]. Reproduced with permission of Royal Society of Chemistry. (c) An optical filter is used to block wavelengths below 500 nm in all the images. Scale bars: 5 mm. Source: Zeng et al. 2018 [40]. Reproduced with permission of John Wiley & Sons.

device with a solar cell panel driven by CLCP nanocomposite actuators doped with SWNT [36]. When facing the sunlight, the CLCP actuators would contract by the photothermal effect from the sunlight, tilting the whole device toward the light. As the angle of the sunlight changed, the artificial heliotropic device would adjust the solar cell panel accordingly, thereby maximizing the use of sunlight.

Zhao and coworkers demonstrated a CLCP-based actuator fabricated with graphene paper, which realized moving on a horizontal or inclined untreated surface. The actuator was composed of three layers – a thin reduced graphene oxide (rGO) top layer, an inactive polymer middle layer, and an active CLCP bottom layer. By shining NIR light on the rGO surface, a moving wave along the strip was generated. It was shown that this light-driven moving wave made the actuator strip an effective caterpillar-type walker to undergo locomotion, on both level and inclined surfaces, controlled by a directional laser scan on it (Figure 9.8b) [37]. Due to the uniform monodomain arrangement in the LC layer, most of the walkers can only move in a straight line or rely on substrate characteristics to determine the walking direction resulting in limited control over the movement. To solve this problem, Zhao and coworkers fabricated a soft and flexible millimeter-scale crawler made of a CLCP doped with NIR dyes with one side cross-linked monodomain and the other cross-linked polydomain along the thickness [38]. Without the need for a support, this crawler could move straight, climb an inclined surface, and turning to right or left guided by light.

Azo compounds are light-absorbing materials that are often used as dyes, so they can be treated as photothermal agents for soft actuators. By incorporating an azo dye, Parmeggiani and Wiersma and coworkers reported a microhand with color recognition ability [39]. Controlled remotely by optical illumination, this microhand could autonomously grab small particles. Objects with an absorption spectrum peak that matches the illumination wavelength create a local heating effect that triggers the microhand to close. The realization of this autonomous hand constituted a crucial element in the development of microscopic creatures that can perform tasks without human intervention and self-organized automation at the micrometer scale. Arri and coworkers reported on a miniature inching robot fabricated from an azo dye containing monolithic CLCP film. Upon visible light excitation, the film was capable of mimicking caterpillar locomotion on different substrates such as a blazed grating and a paper surface. The motion was driven by spatially uniform visible light and was relatively low in intensity, making the robot "human-friendly" and capable of operating on human skin (Figure 9.8c) [40].

A combination of azo-containing CLCP with other photothermal materials will produce more complex motions for soft actuators. Zhao and coworkers reported a biomimetic "flower" using an azobenzene-containing CLCP doped with gold nanorods (AuNRs). This artificial flower could undergo repeated closing and blooming in response to UV and visible lights via the photochemical reaction of azobenzene. Due to the photothermal effect from the surface plasmon resonance of AuNRs, combining the same material with biaxially oriented polypropylene (BOPP) exhibited bending and unbending behaviors to mimic the human "push-ups" and "sit-ups" with the aid of NIR light. Moreover, a polymer "crane" was assembled to demonstrate its potential in other complex motions [41].

9.3.2 Photoinduced Actuation of Soft Actuators

Manipulating soft actuators using light without generating much heat is more biofriendly and may have more practical applications. Thus, light-driven bioinspired actuators prepared from the photomechanical LCPs, especially azobenzene-based LCPs, were summarized in this section. Wasylczyk and coworkers fabricated a soft caterpillar robot to mimic the crawling behavior of the caterpillar, based on a light-sensitive patterned molecular alignment CLCP film, which could perform the light-driven locomotion in the forward and backward directions [42]. In addition to crawling, swimming is another common motion. Tian, Yu, Liu and coworkers prepared a micro swimmer with a CLCP-containing flagellum that moved similar to the microorganism such as choanoflagellate or spermatozoon (Figure 9.9a). Controlled by UV and white light emitting diodes (LEDs), the flexible long flagellum performed a wave-like swing, driving the swimmer to move forward in a liquid-filled tube. Made by the same CLCP incorporated with polyethylene terephthalate (PET), a photoresponsive gripper was fabricated and fixed on the head of this microrobot, introducing a function to the swimmer for grabbing and releasing loads [43]. Additionally, Fischer and coworkers reported a ciliate-like microrobot using photoactive CLCP. Driven by structured monochromatic light, the axial nematic alignment of the microrobot contracted at the axial direction and expanded at radial direction simultaneously, leading to the self-propulsion of the microrobot (Figure 9.9b) [44].

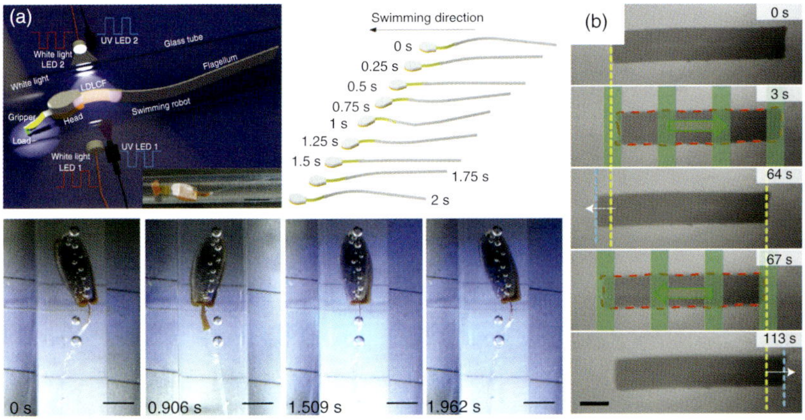

Figure 9.9 (a) Schematics of the micro swimming robot and light irradiation system. Periodic cyclic swinging of the flagellum observed. The light-driven liquid crystal film (LDLCF) periodically bends and drives the flagellum to swing. The scale bar is 2 mm. Source: Huang et al. 2015 [43]. Reproduced with permission of Springer Nature. https://creativecommons.org/licenses/by/4.0/. (b) Back-and-forth swimming of a cylindrical microrobot propelled by traveling-wave deformations (red dashed line: deformed profile). The green overlays and arrows represent the periodic light pattern and its traveling direction, respectively. Yellow and cyan dashed lines represent the initial and final position of the leading edge of the microswimmer, respectively. Source: Palagi et al. 2016 [44]. Reproduced with permission of Springer Nature.

Flying is the eternal dream of humankind. The majority of miniature flying robots are based on electrically powered propeller or piezoactuators driving the wings. When scaled down few centimeters or below, an onboard battery becomes impractical because of the increase in the weight-to-power ratio, whereas LCPs might provide alternative routes toward small-scale flying devices. Through the self-shadowing effect of the photoresponsive polymer devices, such devices will show high-speed oscillation, which demonstrates a possible way to implement the aircraft. White and coworkers reported on the photodriven, fast (30 Hz), and large amplitude (>170°) oscillation of a polymer cantilever consisting of a monodomain azobenzene-containing CLCP (azo-CLCP) [45]. The frequency of the photodriven oscillation is similar to a hummingbird wingbeat, ranging from 20 to 80 Hz. The oscillation of the azo-CLCP cantilever can be turned on and off by switching the polarization direction of the driving laser beam, and the behavior showed little fatigue over 250 000 cycles.

Bleger, Hecht, Debije, and Schenning and coworkers described a CLCP film doped with a visible light-responsive fluorinated azobenzene capable of continuous chaotic oscillatory motion upon non-concentrated sunlight or a combined light of green or blue LEDs. Although the mechanism of erratic oscillatory motion remains unclear and might be attributed to the combination of periodic small changes of cis–trans population and their isomerization rates, this work constituted an important step toward the realization of autonomous, persistently self-propelling machines and self-cleaning surfaces powered by sunlight [46]. With engineering and miniaturized designs, photodriven oscillations of azo-CLCP materials are expected to be applied in photo-fueled miniature flying machines and remotely controlled micro-optomechanical systems.

Inspired by the cilia motion, Broer and coworkers reported a series of photoresponsive fiber arrays for object transport on water surface. They synthesized a cross-linked main-chain LCP functionalized with azobenzene molecules. Reversible complex photoinduced deformation was obtained. When the fiber arrays were placed in liquid and stimulated with light, their cooperative bending motion created a flow that was able to efficiently carry objects (Figure 9.10a) [47].

The iris is an oblate circular structure that exists in the eyes of many animals, which is responsive to the intensity of light. When the eye is illuminated with glare, the purple in the iris contracts to limit the entry of light, while it relaxes when the light is dimmed. Priimagi et al. reported an iris-like CLCP actuator that can shrink automatically according to light intensity (Figure 9.10b). The CLCP film was arranged in a centrally symmetric splayed alignment, with different molecular orientation at the top and bottom interfaces. In dark environment, the CLCPs tended to maintain the iris open because of anisotropic thermal expansion, and the iris will close with light-induced circularly symmetric actuation of the iris segments (Figure 9.10b) [48].

Applying the optical feedback design, a light-sensitive "flytrap" was demonstrated with a photoresponsive CLCP fabricated onto the tip of optical fiber. In response to light, this artificial flytrap was capable of autonomous closure and object recognition. This strategy enabled self-regulated actuation within the fiber-sized architecture, thus opening up avenues toward soft, autonomous

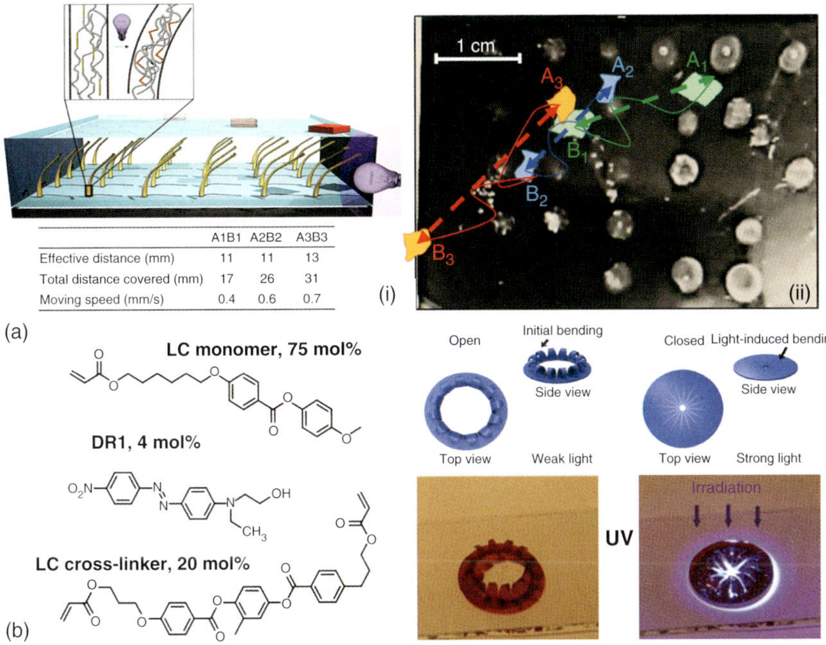

	A1B1	A2B2	A3B3
Effective distance (mm)	11	11	13
Total distance covered (mm)	17	26	31
Moving speed (mm/s)	0.4	0.6	0.7

Figure 9.10 (a) (i) Schematic representation of the collective bending of the photoresponsive fibers leading to the transport of floating objects. (ii) Initial and final position of the objects. The full line represents the path followed by the object from A to B and the double arrows represent the effective distance covered by the floating object. Source: Gelebart et al. 2016 [47]. Reproduced with permission of John Wiley & Sons. (b) Schematic drawing of the light-driven iris at its closed state, where light illumination triggers petal segments to bend inward and to reduce the aperture size. The iris is open under weak/no light conditions, when the anisotropic thermal expansion induces bending in the segments, thus opening the aperture. Source: Zeng et al. 2017 [48]. Reproduced with permission of John Wiley & Sons.

small-scale devices [49]. To simulate a more complex motion like stretching arm, grabbing, and moving the object with hand, Yu and coworkers reported a visible light-driven fully plastic microrobot composed of a "hand," a "wrist," and an "arm" made of the CLCP/polyethylene (PE) bilayer films with different initial shapes and photodeformation modes. Upon visible light irradiation, the bilayer film would bend away from or toward the light source because of the contraction in the surface region of the CLCP layer caused by the trans–cis photoisomerization of the azotolane moieties. Without the aid of any gears, bearings, or contact-based driving systems, the microrobot was manipulated to pick, lift, move, and place the object by irradiating different parts of the microrobot with visible light (Figure 9.11) [50].

In addition to using biomimetic structures to control solids, small amount of liquid can also be modulated to utilize photoresponsive polymers, which has application ranging from biomedical devices to liquid transfer. Superhydrophobic surfaces play wide-ranging roles in self-cleaning materials, lossless droplets transport, and open microreactors. When the adhesive force of the superhydrophobic surface was adjustable, it could work as a switch to control the

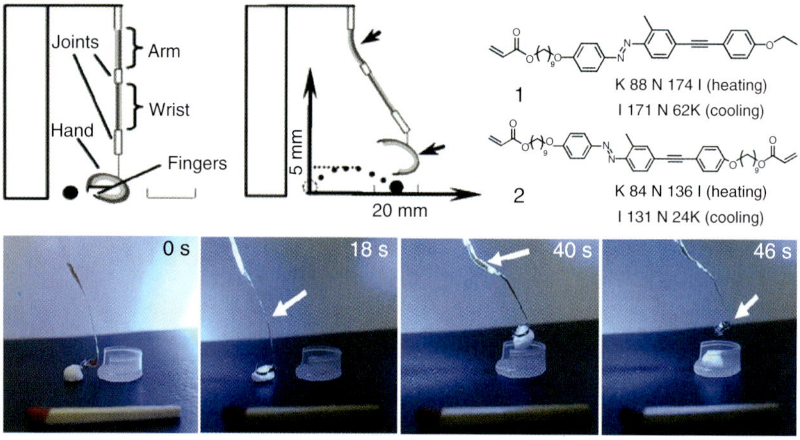

Figure 9.11 Structures of the LC monomer and cross-linker used in this study. Schematic illustration of the microrobot and photographs showing microrobot picking, lifting, moving, and placing an object to a nearby container by turning on and off the blue light. White arrows indicate the parts irradiated with visible light. Source: Cheng et al. 2010 [50]. Reproduced with permission of Royal Society of Chemistry.

movements of water drops. In most of the existing works, the wettability and the adhesive force of a superhydrophobic surface were adjusted at the same time by changing the surface microstructures or chemical composition. It was challenging to regulate the adhesive force while maintaining the superhydrophobicity of a surface. In order to achieve the noncontact regulation of superhydrophobic adhesion, Yu and coworkers prepared a CLCP micro-arrayed surface through a secondary replicating method. The water-sliding angle of the surface could be rapidly and reversibly adjusted through light irradiation while maintaining its large contact angle (Figure 9.12) [51]. Later, using high-throughput etching technique of colloidal lithography, they fabricated two different submicro-arrayed LCP surfaces with average diameters of sub-250 and sub-400 nm, which were easily carried out by modulating different types of etching masks [52]. Owing to the surface chemistry and surface morphology designed to obtain substantial hydrophobicity, the two surfaces exhibited completely different wetting behavior of water adhesion, and mimic a rose petal and a lotus leaf, respectively. Azotolane was used here as the photoresponsive moiety instead of azobenzene to achieve all visible light control, thus the wettability of these films could be reversibly transferred when alternately irradiated by 470 and 530 nm light. This kind of noncontact, real-time photo-controllable switching of microdroplet's adhesion might inspire and facilitate the designs and fabrication of novel microfluidic devices and lossless transportation.

More recently, Yu et al. reported a new strategy to transfer liquids by capillary force caused by asymmetric deformation of photoresponsive microtubes from cylindric to conical. Inspired by the lamellar structure of artery walls, they synthesized a novel azobenzene-based linear liquid crystalline polymer (LLCP), where the flexible backbones and the azobenzene mesogens self-assembled into ordered nanoscale lamellar structures due to the molecular cooperation

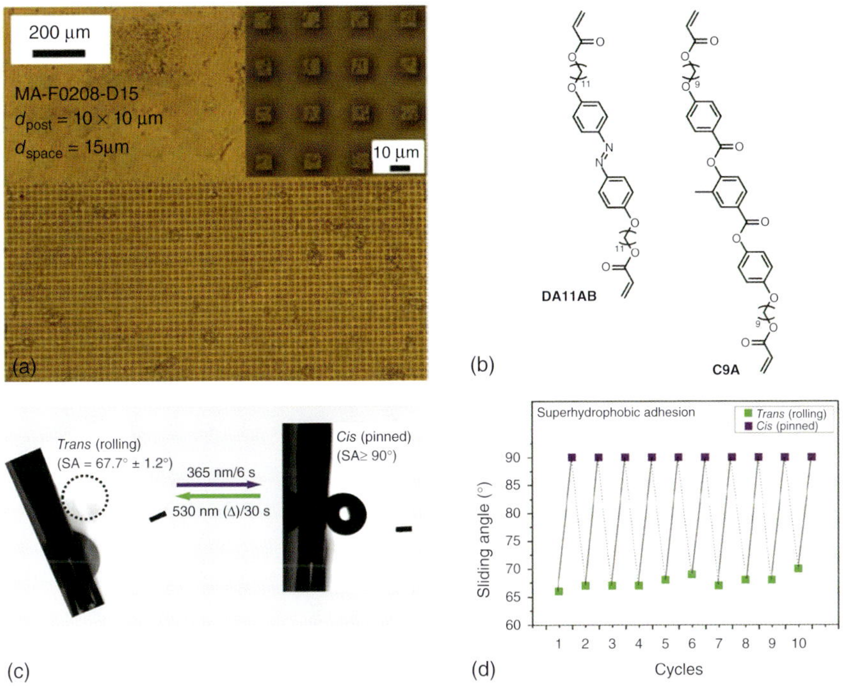

Figure 9.12 (a) Large-area optical microscopic image and local amplified image (inset) of photoresponsive microarrays. (b) Chemical structures of LC monomers (DA11AB and C9A) used to prepare the azobenzene LCP film. (c, d) Light-controlled quick and reversible switching of superhydrophobic adhesion between rolling and pinning on MA-F0208-D15 with a 2 ml water droplet. Source: Li et al. 2012 [51]. Reproduced with permission of Royal Society of Chemistry.

effect. The noncross-linked LLCP was first fabricated into 3D microtubes, which exhibited geometry change from cylindrical to conical upon irradiation of light with gradient intensity, inducing asymmetric capillary force to move the inner water slugs. Compared with the existing photocontrolable water transportation technologies based on wettability gradients or the Marangoni effect, the tubular LLCP microactuator was able to transfer more types of liquids with faster speed. The transportation of multiphase liquids and even mix of different liquids were also achieved with the remote photo-control. Furthermore, the great advantage of water transportation in LLCP microtubes is that, compared with the traditional microfluidics based on the nonresponsive materials, air pumps and complex pipeline are not needed to drive the liquid movements. Thus, our technology has great potential to simplify the microfluidic system and create small volume microreactors, laboratory-on-a-chip contexts, and micro-optomechanical systems (Figure 9.13) [27].

Photonic crystals contain periodic microstructures that reflect light of a particular wavelength. Adjustment of microstructures will induce related reflection change and have potential applications in display, communication, and anti-counterfeiting technologies [53]. Most of the existing phototunable photonic crystals are made of small LC molecules for the convenience

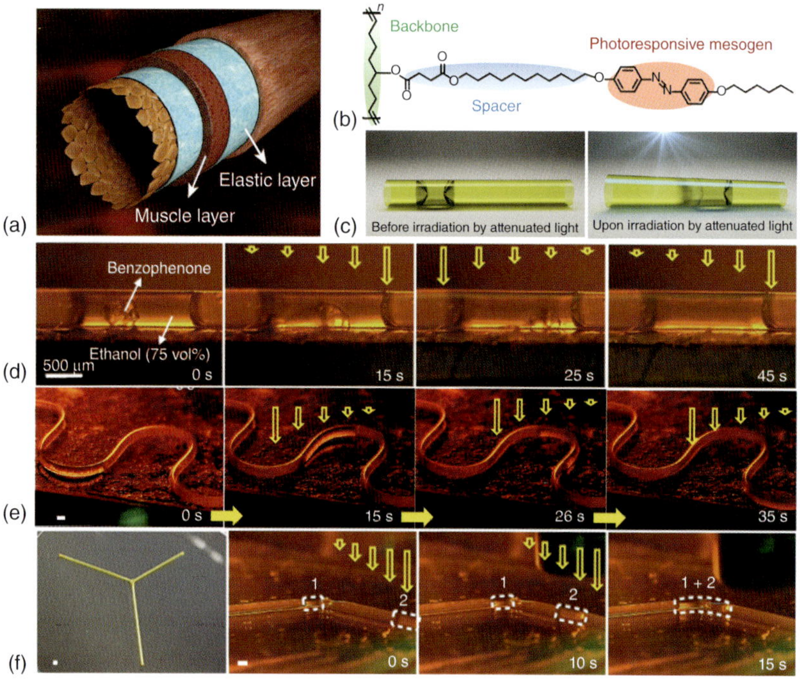

Figure 9.13 (a) Schematic illustration of the structure of artery walls. (b) Molecular structure of a novel linear liquid crystal polymer (LLCP). (c) Schematics showing the motion of a slug of fully wetting liquid confined in a tubular microactuator (TMA) driven by photodeformation. (d) Photocontrolled dissolution of the benzophenone in ethanol in the TMA. (e) Lateral photographs showing light-driven motion of a silicone oil slug in a serpentine TMA. (f) Photocontrolled liquid fusion in the Y-shaped TMA. Source: Lv et al. 2016 [27]. Reproduced with permission of Springer Nature.

to change their arrangement. Yu et al. reported the phototunable CLCP photonic crystals by taking advantage of the photoinduced deformation (Figure 9.14) [54]. A two-dimensional periodic microarray was fabricated with azobenzene-containing CLCP via the replica molding technique. In accordance with the geometry change of the CLCP microarray induced by the photo-isomerization of the azobenzene groups, on–off switch behavior on the reflection spectra of the microarray was observed when irradiated with UV–Vis light alternately. They also prepared a photo and thermal dual-responsive CLCP inverse opal, which was the smallest CLCP photonic structure [55]. Upon alternative irradiation of UV and visible light or temperature variation, the hole size of the inverse opal film changed, leading to rise and fall behavior on the intensity of reflection spectra, which was ascribed to the deformation of CLCP induced by photochemical reactions of the azobenzene moieties or the thermal-induced phase transition. Followed by a similar principle, they manufactured a light micro-regulator with the help of the natural photonic crystal, *Morpho* butterfly wing (MBW), and the abovementioned photoresponsive linear polymer, LLCP. Thanks to the excellent processing behaviors and mechanical

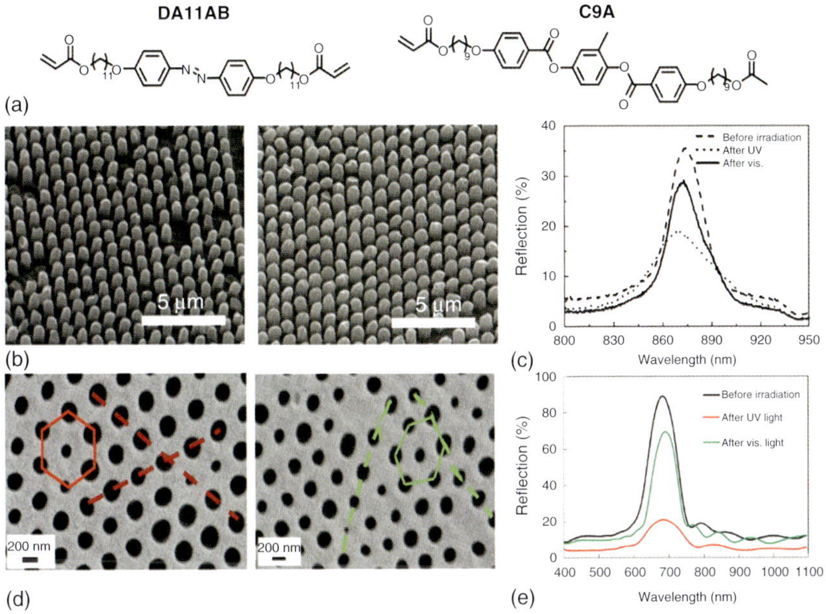

Figure 9.14 (a) Chemical structures of the LC monomers DA11AB and C9A. (b) SEM images of CLCP microarray before (left) and after (right) the irradiation of UV light. After UV irradiation, the mean diameter of the pillars increases from 0.587 to 0.695 μm, showing about 18% diameter increase. (c) Reflection spectra of CLCP microarray under the UV light irradiation and the following visible light irradiation. (d) SEM images of the inverse opal film (left) before and (right) after UV light irradiation. The red and green regular hexagon and straight lines represent the arrangement of the holes before and after UV light irradiation. The distance between every two holes of the inverse opal film is 350 nm. (e) Reflection spectra of CLCP inverse opal film under UV light irradiation and subsequent visible light irradiation. Source: Yan et al. 2012 [54]. Reproduced with permission of John Wiley & Sons.

properties, a LLCP–MBW composite was obtained through electrospinning and demonstrated a phototunable light-regulation performance with a blue shift of its reflection peak (70 nm) and a remarkable change of reflectance (40%). This work provides potentials in a variety of applications ranging from sensors, photochromic pigments, anti-counterfeiting to information storage technologies (Figure 9.15) [56].

9.4 Conclusion

In this chapter, we have described three main actuation mechanisms of photoresponsive materials and summarized the recent development of photomechanical LCPs in bioinspired actuators. With the aid of photoresponisve motifs, especially azobenzene derivatives, various biomimetic porotypes have been fabricated. Controlled by their chemical structures as well as their alignment process, these photoresponsive LCPs exhibit diverse photoinduced performances to mimic the natural behaviors. While most studies focus on the structural mimic from nature, the functional mimic is more crucial to further illustrate the advantages

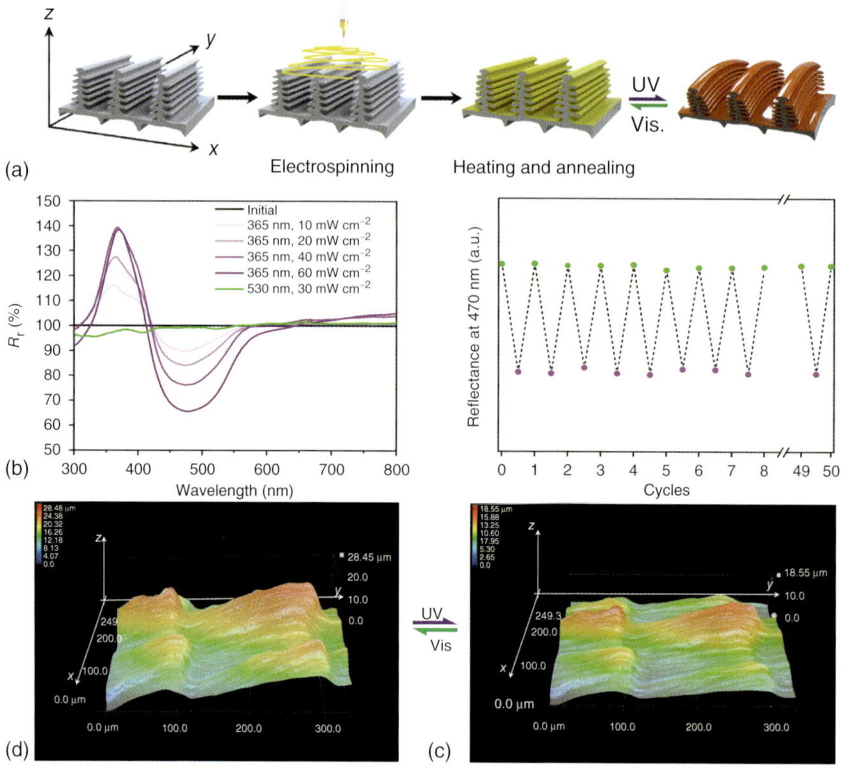

Figure 9.15 (a) Schematic showing the fabrication of LLCP-coated *Morpho* butterfly wing (LLCP–MBW) and the reversible photoinduced deformation. For a better understanding of the placement of the MBW in this and the following figures, a coordinate system was set with *y*-axis parallel to the ridges, *x*-axis perpendicular to the ridges, and *z*-axis vertical to the surface. (b) Relative reflectance (R_r) of LLCP–MBW shows the change ratio of reflection under different lights and intensities. (c) Evaluation on the reversibility of LLCP–MBW in response to alternative irradiation of UV light (365 nm, 60 mW cm^{-2}, 10 seconds) and visible light (530 nm, 30 mW cm^{-2}, 10 seconds) for 50 cycles. (d) 3D reconstruction images show the reversible deformation of the scales on LLCP–MBW upon UV (365 nm, 10 mW cm^{-2}) and visible light (white light from the microscope) irradiations. Source: Qing et al. 2018 [56]. Reproduced with permission of John Wiley & Sons.

of these photomechanical LCPs as smart materials. In recent years, several micro/nano processing technologies including inkjet printing, 3D printing, electrospinning, and so on have evolved to allow the fabrication of actuators in micrometer and even nanometer scales, bringing photoresponsive LCP materials closer to real applications. It is anticipated that the preparation of microscale actuators will become the mainstream for the future development of photodeformable LCPs, which requires not only the guidance of new concepts and theories, but also creative material design and application of novel processing methods. Moreover, further effort is also addressed for integrating LCP actuators into micromechanical systems to fully demonstrate the performance of LCP materials.

References

1 Ikeda, T., Nakano, M., Yu, Y. et al. (2003). Anisotropic bending and unbending behavior of azobenzeneliquid-crystalline gels by light exposure. *Adv. Mater.* 15: 201–205.
2 Liu, Y., Xu, B., Sun, S. et al. (2017). Humidity- and photo-induced mechanical actuation of cross-linked liquid crystal polymers. *Adv. Mater.* 29: 1604792.
3 Zhuang, J., Han, B., Liu, W. et al. (2018). Liposome-amplified photoelectrochemical immunoassay for highly sensitive monitoring of disease biomarkers based on a split-type strategy. *Biosens. Bioelectron.* 99: 230–236.
4 Julie, T., Hugo, O., Olivier, S., and Sebastien, L. (2013). Magnetic responsive polymer composite materials. *Chem. Soc. Rev.*: 7099–7116.
5 Mourran, A., Zhang, H., Vinokur, R., and Möller, M. (2017). Soft microrobots employing nonequilibrium actuation via plasmonic heating. *Adv. Mater.* 29: 1604825.
6 Chen, Z., Hsu, P., Lopez, J. et al. (2016). Fast and reversible thermoresponsive polymer switching materials for safer batteries. *Nat. Energy* 1: 15009.
7 Kocak, G., Tuncer, C., and Bütün, V. (2017). pH-Responsive polymers. *Polym. Chem.* 8: 144–176.
8 Zhang, S., Bellinger, A.M., Glettig, D.L. et al. (2015). A pH-responsive supramolecular polymer gel as an enteric elastomer for use in gastric devices. *Nat. Mater.* 14: 1065–1071.
9 de Gennes, F.G. (1975). One type of nematic polymers. *C. R. Acad. Sci. B. Phys.* 281: 101–103.
10 Fleischmann, E. and Zentel, R. (2013). Liquid-crystalline ordering as a concept in materials science: from semiconductors to stimuli-responsive devices. *Angew. Chem. Int. Ed.* 52: 8810–8827.
11 Ohm, C., Brehmer, M., and Zentel, R. (2010). Liquid crystalline elastomers as actuators and sensors. *Adv. Mater.* 22: 3366–3387.
12 Wei, J. and Yu, Y. (2012). Photodeformable polymer gels and crosslinked liquid-crystalline polymers. *Soft Matter* 8: 8050–8059.
13 White, T.J. and Broer, D.J. (2015). Programmable and adaptive mechanics with liquid crystal polymer networks and elastomers. *Nat. Mater.* 14: 1087–1098.
14 Li, M.H., Keller, P., Yang, J., and Albouy, P.A. (2004). An artificial muscle with lamellar structure based on a nematic triblock copolymer. *Adv. Mater.* 16: 1922–1925.
15 Mol, G.N., Harris, K.D., Bastiaansen, C.W.M., and Broer, D.J. (2005). Thermo-mechanical responses of liquid-crystal networks with a splayed molecular organization. *Adv. Funct. Mater.* 15: 1155–1159.
16 Yu, Y. and Ikeda, T. (2004). Alignment modulation of azobenzene-containing liquid crystal systems by photochemical reactions. *J. Photoch. Photobio. A.* 5: 247–265.
17 Kondo, M., Yu, Y., and Ikeda, T. (2006). How does the initial alignment of mesogens affect the photoinduced bending behavior of liquid-crystalline elastomers? *Angew. Chem. Int. Ed.* 45: 1378–1382.

18 Li, M.H., Keller, P., Li, B. et al. (2003). Light-driven side-on nematic elastomer actuators. *Adv. Mater.* 15: 569–572.

19 Ikeda, T., Nakano, M., Yu, Y. et al. (2003). Anisotropic bending and unbending behavior of azobenzene liquid-crystalline gels by light exposure. *Adv. Mater.* 15: 201–205.

20 Hogan, P.M., Tajbakhsh, A.R., and Terentjev, E.M. (2002). UV manipulation of order and macroscopic shape in nematic elastomers. *Phys. Rev. E* 65: 41720.

21 Finkelmann, H., Nishikawa, E., Pereira, G.G., and Warner, M. (2001). A new opto-mechanical effect in solids. *Phys. Rev. Lett.* 87: 15501.

22 Wu, W., Yao, L., Yang, T. et al. (2011). NIR-light-induced deformation of cross-linked liquid-crystal polymers using upconversion nanophosphors. *J. Am. Chem. Soc.* 133: 15810–15813.

23 Yu, Y., Nakano, M., and Ikeda, T. (2003). Photomechanics: directed bending of a polymer film by light. *Nature* 425: 145–145.

24 Harris, K.D., Cuypers, R., Scheibe, P. et al. (2005). Large amplitude light-induced motion in high elastic modulus polymer actuators. *J. Mater. Chem. A* 15: 5043–5048.

25 Yu, Y., Maeda, T., Mamiya, J., and Ikeda, T. (2007). Photomechanical effects of ferroelectric liquid-crystalline elastomers containing azobenzene chromophores. *Angew. Chem. Int. Ed.* 46: 881–883.

26 Iamsaard, S., Asshoff, S.J., Matt, B. et al. (2014). Conversion of light into macroscopic helical motion. *Nat. Chem.* 6: 229–235.

27 Lv, J., Liu, Y., Wei, J. et al. (2016). Photocontrol of fluid slugs in liquid crystal polymer microactuators. *Nature* 537: 179–184.

28 Tabiryan, N., Serak, S., Dai, X.M., and Bunning, T. (2005). Polymer film with optically controlled form and actuation. *Opt. Express* 13: 7442–7448.

29 Dong, L. and Zhao, Y. (2018). Photothermally driven liquid crystal polymer actuators. *Mater. Chem. Frontiers* 2: 1932–1943.

30 de Haan, L.T., Sanchez-Somolinos, C., Bastiaansen, C.M.W. et al. (2012). Engineering of complex order and the macroscopic deformation of liquid crystal polymer networks. *Angew. Chem. Int. Ed.* 51: 12469–12472.

31 de Haan, L.T., Gimenez-Pinto, V., Konya, A. et al. (2014). Accordion-like actuators of multiple 3d patterned liquid crystal polymer films. *Adv. Funct. Mater.* 24: 1251–1258.

32 Liu, L., Liu, M., Deng, L. et al. (2017). Near-infrared chromophore functionalized soft actuator with ultrafast photoresponsive speed and superior mechanical property. *J. Am. Chem. Soc.* 139: 11333–11336.

33 Wang, M., Lin, B., and Yang, H. (2016). A plant tendril mimic soft actuator with phototunable bending and chiral twisting motion modes. *Nat. Commun.* 7: 13981.

34 Tian, H., Wang, Z., Chen, Y. et al. (2018). Polydopamine-coated main-chain liquid crystal elastomer as optically driven artificial muscle. *ACS Appl. Mater. Interfaces* 10: 8307–8316.

35 Kohlmeyer, R.R. and Chen, J. (2013). Wavelength-selective, IR light-driven hinges based on liquid crystalline elastomer composites. *Angew. Chem. Int.Ed.* 52: 9234–9237.

36 Li, C., Liu, Y., Huang, X., and Jiang, H. (2012). Direct sun-driven artificial heliotropism for solar energy harvesting based on a photo-thermomechanical liquid-crystal elastomer nanocomposite. *Adv. Funct. Mater.* 22: 5166–5174.

37 Dong, L., Tong, X., Zhang, H. et al. (2018). Near-infrared light-driven locomotion of a liquid crystal polymer trilayer actuator. *Mater. Chem. Frontiers* 2: 1383–1388.

38 Ge, F., Yang, R., Tong, X. et al. (2018). A multifunctional dye-doped liquid crystal polymer actuator: light-guided transportation, turning in locomotion, and autonomous motion. *Angew. Chem. Int. Ed.* 57: 11758–11763.

39 Martella, D., Nocentini, S., Nuzhdin, D. et al. (2017). Photonic microhand with autonomous action. *Adv. Mater.* 29: 1704047.

40 Zeng, H., Wani, O.M., Wasylczyk, P., and Priimagi, A. (2018). Light-driven, caterpillar-inspired miniature inching robot. *Macrol. Rap. Comm.* 39: 1700224.

41 Lu, X., Zhang, H., Fei, G. et al. (2018). Liquid-crystalline dynamic networks doped with gold nanorods showing enhanced photocontrol of actuation. *Adv. Mater.* 30: 1706597.

42 Rogóż, M., Zeng, H., Xuan, C. et al. (2016). Light-driven soft robot mimics caterpillar locomotion in natural scale. *Adv. Opt. Mater.* 4: 1689–1694.

43 Huang, C., Lv, J., Tian, X. et al. (2015). Miniaturized swimming soft robot with complex movement actuated and controlled by remote light signals. *Sci. Rep.* 5: 17414.

44 Palagi, S., Mark, A.G., Reigh, S.Y. et al. (2016). Structured light enables biomimetic swimming and versatile locomotion of photoresponsive soft microrobots. *Nat. Mater.* 15: 647–653.

45 White, T.J., Tabiryan, N.V., Serak, S.V. et al. (2008). A high frequency photo-driven polymer oscillator. *Soft Matter* 4: 1796–1798.

46 Kumar, K., Knie, C., Bléger, D. et al. (2016). A chaotic self-oscillating sunlight-driven polymer actuator. *Nat. Commun.* 7: 11975.

47 Gelebart, A.H., Mc Bride, M., Schenning, A.P.H.J. et al. (2016). Photoresponsive fiber array: toward mimicking the collective motion of cilia for transport applications. *Adv. Funct. Mater.* 26: 5322–5327.

48 Zeng, H., Wani, O.M., Wasylczyk, P. et al. (2017). Self-regulating iris based on light-actuated liquid crystal elastomer. *Adv. Mater.* 29: 1701814.

49 Wani, O.M., Zeng, H., and Priimagi, A. (2017). A light-driven artificial flytrap. *Nat. Commun.* 8: 15546.

50 Cheng, F., Yin, R., Zhang, Y. et al. (2010). Fully plastic microrobots which manipulate objects using only visible light. *Soft Matter* 6: 3447–3449.

51 Li, C., Cheng, F., Lv, J. et al. (2012). Light-controlled quick switch of adhesion on a micro-arrayed liquid crystal polymer superhydrophobic film. *Soft Matter* 8: 3730–3733.

52 Zhan, Y., Zhao, J., Liu, W. et al. (2015). Biomimetic submicroarrayed cross-linked liquid crystal polymer films with different wettability via colloidal lithography. *ACS Appl. Mater. Interfaces* 7: 25522–25528.

53 Zhao, Y., Xie, Z., Gu, H. et al. (2012). Bio-inspired variable structural color materials. *Chem. Soc. Rev.* 41: 3297–3317.

54 Yan, Z., Ji, X., Wu, W. et al. (2012). Light-switchable behavior of a microarray of azobenzene liquid crystal polymer induced by photodeformation. *Macromol. Rapid Commun.* 33: 1362–1367.

55 Zhao, J., Liu, Y., and Yu, Y. (2014). Dual responsive inverse opal films based on a crosslinked liquid crystal polymer containing azobenzene. *J. Mater. Chem. C* 2: 10262–10267.

56 Qing, X., Liu, Y., Wei, J. et al. (2018). Phototunable morphobutterfly microstructures modified by liquid crystal polymers. *Adv. Opt. Mater.*: 1801494.

10

Organic–Inorganic Hybrid Materials with Photomechanical Functions

Sufang Guo[1] *and Atsushi Shimojima*[2]

[1] *China University of Geoscience Beijing, Beijing Key Laboratory of Materials Utilization of Nonmetallic Minerals and Solid Wastes, National laboratory of Mineral Materials, School of Materials Science and Technology, 29 Xueyuan Road, Haidian District, Beijing 100083, China*
[2] *Waseda University Department of Applied Chemistry 3-4-1 Ohkubo, Shinjuku-ku, Tokyo 169-8555, Japan*

10.1 Introduction

Researchers have been interested in photoisomerization of molecules to create novel photomechanical materials. Among various chromophore molecules, azobenzene derivatives are most widely utilized because of their easy synthesis, high stability, and excellent photochemical properties. Various azobenzene-containing polymers exhibiting photo-induced motions such as bending, vibration, inchworm movement, and wavelike movement have been reported [1–5]. Photo-induced bending and twisting of the molecular crystals of azobenzene derivatives have also been reported [6, 7].

Enhancement of the thermal stabilities and mechanical strength of these materials is one of the significant issues toward their practical applications. Making composites with high-strength organic layer or with thermally stable polymers is effective to conquer the weakness of azobenzene-containing polymers [8, 9]. Hybridization with inorganic compounds with superior thermal stability and mechanical strength is also a promising approach; however, compared to the organic systems, studies on photomechanical functions of azobenzene–inorganic hybrids have been limited so far.

This chapter focuses on the organic–inorganic hybrid systems, especially on ordered azobenzene–siloxane hybrids, with photomechanical functions. Self-assembly of designed azobenzene-modified silanes leads to the formation of siloxane-based hybrids with ordered nanostructures. Photo-induced variations in the structures at the microscopic scale and bending/curling motions at the macroscopic scale have been observed. In addition, other organic–inorganic hybrids such as azobenzene-intercalated layered inorganic compounds and azobenzene–carbon nanocomposites with photomechanical functions are also briefly introduced. These hybrid materials potentially have higher thermal stability and mechanical strength compared to conventional organic materials, which will be advantageous for practical applications as actuators and other smart devices.

Mechanically Responsive Materials for Soft Robotics, First Edition. Edited by Hideko Koshima.

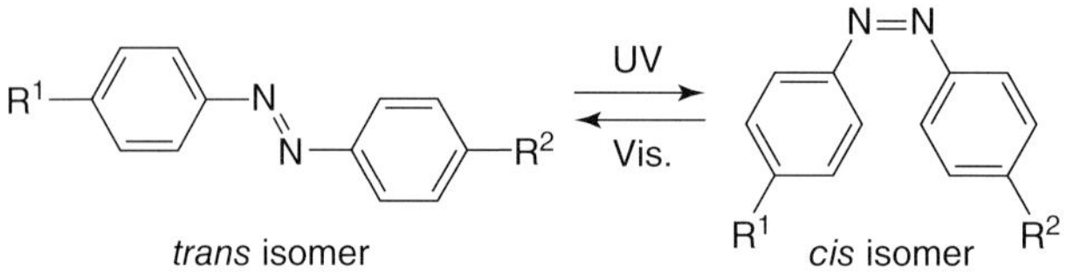

Figure 10.1 Photoisomerization of azobenzene.

10.2 Azobenzene as Organic Components

Azobenzene and its derivatives are known to undergo isomerization between the thermodynamically stable trans isomers and the metastable cis isomers upon ultraviolet (UV) and visible (Vis) light irradiations. These isomers have dramatic differences in their molecular shapes and polarities (Figure 10.1). The isomerization process is completed in picoseconds with high quantum yield and is fully reversible [10].

The mechanism of the trans–cis isomerization has been investigated intensively. It is generally considered that the isomerization takes place through rotation, inversion, concerted inversion, and/or inversion-assisted conversion. For isomerization, enough free volume is necessary (~127 Å^3 for the trans-to-cis isomerization of unsubstituted azobenzene [10]). In the solution state, azobenzene is easy to isomerize, whereas isomerization tends to be inhibited in the solid state due to the lack of enough free volume.

Ordered arrangement of the azobenzene moiety is also crucial to convert the molecular-level shape change or motions into macroscopic responses. In the organic systems, rational design of azobenzene-containing polymers has achieved various photomechanical responses. The most successful example is bending of the liquid crystalline (LC) films along controlled directions by macroscopic ordering of the films on rubbed polymer coatings or by irradiation of linearly polarized light on polydomain LC films [11]. For designing azobenzene–inorganic hybrid materials with photomechanical functions, precise control of the nanostructure of the inorganic moiety as well as the arrangement of the azobenzene moiety should be important.

10.3 Siloxane-Based Organic–Inorganic Hybrids

The sol–gel method provides a convenient route to inorganic oxides under mild conditions [12, 13], which enables the incorporation of thermally unstable organic components to produce organic–inorganic hybrid materials. Incorporation of organic species can endow versatile functions and properties to the inorganic matrices. Siloxane-based hybrids are most extensively studied because of good thermal and chemical stability, diversity of structures, ease in the morphology control (e.g. film, fiber, monolith, nanoparticles, etc.), and tunable physical properties, leading to a wide range of applications such as in optics, optoelectronics, sensors, and catalysis [14, 15]. Typically, alkoxysilanes ($Si(OR)_4$, R = Et, Me, etc.) or organoalkoxysilanes ($R'_nSi(OR)_{4-n}$, R′ = organic substituents) are used as the precursors, and Si—O—Si networks are formed by hydrolysis and polycondensation reactions as shown below.

$$\text{Hydrolysis reaction} \equiv \text{Si—OR} + \text{H}_2\text{O} \rightarrow$$
$$\equiv \text{Si—OH} + \text{ROH (R = Et, Me, etc.)}$$
$$\text{Polycondensation reaction} \equiv \text{Si—OH} + \text{XO—Si} \equiv \rightarrow$$
$$\equiv \text{Si—O—Si} \equiv + \text{XOH (X = H or R)}$$

According to the organic–inorganic interfaces, siloxane-based hybrids prepared by the sol–gel method can be classified into two types – (i) hybrids with noncovalent interactions such as van der Waals force, hydrogen bonding, and electrostatic interactions (Figure 10.2a) and (ii) hybrids with covalent bonds (Figure 10.2b). The hybrids with noncovalent organic–inorganic interfaces can be easily synthesized by hydrolysis and polycondensation of alkoxysilanes in the presence of organic molecules and polymers. The homogeneity of the hybrids depends on the affinity of the siloxane and organic moieties, and phase separation occurs if compatibility is low. Additionally, leaching of the organic components out from the siloxane matrices may occur. On the other hand, starting from organoalkoxysilanes ($R'_n Si(OR)_{4-n}$), organic groups (R′) can be uniformly and stably immobilized on the siloxane networks by Si—C bonds. The products can be obtained as oils, soluble solids, or insoluble resins with various structures such as discrete cage/cyclic structures, linear structures, two-dimensional (2D) layer structures, and three-dimensional (3D) structures, which depend on the types and the number of the R′ groups and the reaction conditions.

The R′ group in organoalkoxysilanes can be varied from simple hydrocarbon groups (e.g. methyl, phenyl, and vinyl) to more complex functional organic groups. The Si—C bonds can be formed by various reactions such as hydrosilylation reaction and Grignard reaction. Furthermore, a wide range

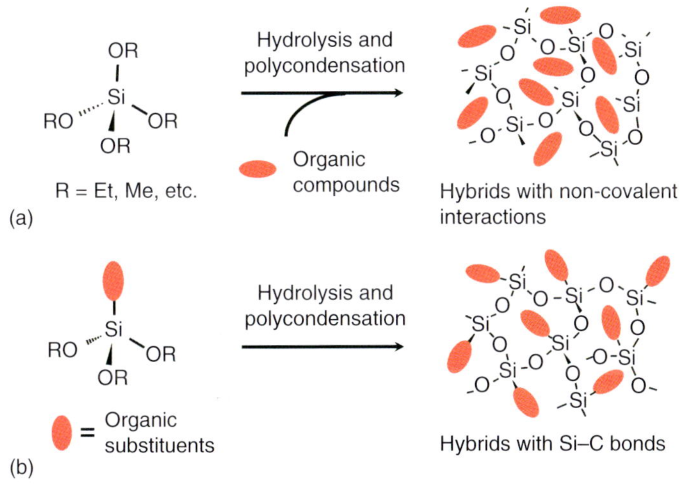

Figure 10.2 Siloxane-based organic–inorganic hybrids prepared by the sol–gel method.

of organic compounds can be introduced by utilizing a grafting reaction to organoalkoxysilanes with an amino group, a thiol group, and a vinyl group. Incorporation of photofunctional organic compounds has been actively researched because of the strong demand for their stabilization in addition to the transparency of siloxane networks in the UV and Vis regions. Various organic chromophores have been incorporated in the amorphous siloxane networks to create photofunctional materials with unique photochromic, luminescent, and nonlinear optical (NLO) properties [16].

Structural control of the siloxane-based hybrid materials can be achieved by self-assembly processes. The surfactant-directed, self-assembly process is the most effective and commonly used (Figure 10.3a). Surfactants such as amphiphilic ionic molecules (e.g. alkyltrimethylammonium bromide) and amphiphilic block copolymers (e.g. polyethylene oxide (PEO)–polypropylene oxide (PPO)–PEO-type triblock copolymers) interact with hydrolyzed alkoxysilanes to self-assemble into ordered nanocomposites with various mesostructures such as 2D/3D hexagonal and 3D cubic structures [17]. For functionalization, organic compounds can be doped within the hydrophobic region of the mesostructured silica–surfactant composites through coassembly (Figure 10.3a, top). For example it is reported that a mesostructured silica–block copolymer film with a 2D hexagonal structure is a good host for laser dyes such as rhodamine 6G [18]. Incorporation of photofunctional molecules into lamellar silica–surfactant mesostructures was also reported. However, these hybrid materials have a drawback of low stability against leaching of organic species.

Covalent linking of organic groups on siloxane networks can be achieved by co-condensation of organotrialkoxysilanes and tetraalkoxysilane in the presence of surfactants (Figure 10.3a, middle) [19]. Lebeau et al. synthesized a transparent thin film of dye-functionalized mesostructured silica by co-condensation method using organoalkoxysilane bearing a NLO chromophore, 3-(2,4-dinitrophenylamino)propyltriethoxysilane [20]. These photofunctional hybrid mesostructures are important for many fields such as optics and sensors. The organic groups can be embedded into the siloxane networks when

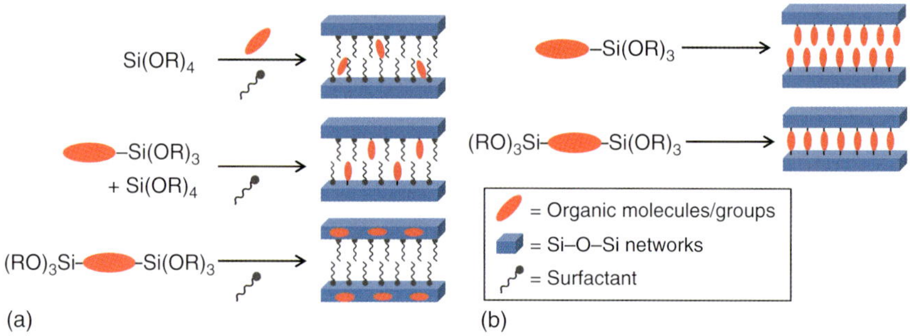

Figure 10.3 Self-assembly processes to produce siloxane-based organic–inorganic hybrids with ordered mesostructures. Lamellar structures are shown as typical examples. (a) Surfactant-directed self-assembly and (b) self-directed assembly of organosilanes.

organoalkoxysilanes with bridging organic groups are used as the precursors (Figure 10.3a, bottom) [19, 21, 22].

Ordered siloxane-based hybrids can also be obtained by the surfactant-free, self-directed assembly of organoalkoxysilanes based on the intermolecular interactions such as van der Waals force, hydrogen bonding, and π–π interaction [23, 24]. It is expected that the organic groups are more densely and homogeneously distributed in the hybrids compared to those by the aforementioned surfactant-directed, self-assembly process. Typically, organoalkoxysilanes with pendant and bridging organic groups ($R'{-}Si(OR)_3$ and $(RO)_3Si{-}R'{-}Si(OR)_3$, respectively) form lamellar mesostructures (Figure 10.3b). Other mesostructures such as hexagonal and cubic structures can be formed by adjusting the packing parameter of the precursors. By introducing a bulky alkoxy-oligosiloxane instead of a single trialkoxysilyl group in $R'{-}Si(OR)_3$, formation of mesostructures with higher surface curvature becomes more favorable [25]. These mesostructures consisting of cylindrical or spherical assembles are useful for producing ordered mesoporous silica by removal of the organic groups.

10.4 Photoresponsive Azobenzene–Siloxane Hybrid Materials

10.4.1 Nanostructural Control by Self-Assembly Processes

As described in Section 10.3, siloxane-based hybrid materials with ordered structures can be obtained by the sol–gel processes in combination with the self-assembly processes. The use of azobenzene-modified organoalkoxysilane precursors allows molecular-level hybridization of azobenzene groups with siloxane networks by Si—C bonds. Ordered arrangement of the azobenzene groups through nanostructural control of the hybrids is the key to realizing novel photoresponsive properties.

The surfactant-directed, self-assembly process has been widely employed for preparing azobenzene-functionalized ordered mesoporous materials. The azobenzene groups can be grafted onto the pore surfaces by using precursors having a pendant azobenzene group ($R'Si(OR)_3$, R′ = azobenzene group) [26–31]. On the other hand, the azobenzene groups can be embedded in the pore walls by using the precursors with a bridging azobenzene group ($(RO)_3Si{-}R'{-}Si(OR)_3$) [32, 33]. Photoresponsive properties such as reversible changes in the pore sizes and photo-controlled release of guest molecules have been achieved. However, photomechanical functions of these hybrids have not been reported because of the rigid 3D siloxane frameworks.

Surfactant-free, self-directed assembly of azobenzene-modified organoalkoxysilane leads to highly ordered and high-density arrangement of azobenzene groups. Liu et al. reported the self-assembly of 4,4′-bis(3-triethoxysilylpropylureido)azobenzene into a highly ordered lamellar structure by intermolecular hydrogen bonding [34]. However, photoisomerization was inhibited possibly due to the strong intermolecular hydrogen bonding between the bis-ureide groups. Manipulation of the intermolecular interactions is significant for the design of photoresponsive materials. To address this issue, we prepared azobenzene–siloxane hybrids by surfactant-free assembly of rationally designed azobenzene-modified organoalkoxysilanes. The synthesis and structures of the hybrids and their photomechanical responses at the microscopic and macroscopic scales are described in the following sections.

10.4.2 Lamellar Siloxane-Based Hybrids with Pendant Azobenzene Groups

In this section, our recent work on the design of azobenzene-functionalized, siloxane-based hybrids with photomechanical functions is described. To obtain azobenzene–siloxane hybrids capable of photoisomerization, we designed and synthesized organoalkoxysilanes having a pendant-type azobenzene group (**P1** and **P2** in Figure 10.4) through hydrosilylation reactions. These precursors have similar structures but with different numbers of ethoxy groups. The $Si(OEt)_3$ group in **P1** can form cross-linked siloxane networks by hydrolysis and polycondensation, whereas the $SiMe(OEt)_2$ group in **P2** can form only linear siloxanes. To avoid strong intermolecular interactions such as hydrogen bonding, the spacers between Si atoms and the azobenzene groups are simple $-O(CH_2)_3-$ chains. In ethanol solutions, both precursors show quick and reversible trans–cis photoisomerization upon UV/Vis irradiations.

When hydrolysis and polycondensation of **P1** and **P2** were performed in an ethanol solvent under an acidic condition, plate-like particles were precipitated (Figure 10.5) [35]. X-ray diffraction (XRD) analysis revealed the formation of lamellar structures, indicating that hydrolyzed **P1** and **P2** have self-assembling abilities. To obtain thin films suitable for photoisomerization experiments, clear hydrolyzed solutions were prepared by using tetrahydrofuran (THF) as the solvent instead of ethanol. Spin coating of these solutions on glass plates gave lamellar thin films (**H1** and **H2**) by evaporation-induced self-assembly and subsequent polycondensation of the hydrolyzed species. The *d*-spacings of the lamellar structures were 3.20 and 2.37 nm for **H1** and **H2**, respectively (Figure 10.6a). The large difference in the *d*-spacings suggested the different tilt angles of azobenzene groups between the siloxane layers (Figure 10.6b). The arrangement of the azobenzene groups was thus influenced by the type of silyl groups. In addition to hydrogen bondings between the silanol (SiOH) groups, the interaction between the azobenzene groups (π–π interactions) might play a role

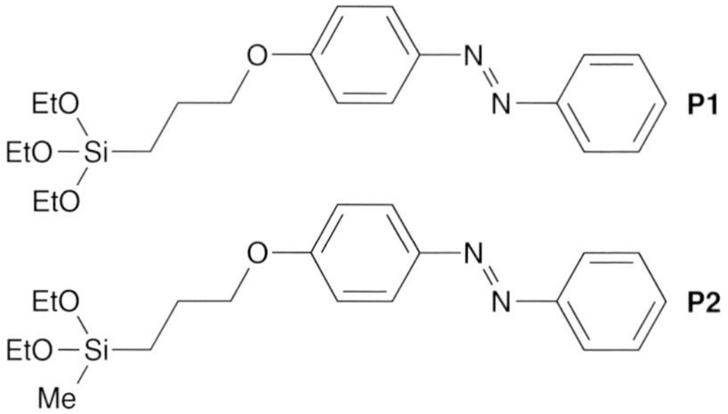

Figure 10.4 Pendant-type azobenzene-modified organoalkoxysilane precursors **P1** and **P2**.

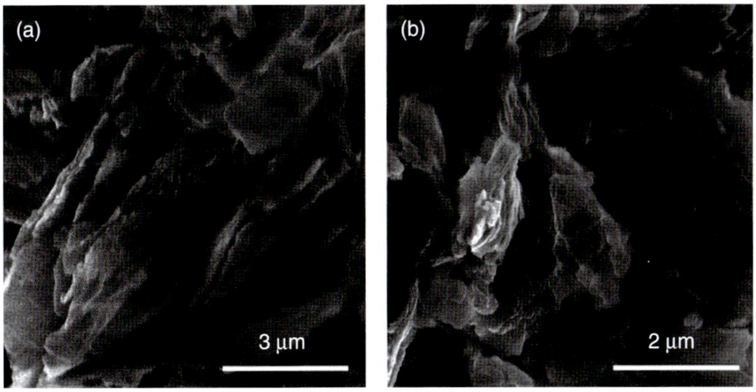

Figure 10.5 Scanning electron microscopy (SEM) images of the precipitates formed from (a) **P1** and (b) **P2**. Source: Guo et al. 2013 [35]. Adapted with permission of The Royal Society of Chemistry.

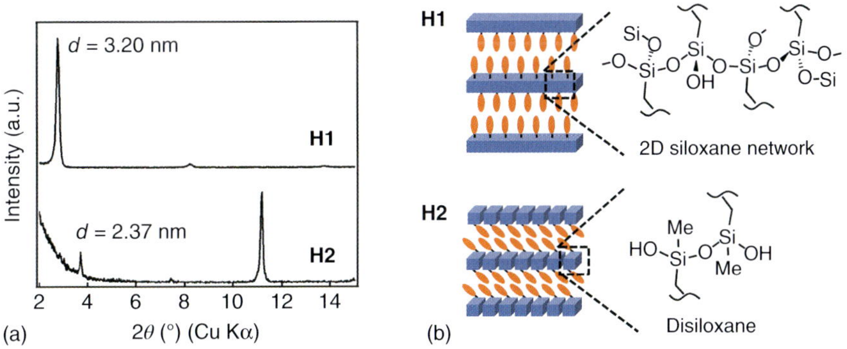

Figure 10.6 (a) XRD patterns and (b) structural models of **H1** and **H2**. Source: Guo et al. 2013 [35]. Adapted with permission of The Royal Society of Chemistry.

in the formation of the lamellar structures. Solid-state ^{29}Si magic-angle spinning (MAS) NMR analysis showed that **H1** had a siloxane network consisting of $R'Si(OSi)_2(OH)$ and $R'Si(OSi)_3$ units (R' = azobenzene group), whereas **H2** was an assembly of dimers ($R'(HO)MeSi–O–SiMe(OH)R'$) (Figure 10.6b).

Photoisomerization of the azobenzene groups in **H1** and **H2** was investigated. Upon UV (340 nm) and subsequent Vis (>420 nm) irradiations, reversible and partial trans–cis photoisomerizations were observed by UV–Vis spectroscopy (Figure 10.7a). Note that the isomerization degree of **H2** is higher than that of **H1**. This is explainable by the difference in the degree of condensation. In contrast to **H1** having a cross-linked siloxane network, the lamellar structure of **H2** is maintained by weak interactions between dimers, which should allow higher mobility of azobenzene groups in **H2**.

Meanwhile, reversible changes of the lamellar periodicities were confirmed by XRD (Figure 10.7b). After UV irradiation, slight decrease (~0.04 nm) and increase (~0.03 nm) in the *d*-spacings were observed for **H1** and **H2**, respectively. Subsequent Vis irradiation recovered the original *d*-spacings. The

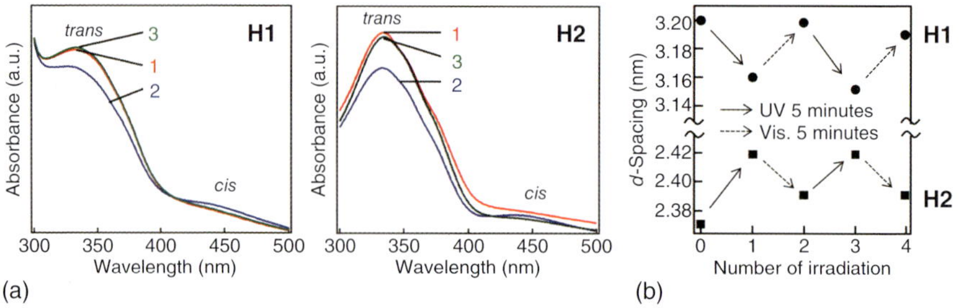

Figure 10.7 (a) UV–Vis absorption spectra of **H1** and **H2** (1, before irradiation; 2, after five minutes of UV irradiation; 3, after subsequent five minutes of Vis irradiation) and (b) reversible changes in their *d*-spacings upon UV and Vis irradiations. Source: Guo et al. 2013 [35]. Adapted with permission of The Royal Society of Chemistry.

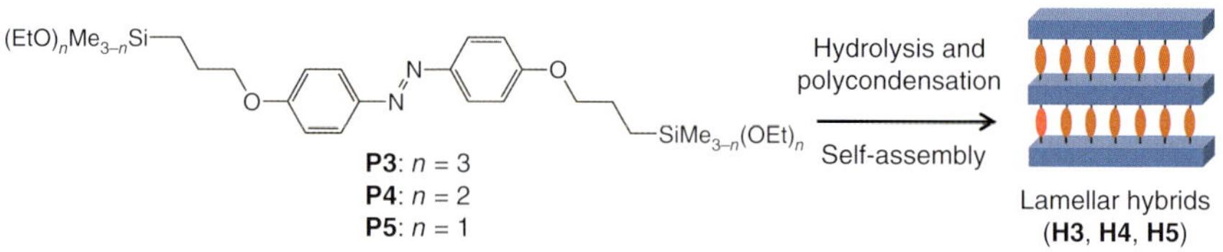

Figure 10.8 Bridging-type, azobenzene-modified organoalkoxysilane precursors **P3**, **P4**, and **P5** and their hydrolysis and polycondensation to form lamellar hybrids **H3**, **H4**, and **H5**, respectively.

opposite changes between **H1** and **H2** are likely associated with the different tilt angles of the azobenzene groups, although the details are unclear. The variations of the *d*-spacings indicate that the film thicknesses are slightly changed by photoirradiation. These behaviors can be regarded as photomechanical responses at the microscopic scale. The ordered arrangements of azobenzene groups in the lamellar structure have enabled their corporative activities to result in changes in the whole structures (changes in *d*-spacing). This is the first report demonstrating the photo-induced structural variation in the ordered azobenzene–siloxane hybrids prepared by the sol–gel method.

10.4.3 Lamellar Siloxane-Based Hybrids with Bridging Azobenzene Groups

As previously discussed, we have successfully obtained lamellar azobenzene–siloxane hybrids showing photo-induced reversible *d*-spacing variations. To understand deeply the molecular factors that affect the self-assembly behaviors of azobenzene-modified alkoxysilanes and photoresponsive properties of the resulting ordered hybrids, organoalkoxysilane precursors having bridging azobenzene groups were investigated [36]. We synthesized bridging-type precursors **P3**, **P4**, and **P5** with different numbers of ethoxy groups ($n = 3$, 2, and 1, respectively) on each Si atom (Figure 10.8). The spacers between Si and azobenzene groups are the same as those for **P1** and **P2**.

Lamellar hybrids **H3**, **H4**, and **H5** were obtained by hydrolysis and polycondensation of **P3**, **P4**, and **P5**, respectively. **P3** formed lamellar precipitates

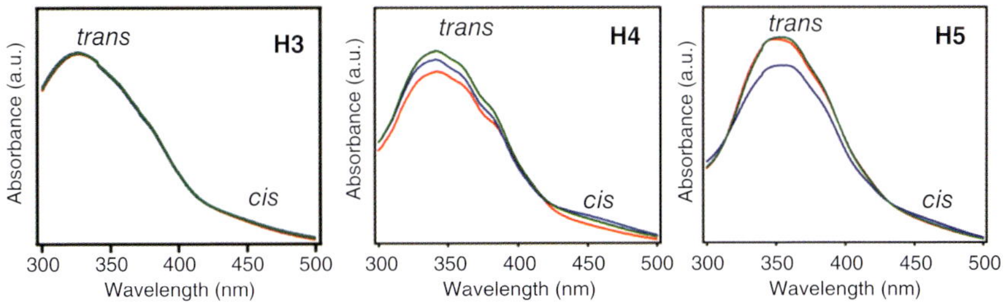

Figure 10.9 UV–Vis absorption spectra of **H3–H5** (red: before irradiation, blue: after five minutes of UV irradiation, green: after subsequent five minutes of Vis irradiation). Source: Guo et al. 2014 [36] Adapted with permission of The Royal Society of Chemistry.

upon hydrolysis in THF, which suggested the strong self-assembling ability of hydrolyzed **P3** due to the large number of Si–OH groups compared to the other two precursors. Self-assembly of hydrolyzed **P4** and **P5** did not occur in THF; therefore, the hydrolyzed solutions were spin coated to obtain lamellar thin films. To investigate the photoisomerization properties of azobenzene in these hybrids, UV–Vis spectra (Figure 10.9) and XRD patterns before and after UV/Vis irradiations were recorded. **H3** showed no changes in the UV–Vis spectra, indicating that isomerization was inhibited probably due to the low mobility of the tightly fixed azobenzene groups. On the other hand, **H4** and **H5** showed partial but reversible photoisomerization, suggesting the increased mobility of the azobenzene groups due to the decreased numbers of Si–O–Si linkages (i.e. lower degree of polycondensation). Note that absorption peaks of these hybrids were blue-shifted compared to their precursors, indicating that H aggregates were formed. This should be responsible for the low degrees of photoisomerization [37]. UV irradiation on **H4** caused disappearance of the XRD peaks; however, the lamellar structure was not recovered after subsequent Vis irradiation. No changes in the XRD patterns of **H3** and **H5** were observed.

10.4.4 Photo-Induced Bending of Azobenzene–Siloxane Hybrid Film

As described in the Sections 10.4.2 and 10.4.3, hydrolyzed organoalkoxysilanes with either pendant or bridging azobenzene groups possessed a self-assembly ability to form lamellar structures. For the hybrids derived from pendant-type precursors, photo-induced structure changes at the microscopic scale were observed, whereas no structural change has been observed for the hybrids derived from bridging-type precursors. In order to realize more dynamic and macroscopic photomechanical responses of the azobenzene–siloxane hybrids, we synthesized new precursors, **P6** (pendant-type) and **P7** (bridging-type), through thiol–ene reactions (Figure 10.10) [38]. Compared with **P1** and **P3**, the spacers between Si atoms and the azobenzene groups are longer and more flexible.

Hybrid films **H6** and **H7** were prepared from **P6** and **P7**, respectively, through the sol–gel processes. Lamellar structures were confirmed by XRD and transmission electron microscopy (TEM) (Figure 10.11a,b). **H6** showed reversible

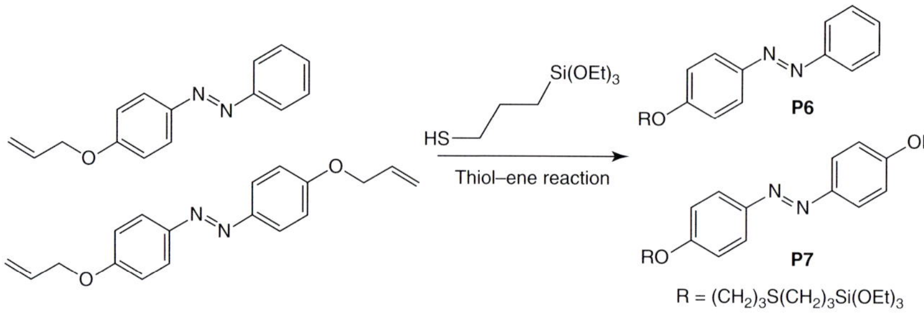

Figure 10.10 Synthesis of precursors **P6** and **P7** through thiol–ene reaction.

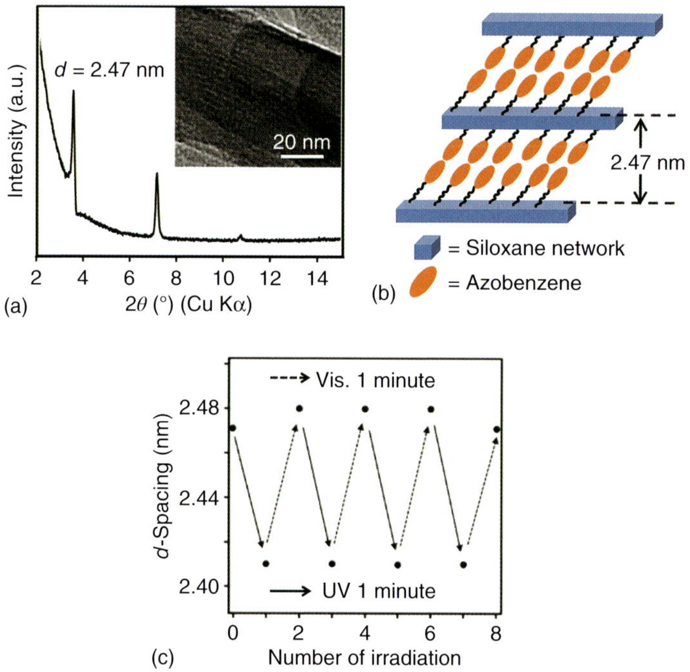

Figure 10.11 (a) XRD pattern of hybrid film **H6** prepared from **P6** (inset: TEM image), (b) a structural model of **H6**, and (c) variation in the *d*-spacing of **H6** upon UV and Vis irradiations.

and partial trans–cis photoisomerization and *d*-spacing changes upon UV/Vis irradiations. Compared to **H1**, **H6** displayed much quicker (within 10 seconds) and larger (~0.06 nm) changes of the *d*-spacing (Figure 10.11c). Reversible photoisomerization of the bridging azobenzene groups was also observed for **H7**, although there was no change in the *d*-spacing. These results suggest the improved mobility of the azobenzene groups owing to the increased length and flexibility of the spacers (cf. **H1** and **H3**).

A flexible, freestanding film **Hm** with a thickness of several micrometers was successfully prepared by co-hydrolysis and polycondensation of a mixture of **P6** and **P7** (Figure 10.12). The hydrolyzed solution was cast on a substrate coated

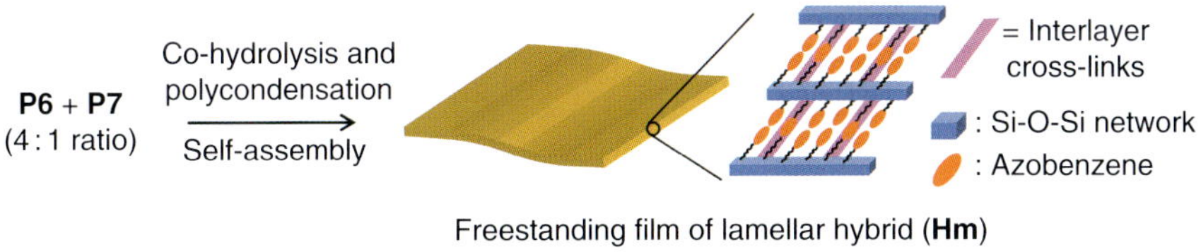

Figure 10.12 Preparation of a freestanding film of lamellar hybrid (**Hm**) by co-hydrolysis and polycondensation of **P6** and **P7**.

with poly(vinyl alcohol) (PVA) as a sacrifice layer so that the hybrid film could be peeled off from the substrate by dissolving PVA. The film consists of a single lamellar phase with a *d*-spacing of 2.64 nm, which is in between those of **H6** and **H7**. This result indicates that both precursors are homogeneously mixed to form a lamellar structure in which adjacent layers are partially cross-linked by bridging azobenzene groups derived from **P7**.

Interestingly, upon UV irradiation on **Hm**, the film bent away from the UV source with a large bending angle (180°) (Figure 10.13a) and the bent shape was maintained after stopping the UV irradiation. The bent film reverted to the original shape upon subsequent Vis irradiation. This reversible bending–unbending motion can be repeated without any detectable fatigue. UV–Vis absorption spectra showed partial but reversible trans–cis isomerizations of the azobenzene groups in **Hm** upon UV/Vis irradiations (Figure 10.13b).

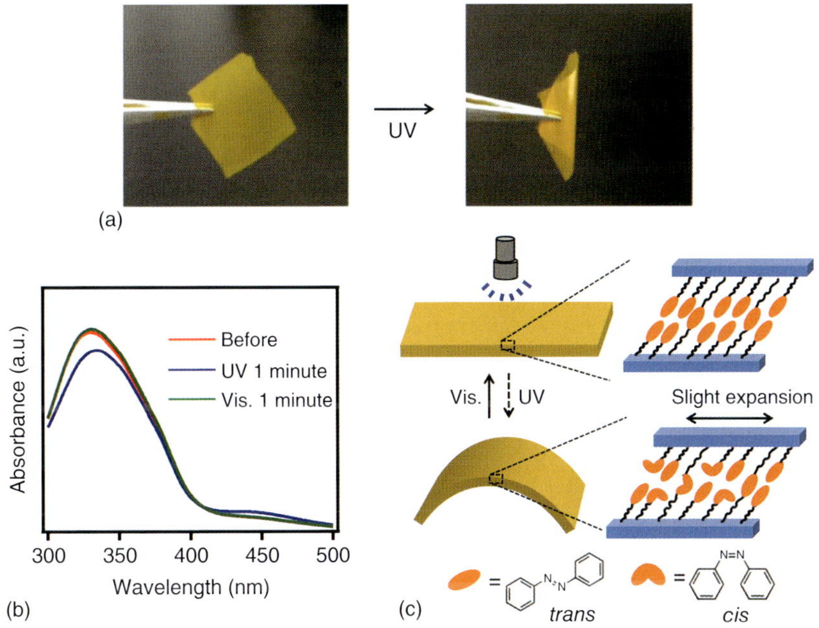

Figure 10.13 (a) Photo-induced bending of **Hm** upon UV irradiation, (b) UV–Vis spectra of a thin film of **Hm** before and after UV/Vis irradiations, and (c) a proposed mechanism of bending. Source: Guo et al. 2015 [38]. Copyright 2015. Reproduced with permission of American Chemical Society.

Figure 10.14 Photo-induced reversible curling motions of **Hm** in water: (a) before irradiation, (b) after one minute of UV irradiation, and (c) after subsequent one minute of Vis irradiation. Source: Guo et al. 2015 [38]. Copyright 2015. Reproduced with permission of American Chemical Society.

The proposed mechanism of the photo-induced bending is shown in Figure 10.13c. Because of the high concentration of azobenzenes and large thickness of the film (**Hm**), UV irradiation can induce trans-to-cis isomerization only in the surface region of the film. The bent cis isomers possess a larger size along the horizontal plane than the trans isomers do, expansion should occur in the horizontal direction and was transduced to the inner part of the film through the bridged azobenzene groups, leading to the bending of the film away from the UV source. It is considered that the thin siloxane layers in **Hm** are flexible enough to allow such an expansion. Note that the bending direction of **Hm** is opposite to that of the conventional azobenzene-containing LC polymers (bend toward UV light source) [2]. This can be explained by the different orientation of the azobenzene groups in the films.

When irradiations were conducted on **Hm** in water, an interesting reversible curling and uncurling motions were observed upon UV/Vis irradiations (Figure 10.14), demonstrating that our materials have potential application in smart aquatic devices. This is the first report on azobenzene–siloxane hybrids capable of bending and curling motions at the macroscopic scale upon UV/Vis irradiations.

The hybrid film **Hm** exhibited excellent thermal and mechanical properties. The thermogravimetry (TG) curve obtained in air showed almost no weight loss up to 200 °C. In the differential scanning calorimetry (DSC) curve, no phase transition was observed in the temperature range of 0–150 °C. The dynamic mechanical analysis of **Hm** at room temperature suggested that the storage modulus was 0.2 GPa. This is much larger than that of liquid crystal (LC) elastomers (<1 MPa). The reversible photo-induced bending–unbending of the film was also observed in a wide temperature range from room temperature up to 130 °C.

10.4.5 Control of the Arrangement of Azobenzene Groups

Although photomechanical responses at microscopic and macroscopic scales have been observed for the lamellar azobenzene–siloxane hybrids, the azobenzene groups are prone to form H aggregates resulting in low degrees of photoisomerization. To address this issue, we designed a new precursor having a bulky cage-type oligosiloxane instead of an alkoxysilyl groups to adjust the

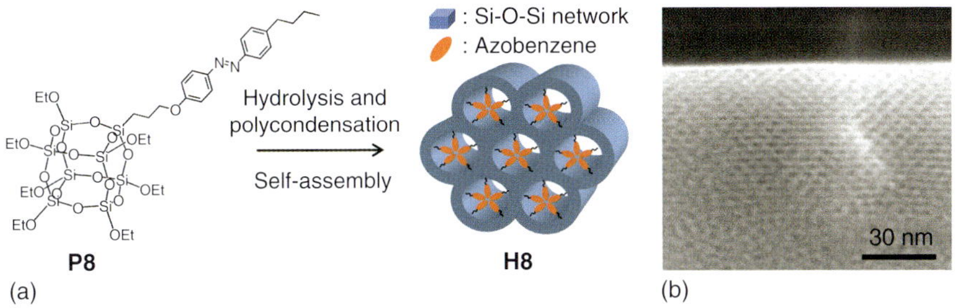

Figure 10.15 (a) Schematic depicting the self-assembly of **P8** into a mesostructure consisting of cylindrical assemblies and (b) HAADF-STEM image of the cross section of the **H8** film.

molecular arrangement. The azobenzene-modified cage-type siloxane precursor (**P8**) consists of an octasiloxane cage with seven hydrolyzable –OEt groups and an azobenzene group bearing flexible alkyl chains at both ends (Figure 10.15a) [39]. A yellow transparent thin film (**H8**) was obtained by spin coating the THF solution of hydrolyzed **P8** on a substrate.

The XRD pattern of **H8** indicated an ordered structure with a *d*-spacing of 3.52 nm. The 2D XRD pattern and the high-angle annular dark-field (HAADF)–scanning transmission electron microscope (STEM) image (Figure 10.15b) revealed a distorted 2D hexagonal structure. It is plausible that the amphiphilic hydrolyzed **P8** were self-assembled into cylinders with the azobenzene chains inside and the siloxane cages outside. When the siloxane cage of **P8** was replaced with a single $Si(OEt)_3$ group (denoted as **P9**), a lamellar structure (**H9**) was obtained, indicating that the bulky siloxane cage contributed to the formation of the cylindrical assemblies.

Compared with the lamellar hybrid film **H9** and **H1**–**H7**, a much higher degree of trans–cis isomerization of azobenzene groups was observed for **H8** (Figure 10.16). Also note that when azobenzene is doped in silica gel matrix or covalently attached on mesoporous silica, photoisomerization is severely inhibited or requires a considerably longer time to occur [40–42]. It is likely

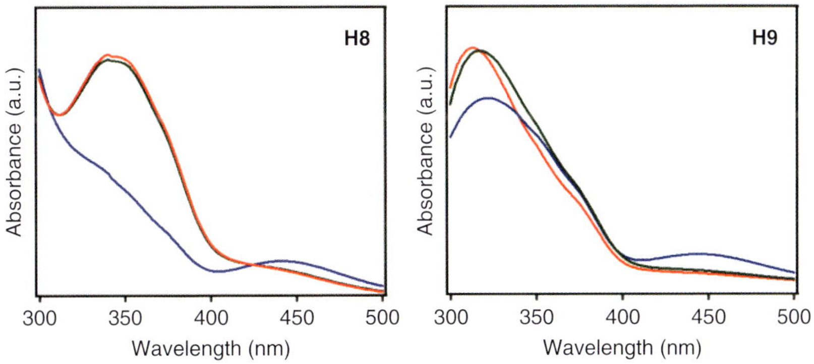

Figure 10.16 UV–Vis spectra of **H8** and **H9** (red: before irradiation; blue: after three minutes of UV irradiation; green: after subsequent three minutes of Vis irradiation). Source: Guo et al. 2017 [39]. Adapted with permission of The Chemical Society of Japan.

that the bulky siloxane cages in **H8** suppressed the aggregation of the attached azobenzene groups and provided a larger free volume for isomerization. However, despite the high degree of photoisomerization in **H8**, no structural change has been observed at the microscopic scale (i.e. variation of the *d*-spacing). This is attributed to the 3D siloxane frameworks of **H8**, which is inherently different from the stacked 2D siloxane frameworks of the lamellar structures (e.g. **H1** and **H6**). Further control of the nanostructure and preparation of a freestanding film will be required for photomechanical responses.

10.5 Other Azobenzene–Inorganic Hybrids

10.5.1 Intercalation Compounds

Intercalation of azobenzene derivatives into layered inorganic compounds is a facile method to prepare ordered photoresponsive hybrid materials. Pioneering work was done by Ogawa et al. They intercalated cationic azobenzene derivatives into layered silicates through ion exchange and observed reversible *d*-spacing changes upon UV/Vis irradiations [43, 44]. More recently, Nabetani et al. reported reversible *d*-spacing changes and sliding motions of the azobenzene-intercalated layered niobate upon UV/Vis irradiations [45, 46]. The morphology change due to the nanosheet sliding mechanism is interesting as it is similar to the phenomenon of human muscle's shrink and expansion. Furthermore, when a cationic azobenzene derivative bearing a polyfluoroalkyl group (Figure 10.17a) was intercalated, nanoscrolls were formed by stirring-induced rolling of the exfoliated niobate nanosheets. The nanoscrolls showed winding and unwinding motions upon photoisomerization of the intercalated azobenzene derivatives (Figure 10.17b) [46]. The nanoscrolls wound more tightly after UV irradiation and reverted to the loosely wound state after subsequent Vis irradiation. Such layered hybrid materials will be a new candidate for the development of materials for actuators and artificial muscle.

10.5.2 Hybridization with Carbon-Based Materials

Carbon-based nanomaterials such as carbon nanotubes, graphene, and graphene oxide (GO) have also been used as inorganic components for preparing azobenzene-functionalized hybrid materials [47–50]. Photo-switchable properties of azobenzene–inorganic hybrids such as photo-controlled dispersion–aggregation behaviors of azobenzene-modified, single-walled carbon nanotubes [49] and photo-regulated conductivity in azobenzene-functionalized reduced GO [50] have been reported.

Furthermore, Yu and coworkers fabricated a nanocomposite film by mixing GO and two LCs (one bearing azobenzene groups) in PVA matrix [51]. The film is easily fabricated and exhibited excellent mechanical properties. The incorporation of GO contributed to the improved mechanical strength, efficient photothermal conversion, and good thermal conductivity. Ordered arrangement of azobenzene groups in the film was achieved by stretching.

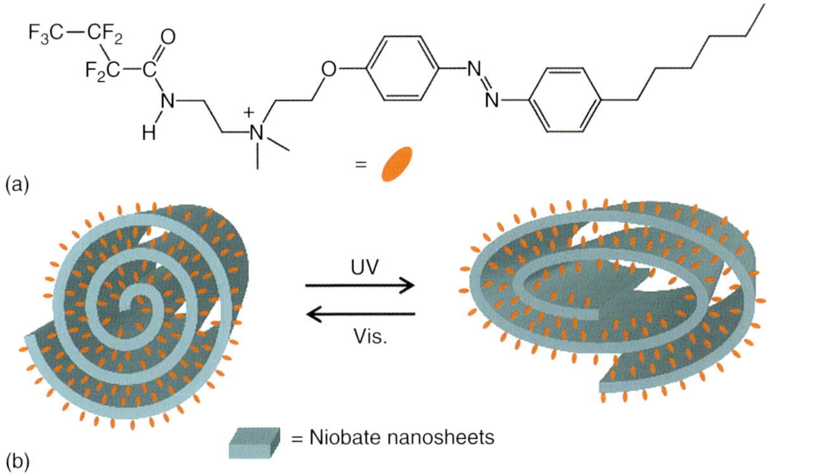

Figure 10.17 (a) The cationic azobenzene derivative used for intercalation. (b) A schematic image for photo-induced winding and unwinding motion of a niobate nanoscroll.

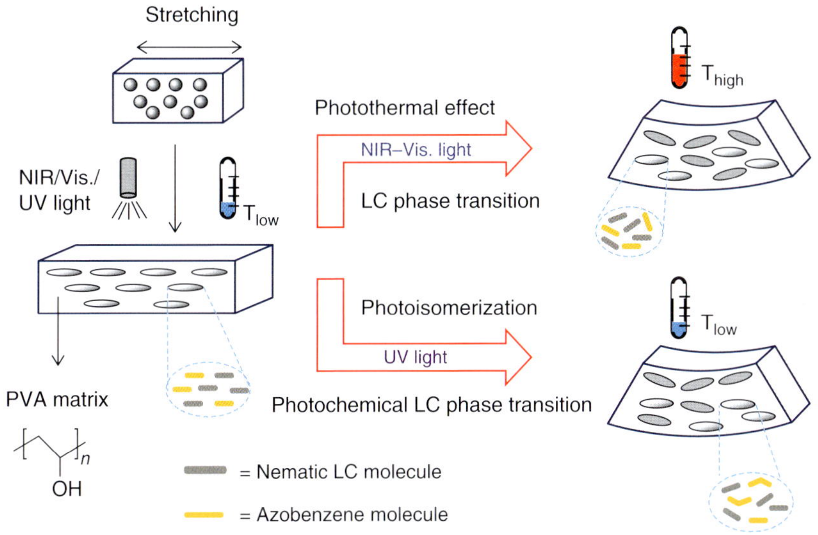

Figure 10.18 UV–NIR–Vis induced bending of an azobenzene/GO/PVA composite film. Source: Cheng et al. 2015 [51]. Copyright 2015, Reproduced with permission of American Chemical Society.

Both photochemical (photoisomerization of azobenzenes under UV irradiation) and photothermal effects (temperature increase upon Vis and near-infrared [NIR] irradiation) induced phase transition of the composite LC film from nematic (N) phase to isotropic (I) phase, causing bending of the film at the macroscopic scale (Figure 10.18).

10.6 Summary and Outlook

In this chapter, ordered azobenzene–siloxane hybrid materials with photomechanical functions were mainly introduced. Oriented lamellar structures were easily formed by the one-step self-assembly process using organoalkoxysilane precursors having pendant or bridging azobenzene groups. Reversible changes of the structures at the microscopic scale and the macroscopic bending and curling motions have been observed. High-density and regular arrangement of azobenzene groups was essential for realizing these photomechanical functions.

Although unique properties and functions have been demonstrated in the organic–inorganic hybrid systems, many challenges still remain. Bending is important as a basic motion, and high bending frequency and generation of large stress are necessary for practical applications. Increase of the types of photomechanical motions is also a significant issue. In addition to bending/curling/winding described in this chapter, other motions such as vibration, folding, and more complicated ones are required for applications in soft robotics. Precise design of the structure of the hybrid at the molecular scale to the mesoscale will enable us to address these issues. Hybridization with upconversion inorganic nanoparticles that can convert NIR or Vis light to the UV light to induce trans-to-cis isomerization of azobenzene is also an important subject of research especially for biomedical applications.

References

1 Barrett, C.J., Mamiya, J., Yager, K.G., and Ikeda, T. (2007). Photo-mechanical effects in azobenzene-containing soft materials. *Soft Matter* 3: 1249–1261.
2 Bisoyi, H.K. and Li, Q. (2016). Light-driven liquid crystalline materials: from photo-induced phase transitions and property modulations to applications. *Chem. Rev.* 116: 15089–15166.
3 Ikeda, T., Mamiya, J., and Yu, Y. (2007). Photomechanics of liquid-crystalline elastomers and other polymers. *Angew. Chem. Int. Ed.* 46: 506–528.
4 Jiang, H., Li, C., and Huang, X. (2013). Actuators based on liquid crystalline elastomer materials. *Nanoscale* 5: 5225–5240.
5 Bushuyev, O.S., Aizawa, M., Shishido, A., and Barrett, C.J. (2017). Shape-shifting azo dye polymers: towards sunlight-driven molecular devices. *Macromol. Rapid Commun.* 39: 1700253.
6 Koshima, H., Ojima, M., and Uchimoto, H. (2009). Mechanical motion of azobenzene crystals upon photoirradiation. *J. Am. Chem. Soc.* 131: 6890–6891.
7 Taniguchi, T., Fujisawa, J., Shiro, M. et al. (2016). Mechanical motion of chiral azobenzene crystals with twisting upon photoirradiation. *Chem. Eur. J.* 22: 7950–7958.
8 Hagaman, D.E., Leist, S., Zhou, J., and Ji, H.-F. (2018). Photoactivated polymeric bilayer actuators fabricated via 3D printing. *ACS Appl. Mater. Interfaces* 10: 27308–27315.

9 Wie, J.J., Chatterjee, S., Wang, D.H. et al. (2014). Azobenzene-functionalized polyimides as wireless actuators. *Polymer* 55: 5915–5923.
10 Klajn, R. (2010). Immobilized azobenzenes for the construction of photoresponsive materials. *Pure Appl. Chem.* 82 (12): 2247–2279.
11 Yu, Y., Nakano, M., and Ikeda, T. (2003). Directed bending of a polymer film by light. *Nature* 425 (6954): 145–145.
12 Brinker, C.J. and Scherer, G.W. (1990). *Sol-Gel Science*. San Diego, CA: Academic Press.
13 Levy, D. and Zayat, M. (eds.) (2015). *The Sol-Gel Handbook: Synthesis, Characterization, and Applications*. Weinheim: Wiley-VCH.
14 Chauhan, B.P.S. (ed.) (2011). *Hybrid Nanomaterials: Synthesis, Characterization, and Application*. NJ: Wiley.
15 Sanchez, C., Julián, B., Belleville, P., and Popall, M. (2005). Applications of hybrid organic–inorganic nanocomposites. *J. Mater. Chem.* 15: 3559–3592.
16 Sanchez, C., Ribot, F., and Lebeau, B. (1999). Molecular design of hybrid organic–inorganic nanocomposites synthesized via sol–gel chemistry. *J. Mater. Chem.* 9: 35–44.
17 Wan, Y. and Zhao, D. (2007). On the controllable soft-templating approach to mesoporous silicates. *Chem. Rev.* 107: 2821–2860.
18 Yang, P., Wirnsberger, G., Huang, H.C. et al. (2000). Mirrorless lasing from mesostructured waveguides patterned by soft lithography. *Science* 287: 465–467.
19 Hoffmann, F., Cornelius, M., Morell, J., and Fröba, M. (2006). Silica-based mesoporous organic–inorganic hybrid materials. *Angew. Chem. Int. Ed.* 45: 3216–3251.
20 Lebeau, B., Fowler, C.E., Hall, S.R., and Mann, S. (1999). Transparent thin films and monoliths prepared from dye-functionalized ordered silica mesostructures. *J. Mater. Chem.* 9: 2279–2281.
21 Fujita, S. and Inagaki, S. (2008). Self-organization of organosilica solids with molecular-scale and mesoscale periodicities. *Chem. Mater.* 20: 891–908.
22 Lu, Y., Fan, H., Doke, N. et al. (2000). Evaporation-induced self-assembly of hybrid bridged silsesquioxane film and particulate mesophases with integral organic functionality. *J. Am. Chem. Soc.* 122: 5258–5261.
23 Shimojima, A. and Kuroda, K. (2006). Designed synthesis of nanostructured siloxane–organic hybrids from amphiphilic silicon-based precursors. *Chem. Rec.* 6: 53–63.
24 Chemtob, A., Ni, L., Croutxé-Barghorn, C., and Boury, B. (2014). Ordered hybrids from template-free organosilane self-assembly. *Chem. Eur. J.* 20: 1790.
25 Kuroda, K., Shimojima, A., Kawahara, K. et al. (2014). Utilization of alkoxysilyl groups for the creation of structurally controlled siloxane-based nanomaterials. *Chem. Mater.* 26: 211.
26 Angelos, S., Johansson, E., Stoddart, J.F., and Zink, J.I. (2007). Mesostructured silica supports for functional materials and molecular machines. *Adv. Funct. Mater.* 17: 2261–2271.
27 Liu, N., Chen, Z., Dunphy, D.R. et al. (2003). Photoresponsive nanocomposite formed by self-assembly of an azobenzene-modified silane. *Angew. Chem. Int. Ed.* 42: 1731–1734.

28 Liu, N., Dunphy, D.R., Atanassov, P. et al. (2004). Photoregulation of mass transport through a photoresponsive azobenzene-modified nanoporous membrane. *Nano Lett.* 4: 551–554.
29 Angelos, S., Choi, E., Vögtle, F. et al. (2007). Photo-driven expulsion of molecules from mesostructured silica nanoparticles. *J. Phys. Chem. C* 111: 6589–6592.
30 Angelos, S., Yang, Y.-W., Khashab, N.M. et al. (2009). Dual-controlled nanoparticles exhibiting AND logic. *J. Am. Chem. Soc.* 131: 11344–11346.
31 Lu, J., Choi, E., Tamanoi, F., and Zink, J.I. (2008). Light-activated nanoimpeller-controlled drug release in cancer cells. *Small* 4: 421–426.
32 Alvaro, M., Benitez, M., Das, D. et al. (2005). Reversible porosity changes in photoresponsive azobenzene-containing periodic mesoporous silicas. *Chem. Mater.* 17: 4958–4964.
33 Besson, E., Mehdi, A., Lerner, D.A. et al. (2005). Photoresponsive ordered hybrid materials containing a bridged azobenzene group. *J. Mater. Chem.* 15: 803–809.
34 Liu, N., Yu, K., Smarsly, B. et al. (2002). Self-directed assembly of photoactive hybrid silicates derived from an azobenzene-bridged silsesquioxane. *J. Am. Chem. Soc.* 124: 14540–14541.
35 Guo, S., Sugawara-Narutaki, A., Okubo, T., and Shimojima, A. (2013). Synthesis of ordered photoresponsive azobenzene–siloxane hybrids by self-assembly. *J. Mater. Chem. C* 1 (42): 6989–6995.
36 Guo, S., Chaikittisilp, W., Okubo, T., and Shimojima, A. (2014). Azobenzene–siloxane hybrids with lamellar structures from bridge-type alkoxysilyl precursors. *RSC Adv.* 4: 25319–25325.
37 Deng, Y., Yuan, W., Jia, Z., and Liu, G. (2014). H- and J-aggregation of fluorene-based chromophores. *J. Phys. Chem. B* 118: 14536–14545.
38 Guo, S., Matsukawa, K., Miyata, T. et al. (2015). Photoinduced bending of self-assembled azobenzene–siloxane hybrid. *J. Am. Chem. Soc.* 137: 15434–15440.
39 Guo, S., Sasaki, J., Tsujiuchi, S. et al. (2017). Role of cubic siloxane cages in mesostructure formation and photoisomerization of azobenzene–siloxane hybrid. *Chem. Lett.* 46 (8): 1237–1239.
40 Ide, T., Ozama, Y., and Matsui, K. (2011). Photochemistry of azobenzene in sol–gel systems. *J. Non-Cryst. Solids* 357: 100–104.
41 Ueda, M., Kim, H.-B., and Ichimura, K. (1994). Photochemical and thermal isomerization of azobenzene derivatives in sol–gel bulk materials. *Chem. Mater.* 6: 1771–1775.
42 Ueda, M., Kim, H.-B., Ikeda, T., and Ichimura, K. (1993). Photoisomerizability of an azobenzene covalently attached to silica-gel matrix. *J. Non-Cryst. Solids* 163: 125–132.
43 Ogawa, M., Ishii, T., Miyamoto, N., and Kuroda, K. (2001). Photocontrol of the basal spacing of azobenzene–magadiite intercalation compound. *Adv. Mater.* 13: 1107–1109.
44 Okada, T., Nozaki, N., Seo, J. et al. (2017). Photoinduced structural changes of cationic azo dyes confined in a two dimensional nanospace by two different mechanisms. *RSC Adv.* 7 (13): 8077–8081.

45 Nabetani, Y., Takamura, H., Hayasaka, Y. et al. (2013). An artificial muscle model unit based on inorganic nanosheet sliding by photochemical reaction. *Nanoscale* 5 (8): 3182–3193.
46 Nabetani, Y., Takamura, H., Uchikoshi, A. et al. (2016). Photo-induced morphological winding and unwinding motion of nanoscrolls composed of niobate nanosheets with a polyfluoroalkyl azobenzene derivative. *Nanoscale* 8: 12289–12293.
47 Feng, W., Luo, W., and Feng, Y. (2012). Photo-responsive carbon nanomaterials functionalized by azobenzene moieties: structures, properties and application. *Nanoscale* 4: 6118–6134.
48 Ozawa, T., Kondo, M., Mamiya, J., and Ikeda, T. (2014). Enhancement of mechanical stability in hydrogen-bonded photomobile materials with chemically modified single-walled carbon nanotubes. *J. Mater. Chem. C* 2: 2313–2315.
49 Guo, Z., Feng, Y., Zhu, D. et al. (2013). Light-switchable single-walled carbon nanotubes based on host-guest chemistry. *Adv. Funct. Mater.* 23: 5010–5018.
50 Park, J.Y., Male, U., and Huh, D.S. (2017). Photo-regulated conductivity of polycaprolactone honeycomb-patterned porous films containing azobenzene-functionalized reduced graphene oxide. *Macromol. Res.* 25 (8): 849–855.
51 Cheng, Z., Wang, T., Li, X. et al. (2015). NIR–Vis–UV light-responsive actuator films of polymer-dispersed liquid crystal/graphene oxide nanocomposites. *ACS Appl. Mater. Interfaces* 7: 27494–27501.

11

Multi-responsive Polymer Actuators by Thermo-reversible Chemistry

Antoniya Toncheva[1,2], Loïc Blanc[3], Pierre Lambert[3], Philippe Dubois[2], and Jean-Marie Raquez[2]

[1] *University of Mons Laboratory of Polymeric and Composite Materials 23 Place du Parc, Mons 7000, Belgium*
[2] *Bulgarian Academy of Sciences Laboratory of Bioactive Polymers, Institute of Polymers 103A Academik G. Bonchev Street, Sofia 1113, Bulgaria*
[3] *Université Libre de Bruxelles Transfers, Interfaces and Processes Department 50 Avenue F.D. Roosevelt CP 165/67, Brussels 1050, Belgium*

11.1 Introduction

From the initial raw materials to the fabrication of a new generation functional polymer devices, chemistry would be playing a crucial role in the discoveries of tomorrow. To define the next wave of innovative directions, material scientists and chemists need to build an efficient collaboration network with various industry partners, process and hardware engineers, and device designers. This network should be well equipped to deal with the transfer from laboratory settings to the industrial scale, in relatively short periods of time, and elevated process yields.

Latest trends in the materials science focus on the design of robotic systems with biomimetic characteristics made from lightweight, low-cost, and high-efficient materials [1]. Generally, such systems must address requirements such as compactness, accuracy, safety, dynamics, and compatibility in specific environments. Naturally, the development of flexible mechanics or soft robotics, where the soft deformable materials being the primary building blocks, are of particular interest [2]. In contrast to conventional robots made of hard materials that limit their ability to elastically deform and adapt their shape to external constraints and obstacles, the soft robots are composed of fluids, gels, functional polymers, and other easily deformable matter. They are also characterized, to large extend, by the same elastic and rheological properties of soft biological matter, allowing the robot to remain operational even as it is stretched and squeezed (e.g. silicone materials). Another advantage of the soft robotic systems is their safety (reduced stress concentration during conformal contacts, continuous motion, and development of low-impact forces), making them safely cooperating with humans and steering through constrained environments [3]. Despite the indisputable advantages (low Young's modulus and high elastic and fatigue resistance values), silicone materials are often produced through molding and casting, and are subject to limit actuating stimuli such as pneumatic

Mechanically Responsive Materials for Soft Robotics, First Edition. Edited by Hideko Koshima.

pressure with lack of additional functionalities (self-healing, recyclability, or shape-memory properties), preventing them from becoming true "active matter." One of the most appropriate candidates for soft robotic fabrication are the polymers due to their high elastic limit to Young's modulus ratio, lightweight, one- to two-way shape-memory properties, and the possibility to be remotely and reversibly actuated. With the emergence of "smart" polymers, it is becoming easier to convert the chemical energy into mechanical work by enlarging the choice of the actuating stimuli (heat, solvent, humidity, electric-magnetic field, light) [4]. Consequently, the soft actuator deformations (anisotropic shape or volume change) were characterized by reduced friction, wear, and hysteresis, in opposition to the conventional motors (electric motors, combustion engines, and hydraulic pumps) where the motion is generated via changes of relative positions between their components.

Nowadays, polymers are also able to fulfill these requirements while allowing to be printed (Figure 11.1) in devices with complex three-dimensional (3D) geometries and their mechanical characteristics depend directly on the polymer macromolecules architecture (linear, graft, dendrimer-like, etc.). The advantage herein is that by changing the macromolecular architecture it is possible to generate superior supramolecular interactions, explaining why polymers with the same chemical composition can possess variant mechanical properties. However, most of the commercially available polymers cannot respond to any desired structure–processing–property relationship, thus limiting their manufacturing and possible implementation in additive manufacturing (AM) technologies such as 3D printing. A sensible approach toward solving this

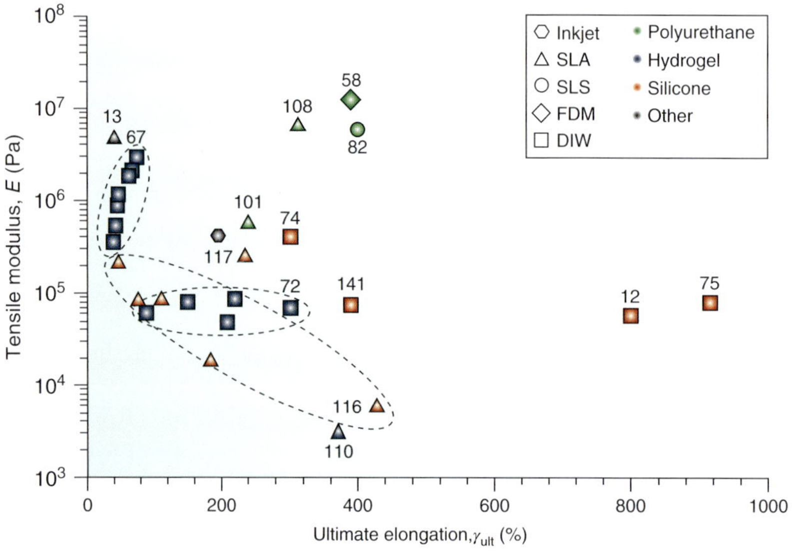

Figure 11.1 General presentation of polymeric materials mechanical properties printed using stereolithography (SLA), selective laser sintering (SLS), fused deposition modeling (FDM), or direct ink write (DIW). Source: Wallin et al. 2018 [2]. Adapted with permission of Springer Nature.

difficulty could be the focus on polymer networks based on thermo-reversible covalent adaptable networks (CANs), for their good processing, recyclability, and actuation upon different environmental stimuli.

In the present chapter the focus will be on using thermo-reversible CANs as unique strategy for the design and fabrication of smart polymer materials in soft robotics application. A special focus on Diels–Alder (DA) reactions as an excellent way to create the dynamic polymer networks is presented for their high efficiency, selectivity, and versatile approaches in the fabrication of nonconventional polymer architectures and materials topologies. Different aspects in term of materials with shape-memory and self-healing properties, and stimuli-responsiveness have been discussed. A section is dedicated to the latest progress in the field, revealing the great potential of the thermo-reversible chemistry as an elegant solution for fast soft robotics prototyping via 3D printing. At the very end, the reader can find a short list of conclusive remarks on the various trends in the domain, as well as some tips to consider when looking to meet the demands of the constantly evolving AM technologies.

11.2 Covalent Adaptive Networks

In the materials world, thermosets find various applications due to their good chemical stability, mechanical properties, and creep resistance. Regardless of these advantages, they cannot be easily reprocessed, reshaped, or recycled, which is limiting their life cycle. To face such a difficulty, scientists proposed a successful approach to increase the materials plasticity through the building of dynamic reversible and exchangeable chemical bonds within the polymer matrix. The advantage here is that such covalent bonds can occur between different locations within the polymer chains, while preserving the material structural integrity and properties. In the literature such polymer networks are known to be defined as CANs [5, 6] and classified into two groups depending on the exchange mechanism: (i) CANs based on associative bond exchange between the macromolecular chains with destruction of the cross-link only when a new covalent bond to another position is build (Figure 11.2a) and (ii) CANs based on dissociative cross-link exchange mechanism, where the cross-link is destroyed and reform at another place (Figure 11.2b).

11.2.1 Associative CANs

Generally, in the associative CANs the polymer network is not undergoing depolymerization. The material is maintaining its integrality without any drop of the viscosity state, due to the almost constant level of cross-linking density (the old covalent bonds are only broken when new ones are formed). Such permanent and dynamic state networks were developed at the beginning of 2000s where allyl sulfides were used as photo-mediated free radical reactions [7, 8] or radical generators such as trithiocarbonates [9–11]. Despite of the promising aspects of these networks (flow rate and stress relaxation) their dynamic lifetime is limited mainly because of the radical nature of the involved reactions and the occurrence

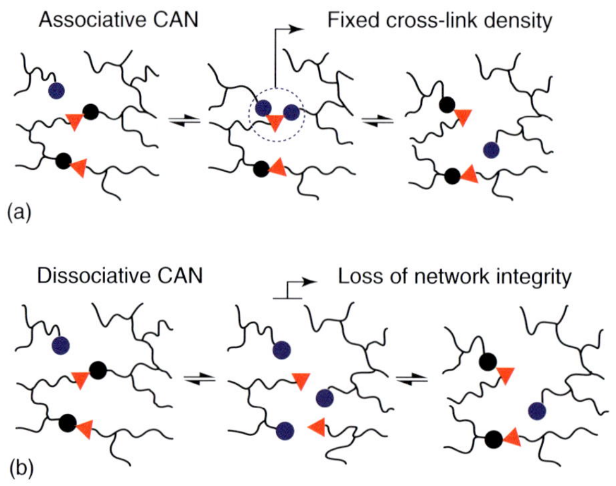

Figure 11.2 Schematic presentation of CAN types, depending on the exchange reactions without or with a temporary net loss of the cross-link density: associative (a) and dissociative (b) CANs. Source: Denissen et al. 2016 [14]. Adapted with permission of RSC.

of unavoidable termination reactions. Some efforts were made in the exploration of associative CANs by the development of thermally triggered transesterification catalyst for epoxy/acid or epoxy/anhydride polyester-based networks [12]. Interestingly in the particular case of the permanent polyester/polyol networks a gradual viscosity decrease upon direct heating was observed (Arrhenius law) [13]. This phenomenon gave birth to polymer materials named "vitrimers" [14].

11.2.2 Dissociative CANs

Well-explored chemical systems for the design of dissociative CANs are the polymer networks capable of thermo-reversible DA cycloaddition reactions. The thermal reversibility is occurring for temperatures below 50 °C and above 110 °C for complete DA and retro Diels–Alder (rDA) reactions, respectively. DA-based materials easily undergo fast topology rearrangements at high temperatures related to the fast drop in the viscosity values upon heating and the cross-links reform upon cooling while reaching the thermosets characteristics (stiffness and insolubility). Today DA chemistry presents a suitable tool for the development of polymer materials with remarkable dynamic features: shape-memory [15] and self-healing [16] or reconfigurable and recyclable polymer materials [17] with sufficient mechanical properties.

11.3 Thermo-reversible Chemistry

In the literature, different thermo-reversible reactions are used as cross-links for the fabrication of dissociative thermo-reversible networks (Figure 11.3). An interesting strategy is the nucleophilic addition reaction (nucleophile with characteristics of good leaving group) for the absence of produced condensate.

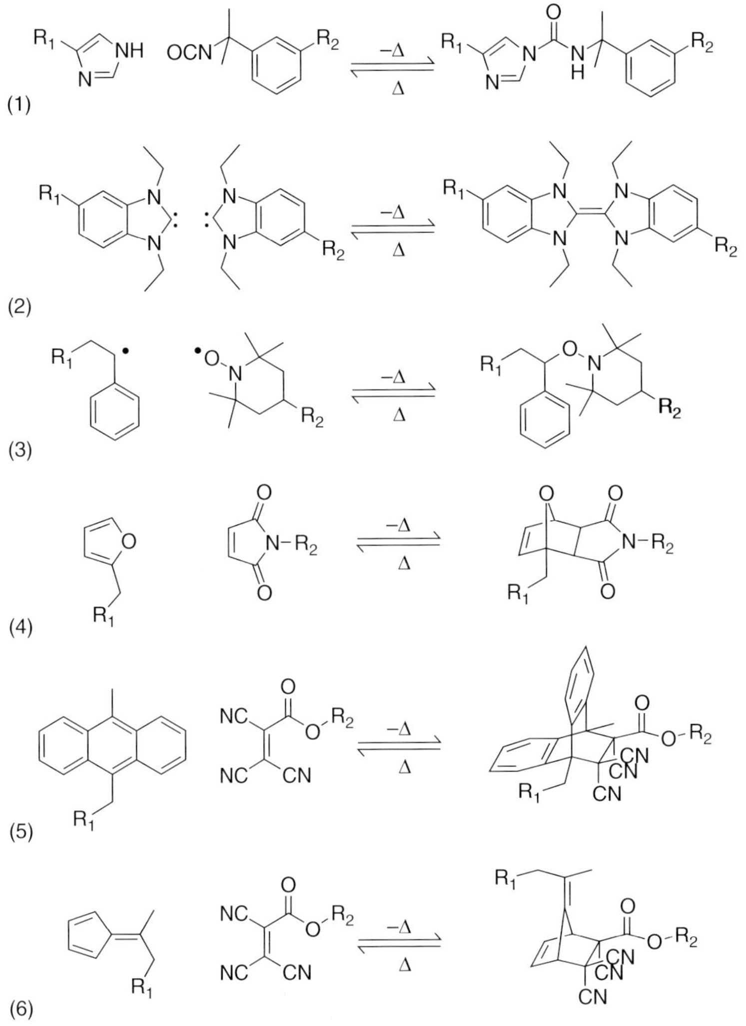

Figure 11.3 A short list of the most applied thermo-reversible reactions for polymer networks production. Source: Kloxin et al. 2010 [5]. Adapted with permission of ACS.

Such systems rely on the reactions of isocyanate with nucleophile (alcohol or amine) under inert conditions (reaction 1 in Figure 11.3) [18]. However, a limitation of these systems is the high moisture sensitivity of the isocyanates restricting the materials practical application. Thermo-reversible reactions can be achieved also by developing ring-opening chemistry – polymer chains carrying maleate [19] or azlactone [20] moieties reacting with bis-alcohol. Somehow, in this case the authors described limited reaction reversibility as a result of the presented various side reactions [21]. In controlled conditions thermally reversible CANs can be produced by a direct dimerization of otherwise oxygen-sensitive carbene (reaction 2 in Figure 11.3) [22]. Active radicals are making also a part of the agents capable of thermo-reversible reactions. One good candidate is the thermo-reversible homolytic cleavage of the alkoxyamine bond in nitroxide-mediated living/controlled radical polymerizations (reaction 3 in Figure 11.3). This concept was successfully used in studies for networks swollen in monomer solution [23], or to cross-link linear polymers [24]. Here, the main limitation is the biomolecular termination of the active radical, leading to a decrease in the network reversible character.

Thermally reversible CANs can be easily produced by [4+2] cycloaddition (DA cycloaddition reaction) between conjugated diene and an electronically activated compound carrying a double bond (dieneophile). This chemical process might occur in a wide range of temperatures through changes in the macromolecular functional reactivity with relatively low water and oxygen sensibility. Some examples are shown in Figure 11.3 (reactions 4–6), demonstrating the concept of producing thermo-reversible networks from linear or starlike polymers functionalized with bi-, tri-, or more active moieties (e.g. maleimide and furan moieties, cyclopentadiene as cross-linking dimer and others) [25–31]. Practical limitations here are the required high working temperatures to activate the rDA reactions, imparting the formation of irreversible side reactions, as well as the stability of the temperature depending kinetics.

11.4 DA Reactions for Thermo-reversible Networks

11.4.1 Basic Definitions

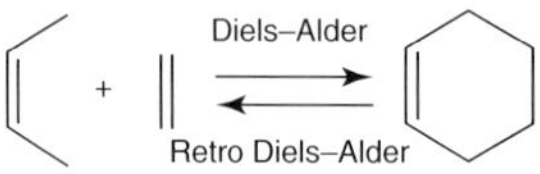

Figure 11.4 General presentation of DA pair involved in thermo-reversible reaction.

Between the previously listed thermo-reversible reactions, DA are among the most common ones used for the synthesis of materials with complex macromolecular architectures and supramolecular interaction. By definition, DA reactions involve [4+2] cycloaddition process between an electron-rich diene (e.g. furan, 1,3 cyclopentadiene, and others) and an electron-poor dienophile (e.g. maleic acid, vinyl ketone, and others) to form a stable cyclohexene adduct (Figure 11.4). The chemical mechanisms were studied in detail by Otto Diels and Kurt Alder who received Nobel Prize in 1950 for their research. DA reactions allow not only the creation of carbon–carbon covalent bonds but also that of heteroatom–heteroatom-type bonds (hetero-Diels–Alder, HDA) widely used synthetically for the preparation of six-membered rings.

11.4.2 DA Reactions for Polymer Synthesis

Till date, various strategies have been proposed to produce polymers starting from DA functionalized monomers or oligomers. Some of them deal with the synthesis of homopolymers: (i) linear thermoplastic [32] and thermosetting polymers [33] (polyimides [34–37], polyphenylenes [38–40]); (ii) telechelic polymers [41–43] for macromolecules containing reactive end groups with the capacity to participate in additional polymerization or reactions avoiding potential cross-linking side reactions; (iii) block copolymers by DA [44] and by HDA [45–47] as an access for more complex macromolecular architectures; (iv) graft copolymers [48]; (v) complex macromolecular architectures applying DA [49, 50] and HDA chemistry [47, 51] for cyclic homo and block copolymers [52]; (vi) dendrimer and dendronized polymers [53, 54]; and (vii) side chain functionalization [55–57] (Figure 11.5). Having in mind all of these possibilities, the idea to build networks based on thermo-reversible DA reaction did not wait to become reality.

11.4.3 DA Reactions for Thermo-reversible Polymer Network

In current materials science, DA reactions have been used as solid tool for the successful synthesis of cross-linked and mendable polymers. Inspired by this progress, smart materials with shape-memory and self-healing properties have been designed with potential applications such as adaptive coatings, actuators, sensors, and devices for soft robotics. Nowadays, three main synthetic strategies are used in the field: (i) direct DA cycloaddition reactions involving multifunctional monomers [35, 59–62], (ii) DA cycloaddition of linear polymers bearing pendant active moieties [25, 29, 31, 63–68], or (iii) cross-linker or initiator containing DA linkage ensuring the polymerization [69, 70] (Figure 11.6). It is precisely the reversible character of the CANs that enables reshaping reversibly the materials to confer them desired initial shapes and to program their life cycle actuation with low degree of fatigue. In the case of the thermo-modulable networks, the rate of the bond rearrangement directly depends on the conversion temperatures situated in the solid-to-liquid transition (static dormant CANs into dynamic activated one) with impact on the time of macromolecules stress relaxation. Taking advantage of these considerations, the focus has been moved to the design of materials with remarkable thermo-adaptive materials exhibiting self-healing properties and biocompatibility (hydrogels).

11.4.3.1 Self-healing Materials

From a practical point of view, polymer materials and composites with long lifetime have become a topic of considerable interest inspiring researchers over the world to tackle the problem of restoring damaged materials mechanical properties. Subsequently, two main strategies were defined: (i) production of polymer networks loaded with microcapsules releasing a healing agent upon fracture and (ii) fabrication of reversible CANs activated upon heat, light, or another stimulus.

In the study of Willocq et al. the authors proposed an interesting synthetic pathway for single-component, self-healing polyurethanes (PUs) based on DA chemistry [71]. The polymer matrix was made of thermo-reversible furfuryl and maleimide moieties obtained from 2,2-bis(hydroxymethyl)propionic acid and thermo-responsive monoalcohol coupled with 1,6-hexamethylene diisocyanate and polypropylene glycol. To ensure a good chain mobility for the self-healing processes of importance was the production of amorphous PUs with glass transition temperature (T_g) values below the room temperature. The materials demonstrated good post-healing mechanical properties by applying on the samples a load of 250 g representing a shear stress of approximately

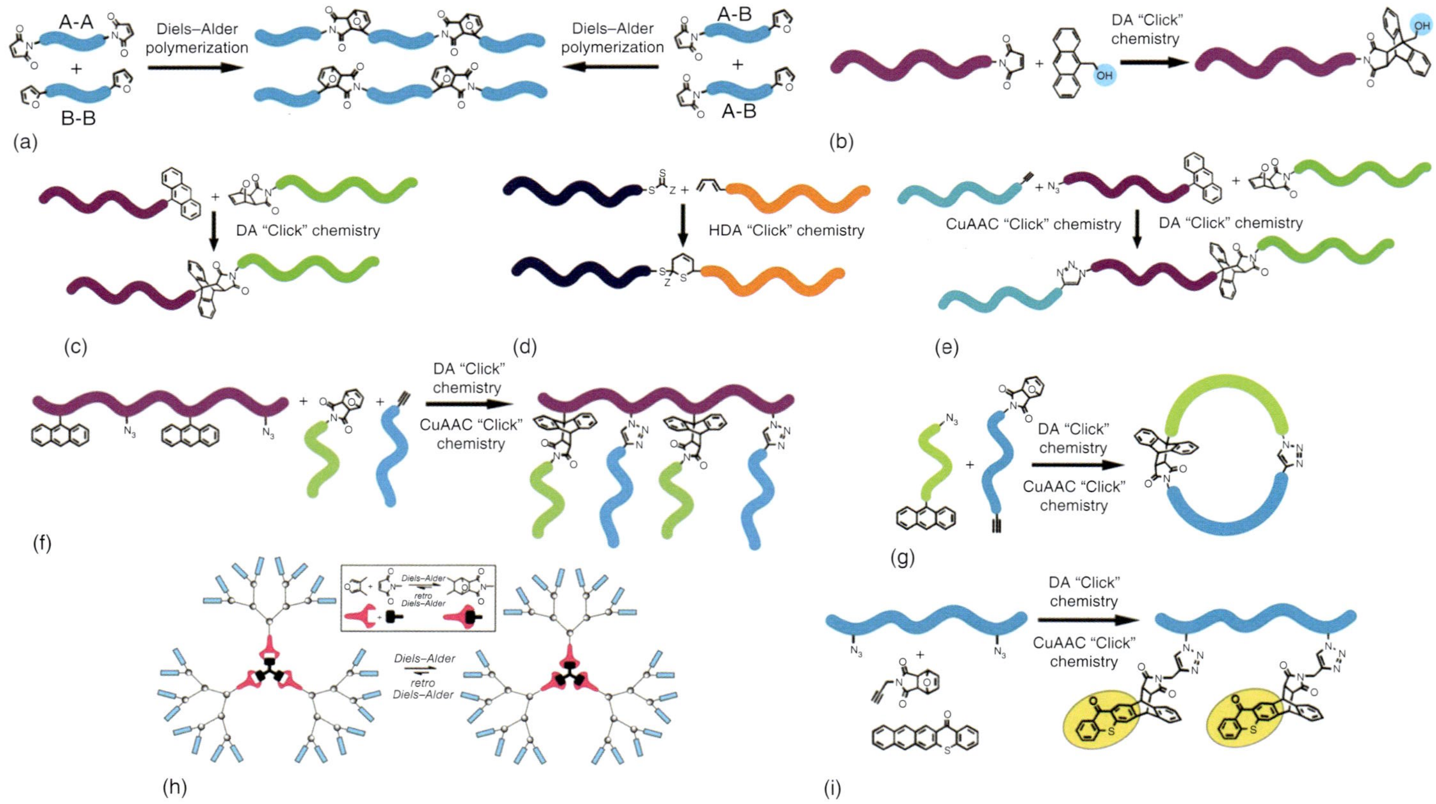

Figure 11.5 General presentation of the polymerization strategies based on DA chemistry from DA functionalized monomers (a), telechelic polymers (b), DA block copolymers (c), HDA block copolymers (d), triblock copolymers (e), hetero-graft copolymers (f), monomers for cyclic bock copolymers (g), dendrimers (h), and side chain DA functionalized polymers (i). Source: Tasdelen et al. 2011 [58]. Adapted with permission of RSC.

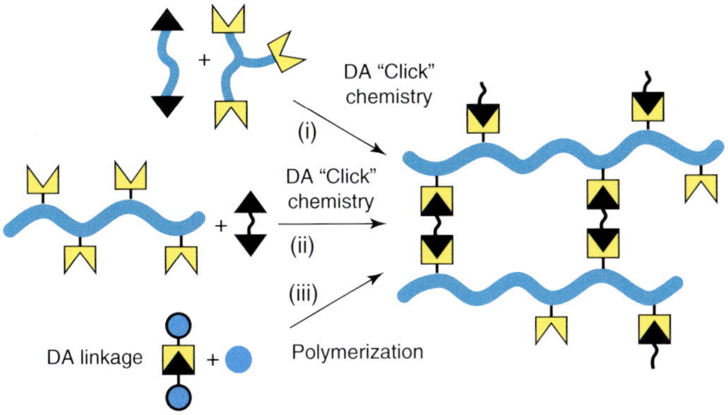

Figure 11.6 Schematic presentation of the generally used strategies for polymer network synthesis by DA cycloaddition reactions. Source: Tasdelen et al. 2011 [58]. Adapted with permission of RSC.

2.5×10^4 Pa (Figure 11.7a). It was reported that DA reactions can be used in the improvement of long-time service life of poly(siloxane-urethane) elastomers (PDMS-DA-PU and PDMS/PCL-DA-PU with PCL for poly(ϵ-caprolactone) and PDMS – polydimethylsiloxane) [72]. In addition, the materials acquired thermal self-healing ability and excellent recyclability, recovering their properties after impact (Figure 11.7b). Despite these advantages, the number of suitable DA pairings occurring at reasonable and at acceptable processable temperatures (below the materials degradation temperature) are barely investigated. Somehow, new directions in this aspect were proposed by Oehlenschlaeger et al., where reversible HDA reactions were developed from activated dithioesters as dienophiles and cyclopentadiene (Cp, slow self-healing kinetics) [73] or cyanodithioester (CDTE, hysteresis-free repetitive cyclability between 40 and 120 °C for less than five minutes) [74]. By combining the CDTE/Cp HDA as multifunctional building blocks it was found that full healing process can be reached in short time at relatively low temperatures. The strategy of the synthesis consists of several steps allowing the protection and deprotection of the C=S-double bond of a tetra-functional linker moiety in order to ensure the DA-based network formation with good healing efficiency and mechanical properties (Figure 11.7c). Such materials can find potential application as adhesives, coatings, or sealants.

The incorporation of nanofillers within the polymer matrix presented also some great perspectives for the generation of multi-responsive materials. Several studies were reported for the synthesis of polymer hybrid materials by DA "click" reactions from organic molecules onto gold nanoparticles [75–77] or tetraethoxysilanes [78, 79]. The cycloaddition reaction between organic and inorganic active moieties was a result of the interactions involving π-electrons of carbon-rich materials (graphene planes of carbon black [80], carbon nanotubes (CNTs) [81–85], or selective modification of fullerenes [46, 86]). For example, CNTs were activated to exhibit DA reactivity within the furan-modified poly(styrene-*block*-butadiene-*block*-styrene) as presented in Figure 11.7d [87].

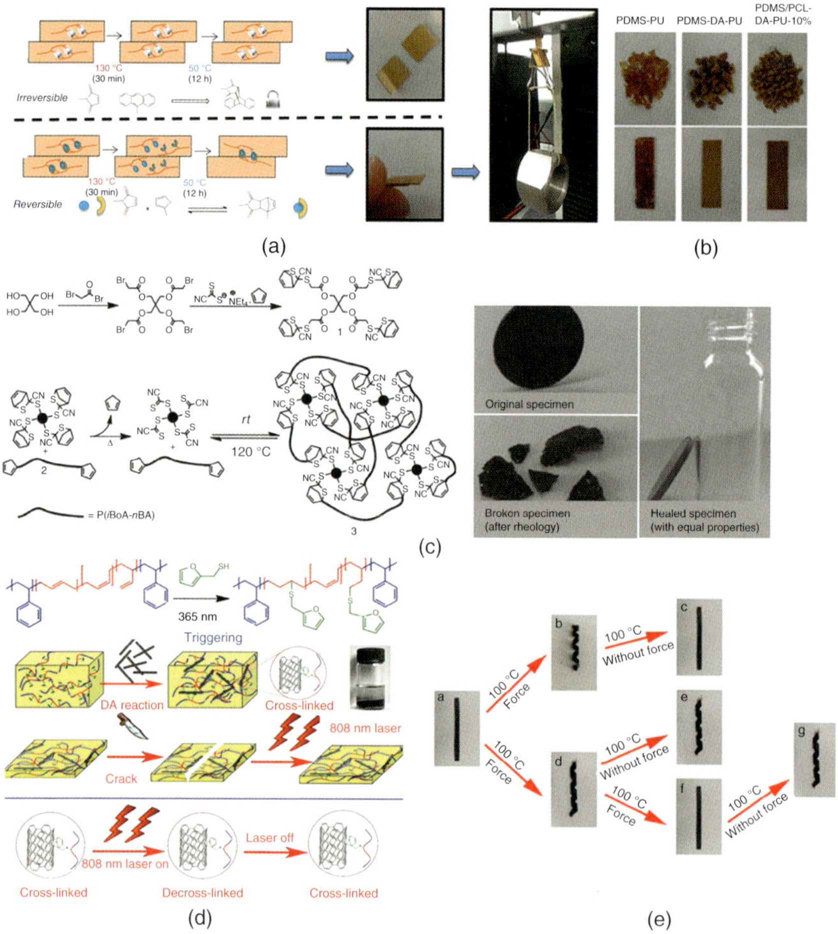

Figure 11.7 Polymer systems capable of DA pathways for materials with heat-induced self-healing properties (a), recyclability (b and c), light-induced self-healing (d), and shape memory properties (e). Source: (A) Willocq et al. 2017 [71]. Adapted with permission of RSC. (B) Zhao et al. 2016 [72]. Adapted with permission of RSC. Oehlenschlaeger et al. 2014 [73]. Adapted with permission of John Wiley & Sons. (d and e) Bai and Shi 2017 [87]. Adapted with permission of ACS.

The covalent strategy to incorporate 1 wt% CNT not only increased the materials mechanical properties (16.5 MPa or 9.4 times compared to the stress of the pristine sample) but ensured their homogeneous incorporation resolving the problem with the nanofiller dispersion within the composite material. The system was characterized with excellent recyclability and exhibited self-healing properties. In addition, the nanofiller photothermal effect favor the activation of the rDA by increasing locally the temperature above 200 °C and repairing the surface crack for less than 10 seconds for almost unchanged post-process mechanical properties (Figure 11.7d). Interestingly, this system was subject of successful re-programming offering a possibility to tune locally or remotely the shape-memory effect (Figure 11.7e).

11.4.3.2 Hydrogels

Hydrogels, i.e. 3D polymer networks swollen in aqueous solutions, are an interesting type of materials because of their reversible change in volume or shapes upon triggered environmental stimuli such as temperature, pH, and ionic strength [88, 89]. Often, they are made of biocompatible and biodegradable polymers with tunable water absorption and tissue-like architecture. Hydrogels have wide range of applications, such as injectable drug delivery systems, liquid-absorbing wound dressings, scaffolds for tissue engineering, biosensors, food packaging, and many others [90, 91]. Their successful application directly depends on the material cross-linking density, the swelling capacity, mechanical properties, and degradability [92]. Some of the latest progress in this field established that DA cycloaddition reactions are another powerful tool for the hydrogel formation providing mild reaction conditions, absence of initiator or catalyst, tunable gelation rate, and thermal reversibility.

Zhan et al. reported hydrogels-based drug delivery systems from furfural functionalized chitosan and maleimide functionalized hydroxypropyl β-cyclodextrin (CD) as a cross-linker [93]. The CD was used simultaneously as a cross-linking agent and as a key element in the guest–host chemistry for model substances release. The network stability during the gelation process revealed that the G' and G'' values sharply declined in the mixture and slowly increased in the temperature range between 25 and 45 °C as an evidence of the formation of supramolecular physical hydrogels (Figure 11.8a,b). In contrast, with the increase of the temperature a drop in the elastic modulus with pronounced strength of the hydrogels was noticed due to the formation of the covalent cross-linked network based on the DA reactions. In another study double, cross-linked hydrogel network was prepared by combining DA click reaction and acylhydrazone bond for functional tissue-engineered hydrogel for microenvironment cell survival [94]. The hyaluronic acid networks structural integrity and mechanical strength were maintained by the DA reactions and the dynamic covalent acylhydrazone bond ensured self-healing property by adapting the cross-link density. As presented in Figure 11.8c,d, the materials mechanical properties were well preserved after healing, revealing the effectiveness of the process.

As biomaterials, interesting in vivo injectable thermo-responsive hydrogels were produced from poly(*N*,*N*-dimethylacrylamide-*co*-furfuryl methacrylate) and *N*-[4-(formyl polyethylene glycol ester)] bismaleimide [95]. The authors underline the possibility to accelerate the DA reactions in aqueous solution and favor the rDA upon immersion of the materials in *N*,*N*-dimethylformamide. Here, a clear relationship was found between the gelation time and the working temperature: increasing the temperature led to a decrease in the gel fraction. More complex structures were built from starlike polymers. For example, Gregoritza et al. developed hydrogels for controlled antibody delivery by dual approach dealing with both thermal gelation and the DA cross-linking [96]. The starting building blocks were amine-modified four- and eight-armed poloxamines functionalized with maleimide or furyl groups. Once the gel was formed in situ at body temperature the DA reactions gradually took place conferring greater degree of reticulation. Similar cross-linking mechanism were described in the study of Kirchhof et al., where DA reactions were used for the

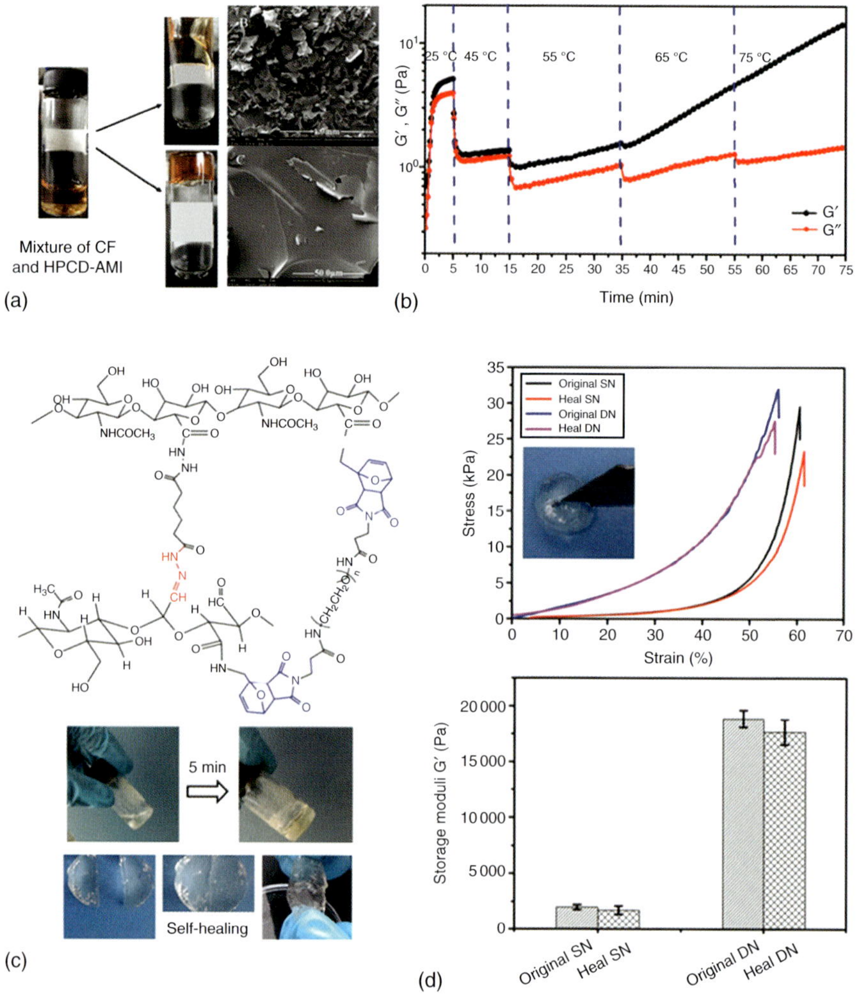

Figure 11.8 Examples of hydrogels produced via DA chemistry (a), gel formation procedure upon measuring its mechanical properties (b), hydrogel self-healing properties (c), and post-healing mechanical properties recovery (d). Source: (a and b) Zhang et al. 2018 [93]. Adapted with permission of Elsevier. (c and d) Yu et al. 2015 [94]. Adapted with permission of ACS.

synthesis of four- and eight-arm poly(ethylene glycol)-based hydrogels [92]. The performed rheological tests revealed that the gelation time and mechanical properties depended on the polymer concentration, the branching degree, and the macromonomers molecular weight.

In some cases, the DA-based gelation period is relatively slow (more than 10 minutes) and additional hydrophobic association [97] or branching structures [98] are required. This last point can be easily achieved by using stimuli-responsive polymers activated upon heating exposure, temperature, pH, or ionic strength [98–101]. Such hydrogels forming polymers are poly(*N*-isopropylacrylamide), poloxamer (Pluronic®), and poloxamine (Tetronic®) known for their lower critical solution temperature (reduction in solubility above a certain temperature) [102].

While focusing on the advantages of the thermally triggered CANs chemistry and effectiveness of production, another critical consideration must be taken into account regarding their practical application. In general aspect, DA-based CANs should respond simultaneously to sufficient mechanical properties at room temperature and be dynamic in thermal range above 100 °C. This limit the application to only high thermally stable systems. In this aspect, alternative solution can be found in the combination of the shape-memory polymers (SMPs) and the thermo-sensible systems, where CANs will serve as skeleton ensuring the polymers permanent domain. Here the great advantage is that the material T_g and melting temperature (T_m) can be modulated with respect to the network architecture, cross-linking points, and the molar mass of the building oligomers/polymers.

11.5 Soft Actuators

In materials science, latest trends show an increased interest in the fabrication of materials with complex deployable structures and functionalities described as actuators. Generally, they are made of materials capable of reversibly changing their shape and size in response to various stimuli. Among potential applications of such actuators, one can mention sensors, artificial muscles, controlled encapsulation/delivery systems, or biomimetic actuating systems as well as soft robots [103]. Basically, the function of an actuator is to provide mechanical energy (force and displacement) along different degrees of freedom (contraction, bending, torsion, expansion, swelling, etc.). It is usually centered on preexisting anisotropic structuration in the material (gradient-like structure, bilayer with different cross-links, and rigidity) responsible for the system expansion, contraction, or bending upon appropriate stimulation (compressed air, remote light irradiation, direct heating, solvent exposure, etc.). The range of actuation depends on the properties of each component with the relative thermal expansion coefficients, swelling abilities, layer thickness, and size. Among many, actuators based on SMPs are very promising, since they are able to memorize their original shape, to acquire a metastable temporary configuration upon deformation (even with large strains), and to revert to the permanent shape when exposed to an appropriate stimulus. The ability of the material to achieve shape recovery from the fixed temporary shape to the initial one is defined as shape-memory effect. This property can be quantified using the shape fixity ratio (R_f, Eq. (11.1)) or the ability of the material to fix the temporary shape and the shape-recovery ratio (R_r, Eq. (11.2)), which represents the extent to revert to the original shape.

$$R_f = \epsilon_{un}/\epsilon_{ext} \tag{11.1}$$

$$R_r = (\epsilon_{un} - \epsilon_f)/\epsilon_{un} \tag{11.2}$$

where ϵ_{un} is the strain directly obtained before constant loading was released and ϵ_f is the strain obtained directly after cooling and unloading, and ϵ_{ext} is the strain after heating in the recovery step.

Usually, the major actuation is directly related to the thermal transitions (T_g for amorphous regions mobility or T_m for crystalline phase melting) of the polymeric matrix, but sometimes other stimuli (solvent, vapor, illumination, electro-magnetic field, and others) can be also used after incorporation of nanostructured fillers (cellulose nanocrystals (CNCs), metal oxide nanoparticles (NPs), CNTs, and others). In 2011 Raquez et al. reported the synthesis of thermo-reversible cross-linked PCL-based PU networks by reactive extrusion technology [15, 17]. The proposed solution was characterized by having one- and two-way shape-memory properties as a result of the cycloaddition reaction between furan and maleimide moieties, enabling the final control of the microstructure. The PCL-based PU matrix was made out of linear furfuryl-modified α,ω-diol PCL coupled with methylene diphenyl 4,4′-diisocyanate and subsequently subjected to DA reactions with *N*,*N*-phenylenedimaleimide. A key element was the possibility to modulate the cross-linking density by partially replacing the linear PCL diol with starlike one PCL-tetraol with excellent shape-memory properties. With this it was clear that DA can be used as a versatile way to prepare SMP with controllable cross-link density and multi-shape-memory effects. For this purpose films from poly(2,5-furandimethylene succinate) (PFS) and 1,8-bismaleimidotriethyleneglycol (M2) were produced via DA pathways [104]. The quintuple shape-memory effect was demonstrated by star-shaped like PFS/M2 film with four sections each with distinct T_g ensuring the number of temporary and recovery steps (T_g transitions created by different cross-linking degree – the higher the cross-linking density the greater the T_g values; Figure 11.9a). This highly flexible design capability allowed the fabrication of materials with desired mechanical properties with Young's modulus between 20 and 2000 MPa, while selectively actuating the sections (stepwise increase of the activating temperature from 0 °C to 25, 35, 45, and 65 °C Figure 11.9b).

In most of the studies the DA reaction reversibility is dedicated to the design of thermosets that can be liquefied for reprocessing and self-healing, while neglecting their dynamic equilibrium nature. Going in this direction Zhang et al. [105] demonstrated that in addition to their remoldability at the liquid state (temperatures above 110 °C), such thermo-reversible network can undergo plastic deformation in its solid state at intermediate temperatures (60–100 °C) as a result of the occurred dynamic equilibrium making possible the network topological rearrangement. An example for such transformations are presented in Figure 11.9c: (i) classical SMP behavior with deformation and recovery step at 70 °C followed by (ii) sample plastic deformation (100 °C, 30 minutes) to achieve different umbrella-like permanent shape involved in (iii) next shape-memory cycle at higher temperature (140 °C, part of the material was remolded into a coin while maintaining its shape-memory characteristics). With this concept the liquid state remoldability and solid-state plasticity were studied as two distinct yet complementary mechanisms in the manipulation of the SMP permanent shape opening the door to a variety of future applications.

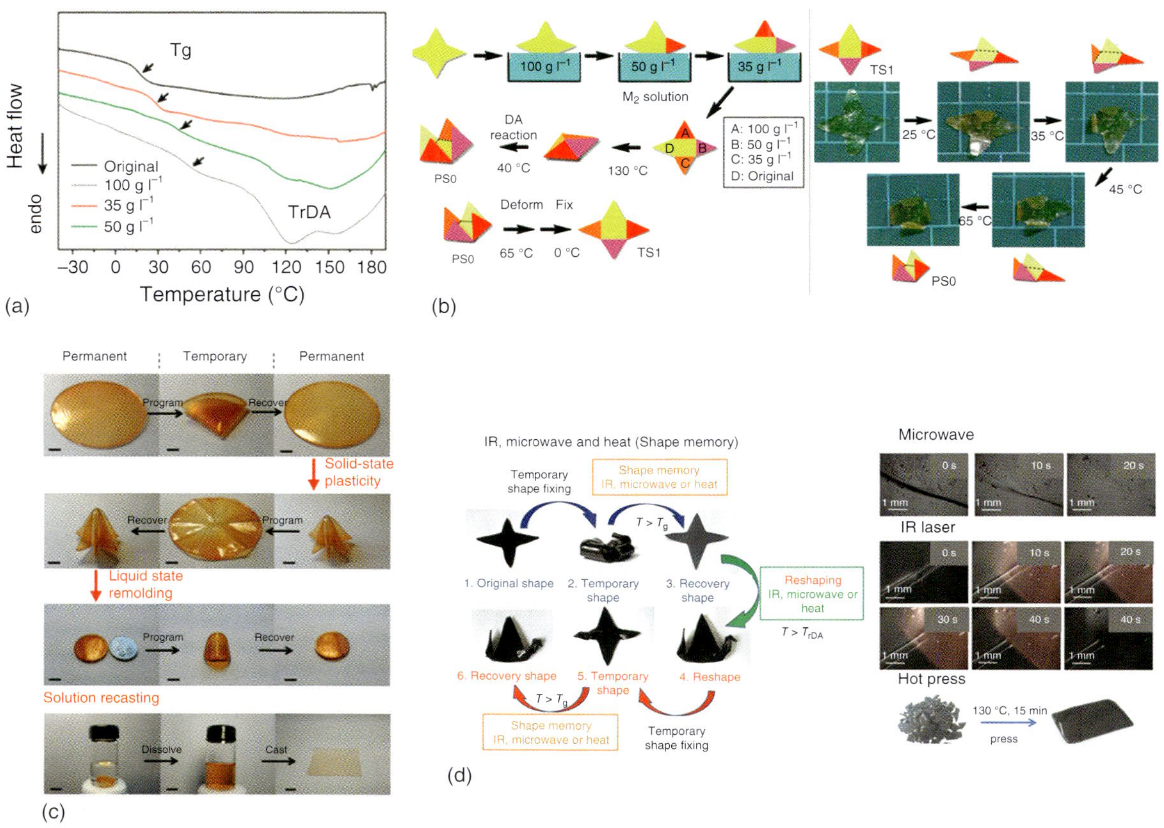

Figure 11.9 SMP materials produced via DA chemistry: differential scanning calorimetry (DSC) thermograms revealing the possibility to produce materials with different T_g (a); films with multistep actuation upon direct heating (b); digital images of shape memory films programming, activating, and recovery (c); and IR, microwave, or thermal activation of graphene-containing materials (d). Source: (a and b) Zeng et al. 2014 [104]. Adapted with permission of ACS. (c) Zhang et al. 2016 [105]. Adapted with permission of ACS. (d) Cai et al. 2018 [106]. Adapted with permission of RSC.

DA chemistry was successfully used for the fabrication of nanocomposite such as cross-linked polyacrylonitrile/graphene nanocomposite for materials with multi-responsiveness (IR light, microwave stimulation, and heat) and enhanced mechanical properties [106]. Here the dispersed graphene sheets (<1 wt%) were used as intrinsic thermal sources by converting absorbed environmental IR light/microwave energy to heat. The main challenge was to trigger the T_g (elasticity-based shape-memory properties) in addition to the rDA reactions ensuring solid-state plasticity through topological network rearrangement of the materials. Digital images of the healing process for different stimuli and material reprocessing are presented in Figure 11.9d.

11.6 DA-based SMPs for Soft Robotics Application

Within its perfection Nature often inspires scientists in the development of bio-inspired robotic structures. As discussed, soft robots are almost entirely made out of flexible soft materials, making them suitable for applications in dynamic task environments, including safe human–robot interactions. However, softness comes at a price: the actuators can be affected by cuts, shears, and perforation while executing complex tasks in unpredictable environments. In a recent study, Terryn et al. [107] proposed an interesting solution for self-healing, soft pneumatic actuators made of Jeffamine-based elastomers capable of DA pathways. The authors demonstrated the successful development of self-healing, soft pneumatic hand, gripper, and artificial muscles model (Figure 11.10a). Of importance for the materials practical application were the mechanical properties after several self-healing cycles (30 cycles), which were found to be slightly changed with contraction force values around 15 N, storage and loss modulus of 10 and 1 MPa, respectively (Figure 11.10b) [108].

In our recent studies, we proposed a simple and straightforward approach to produce solvent, vapor, and IR light actuated materials by reactive extrusion technology [109, 110]. These actuators were based on rigidity gradients created in bilayer architectures of reversibly cross-linked PCL networks (PCL diol and tetraol funcionalized with furan and maleimide moieties) into which functional nanofillers, i.e. multi-walled CNTs were incorporated. A key element of the bilayer functionality depended on the good adhesion between the layers. Thereby, the produced material instantaneously swells in an anisotropic way due to the rigidity gradient, resulting in reproducible bending actuations for object grabbing and transportation (Figure 11.10c). Additionally, we designed plasmonic silver nanoparticles (AgNPs) grafted onto CNCs as an efficient plasmonic system for fast and remote actuation. Such CNC-*g*-AgNPs "nanorod-like" structures thereby allowed long-distance and strong coupling plasmonic effect between the NPs. This allowed fast photothermal shape-recovery upon non-invasive IR light illumination (at wavelength 780–1400 nm) by transporting objects 500 times the samples weight (Figure 11.10d). The materials could find applications in the realm of soft robotics for remote object transportation or as smart biomaterials such as self-tightening knots with antibacterial properties.

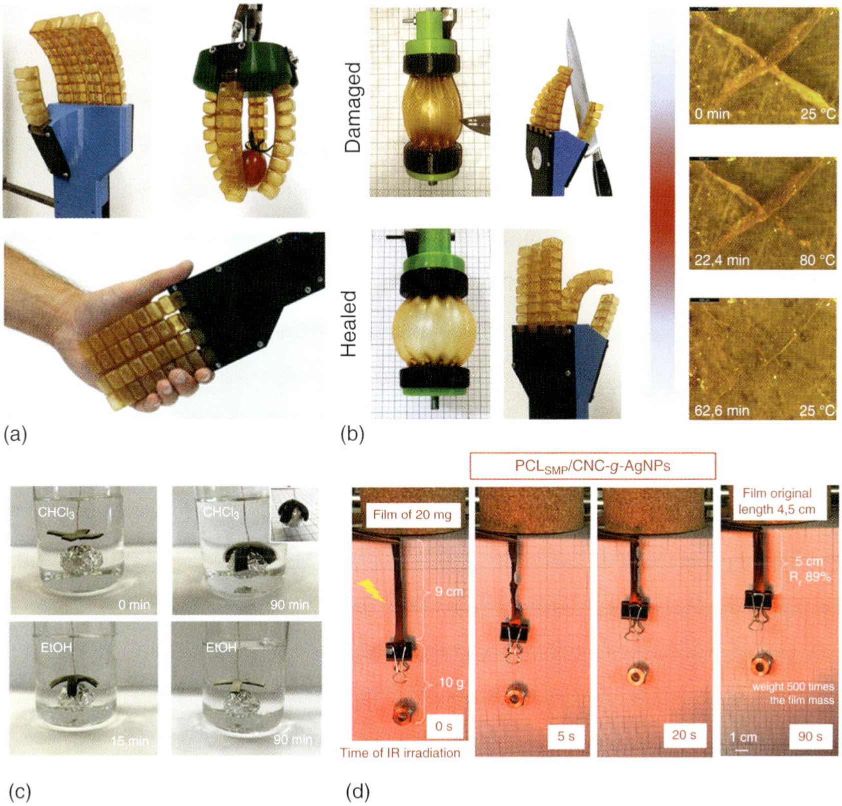

Figure 11.10 Soft robotic DA-based thermo-reversible networks with self-healing properties (a and b) and for object catching and transportation upon solvent (c) or remote IR light (d) actuation. Source: (a and b) Terryn et al. 2017 [107]. Adapted with permission of Science Robotics. (c) Toncheva et al. 2017 [109]. Adapted with permission of RSC. (d) Toncheva et al. 2018 [110]. Adapted with permission of ACS.

11.7 On the Road to 3D Printing

Recent developments in the field of AM technologies have started a new era in polymer materials fabrication. The technological advancement made possible the fast prototyping process with full control of the object structuration, while conferring it precise environmental responsiveness. There are three main AM technologies using polymeric materials including selective laser sintering (starting materials – powders), stereolithography (SLA), and two-photon polymerization (2PP) (polymer solutions) and fused deposition modeling (FDM; polymer melt). While during the conventional SLA and 2PP methods, the polymer solutions undergo photochemical reactions in the presence of photo-initiator with appropriate absorption wavelength window; in the case of the FDM the polymer thermal properties are the prominent. Up to date, a limited number of studies discuss the building of dual networks by combining photosensible and thermo-reversible chemistry (DA network containing dissolved multiacrylate monomers) applying SLA (Figure 11.11a,b) [111] or for controlling the photoinitiation and thiol–ene reaction for selective conversion of DA adducts into irreversible cross-links through 2PP (Figure 11.11c,d) [112]. In this section the reader's attention is drawn to the FDM technique, as the more explored, solvent–free, and appropriate 3D printing method to be used in the case of thermoplastic systems.

Generally, during an FDM layer-by-layer deposition, the polymer filament is subjected to direct heating trough melting and subsequently exposed to several cooling steps. The use of small molecules, monomers or oligomers, capable of thermo-reversible pathways offered new possibilities to produce polymer networks with desired levels of cross-linking [113, 114]. This is giving the advantage not only to produce complex gradient structures, but also the possibility to increase the interfilamentous adhesion through the creation of DA covalent bonds. With this, one more step forward was made in the improvement of the mechanical properties of the printed devices, avoiding their anisotropic behavior problem (applied stress with respect to the printing direction). These aspects were found to have direct impact on the devices mechanical properties and can be improved by applying some thermo-reversible chemistry strategies, where the networks will "de-polymerize" during the print process and re-polymerizes during the cooling phase, solving the weak welding between the printed filaments.

In a recent study, polylactic acid (PLA)-based actuator was designed for object catching and transportation through the polymerization of maleimide moieties and furan-containing mending agent as cross-linkers (Figure 11.12a) [113]. In this case, the degree of interfilamentous adhesion in the final device was additionally increased and the anisotropic mechanical behavior was prevented (strength and toughness end up being improved with 130% and 460%, respectively along the z-axis). Self-healing mechanism have been described by Yang et al., with multi-furan monomer and bismaleimide as initial compounds resulting in more than 95% interlayer adhesion [115]. These approaches ensure a good adhesion between the printed filaments, ensuring the controlled producing of 3D-printed parts with desired design, sufficient mechanical properties, and reproducible functionalities.

New types of filaments from noncommercially available polymers were also produced from SMP materials. For example, Chen et al. discussed the fabrication of monofilaments from ternary blends between styrene-*b*-(ethylene-*co*-butylene)-*b*-styrene, polyethylene wax, and low-density polyethylene [116]. The smart materials were described with at least three different temporary shapes with specific temperatures (dual, triple, and quadruple heating processes). This gradual shape recovery was explained with the presence of multiple individual melting peaks (heterogeneous microstructuration), offering broad thermally switching transitions. In another study, hydroxyapatite (HA, 15 wt%) was used as nanofiller in 3D SMP filaments [117]. The main role of the HA particles was to reinforce the porous scaffolds acting as nucleation agents on PLA phase, by inhibiting the growth of cracks during compression–heating cycles ensuring the devices self-healing properties. The authors discussed also the feasibility to print personalized scaffolds and implants for self-fitting bone defect replacements with Rr of 98% and absence of delamination. Additionally, thermally triggered parts can be made from filaments made of an ionomeric thermoplastic SMP-zinc-neutralized poly(ethylene-*co*-methacrylic acid) [118], by combining two networks: nanodomains with long relaxation times of supramolecular cross-links by ionic bonds (permanent network) and ethylene crystals (temporary network). PU-based filaments (T_g in the range from 145 to 195 °C) were suitable materials to print parts with complex geometry for actuators capable of catching and transporting objects (Figure 11.12b) [119]. FDM finds also application in the design of light-triggered devices. Therefore, photo-responsive PU-based (nano)composites loaded with carbon black were produced by Yang et al. The materials were characterized with excellent shape-memory properties and photo-thermal conversion efficiency [120]. It was possible to remotely control (light source or sunshine) the printed parts shaped in complex geometric shapes – cubic frame or sunflower-shaped structure (Figure 11.12c). The original shape had a very fast recovery process (within few minutes) due to the excellent photothermal conversion efficiency of the carbon black, revealing the possibility to 3D-print materials as biomimetic solar tracking sensors or smart solar cell systems by FDM.

11.8 Perspectives and Challenges

Over the past years, DA chemistry has shown significant potential in boosting the creation of the next generation of materials with remarkable macromolecular architectures and composition. Polymer networks with complex geometry, desired mechanical properties, and functionalities (self-healing, recyclability, and stimuli-responsiveness) were therefore produced via 3D printing methodologies. Despite the dynamic perspectives and the already existing applications in soft robotics, some major challenges remain unsolved:

(i) From a technological point of view, a close collaboration between chemists and mechanical engineers would be crucial for tackling some of the challenges ahead. While currently available 3D printers (particularly FDM) offer the possibility to print commercially available polymers or some custom-made filaments, there are opportunities for improvements in various aspects – the working temperature range, the diameter of the printable filaments, improvement of open source software, increase in printing resolution, increase in the repeatability of the process, and others. For the moment, biomimetic engineering does not always achieve the best technological solution, but it still stimulates scientific curiosity, imagination, and creativity. As the FDM technology is currently showing limitations (the resolution is limited by the size of the nozzle diameter 0.2 mm), the study of smart materials should be extended to other AM technologies.

(ii) Recently the scientific community released formulas for active inks that can exploit the thermodynamically driven self-assembly of block copolymers and generate the next level of ordered structures. A difficulty to overcome in this field is to maintain the generated assemblies during the 3D printing

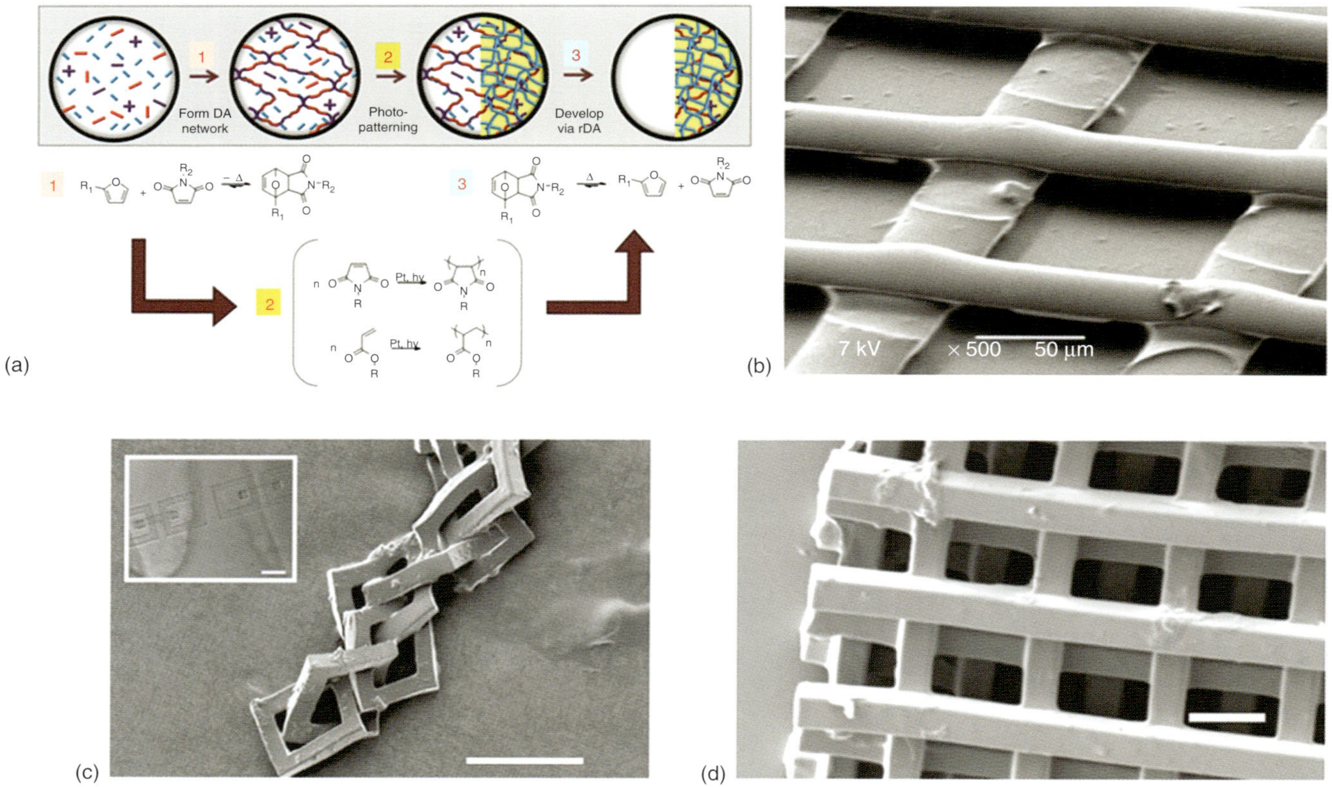

Figure 11.11 DA chemistry for 3D printed objects: dual networks by combining photosensible and thermo-reversible systems applying SLA technique (a) and related SEM micrographes (b). Eight freely rotating rings printed through 2PP (SEM micrographs – 200 μm scale bar, c) and five-layer log pile type structure (40 μm scale bar, d). Source: (a and b) Berg et al. 2014 [111]. Adapted with permission of ACS. (c and d) Adzima et al. 2012 [112]. Adapted with permission of John Wiley & Sons.

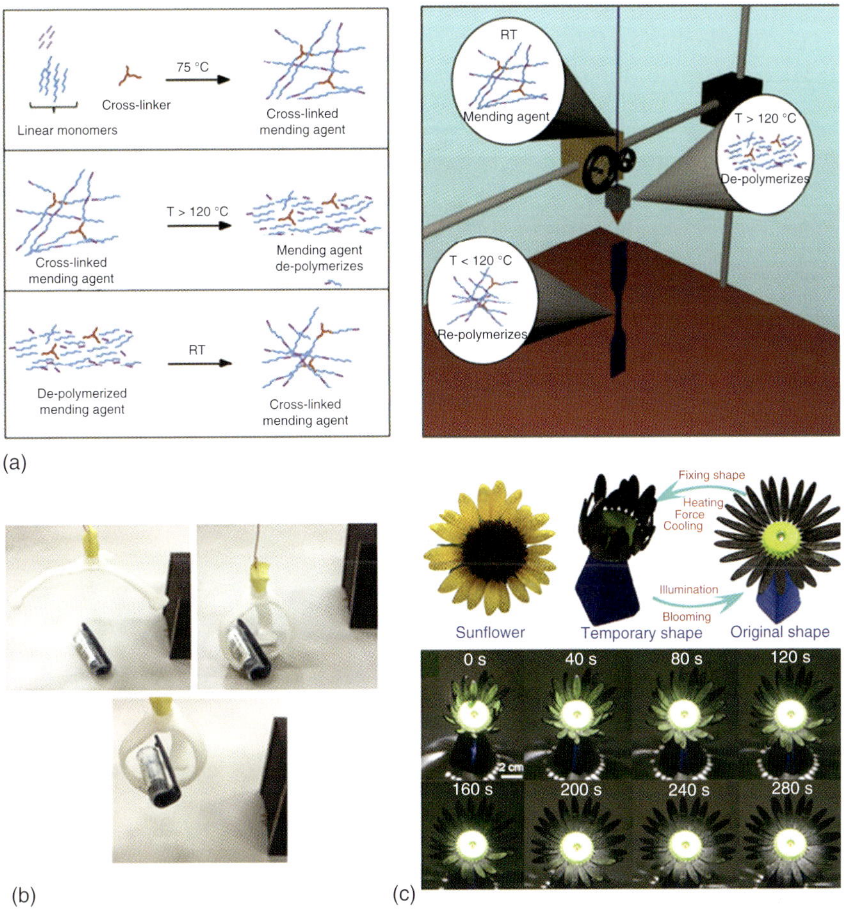

Figure 11.12 Schematic presentation of FDM procedure using DA-activated monomers (a) and digital images of soft robotics devices for object catching and transportation (b) or for light-triggered system (c). Source: (a) Davidson et al. 2016 [113]. Adapted with permission of ACS. (b) Yang et al. [119]. Adapted with permission of Springer Nature. (c) Yang et al. 2017 [120]. Adapted with permission of John Wiley & Sons.

process, as this requires deep understanding of the process thermodynamics, kinetics, phase space, and morphological transitions occurring in the polymer system. On a more practical aspect, during the printing process it is of importance to minimize the occurrence of side reactions and chemical decomposition, while considering the materials application and at acceptable timescale. This creates additional challenges for both chemists and engineers to successfully adjust high temperature activated chemical process for the safety of working environment and protective apparatus conditions. Improvement of the 3D printer parameters in terms of resolution, precise control of the filament melt and multi-head printing (FDM), and greater choice of wavelengths and organic solvents (2PP, SLA, bioprinting) will additionally help and inspire the scientific community in the development of new concepts and know-how in the materials making.

(iii) As discussed in the chapter, DA chemistry has a great potential in solving the problem with the anisotropic mechanical properties of 3D printed parts. However, increasing in situ the interfilamentous adhesion by newly generated covalent bonds might turn out insufficient. Additional posttreatment of the surface by plasma, coatings, nanofillers loading, or solvent vapor exposure could be needed.

(iv) Modification of biomolecules via DA "click" chemistry is another possibility that would need further exploration. Some of the studies in this direction are already presenting encouraging results, including increased stability and solubility of the bioconjugates as well as the ability to modulate the protein activity and their bioavailability. The possibility to immobilize oligonucleotides, peptides, and antibodies via DA strategies is also a portal toward interesting therapeutic applications in the field.

Acknowledgments

A. T. thanks the BEWARE (BElgium WAllonia REsearch, project convention no. 410161) Fellowships Academia program cofunded by the COFUND program of the European Union (FP7-Marie Curie Actions) for financial support. J.-M. R. as a F.R.S.-FNRS research associate and the LPCM thank the Belgian Federal Government Office of Science Policy (SSTC-PAI 6/27) for general support, and LCFM-BIOMAT project is gratefully acknowledged. This work was also supported by the FNRS through an FRIA grant as well as the PREDICTION ARC project between Université Libre de Bruxelles and Université de Mons (2012–17) and FNRS PDR (project convention no. 19471061, 2014-2018). The authors would like to acknowledge Erasmus Mundus EASED (2014–17) for enabling the collaboration with Prof. Kanda in Okayama.

References

1 Momeni, F., Mehdi Hassani, S.M., Liu, X., and Ni, J. (2017). A review of 4D printing. *Mater. Des.* 122: 42–79.

2 Wallin, T.J., Pikul, J., and Shepherd, R.F. (2018). 3D printing of soft robotic systems. *Nat. Rev. Mater.* 3 (6): 84–100.

3 Polygerinos, P., Wang, Z., Galloway, K.C. et al. (2015). Soft robotic glove for combined assistance and at-home rehabilitation. *Rob. Autom. Syst.* 73: 135–143.

4 Pilate, F., Toncheva, A., Dubois, P., and Raquez, J.-M. (2016). Shape-memory polymers for multiple applications in the materials world. *Eur. Polym. J.* 80: 268–294.

5 Kloxin, C.J., Scott, T.F., Adzima, B.J., and Bowman, C.N. (2010). Covalent adaptable networks (CANs): a unique paradigm in cross-linked polymers. *Macromolecules* 43 (6): 2643–2653.

6 Bowman, C.N. and Kloxin, C.J. (2012). Covalent adaptable networks: reversible bond structures incorporated in polymer networks. *Angew. Chem. Int. Ed.* 51 (18): 4272–4274.
7 Scott, T.F., Schneider, A.D., Cook, W.D., and Bowman, C.N. (2005). Photoinduced plasticity in cross-linked polymers. *Science* 308 (5728): 1615–1617.
8 Kloxin, C.J., Scott, T.F., Park, H.Y., and Bowman, C.N. (2011). Mechanophotopatterning on a photoresponsive elastomer. *Adv. Mater.* 23 (17): 1977–1981.
9 Nicolaÿ, R., Kamada, J., Van Wassen, A., and Matyjaszewski, K. (2010). Responsive gels based on a dynamic covalent trithiocarbonate cross-linker. *Macromolecules* 43 (9): 4355–4361.
10 Amamoto, Y., Kamada, J., Otsuka, H. et al. (2011). Repeatable photoinduced self-healing of covalently cross-linked polymers through reshuffling of trithiocarbonate units. *Angew. Chem. Int. Ed.* 50 (7): 1660–1663.
11 Amamoto, Y., Otsuka, H., Takahara, A., and Matyjaszewski, K. (2012). Self-healing of covalently cross-linked polymers by reshuffling thiuram disulfide moieties in air under visible light. *Adv. Mater.* 24 (29): 3975–3980.
12 Montarnal, D., Capelot, M., Tournilhac, F., and Leibler, L. (2011). Silica-like malleable materials from permanent organic networks. *Science* 334 (6058): 965–968.
13 Angell, C.A. (1995). Formation of glasses from liquids and biopolymers. *Science* 267 (5206): 1924–1935.
14 Denissen, W., Winne, J.M., and Du Prez, F.E. (2016). Vitrimers: permanent organic networks with glass-like fluidity. *Chem. Sci.* 7 (1): 30–38.
15 Raquez, J.-M., Vanderstappen, S., Meyer, F. et al. (2011). Design of cross-linked semicrystalline poly(ϵ-caprolactone)-based networks with one-way and two-way shape-memory properties through Diels–Alder reactions. *Chem. Eur. J.* 17 (36): 10135–10143.
16 Willocq, B., Bose, R.K., Khelifa, F. et al. (2016). Healing by the Joule effect of electrically conductive poly(ester-urethane)/carbon nanotube nanocomposites. *J. Mater. Chem. A* 4 (11): 4089–4097.
17 Defize, T., Riva, R., Raquez, J.-M. et al. (2011). Thermoreversibly crosslinked poly(ϵ-caprolactone) as recyclable shape-memory polymer network. *Macromol. Rapid Commun.* 32 (16): 1264–1269.
18 Chang, J.Y., Do, S.K., and Han, M.J. (2001). A sol–gel reaction of vinyl polymers based on thermally reversible urea linkages. *Polymer* 42 (18): 7589–7594.
19 Sun, C.X., van der Mee, M.A.J., Goossens, J.G.P., and van Duin, M. (2006). Thermoreversible cross-linking of maleated ethylene/propylene copolymers using hydrogen-bonding and ionic interactions. *Macromolecules* 39 (9): 3441–3449.
20 Wagener, K.B. and Engle, L.P. (1991). Thermally reversible polymer linkages. 3. Covalently crosslinked poly(azlactone). *Macromolecules* 24 (26): 6809–6815.
21 van der Mee, M.A.J., Goossens, J.G.P., and van Duin, M. (2008). Thermoreversible covalent crosslinking of maleated ethylene/propylene copolymers with diols. *J. Polym. Sci., Part A: Polym. Chem.* 46 (5): 1810–1825.

22 Kamplain, J.W. and Bielawski, C.W. (2006). Dynamic covalent polymers based upon carbene dimerization. *Chem. Commun.* (16): 1727–1729.
23 Amamoto, Y., Kikuchi, M., Masunaga, H. et al. (2009). Reorganizable chemical polymer gels based on dynamic covalent exchange and controlled monomer insertion. *Macromolecules* 42 (22): 8733–8738.
24 Higaki, Y., Otsuka, H., and Takahara, A. (2006). A thermodynamic polymer cross-linking system based on radically exchangeable covalent bonds. *Macromolecules* 39 (6): 2121–2125.
25 Gheneim, R., Perez-Berumen, C., and Gandini, A. (2002). Diels–Alder reactions with novel polymeric dienes and dienophiles: synthesis of reversibly cross-linked elastomers. *Macromolecules* 35 (19): 7246–7253.
26 Goiti, E., Heatley, F., Huglin, M.B., and Rego, J.M. (2004). Kinetic aspects of the Diels–Alder reaction between poly(styrene-*co*-furfuryl methacrylate) and bismaleimide. *Eur. Polym. J.* 40 (7): 1451–1460.
27 Goiti, E., Huglin, M.B., and Rego, J.M. (2004). Some properties of networks produced by the Diels–Alder reaction between poly(styrene-*co*-furfuryl methacrylate) and bismaleimide. *Eur. Polym. J.* 40 (2): 219–226.
28 Liu, Y.-L. and Hsieh, C.-Y. (2006). Crosslinked epoxy materials exhibiting thermal remendablility and removability from multifunctional maleimide and furan compounds. *J. Polym. Sci., Part A: Polym. Chem.* 44 (2): 905–913.
29 Liu, Y.-L., Hsieh, C.-Y., and Chen, Y.-W. (2006). Thermally reversible cross-linked polyamides and thermo-responsive gels by means of Diels–Alder reaction. *Polymer* 47 (8): 2581–2586.
30 Zhang, Y., Broekhuis, A.A., and Picchioni, F. (2009). Thermally self-healing polymeric materials: the next step to recycling thermoset polymers? *Macromolecules* 42 (6): 1906–1912.
31 Imai, Y., Itoh, H., Naka, K., and Chujo, Y. (2000). Thermally reversible IPN organic–inorganic polymer hybrids utilizing the Diels–Alder reaction. *Macromolecules* 33 (12): 4343–4346.
32 Kennedy, J.P. and Carlson, G.M. (1983). Synthesis, characterization, and Diels–Alder extension of cyclopentadiene telechelic polyisobutylene. IV. α,ω-di(3-Cyclopentadienylpropyldimethylsilyl)polyisobutylene. *J. Polym. Sci.: Polym. Chem.* 21 (12): 3551–3561.
33 Goodall, G.W. and Hayes, W. (2006). Advances in cycloaddition polymerizations. *Chem. Soc. Rev.* 35 (3): 280–312.
34 Mikroyannidis, J.A. (1992). Synthesis and Diels–Alder polymerization of furfurylidene and furfuryl-substituted maleamic acids. *J. Polym. Sci., Part A: Polym. Chem.* 30 (1): 125–132.
35 Goussé, C. and Gandini, A. (1999). Diels–Alder polymerization of difurans with bismaleimides. *Polym. Int.* 48 (8): 723–731.
36 Kamahori, K., Tada, S., Ito, K., and Itsuno, S. (1999). Optically active polymer synthesis by Diels–Alder polymerization with chirally modified lewis acid catalyst. *Macromolecules* 32 (3): 541–547.
37 Mallakpour, S.E., Hajipour, A.-R., Mahdavian, A.-R., and Khoee, S. (2000). Synthesis and characterization of novel optically active and flame-retardant heterocyclic polyimides. *J. Appl. Polym. Sci.* 76 (2): 240–248.

38 Schilling, C.L., Reed, J.A., and Stille, J.K. (1969). Diels–Alder polymerization. VI. Phenylated polyphenylenes from bis-2-pyrones and *p*-diethynylbenzene. *Macromolecules* 2 (1): 85–88.
39 Stille, J.K., Rakutis, R.O., Mukamal, H., and Harris, F.W. (1968). Diels–Alder polymerizations. IV. Polymers containing short phenylene blocks connected by alkylene units. *Macromolecules* 1 (5): 431–436.
40 Mukamal, H., Harris, F.W., and Stille, J.K. (1967). Diels–Alder polymers. III. Polymers containing phenylated phenylene units. *J. Polym. Sci., Part A: Polym. Chem.* 5 (11): 2721–2729.
41 Durmaz, H., Karatas, F., Tunca, U., and Hizal, G. (2006). Heteroarm H-shaped terpolymers through the combination of the Diels–Alder reaction and controlled/living radical polymerization techniques. *J. Polym. Sci., Part A: Polym. Chem.* 44 (13): 3947–3957.
42 Li, M., De, P., Gondi, S.R., and Sumerlin, B.S. (2008). End group transformations of RAFT-generated polymers with bismaleimides: functional telechelics and modular block copolymers. *J. Polym. Sci., Part A: Polym. Chem.* 46 (15): 5093–5100.
43 Akat, H., Gacal, B., Balta, D.K. et al. (2010). Poly(ethylene glycol)-thioxanthone prepared by Diels–Alder click chemistry as one-component polymeric photoinitiator for aqueous free-radical polymerization. *J. Polym. Sci., Part A: Polym. Chem.* 48 (10): 2109–2114.
44 Masutani, K., Kawabata, S., Aoki, T., and Kimura, Y. (2010). Efficient formation of stereocomplexes of poly(*L*-lactide) and poly(*D*-lactide) by terminal Diels–Alder coupling. *Polym. Int.* 59 (11): 1526–1530.
45 Jørgensen, K.A. (2000). Catalytic asymmetric hetero-Diels–Alder reactions of carbonyl compounds and imines. *Angew. Chem. Int. Ed.* 39 (20): 3558–3588.
46 Nebhani, L., Sinnwell, S., Lin, C.Y. et al. (2009). Strongly electron deficient sulfonyldithioformate based RAFT agents for hetero Diels–Alder conjugation: computational design and experimental evaluation. *J. Polym. Sci., Part A: Polym. Chem.* 47 (22): 6053–6071.
47 Sinnwell, S., Synatschke, C.V., Junkers, T. et al. (2008). A study into the stability of 3,6-dihydro-2H-thiopyran rings: key linkages in the RAFT hetero-Diels–Alder click concept. *Macromolecules* 41 (21): 7904–7912.
48 Bousquet, A., Boyer, C., Davis, T.P., and Stenzel, M.H. (2010). Electrostatic assembly of functional polymer combs onto gold nanoparticle surfaces: combining RAFT, click and LbL to generate new hybrid nanomaterials. *Polym. Chem.* 1 (8): 1186–1195.
49 Dag, A., Durmaz, H., Hizal, G., and Tunca, U. (2008). Preparation of 3-arm star polymers (A3) via Diels–Alder click reaction. *J. Polym. Sci., Part A: Polym. Chem.* 46 (1): 302–313.
50 Durmaz, H., Karatas, F., Tunca, U., and Hizal, G. (2006). Preparation of ABC miktoarm star terpolymer containing poly(ethylene glycol), polystyrene, and poly(*tert*-butylacrylate) arms by combining Diels–Alder reaction, atom transfer radical, and stable free radical polymerization routes. *J. Polym. Sci., Part A: Polym. Chem.* 44 (1): 499–509.
51 Sinnwell, S., Inglis, A.J., Stenzel, M.H., and Barner-Kowollik, C. (2008). Access to three-arm star block copolymers by a consecutive combination

of the copper(I)-catalyzed azide–alkyne cycloaddition and the RAFT hetero Diels–Alder concept. *Macromol. Rapid Commun.* 29 (12–13): 1090–1096.

52 Durmaz, H., Dag, A., Hizal, G., and Tunca, U. (2010). Cyclic homo and block copolymers through sequential double click reactions. *J. Polym. Sci., Part A: Polym. Chem.* 48 (22): 5083–5091.

53 Szalai, M.L., McGrath, D.V., Wheeler, D.R. et al. (2007). Dendrimers based on thermally reversible furan–maleimide Diels–Alder adducts. *Macromolecules* 40 (4): 818–823.

54 McElhanon, J.R. and Wheeler, D.R. (2001). Thermally responsive dendrons and dendrimers based on reversible furan-maleimide Diels–Alder adducts. *Org. Lett.* 3 (17): 2681–2683.

55 Kim, T.-D., Luo, J., Tian, Y. et al. (2006). Diels–Alder "click chemistry" for highly efficient electrooptic polymers. *Macromolecules* 39 (5): 1676–1680.

56 Shi, Z., Luo, J., Huang, S. et al. (2009). Controlled Diels–Alder reactions used to incorporate highly efficient polyenic chromophores into maleimide-containing side-chain polymers for electro-optics. *Macromolecules* 42 (7): 2438–2445.

57 Gacal, B., Akat, H., Balta, D.K. et al. (2008). Synthesis and characterization of polymeric thioxanthone photoinitatiors via double click reactions. *Macromolecules* 41 (7): 2401–2405.

58 Tasdelen, M.A. (2011). Diels–Alder "click" reactions: recent applications in polymer and material science. *Polym. Chem.* 2: 2133–2145.

59 Ghezzo, F., Smith, D.R., Starr, T.N. et al. (2010). Development and characterization of healable carbon fiber composites with a reversibly cross linked polymer. *J. Compos. Mater.* 44 (13): 1587–1603.

60 Inoue, K., Yamashiro, M., and Iji, M. (2009). Recyclable shape-memory polymer: poly(lactic acid) crosslinked by a thermoreversible Diels–Alder reaction. *J. Appl. Polym. Sci.* 112 (2): 876–885.

61 Adzima, B.J., Aguirre, H.A., Kloxin, C.J. et al. (2008). Rheological and chemical analysis of reverse gelation in a covalently cross-linked Diels–Alder polymer network. *Macromolecules* 41 (23): 9112–9117.

62 Nemoto, T. and Konishi, G.-i. (2008). Synthesis and properties of new organosoluble alkoxylated naphthalene-based novolacs preparaed by addition-condensation of mono- or di-alkoxynaphthalene with formaldehyde. *Polym. J.* 40: 651.

63 Kavitha, A.A. and Singha, N.K. (2010). Smart "all acrylate" ABA triblock copolymer bearing reactive functionality via atom transfer radical polymerization (ATRP): demonstration of a "click reaction" in thermoreversible property. *Macromolecules* 43 (7): 3193–3205.

64 Watanabe, M. and Yoshie, N. (2006). Synthesis and properties of readily recyclable polymers from bisfuranic terminated poly(ethylene adipate) and multi-maleimide linkers. *Polymer* 47 (14): 4946–4952.

65 Shi, Z., Liang, W., Luo, J. et al. (2010). Tuning the kinetics and energetics of Diels–Alder cycloaddition reactions to improve poling efficiency and thermal stability of high-temperature cross-linked electro-optic polymers. *Chem. Mater.* 22 (19): 5601–5608.

66 Adzima, B.J., Kloxin, C.J., and Bowman, C.N. (2010). Externally triggered healing of a thermoreversible covalent network via self-limited hysteresis heating. *Adv. Mater.* 22 (25): 2784–2787.
67 Peterson, A.M., Jensen, R.E., and Palmese, G.R. (2010). Room-temperature healing of a thermosetting polymer using the Diels–Alder reaction. *ACS Appl. Mater. Interfaces* 2 (4): 1141–1149.
68 Amalin Kavitha, A. and Singha, N.K. (2007). A tailor-made polymethacrylate bearing a reactive diene in reversible Diels–Alder reaction. *J. Polym. Sci., Part A: Polym. Chem.* 45 (19): 4441–4449.
69 Heath, W.H., Palmieri, F., Adams, J.R. et al. (2008). Degradable cross-linkers and strippable imaging materials for step-and-flash imprint lithography. *Macromolecules* 41 (3): 719–726.
70 Syrett, J.A., Mantovani, G., Barton, W.R.S. et al. (2010). Self-healing polymers prepared via living radical polymerisation. *Polym. Chem.* 1 (1): 102–106.
71 Willocq, B., Khelifa, F., Brancart, J. et al. (2017). One-component Diels–Alder based polyurethanes: a unique way to self-heal. *RSC Adv.* 7 (76): 48047–48053.
72 Zhao, J., Xu, R., Luo, G. et al. (2016). Self-healing poly(siloxane-urethane) elastomers with remoldability, shape memory and biocompatibility. *Polym. Chem.* 7 (47): 7278–7286.
73 Oehlenschlaeger, K.K., Mueller, J.O., Brandt, J. et al. (2014). Adaptable hetero Diels–Alder networks for fast self-healing under mild conditions. *Adv. Mater.* 26 (21): 3561–3566.
74 Oehlenschlaeger, K.K., Guimard, N.K., Brandt, J. et al. (2013). Fast and catalyst-free hetero-Diels–Alder chemistry for on demand cyclable bonding/debonding materials. *Polym. Chem.* 4 (16): 4348–4355.
75 Liu, X., Zhu, M., Chen, S. et al. (2008). Organic–inorganic nanohybrids via directly grafting gold nanoparticles onto conjugated copolymers through the Diels–Alder reaction. *Langmuir* 24 (20): 11967–11974.
76 Liu, X., Liu, H., Zhou, W. et al. (2010). Thermoreversible covalent self-assembly of oligo(*p*-phenylenevinylene) bridged gold nanoparticles. *Langmuir* 26 (5): 3179–3185.
77 Zhu, J., Kell, A.J., and Workentin, M.S. (2006). A retro-Diels–Alder reaction to uncover maleimide-modified surfaces on monolayer-protected nanoparticles for reversible covalent assembly. *Org. Lett.* 8 (22): 4993–4996.
78 Gotsmann, B., Duerig, U., Frommer, J., and Hawker, C.J. (2006). Exploiting chemical switching in a Diels–Alder polymer for nanoscale probe lithography and data storage. *Adv. Funct. Mater.* 16 (11): 1499–1505.
79 Adachi, K., Achimuthu, A.K., and Chujo, Y. (2004). Synthesis of organic–inorganic polymer hybrids controlled by Diels–Alder reaction. *Macromolecules* 37 (26): 9793–9797.
80 Zhou, X., Li, Q., and Wu, C. (2008). Grafting of maleic anhydride onto carbon black surface via ultrasonic irradiation. *Appl. Organomet. Chem.* 22 (2): 78–81.
81 Ménard-Moyon, C., Dumas, F., Doris, E., and Mioskowski, C. (2006). Functionalization of single-wall carbon nanotubes by tandem

high-pressure/$Cr(CO)_6$ activation of Diels–Alder cycloaddition. *J. Am. Chem. Soc.* 128 (46): 14764–14765.
82 Delgado, J.L., de la Cruz, P., Langa, F. et al. (2004). Microwave-assisted sidewall functionalization of single-wall carbon nanotubes by Diels–Alder cycloaddition. *Chem. Commun.* (15): 1734–1735.
83 Mercuri, F. and Sgamellotti, A. (2009). First-principles investigations on the functionalization of chiral and non-chiral carbon nanotubes by Diels–Alder cycloaddition reactions. *Phys. Chem. Chem. Phys.* 11 (3): 563–567.
84 Hayden, H., Gun'ko, Y.K., and Perova, T.S. (2007). Chemical modification of multi-walled carbon nanotubes using a tetrazine derivative. *Chem. Phys. Lett.* 435 (1): 84–89.
85 Chang, C.-M. and Liu, Y.-L. (2009). Functionalization of multi-walled carbon nanotubes with furan and maleimide compounds through Diels–Alder cycloaddition. *Carbon* 47 (13): 3041–3049.
86 Vida, Y., Suau, R., Casado, J. et al. (2007). Electropolymerizable terthiophene *S,S*-dioxide-fullerene Diels–Alder adduct for donor/acceptor double-cable polymers. *Macromol. Rapid Commun.* 28 (12): 1345–1349.
87 Bai, J. and Shi, Z. (2017). Dynamically cross-linked elastomer hybrids with light-induced rapid and efficient self-healing ability and reprogrammable shape memory behavior. *ACS Appl. Mater. Interfaces* 9 (32): 27213–27222.
88 Lim, H.L., Hwang, Y., Kar, M., and Varghese, S. (2014). Smart hydrogels as functional biomimetic systems. *Biomater. Sci.* 2 (5): 603–618.
89 Stuart, M.A.C., Huck, W.T.S., Genzer, J. et al. (2010). Emerging applications of stimuli-responsive polymer materials. *Nat. Mater.* 9: 101–113.
90 Varaprasad, K., Raghavendra, G.M., Jayaramudu, T. et al. (2017). A mini review on hydrogels classification and recent developments in miscellaneous applications. *Mater. Sci. Eng., C* 79: 958–971.
91 Gandini, A. and Lacerda, T.M. (2015). From monomers to polymers from renewable resources: recent advances. *Prog. Polym. Sci.* 48: 1–39.
92 Kirchhof, S., Brandl, F.P., Hammer, N., and Goepferich, A.M. (2013). Investigation of the Diels–Alder reaction as a cross-linking mechanism for degradable poly(ethylene glycol) based hydrogels. *J. Mater. Chem. B* 1 (37): 4855–4864.
93 Zhang, M., Wang, J., and Jin, Z. (2018). Supramolecular hydrogel formation between chitosan and hydroxypropyl β-cyclodextrin via Diels–Alder reaction and its drug delivery. *Int. J. Biol. Macromol.* 114: 381–391.
94 Yu, F., Cao, X., Du, J. et al. (2015). Multifunctional hydrogel with good structure integrity, self-healing, and tissue-adhesive property formed by combining Diels–Alder click reaction and acylhydrazone bond. *ACS Appl. Mater. Interfaces* 7 (43): 24023–24031.
95 Wei, H.-L., Yang, J., Chu, H.-J. et al. (2011). Diels–Alder reaction in water for the straightforward preparation of thermoresponsive hydrogels. *J. Appl. Polym. Sci.* 120 (2): 974–980.
96 Gregoritza, M., Messmann, V., Abstiens, K. et al. (2017). Controlled antibody release from degradable thermoresponsive hydrogels cross-linked by Diels–Alder chemistry. *Biomacromolecules* 18 (8): 2410–2418.

97 Gregoritza, M., Messmann, V., Goepferich, A.M., and Brandl, F.P. (2016). Design of hydrogels for delayed antibody release utilizing hydrophobic association and Diels–Alder chemistry in tandem. *J. Mater. Chem. B* 4 (19): 3398–3408.

98 Kirchhof, S., Gregoritza, M., Messmann, V. et al. (2015). Diels–Alder hydrogels with enhanced stability: first step toward controlled release of bevacizumab. *Eur. J. Pharm. Biopharm.* 96: 217–225.

99 He, C., Kim, S.W., and Lee, D.S. (2008). In situ gelling stimuli-sensitive block copolymer hydrogels for drug delivery. *J. Controlled Release* 127 (3): 189–207.

100 Koetting, M.C., Peters, J.T., Steichen, S.D., and Peppas, N.A. (2015). Stimulus-responsive hydrogels: theory, modern advances, and applications. *Mater. Sci. Eng. R. Rep.* 93: 1–49.

101 Qiu, Y. and Park, K. (2001). Environment-sensitive hydrogels for drug delivery. *Adv. Drug Delivery Rev.* 53 (3): 321–339.

102 Ruel-Gariépy, E. and Leroux, J.-C. (2004). In situ-forming hydrogels-review of temperature-sensitive systems. *Eur. J. Pharm. Biopharm.* 58 (2): 409–426.

103 Lee, C., Kim, M., Kim, Y.J. et al. (2017). Soft robot review. *Int. J. Control Autom. Syst.* 15 (1): 3–15.

104 Zeng, C., Seino, H., Ren, J., and Yoshie, N. (2014). Polymers with multishape memory controlled by local glass transition temperature. *ACS Appl. Mater. Interfaces* 6 (4): 2753–2758.

105 Zhang, G., Zhao, Q., Yang, L. et al. (2016). Exploring dynamic equilibrium of Diels–Alder reaction for solid state plasticity in remoldable shape memory polymer network. *ACS Macro Lett.* 5 (7): 805–808.

106 Cai, C., Zhang, Y., Li, M. et al. (2018). Multiple-responsive shape memory polyacrylonitrile/graphene nanocomposites with rapid self-healing and recycling properties. *RSC Adv.* 8 (3): 1225–1231.

107 Terryn, S., Brancart, J., Lefeber, D. et al. (2017). Self-healing soft pneumatic robots. *Sci. Rob.* 2 (9): eaan4268.

108 Seppe, T., Glenn, M., Joost, B. et al. (2015). Development of a self-healing soft pneumatic actuator: a first concept. *Bioinspiration Biomimetics* 10 (4): 1–17.

109 Toncheva, A., Willocq, B., Khelifa, F. et al. (2017). Bilayer solvent and vapor-triggered actuators made of cross-linked polymer architectures via Diels–Alder pathways. *J. Mater. Chem. B* 5 (28): 5556–5563.

110 Toncheva, A., Khelifa, F., Paint, Y. et al. (2018). Fast IR-actuated shape-memory polymers using in situ silver nanoparticle-grafted cellulose nanocrystals. *ACS Appl. Mater. Interfaces* 10 (35): 29933–29942.

111 Berg, G.J., Gong, T., Fenoli, C.R., and Bowman, C.N. (2014). A dual-cure, solid-state photoresist combining a thermoreversible Diels–Alder network and a chain growth acrylate network. *Macromolecules* 47 (10): 3473–3482.

112 Adzima, B.J., Kloxin, C.J., DeForest, C.A. et al. (2012). 3D photofixation lithography in Diels–Alder networks. *Macromol. Rapid Commun.* 33 (24): 2092–2096.

113 Davidson, J.R., Appuhamillage, G.A., Thompson, C.M. et al. (2016). Design paradigm utilizing reversible Diels–Alder reactions to enhance the mechanical properties of 3D printed materials. *ACS Appl. Mater. Interfaces* 8 (26): 16961–16966.

114 Appuhamillage, G.A., Reagan, J.C., Khorsandi, S. et al. (2017). 3D printed remendable polylactic acid blends with uniform mechanical strength enabled by a dynamic Diels–Alder reaction. *Polym. Chem.* 8 (13): 2087–2092.

115 Yang, K., Grant, J.C., Lamey, P. et al. (2017). Diels–Alder reversible thermoset 3D printing: isotropic thermoset polymers via fused filament fabrication. *Adv. Funct. Mater.* 27 (24): 1–11.

116 Chen, S., Zhang, Q., and Feng, J. (2017). 3D printing of tunable shape memory polymer blends. *J. Mater. Chem. C* 5 (33): 8361–8365.

117 Senatov, F.S., Niaza, K.V., Zadorozhnyy, M.Y. et al. (2016). Mechanical properties and shape memory effect of 3D-printed PLA-based porous scaffolds. *J. Mech. Behav. Biomed. Mater.* 57: 139–148.

118 Zhao, Z., Peng, F., Cavicchi, K.A. et al. (2017). Three-dimensional printed shape memory objects based on an olefin ionomer of zinc-neutralized poly(ethylene-*co*-methacrylic acid). *ACS Appl. Mater. Interfaces* 9 (32): 27239–27249.

119 Yang, Y., Chen, Y., Wei, Y., and Li, Y. (2016). 3D printing of shape memory polymer for functional part fabrication. *Int. J. Adv. Manuf. Technol.* 84 (9): 2079–2095.

120 Yang, H., Leow, W.R., Wang, T. et al. (2017). 3D Printed photoresponsive devices based on shape memory composites. *Adv. Mater.* 29 (33): 1–7.

12

Mechanochromic Polymers as Stress-sensing Soft Materials

Daisuke Aoki and Hideyuki Otsuka

Tokyo Institute of Technology, Department of Chemical Science and Engineering, 2-12-1-S1-6 Ookayama, Meguro-ku, Tokyo 152-8550, Japan

12.1 Introduction

"Mechanochromism" refers to changes in the optical properties of compounds or materials under mechanical stress such as grinding, compressing, stretching, or rubbing. These phenomena have attracted much attention due to their potential applications in probes that allow the detection of stress in materials and provide insight into their fracture mechanisms, thus facilitating the prediction of risks and an estimation of the lifetime of such materials. Materials that exhibit mechanochromism are therefore called "mechanochromic" or "mechanoresponsive," as they represent functional materials whose optical properties change upon exposure to external stimuli in a manner similar to that of thermochromic, photochromic, and electrochromic materials. Mechanochromic materials that consist of polymer chains are called "mechanochromic polymers" and these have attracted substantial attention given that their mechanical properties can be readily tuned by a judicious choice of the appropriate polymer composition. Several methods have been explored for the design of mechanochromic polymers. In this chapter, different approaches for the development of mechanochromic polymers and their characteristics are discussed with a focus on the dynamic covalent chemistry of mechanochromic polymers, which is described in detail.

12.2 Classification of Mechanochromic Polymers

The mechanochromic polymers reported so far have been mainly classified into three categories: structural-color-based systems, fluorescent-dye-based systems, and chemical-reaction-based systems (Figure 12.1). Structural color can be observed when the periodicity of regular arrays is on a length scale of about half the wavelength of visible light [1–5]. A number of applications, including full-color displays, colorimetric sensors, coatings, and paints, have been reported so far based on such structural-color-based systems. Most of these are obtained by arraying silica particles, polystyrene particles, and lamellar structures into gels

Mechanically Responsive Materials for Soft Robotics, First Edition. Edited by Hideko Koshima.

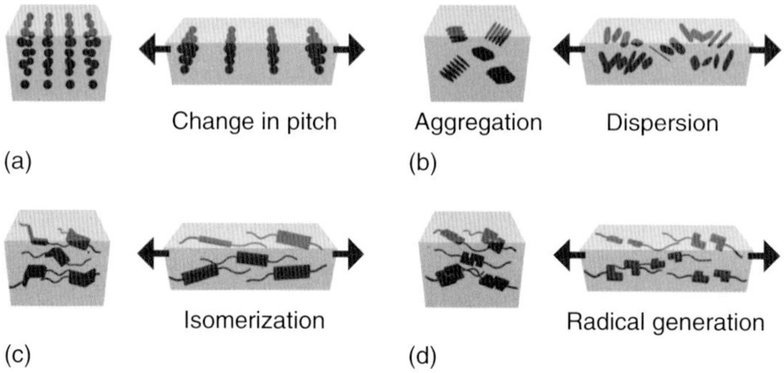

Figure 12.1 Classification of mechanochromic polymers. (a) Structural-color-based system, (b) fluorescent-dye-based system, (c) chemical-reaction-based system (isomerization), and (d) chemical-reaction-based system (bond cleavage).

or elastomers that respond to light of specific wavelengths. The stress-induced deformation of these materials changes the pitch of the ordered structures, which results in changes of the structural color (Figure 12.1a). Although some limitations remain to be solved in structural-color-based systems (e.g. color contrast and angle dependency), most of them show reversible color changes, which is an attractive feature for the practical application of mechanochromism.

Systems based on fluorescent dyes are obtained by dispersing fluorescent molecules into polymer matrices. Flat fluorescent molecules, typically conjugated fluorescent molecules, tend to form aggregates in polymer matrices (Figure 12.1b). When a mechanical force is applied to such materials, the aggregated fluorescent molecules become dispersed, which is accompanied by a change in the fluorescence color [6–8]. Although the color change observable by the naked eye in systems based on fluorescent dyes is not very drastic, the fluorescent colors of the aggregated and dispersed states can be tuned by choosing appropriate fluorescent molecules.

Mechanochromic polymers based on chemical reactions were initially reported by Moore and coworkers [9]. The introduction of spiropyran, a dye molecule, in the primary structure of linear polymer chains endows these with mechanochromic properties due to an isomerization of spiropyran into merocyanine in response to mechanical stress such as stretching and compressing (Figure 12.1c) [10–12]. Molecules that respond to mechanical stimuli have been referred to as "mechanophores" by Moore and coworker [13]. Some of these change color in response to mechanical stimuli, allowing the visualization of mechanical stress in the polymeric materials. However, the number of reported examples of mechanophores remains limited, making this research area a field that is still under development. In addition to the isomerization in spiropyran derivatives, metal complexes [14, 15] whose structure changes upon exposure to a mechanical force and bulky dioxetane derivatives [16], which act as chemiluminescent mechanophores, have also been reported. Systems based on the latter can be employed to elucidate the fracture mechanisms of polymeric materials [17] (Figure 12.1d).

12.3 Mechanochromophores Based on Dynamic Covalent Chemistry

Our group has previously reported polymer reactions based on dynamic covalent chemistry using chemical equilibrium systems, i.e. those exhibiting a reversible cleavage of covalent bonds [18–22]. Among these, diarylbibenzofuranone (DABBF) derivatives, synthesized by the dimerization of the corresponding arylbenzofuranone (ABF) derivatives, contain relatively weak C—C bonds. DABBF and its radicals are in equilibrium at room temperature, by-products are not generated, and the amount of radicals formed is quite small (Figure 12.2a). We have discovered that DABBF works as a bond-exchanging unit via the generated radicals, endowing DABBF-containing materials with self-healing ability at room temperature, which is derived from bond-exchange reactions that occur spontaneously at room temperature [23–25]. Owing to the stability of the ABF radicals toward oxygen, the equilibrium between DABBF and ABF radicals enables a quantitative in situ evaluation of the cleavage of C—C bonds using electron paramagnetic resonance (EPR) measurements. The bond dissociation energy (BDE) of the central C—C bond in DABBF has been estimated as 85.4–104.8 kJ mol^{-1}, which is lower than that of typical C—C bonds (e.g. 380 kJ mol^{-1} in ethane), resulting in an easy dissociation of the DABBF unit. The fact that the ABF radical generated upon heat-induced dissociation of DABBF is blue and exhibits considerable stability in air led us to the idea that DABBF may work as a mechanochromophore. As a preliminary test, a DABBF derivative with two hydroxy groups was ground, resulting in a color change from white to blue, i.e. DABBF indeed exhibits mechanochromism (Figure 12.2a). The peak intensity in the EPR spectrum after grinding increased dramatically compared to that before grinding, indicating that the central C—C bond of DABBF had been cleaved and confirming that radicals are generated upon grinding (Figure 12.2b)

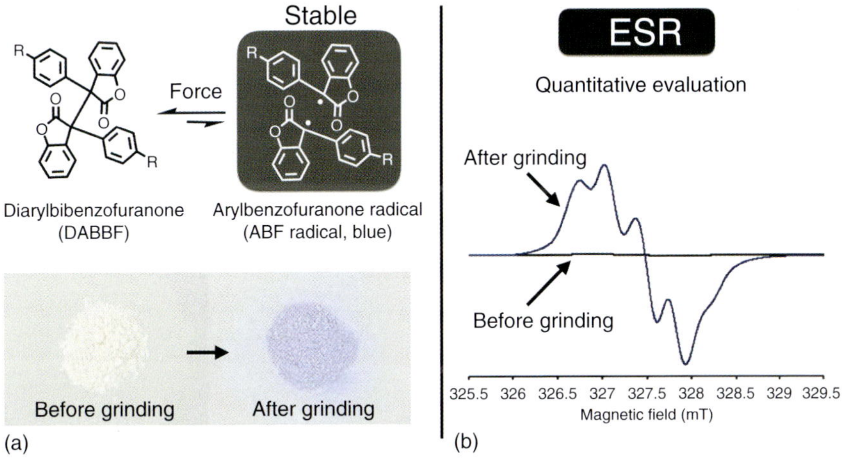

Figure 12.2 (a) Chemical structure of DABBF and equilibrium between DABBF and ABF radicals, as well as (b) EPR spectra of a DABBF derivative before and after grinding.

[26]. The blue color of the sample after grinding gradually faded with time to return to the original white color, which further supports the notion that DABBF derivatives serve as "dynamic" mechanochromophores.

12.4 Mechanochromic Polymers Based on Dynamic Covalent Chemistry

12.4.1 Polystyrenes with Mechanochromophores at the Center of the Polymer Chain

The incorporation of a mechanochromophore at the center of the polymer backbone is the simplest approach for the preparation of mechanochromic polymers based on dynamic covalent chemistry. For example, a DABBF moiety was introduced at the center of polystyrenes with linear or star-shaped structures [27]. Atom transfer radical polymerization and living anionic polymerization techniques were employed to synthesize polystyrene samples with narrow polydispersity indexes (PDIs), followed by conversion of the terminal structure to afford azide-terminated polystyrene precursors. Linear, four-armed star-shaped, and eight-armed star-shaped polymers bearing the DABBF structure at the center were successfully synthesized as white powders by click reactions between multifunctional alkyne–DABBF compounds and the polymer precursors (Figure 12.3).

All the obtained polymers exhibited mechanochromic properties, i.e. a color change from white to blue was confirmed. To quantitatively evaluate these mechanochromic properties, EPR measurements were carried out after grinding the polystyrene samples in a ball mill at specific grinding frequencies and times. The DABBF dissociation ratios in these samples were employed as an index of the force-transmitting efficiency, estimated by the peak intensity using 4-hydroxy-2,2,6,6-tetramethylpiperidine-1-oxyl (TEMPOL) as a radical standard. The dissociation ratio of DABBF with the polymer chains was >10%, i.e. much higher than that of DABBF without the polymer chain (<0.2%). Figure 12.4 summarizes the dissociation ratios of DABBF diol and DABBF-centered polystyrenes, revealing that the dissociation ratio increases with the molecular weight and the number of branches. Surprisingly, the DABBF dissociation ratio was approximately 46% in the eight-armed star-shaped polymer. This remarkable mechanochromic behavior in all the DABBF-based polymer chains demonstrates that the mechanical force is effectively transmitted to the central DABBF unit through the polystyrene chains.

12.4.2 Polyurethane Elastomers with Mechanophores in the Repeating Units

Mechanochromic polymers that exhibit elastic properties are discussed in this (12.4.2) and the following Section (12.4.3). Segmented polyurethanes (SPUs) have been widely used in various fields of materials science and engineering, including biocompatible materials, owing to their excellent elastic properties, thermoplasticity (i.e. processability into various shapes), and durability. Furthermore,

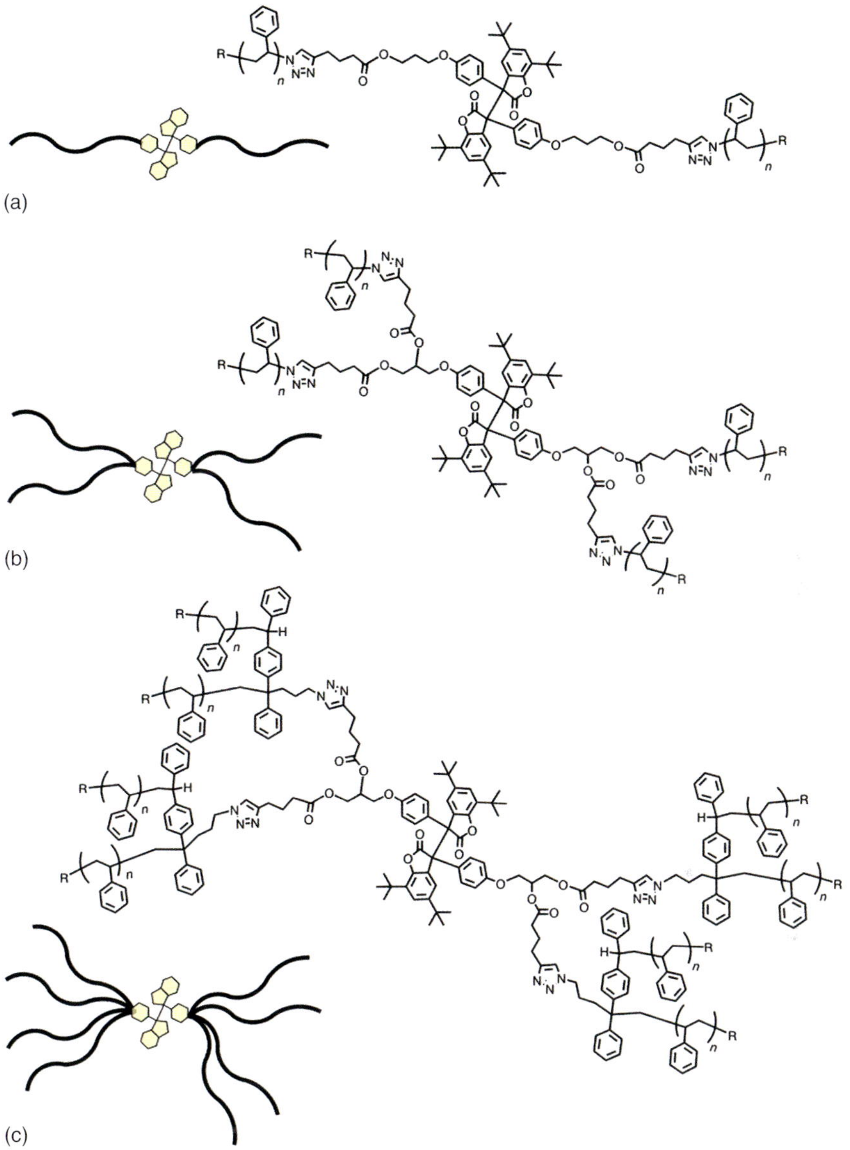

Figure 12.3 (a) Linear, (b) four-armed star-shaped, and (c) eight-armed star-shaped polymers containing a central DABBF unit.

these properties can be easily tuned by controlling the chemical structure and composition of the SPUs. To endow mechanochromic materials with elastic properties, which is an important feature in the design of strain-sensing materials, DABBF-containing SPUs were synthesized from a DABBF derivative with two hydroxy groups, a polyol, and a diisocyanate (Figure 12.5) [28, 29]. As expected, the obtained SPUs, which bear a DABBF moiety in the polymer chain, exhibited mechanochromism upon stretching and their behavior could be tuned

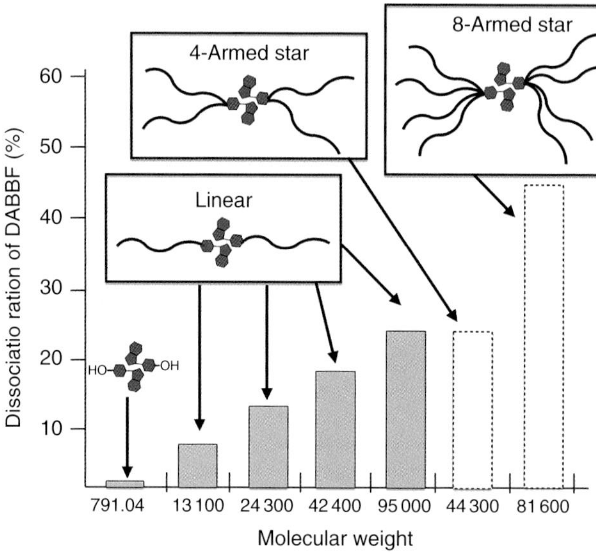

Figure 12.4 Dissociation ratio of the DABBF moiety in polymer chains.

by changing their chemical structure and composition. The blue color of the mechanically activated polyurethanes gradually faded during several hours at room temperature until disappearing completely under almost full restoration of the original dimensions.

These observations were confirmed by EPR spectroscopy, where the spectral intensity derived from the ABF radical greatly increased after stretching. Upon removing the applied force, the spectral intensity decreased together with a fading of the color (Figure 12.6). Throughout repeated cycling, the spectral intensity remained almost unchanged, demonstrating repeated mechanical activation and regeneration based on the dynamic covalent chemistry of DABBF.

12.4.3 Mechanochromic Elastomers Based on Polymer–Inorganic Composites with Dynamic Covalent Mechanochromophores

In this section, an example based on chemical cross-linking for mechanochromic materials with elastic properties is introduced [30]. A DABBF alkoxysilane derivative (DABBF-bis(trimethoxysilane)) was synthesized and applied to the fabrication of polymer–inorganic composite materials via a sol–gel method, which led to the incorporation of DABBF moieties into highly cross-linked rigid silica network domains. For the synthesis of DABBF-bis(trimethoxysilane), a DABBF derivative with a hydroxy group was reacted with 3-isocyanato-propyltrimethoxysilane (IPTMS) in the presence of di-*n*-butyltin dilaurate (DBTDL) as the catalyst. As shown in Figure 12.7, the DABBF-bis(trimethoxysilane) synthesized in situ was directly reacted with commercially available poly(butyl acrylate) that contained dimethoxymethylsilyl functionalities at both chain ends (Si-PBA) as the precursor for the soft domain, thus affording a series of

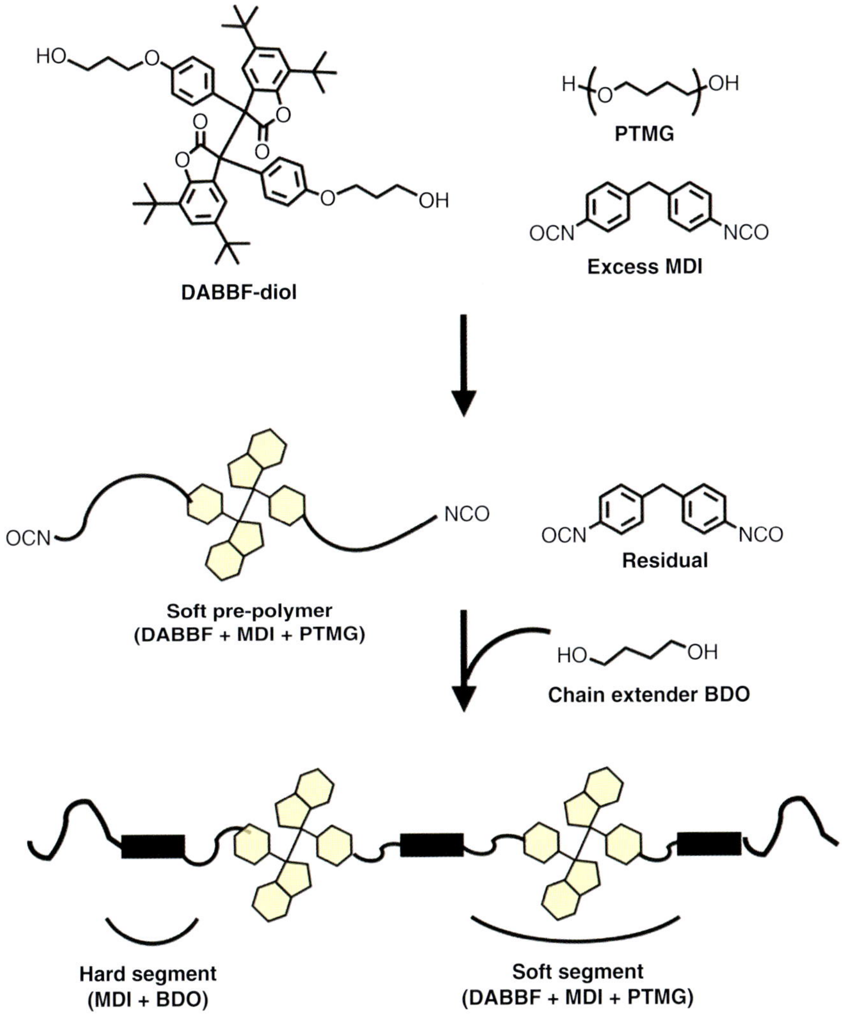

Figure 12.5 Synthetic scheme for the formation of segmented polyurethanes containing DABBF.

Figure 12.6 Photographic images of a dumbbell-shaped segmented polyurethane elastomer before and after manual elongation.

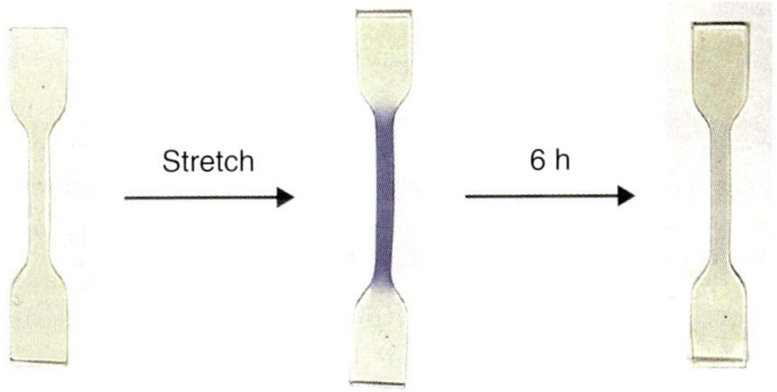

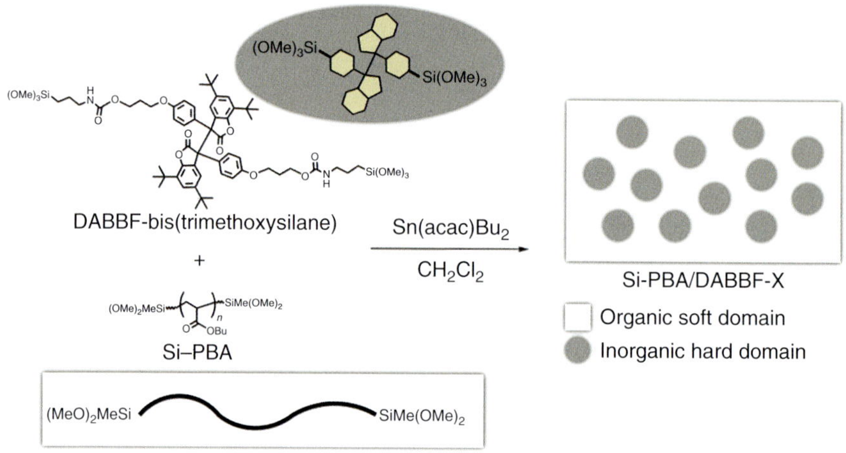

Figure 12.7 Synthetic route to polymer–inorganic composites containing DABBF moieties.

organic–inorganic hybrid elastomers (Si-PBA/DABBF-X) with different contents of DABBF-bis-(trimethoxysilane) (indicated by X in wt%).

Si-PBA/DABBF-65, which is the hybrid with the highest content of DABBF mechanophores, is brittle and inflexible, but very sensitive to mechanical stress, i.e. the sample turns blue upon grinding, scratching, and compressing. On the other hand, other samples with lower DABBF contents (Si-PBA/DABBF-0, 5, 16, 25, and 35) are flexible and could be molded into films. Tensile tests of the obtained elastic films (Si-PBA/DABBF-0, 5, 16, 25, and 35) revealed that they become tougher with increasing DABBF alkoxysilane content, albeit that their transparency is low. Transmission electron microscopy (TEM) images of the samples showed a discontinuous sphere-like morphology, whose average sphere size increases with the DABBF-bis(trimethoxysilane) weight ratio (Figure 12.8).

The DABBF-containing silica domains appear as darker regions, while the poly(butyl acrylate) matrix appears bright. The stress–strain curves and TEM images of these samples indicated that the cross-linking density increases with increasing silica domains, resulting in an improvement of the mechanical properties (Figure 12.9).

Subsequently, we examined the mechanoresponsiveness of Si-PBA/DABBF elastomer films to tensile elongation. Color changes were observed upon stretching Si-PBA/DABBF-16, 25, and 35 films, where the onset of the color change is clearly related to the mechanical properties, i.e. the strain and applied stress. We found that the mechanoresponsiveness strongly depends on the content of DABBF alkoxysilane, the phase-separation structure, and the mechanical properties. It was thus presumed that the DABBF moieties from the DABBF alkoxysilane derivatives could be incorporated in the hard silica domain, soft polymer domain, or at their interface. Although we will not go into details about a plausible mechanism for the stretching-triggered DABBF mechanochromophore here, at the interfaces, the DABBF moieties are selectively dissociated upon stretching (Figure 12.10). Therefore, Si-PBA/DABBF-35 exhibits remarkable mechanochromic properties, as the large silica domain in the film promotes an

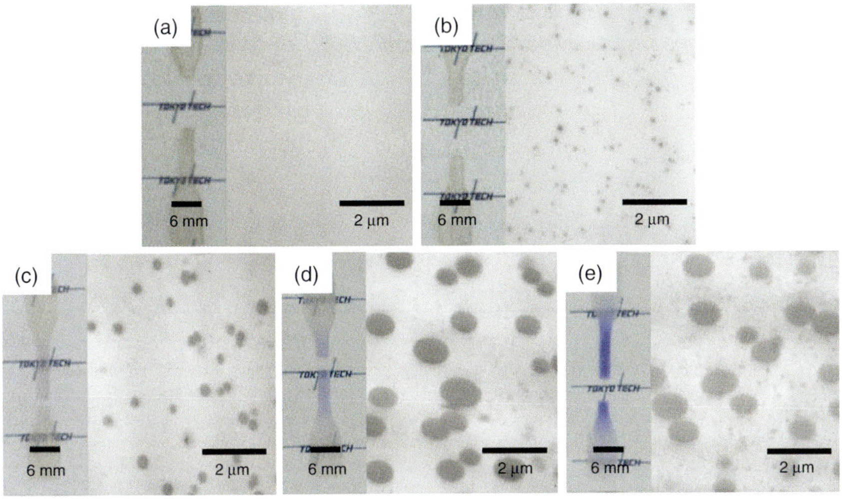

Figure 12.8 Photographic images of broken dumbbell specimens after stretching and TEM images of Si-PBA/DABBF-X at (a) X = 0, (b) X = 5, (c) X = 16, (d) X = 25, and (e) X = 35 wt% DABBF alkoxysilane. Source: Kosuge et al. 2016 [30]. Copyright 2016, Reproduced with permission of American Chemical Society.

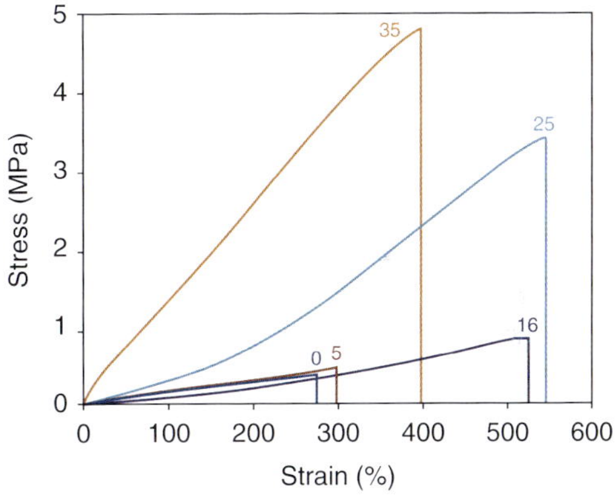

Figure 12.9 Stress–strain curves for Si-PBA/DABBF-0, 5, 16, 25, and 35. Source: Kosuge et al. 2016 [30]. Copyright 2016, Reproduced with permission of American Chemical Society.

effective transmission of the applied mechanical force to the DABBF moieties at the interface.

12.5 Mechanochromic Polymers Exhibiting Mechanofluorescence

In the last sections of this chapter, other radical-type mechanophores with relatively weak central C—C bonds will be discussed. Mechanofluorescence, which exhibits simple and/or fluorescence color changes in response to external mechanical stimuli such as shearing, stretching, and compressing has attracted considerable research interest due to its potential applications in optoelectronic

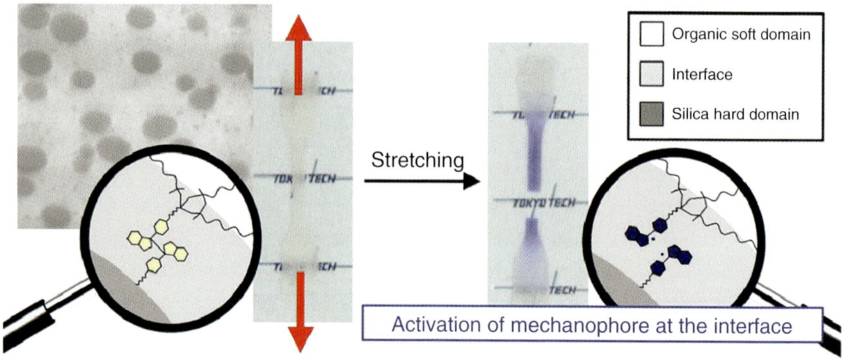

Figure 12.10 Plausible activation mechanism of DABBF at the interface of Si-PBA/DABBF films. Source: Kosuge et al. 2016 [30]. Copyright 2016, Reproduced with permission of American Chemical Society.

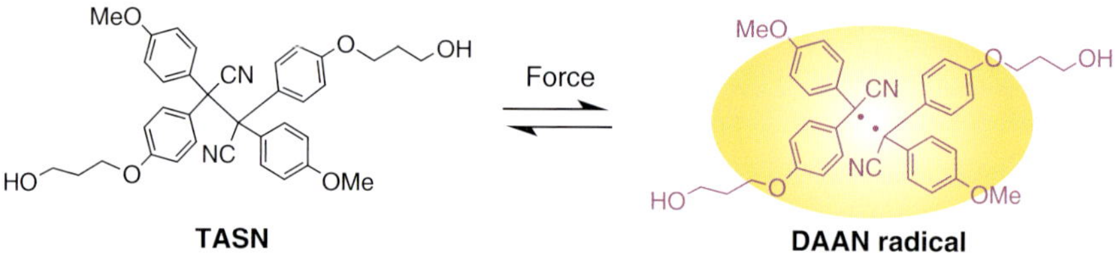

Figure 12.11 Chemical structure of TASN and the generation of its radicals.

devices, sensors, probes, and optical data storage devices. Recently, we have developed tetraarylsuccinonitrile (TASN) as a radical-type mechanochromophore that can be cleaved in response to mechanical stress to generate pink diarylacetonitrile (DAAN) radicals, as well as the emission of yellow light under UV irradiation (Figure 12.11) [31].

The absolute photoluminescence quantum yield of a TASN derivative with two hydroxy groups after grinding is 39%. Polymeric materials exhibiting mechanoluminescence in response to mechanical stress were subsequently developed by introducing TASN derivatives in polystyrene using click chemistry. The TASN units located at the center of a polystyrene polymer chain not only exhibit a color change but also emit yellow light under UV irradiation (Figure 12.12). Owing to such light-emission properties in response to mechanical stimuli, TASN is expected to be an appropriate molecular probe to detect stress and damage in polymeric materials with both high visibility and sensitivity [32].

12.6 Rainbow Mechanochromism Based on Three Radical-type Mechanochromophores

In the previous section, the two radical-type mechanochromophores DABBF (Sections 12.3 and 12.4) and TASN (Section 12.5) were introduced. Homolytic cleavage of the central C—C bond in the DABBF and TASN units in response to

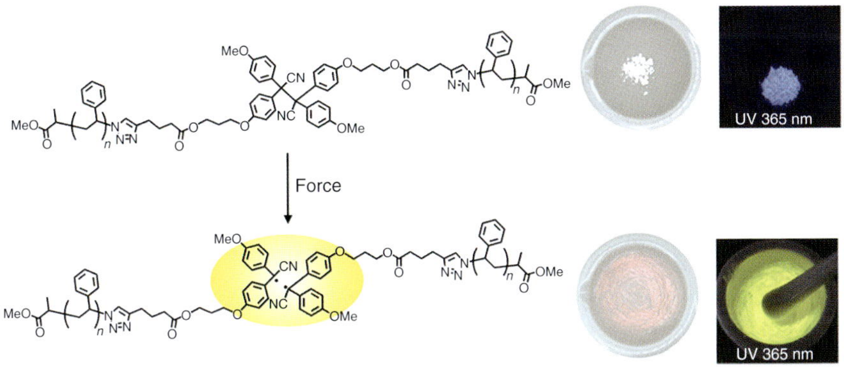

Figure 12.12 Chemical structure of TASN-based polystyrene and photographic images before and after grinding.

mechanical stress afforded ABF (blue) and DAAN (pink) radicals, respectively. As the two radical-type mechanochromophores display different colors, we envisioned that the development of a green radical-type mechanochromophore would complete a set of three primary colors, which would allow the subsequent development of "rainbow mechanochromism," i.e. a novel concept providing combinatorial colors for mechanochromic polymers (Figure 12.13). It has already been reported that arylbenzothiophenonyl (ABT) radicals generated

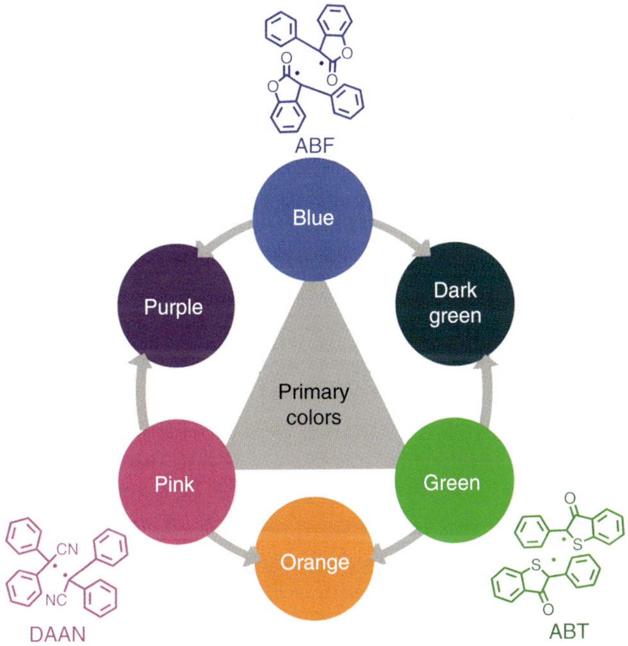

Figure 12.13 Combinatorial colors ("color palette") for mechanochromic polymers.

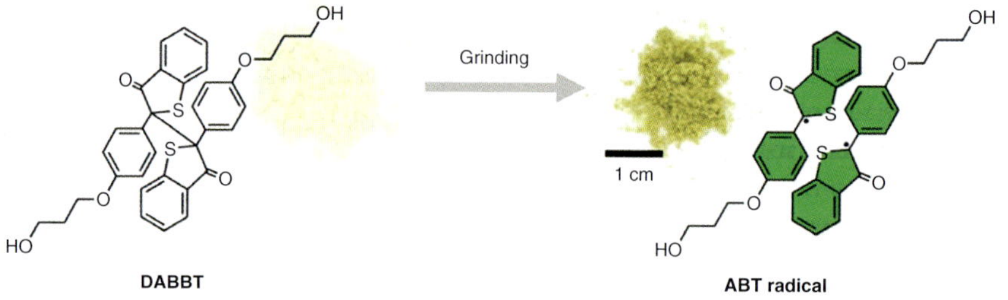

Figure 12.14 Chemical structure of DABBT and ABT radicals as well as photographic images before and after grinding.

from diarylbibenzothiophenonyl (DABBT) by homolytic cleavage in response to mechanical stress exhibit green color [33].

We thus focused on the DABBT skeleton, which can generate green ABT radicals in response to mechanical stress, and designed a DABBT-based polystyrene as a mechanochromic polymer [34]. As expected, a DABBT derivative with hydroxy groups appeared green upon exposure to mechanical stress, and the mechanochromic properties of DABBT were amplified upon incorporation in polystyrene, i.e. polymer chains with a relatively high glass-transition temperature (Figure 12.14).

The three radical-type mechanochromophores with different colors, i.e. blue, pink, and green, enabled the development of rainbow mechanochromism by simply mixing these mechanochromophores like paint. Three radical-type mechanochromic polystyrenes, namely PS-DABBT-PS, PS-DABBF-PS, and PS-TASN-PS, were mixed and ground by ball milling (30 Hz, five minutes). As shown in Figure 12.15, the mixture containing PS-DABBT-PS and PS-DABBF-PS (1 : 1, w/w) exhibited a color change from white to dark green after grinding, which is a color different from those of the individual polymers. This color change can be explained by the fact that the individual polymers in the mixture absorb light at different wavelengths and reflect or transmit other colors, as observed in the case of paint. Similarly, mixtures of these polymers display different colors depending on the mixing ratio of the three primary colors (Figure 12.15).

12.7 Multicolor Mechanochromism Based on Radical-type Mechanochromophores

The mechanochromic polymers featured in this chapter display simple changes between two colors upon exposure to mechanical stimuli. Although many types of mechanochromic polymers have been reported so far, only a few examples exist on further functionalization based on multiple color changes (multicolor mechanochromism) (Figure 12.16) [35–38]. The rational molecular design of such multicolor mechanochromic polymers remains challenging, predominantly due to the lack of insight on the relationship between the chemical structure and

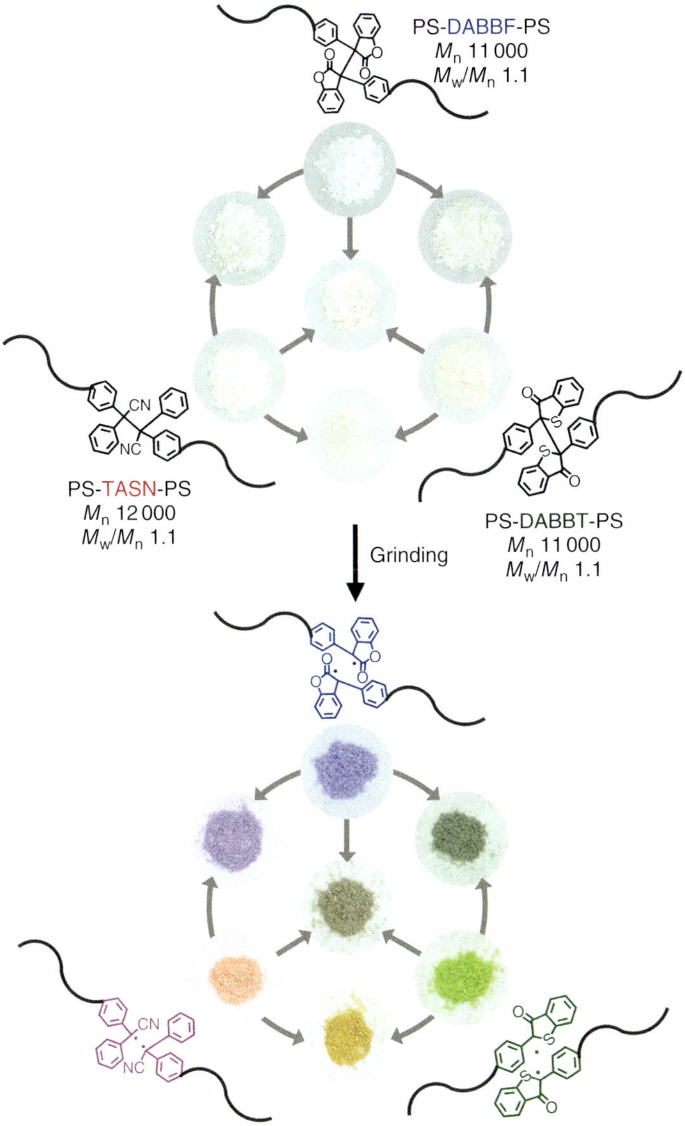

Figure 12.15 Photographic images of polystyrene mixtures containing DABBF, TASN, and DABBT before and after grinding.

mechanochromic behavior. In this section, multicolor mechanochromism based on rainbow mechanochromism, i.e. the generation of combinatorial colors in mechanochromic polymers akin to a color palette, is introduced as a preliminary finding [39].

To induce multicolor mechanochromism, we designed a polymeric material comprising soft and hard domains, where each domain contains different mechanochromophores displaying different colors. The idea is that under

Figure 12.16 Multicolor mechanochromism.

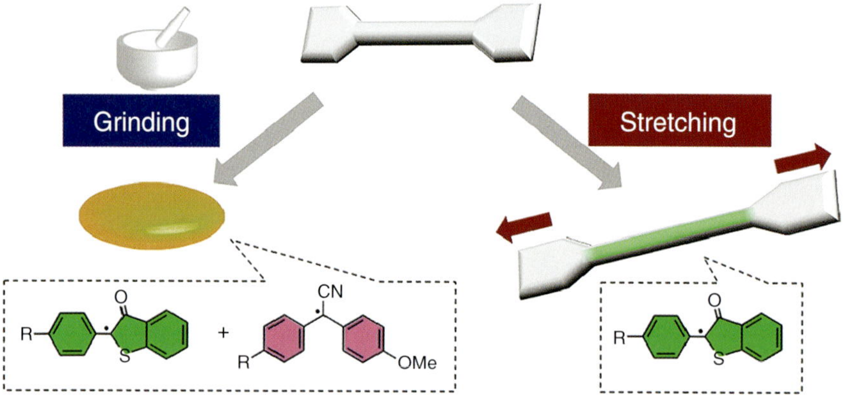

Figure 12.17 Conceptual illustration of multicolor mechanochromism based on two radical-type mechanophores that is able to discriminate between different mechanical stimuli. Source: Ishizuki 2018 [39]. Copyright 2018, Reproduced with permission of American Chemical Society.

mechanical force at low shear rates (e.g. stretching), the mechanical stress resulting from the large deformation accumulates in the soft domains, selectively activating the mechanochromophores therein. On the other hand, at high shear rates (e.g. grinding), both soft and hard domains experience mechanical stress, resulting in activation of both mechanochromophores (Figure 12.17). It is well known that block polymers can exhibit microphase separation, thus the combination of a block copolymer and polymer blend technique allowed us to realize this novel concept (Figure 12.18).

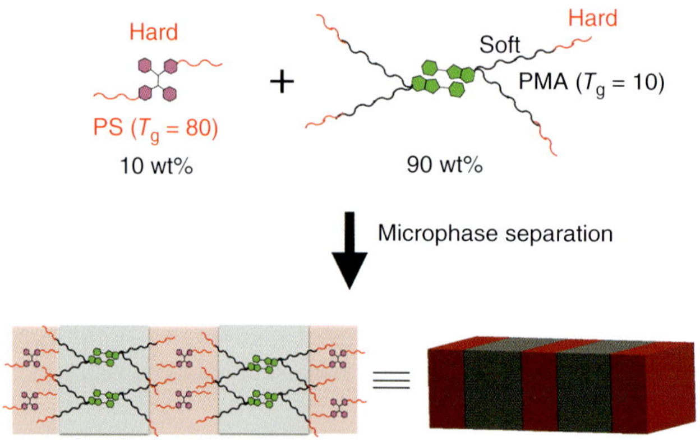

Figure 12.18 Conceptual illustration for the introduction of two mechanophores in different domains (soft and hard) by microphase separation.

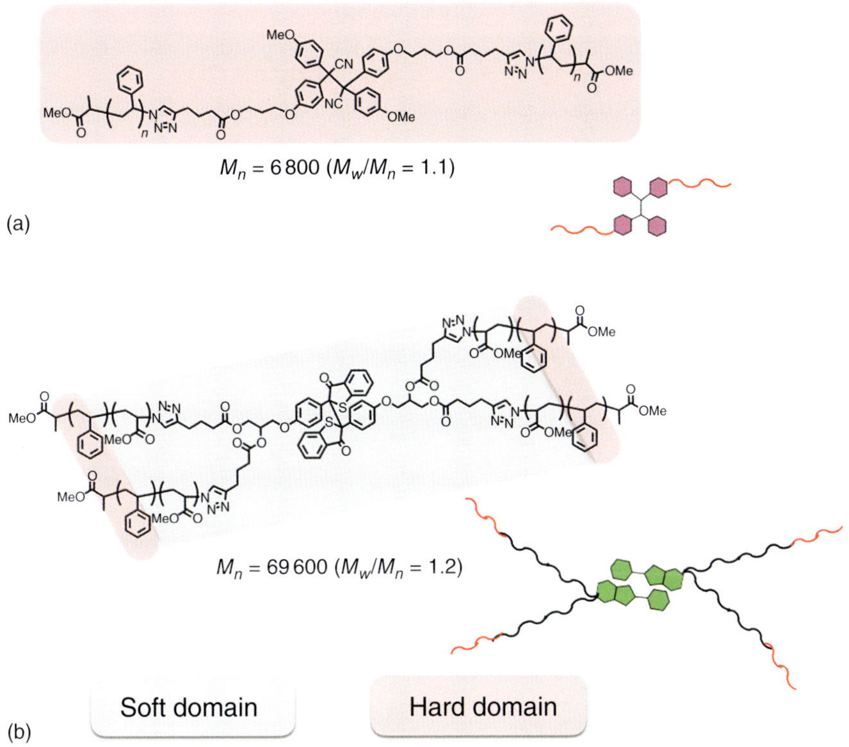

Figure 12.19 Chemical structures of (a) PS-TASN-PS and (b) (PS-*b*-PMA)$_4$–DABBT in the polymer blends.

We designed two mechanochromic polymers with different mechanochromophores, i.e. a linear polystyrene with a TASN moiety at the center of the PS chain (PS-TASN-PS; Figure 12.19a) and a star-shaped block copolymer with a DABBT moiety at the center and four polystyrene-*b*-poly(methyl acrylate) arms ((PS-*b*-PMA)$_4$-DABBT; Figure 12.19b). The star-shaped polymer structure was employed to amplify the mechanical stress on the DABBT moiety as much as possible. As shown in Figure 12.20a, two different mechanical forces, i.e. stretching and grinding, were applied to the polymer blend, resulting in different colors. EPR and solid-state UV–Vis measurements supported the mechanism proposed for this multicolor mechanochromism, which is able to discriminate different types of mechanical stimuli (Figure 12.20b).

12.8 Foresight

In this chapter, recent approaches to the development of mechanochromic polymers and their characterization have been summarized. Although the mechanochromic polymers based on dynamic covalent chemistry featured in this chapter exhibit their properties through radical processes, they can be used

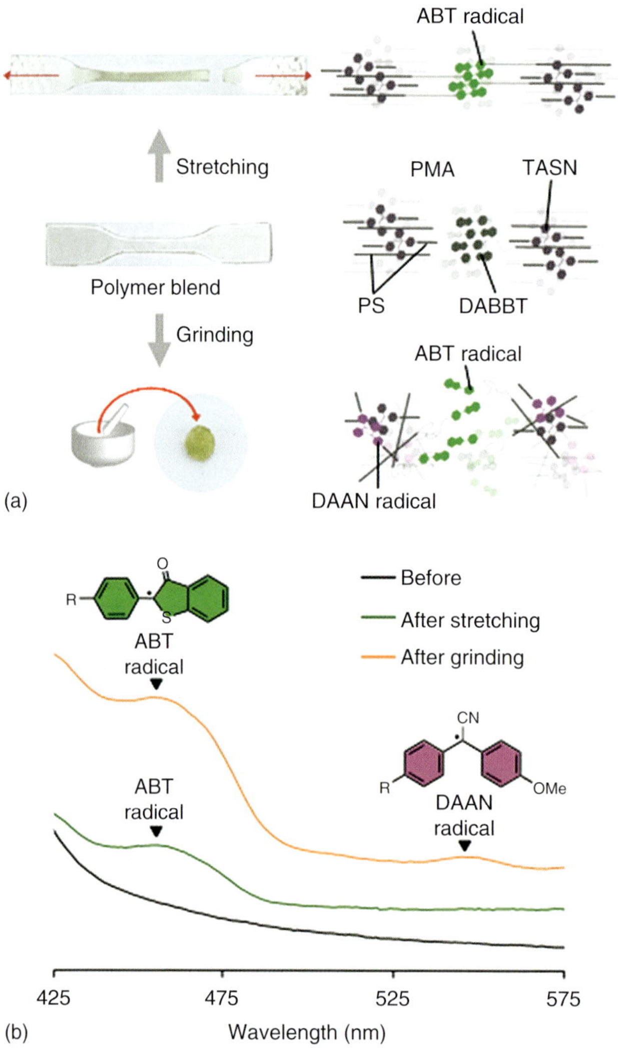

Figure 12.20 (a) Photographic images of the polymer blend and schematic illustration of the origin of the different colors upon exposure to stretching or grinding as different types of mechanical stimuli. (b) Solid-state UV–Vis spectra of the polymer blend before and after stretching and grinding. Source: Ishizuki et al. [39]. Copyright 2018, Reproduced with permission of American Chemical Society.

in air due to the considerable stability of the generated radicals. Furthermore, a system that can be used repeatedly was realized by exploiting the dynamic nature of covalent chemistry. Based on these systems, the visualization and elucidation of the fracture behavior in polymeric materials and the development of novel mechanochromophores will be possible in the near future. We believe that the fundamental research shown here will contribute to the development of advanced smart materials.

References

1 Asher, S.A., Holtz, J., Liu, L., and Wu, Z. (1994). Self-assembly motif for creating submicron periodic materials. Polymerized crystalline colloidal arrays. *J. Am. Chem. Soc.* 116: 4997–4998.
2 Sunkara, H.B., Jethmalani, J.M., and Ford, W.T. (1994). Composite of colloidal crystals of silica in poly(methyl methacrylate). *Chem. Mater.* 6: 362–364.
3 Haque, M.A., Kamita, G., Kurokawa, T. et al. (2010). Unidirectional alignment of lamellar bilayer in hydrogel: one-dimensional swelling, anisotropic modulus, and stress/strain tunable structural color. *Adv. Mater.* 22: 5110–5114.
4 Schäfer, C.G., Gallei, M., Zahn, J.T. et al. (2013). Reversible light-, thermo-, and mechano-responsive elastomeric polymer opal films. *Chem. Mater.* 25: 2309–2318.
5 Williams, G.A., Ishige, R., Cromwell, O.R. et al. (2015). Mechanically robust and self-healable superlattice nanocomposites by self-assembly of single-component "sticky" polymer-grafted nanoparticles. *Adv. Mater.* 27: 3934–3941.
6 Löwe, C. and Weder, C. (2002). Oligo(*p*-phenylene vinylene) excimers as molecular probes: deformation-induced color changes in photoluminescent polymer blends. *Adv. Mater.* 14: 1625–1629.
7 Crenshaw, B.R. and Weder, C. (2003). Deformation-induced color changes in melt-processed photoluminescent polymer blends. *Chem. Mater.* 15: 4717–4724.
8 Kinami, M., Crenshaw, B.R., and Weder, C. (2006). Polyesters with built-in threshold temperature and deformation sensors. *Chem. Mater.* 18: 946–955.
9 Davis, D.A., Hamilton, A., Yang, J. et al. (2009). Force-induced activation of covalent bonds in mechanoresponsive polymeric materials. *Nature* 459: 68–72.
10 Lee, C.K., Davis, D.A., White, S.R. et al. (2010). Force-induced redistribution of a chemical equilibrium. *J. Am. Chem. Soc.* 132: 16107–16111.
11 Kingsbury, C.M., May, P.A., Davis, D.A. et al. (2011). Shear activation of mechanophore-crosslinked polymers. *J. Mater. Chem.* 21: 8381–8388.
12 Gossweiler, G.R., Hewage, G.B., Soriano, G. et al. (2014). Mechanochemical activation of covalent bonds in polymers with full and repeatable macroscopic shape recovery. *ACS Macro Lett.* 3: 216–219.
13 Li, J., Nagamani, C., and Moore, J.S. (2015). Polymer mechanochemistry: from destructive to productive. *Acc. Chem. Res.* 48: 2181–2190.
14 Balkenende, D.W., Coulibaly, S., Balog, S. et al. (2014). Mechanochemistry with metallosupramolecular polymers. *J. Am. Chem. Soc.* 136: 10493–10498.
15 Filonenko, G.A. and Khusnutdinova, J.R. (2017). Dynamic phosphorescent probe for facile and reversible stress sensing. *Adv. Mater.* 29: 1700563.
16 Chen, Y., Spiering, A.J.H., Karthikeyan, S. et al. (2012). Mechanically induced chemiluminescence from polymers incorporating a 1,2-dioxetane unit in the main chain. *Nat. Chem.* 4: 559–562.
17 Ducrot, E., Chen, Y., Bulters, M. et al. (2014). Toughening elastomers with sacrificial bonds and watching them break. *Science* 344: 186–189.

18 Otsuka, H., Aotani, K., Higaki, Y., and Takahara, A. (2003). Polymer scrambling: macromolecular radical crossover reaction between the main chains of alkoxyamine-based dynamic covalent polymers. *J. Am. Chem. Soc.* 125: 4064–4065.
19 Amamoto, Y., Kamada, J., Otsuka, H. et al. (2011). Repeatable photoinduced self-healing of covalently cross-linked polymers through reshuffling of trithiocarbonate units. *Angew. Chem. Int. Ed.* 50: 1660–1663.
20 Takahashi, A., Goseki, R., and Otsuka, H. (2017). Thermally adjustable dynamic disulfide linkages mediated by highly air-stable 2,2,6,6-tetramethylpiperidine-1-sulfanyl (TEMPS) radicals. *Angew. Chem. Int. Ed.* 56: 2016–2021.
21 Imato, K. and Otsuka, H. (2018). Reorganizable and stimuli-responsive polymers based on dynamic carbon–carbon linkages in diarylbibenzofuranones. *Polymer* 137: 395–413.
22 Suzuki, N., Takahashi, A., Ohishi, T. et al. (2018). Enhancement of the stimuli-responsiveness and photo-stability of dynamic diselenide bonds and diselenide-containing polymers by neighboring aromatic groups. *Polymer* 154: 281–290.
23 Imato, K., Nishihara, M., Kanehara, T. et al. (2012). Self-healing of chemical gels cross-linked by diarylbibenzofuranone-based trigger-free dynamic covalent bonds at room temperature. *Angew. Chem. Int. Ed.* 51: 1138–1142.
24 Imato, K., Takahara, A., and Otsuka, H. (2015). Self-healing of a cross-linked polymer with dynamic covalent linkages at mild temperature and evaluation at macroscopic and molecular levels. *Macromolecules* 48: 5632–5639.
25 Imato, K., Ohishi, T., Nishihara, M. et al. (2014). Network reorganization of dynamic covalent polymer gels with exchangeable diarylbibenzofuranone at ambient temperature. *J. Am. Chem. Soc.* 136: 11839–11845.
26 Imato, K., Irie, A., Kosuge, T. et al. (2015). Mechanophores with a reversible radical system and freezing-induced mechanochemistry in polymer solutions and gels. *Angew. Chem. Int. Ed.* 54: 6168–6172.
27 Oka, H., Imato, K., Sato, T. et al. (2016). Enhancing mechanochemical activation in the bulk state by designing polymer architectures. *ACS Macro Lett.* 5: 1124–1127.
28 Imato, K., Kanehara, T., Ohishi, T. et al. (2015). Mechanochromic dynamic covalent elastomers: quantitative stress evaluation and autonomous recovery. *ACS Macro Lett.* 4: 1307–1311.
29 Imato, K., Kanehara, T., Nojima, S. et al. (2016). Repeatable mechanochemical activation of dynamic covalent bonds in thermoplastic elastomers. *Chem. Commun.* 52: 10482–10485.
30 Kosuge, T., Imato, K., Goseki, R., and Otsuka, H. (2016). Polymer–inorganic composites with dynamic covalent mechanochromophore. *Macromolecules* 49: 5903–5911.
31 Sumi, T., Goseki, R., and Otsuka, H. (2017). Tetraarylsuccinonitriles as mechanochromophores to generate highly stable luminescent carbon-centered radicals. *Chem. Commun.* 53: 11885–11888.
32 Kato, S., Ishizuki, K., Aoki, D. et al. (2018). Freezing-induced mechanoluminescence of polymer gels. *ACS Macro Lett.* 7: 1087–1091.

33 Mori, Y., Yamada, N., Kanazawa, M. et al. (1996). Radical dissociation of 2,2′-bis(2-aryl-3-benzothiophenonyl)s by mechanical energy. *Bull. Chem. Soc. Jpn.* 69: 2355–2359.
34 Ishizuki, K., Oka, H., Aoki, D. et al. (2018). Mechanochromic polymers that turn green upon the dissociation of diarylbibenzothiophenonyl: the missing piece toward rainbow mechanochromism. *Chem Eur J* 24: 3170–3173.
35 Zhang, Y., Wang, K., Zhuang, G. et al. (2015). Multicolored-fluorescence switching of ICT-type organic solids with clear color difference: mechanically controlled excited state. *Chem Eur J* 21: 2474–2479.
36 Meng, X., Qi, G., Li, X. et al. (2016). Spiropyran-based multi-colored switching tuned by pressure and mechanical grinding. *J. Mater. Chem. C* 4: 7584–7588.
37 Zhao, Z., Chen, T., Jiang, S. et al. (2016). The construction of a multicolored mechanochromic luminogen with high contrast through the combination of a large conjugation core and peripheral phenyl rings. *J. Mater. Chem. C* 4: 4800–4804.
38 Wang, T., Zhang, N., Dai, J. et al. (2017). Novel reversible mechanochromic elastomer with high sensitivity: bond scission and bending-induced multicolor switching. *ACS Appl. Mater. Interfaces* 9: 11874–11881.
39 Ishizuki, K., Aoki, D., Goseki, R., and Otsuka, H. (2018). Multicolor mechanochromic polymer blends that can discriminate between stretching and grinding. *ACS Macro Lett.* 7: 556–560.

Part III

Application of Mechanically Responsive Materials to Soft Robots

13

Soft Microrobots Based on Photoresponsive Materials

Stefano Palagi[1,2]

[1] *Istituto Italiano di Tecnologia, Center for Micro-BioRobotics, viale R. Piaggio 34, 56025 Pontedera (PI), Italy*
[2] *Max Planck Institute for Intelligent Systems, Micro, Nano, and Molecular Systems, Heisenbergstraße 3, 70569 Stuttgart, Germany*

A new approach known as *soft robotics* has emerged for the design, fabrication, and operation of robots at the *macro*scale. Inspired by natural organisms such as invertebrate animals, researchers are developing unconventional robots consisting of soft materials such as elastomers and mechanically responsive polymers. In this way, robots with advanced biomimetic functionalities and spontaneous adaptability can be obtained. In this chapter I will show that adopting soft and smart materials is essential to achieve advanced robotic functionalities in robots at the *micro*scale and I will introduce the emergent approach of *soft microrobotics*.

The chapter is organized as follows. Section 13.1 gives a brief introduction to the field of microrobotics and its open challenges, followed by an overview of soft microrobots and of the materials used to build them. Section 13.2 is dedicated to the description of the behavior of liquid-crystal elastomers (LCEs) and, in particular, of the photothermal actuation of photoresponsive LCEs in fluidic environments. The use of such photothermal LCEs as light-controlled soft microrobots was first introduced in [1]. Finally, a brief outlook on the future of soft microrobotics is given at the end of the chapter (Section 13.4).

13.1 Soft Robotics at the Micro Scale

The term "microrobots" refers to submillimeter robotic devices that are untethered, mobile, self-propelled, and/or wirelessly controlled by external fields [2]. Microrobots are expected to revolutionize medicine by enabling groundbreaking minimally-invasive diagnostic or therapeutic procedures, such as targeted delivery of drugs and cells [3].

Realizing microrobots is not simply a matter of miniaturizing existing robotic technologies, as sensors, actuators, and controllers are extremely challenging to fabricate and integrate in robotic devices at the microscale. The two main approaches to develop mobile microrobots are based on:

Mechanically Responsive Materials for Soft Robotics, First Edition. Edited by Hideko Koshima.

- chemical reactions at the microrobot's surface that result in self-propulsion of the microrobot in a liquid, e.g. catalytic janus particles that consume a chemical fuel (often hydrogen peroxide – H_2O_2) and show *active* movement or diffusion (these microrobots are also known as *active particles* or *micro/nanomotors*) [4];
- external fields (mainly magnetic, but also acoustic, optical, etc.) that exert controlled actions (forces and/or torques) on microscopic rigid or flexible structures; an example is given by helical magnetic microrobots, which propel in fluids by being rotated by a homogeneous rotating magnetic field [5, 6].

The two approaches can also be combined to obtain self-propelled microrobots whose movement is controlled by external fields.

Recently a different approach has been proposed that involves using soft and stimuli-responsive (*smart*) materials to provide microrobots with novel functionalities and improve their performance [7]. Since the introduction of the first magnetic microrobots based on soft materials [8], the "soft microrobotics" approach has lead to microrobots with increased adaptability and flexibility [9]. In particular, the adoption of stimuli-responsive polymers and gels has enabled the implementation of basic actuation and sensing functionalities *on board* the microrobots [10]. For example soft microrobots have been reported that are able to excise cells [11], release drugs [12], or reconfigure their shape [13, 14].

The soft smart materials adopted in microrobotics include, for instance, hydrogels based on poly(*N*-isopropylacrylamide) (PNIPAM). This has a low critical solution temperature (LCST) in water of about 32 °C, above which it undergoes a transition from a hydrated to a dehydrated state. This transition leads to a macroscopic *volume change* (de-swelling) of the hydrogels based on PNIPAM. Photoresponsive LCEs have also been used in mobile microrobots [15]. Photoresponsive LCEs were also the constituent materials of the light-controlled soft microrobots we presented in [1]. In the following section I will therefore describe in details the response of LCEs for microrobotics applications.

13.2 LCEs for Microrobotics

Soft stimuli-responsive materials can provide microrobots with advanced functionalities and enable new locomotion strategies. To do so, such smart materials must be capable of being controlled or triggered by suitable external stimuli, such as magnetic fields or light. LCEs consist of rubber-like polymeric networks of liquid-crystal monomers that respond to temperature, and can respond to light (if the material is loaded with a proper light-absorbing molecule), with an isovolumetric *shape change*. This section is dedicated to the analysis of the behavior of LCEs from a microrobotics perspective.

13.2.1 Thermal Response of LCEs

The peculiar actuation capabilities of LCEs stem from the physical behavior of liquid crystals. At low temperature, liquid crystals molecules are in an ordered state:

for instance, in the so-called *nematic* phase, the molecules are aligned along a common direction (the *nematic director*, **n**). When a liquid crystal is heated above a certain threshold temperature (known as *clearing* temperature, or transition temperature) the liquid-crystal order is lost and the molecules enter a disordered, isotropic phase. In LCEs, the molecular reorientation stretches the polymeric networks and generates a macroscopic deformation. In the case of nematic LCEs, the phase transition results in a contraction along the nematic director **n** whose amplitude depends on the initial ordering:

$$\Lambda_{\mathrm{NI}} = \frac{L_{\mathrm{I}}}{L_{\mathrm{N}}} = \left(\frac{1+2Q_{\mathrm{N}}}{1-Q_{\mathrm{N}}}\right)^{-\frac{1}{3}}, \tag{13.1}$$

where Λ_{NI} is the longitudinal (along **n**) stretch of the LCE across the nematic-to-isotropic transition, L_{I} and L_{N} are the lengths (along **n**) of the LCE in the isotropic and nematic phase, respectively, and Q_N is the order parameter in the nematic phase [16]. The order parameter Q synthetically describes the orientation of the liquid-crystal molecules with respect to the nematic director **n**. It has values ranging from 0 (disorder, random orientation) to 1 (perfect order, all molecules aligned along the director). Therefore, in the nematic phase $Q = Q_{\mathrm{N}} \leq 1$, whereas in the isotropic phase $Q \approx 0$.

Figure 13.1 reports experimental data obtained by the small angle X-ray scattering (SAXS) characterization of a cylindrical LCE sample performed in [1]. It can be noticed that the shape change of the material almost exactly matches the change in the order parameter. The LCE order parameter data are fitted by a sigmoid function, namely:

$$Q(T) = Q_{\mathrm{N}}\left(1 + \mathrm{e}^{\frac{T-T_{\mathrm{NI}}}{\gamma}}\right)^{-1}, \tag{13.2}$$

where T_{NI} is the nematic-to-isotropic transition temperature and γ is a constant related to the width of the transition region. The parameter values obtained from the fitting are: $Q_{\mathrm{N}} = 0.396$, $T_{\mathrm{NI}} = 83.2\,°\mathrm{C}$, $\gamma = 4.32\,°\mathrm{C}$. The fitted function is then used to calculate the longitudinal stretch $\Lambda(T)$ by the following relation:

$$\Lambda(T) = \frac{L(T)}{L_{\mathrm{N}}} = \left(\frac{1+2Q(T)}{1-Q(T)}\right)^{\frac{1}{3}} \left(\frac{1+2Q_{\mathrm{N}}}{1-Q_{\mathrm{N}}}\right)^{-\frac{1}{3}}. \tag{13.3}$$

The above relation is obtained by considering that

$$\Lambda(T) = \frac{L(T)}{L_{\mathrm{N}}} = \frac{L(T)/L_{\mathrm{I}}}{L_{\mathrm{N}}/L_{\mathrm{I}}} = \left(\frac{1+2Q(T)}{1-Q(T)}\right)^{\frac{1}{3}} \Big/ \left(\frac{1+2Q_{\mathrm{N}}}{1-Q_{\mathrm{N}}}\right)^{\frac{1}{3}}. \tag{13.4}$$

Fitting the relation $L(T) = L_{\mathrm{N}}\,\Lambda(T)$ to the original measurements of the length of the LCE sample (see supplementary information of [1]), the nematic length $L_{\mathrm{N}} = 2.06\,\mathrm{mm}$ is obtained and used to calculate the stretch values in Figure 13.1.

13.2.2 Photothermal Actuation of LCEs

The LCEs we have used for light-controlled soft microrobots in [1] and subsequent works contain a covalently linked azo-benzene dye, modified to have an

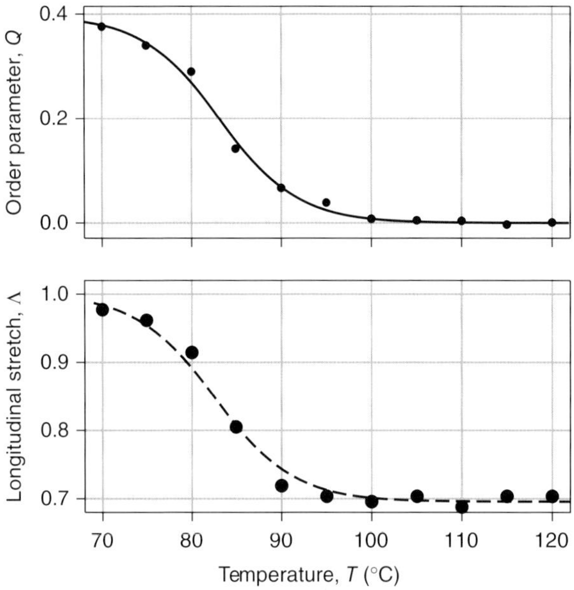

Figure 13.1 Dependence of order parameter Q and longitudinal stretch Λ on the temperature T. The graphs report experimental data obtained by SAXS for a cylindrical LCE sample. The solid line in the top graph represents the fitting of Eq. (13.2) to the experimental data on the order parameter Q. Parameter values obtained from fitting are: $Q_N = 0.396$, $T_{NI} = 83.2\,°C$, $\gamma = 4.32\,°C$. The dashed line in the bottom graph represents the longitudinal stretch from Eq. (13.3), calculated with the fitted order parameter. The nematic length $L_N = 2.06$ mm is obtained by fitting the function $L(T) = L_N \cdot \Lambda(T)$ to the original length data and used to calculate the stretch $\Lambda(T)$. Source: Palagi et al. 2016 [1]. Experimental data from Springer Nature.

absorption peak in the visible green region [17]. When exposed to light near the absorption peak, azo dyes undergo a *trans–cis* photoisomerization, which means that the dye molecule deforms from a rodlike shape to a bent shape. The azo dyes commonly adopted in LCEs have the *trans* absorption peak in the UV and present a quite stable *cis* state, which absorbs light in the yellow region. The buildup of a population of bent *cis* molecules by UV light exposure triggers the nematic-to-isotropic transition of the host LCE, effectively lowering the transition temperature [16]. LCEs based on these dyes are activated by UV light and need an additional yellow light signal to switch back to the original shape. On the contrary, the *cis* state of the green light-absorbing azo dye adopted in our LCE-based microrobots is not stable and switches back to the *trans* conformation spontaneously and quickly. This has three main consequences:

(i) the dye triggers the material response by a *photothermal* mechanism (rather than by an *optical* one);
(ii) the light intensity required to activate the LCE is higher (than in the purely *optical* case) and depends on the thermal properties of the environment;
(iii) the LCE response is reversible and requires only one light source: the material deforms in response to green light and spontaneously relaxes back to the initial shape when the light is turned off.

In particular, it has been shown that microscopic LCEs based on this azo dye and powered by a 532 nm laser can operate up to frequencies of almost 2 kHz [15]. These aspects are particularly important in the design and operation of light-controlled LCE-based microrobots.

To a first approximation, the photothermal response of an LCE microrobot in a fluid is given by:

$$\frac{\partial}{\partial t}[\rho c V(T - T_0)] = \eta I A_{\mathrm{pr}} - h A_{\mathrm{sur}}(T - T_{\mathrm{f}}), \tag{13.5}$$

where ρ, c, and V are the mass density, specific heat capacity, and volume of the LCE, respectively, T_{f} is the equilibrium fluid temperature, I is the light intensity, and h is the heat transfer coefficient of the LCE in the fluid. The coefficient η represents the light-absorption efficiency of the LCE ($\eta = 0$: no absorption; $\eta = 1$: full absorption). A_{pr} and A_{sur} are the area of the LCE's projection on the workspace plane and the area of the surface of the LCE, respectively. Moreover, the microrobot is assumed to be a cylindrical LCE rod, placed horizontally in the workspace, such that $V = \pi R^2 L$, $A_{\mathrm{pr}} = 2RL$, and $A_{\mathrm{sur}} = 2\pi RL$ (the heat transfer at the two ends of the rods is neglected, as their surface is much smaller than the lateral surface), and to initially be in equilibrium with the fluid ($T_0 \equiv T_{\mathrm{f}}$). Equation (13.5) can thus be rewritten as

$$\frac{\partial}{\partial t}[\rho c V(T - T_{\mathrm{f}})] = 2RL[\eta I - \pi h(T - T_{\mathrm{f}})]. \tag{13.6}$$

Assuming the physical properties of the system to be constant, Eq. (13.6) can be further simplified to

$$\begin{aligned}\frac{\partial T}{\partial t} &= \frac{2RL}{\rho c V}[\eta I - \pi h(T - T_{\mathrm{f}})] \\ &= \frac{2}{\pi R \rho c}[\eta I - \pi h(T - T_{\mathrm{f}})],\end{aligned} \tag{13.7}$$

where $T(t = 0) = T_{\mathrm{f}}$. The heat transfer equation used to calculate the cooling term $\dot{q}_{\mathrm{out}} = h A_{\mathrm{sur}}(T - T_{\mathrm{f}})$ in Eq. (13.5) is based on the assumption that a thin thermal boundary layer δ_{T} forms around the object being cooled, in which there is only heat conduction and the temperature is $T_{\delta_{\mathrm{T}}} \sim (T + T_{\mathrm{f}})/2$ (with T being the temperature of the body and T_{f} the temperature of the fluid at infinite). The size of thermal boundary layer can be calculated as $\delta_{\mathrm{T}} = D/Nu_D$, where $D = 2R$ is the diameter of the cylindrical LCE and Nu_D is its average Nusselt number in the fluid (calculated for the case of natural convection from horizontal isothermal cylinders) [18]. Whereas in most common cases the boundary layer is very thin compared to the body size ($\delta_{\mathrm{T}} \ll D$), in the case of our microscopic LCE $\delta_{\mathrm{T}} \gtrsim D$. To take into account the non-negligible heating inertia of the boundary layer, Eq. (13.6) can be modified as follows:

$$\begin{aligned}&\frac{\partial}{\partial t}\left[\rho c V(T - T_{\mathrm{f}}) + \rho_{\mathrm{f}} c_{\mathrm{f}} V_{\delta_{\mathrm{T}}}\left(\frac{T + T_{\mathrm{f}}}{2} - T_{\mathrm{f}}\right)\right] = \\ &= \frac{\partial}{\partial t}\left[\left(\rho c + 2\rho_{\mathrm{f}} c_{\mathrm{f}}\left(\frac{1}{Nu_D} + \frac{1}{Nu_D^{\,2}}\right)\right)\pi R^2 L(T - T_{\mathrm{f}})\right] \\ &= \frac{\partial}{\partial t}[(\rho c)^* V(T - T_{\mathrm{f}})] = 2RL[\eta I - \pi h(T - T_{\mathrm{f}})],\end{aligned} \tag{13.8}$$

where ρ_f and c_f are the mass density and specific heat capacity of the fluid, respectively, and V_{δ_T} is the volume of the boundary layer around the body. The term $(\rho c)^* = \rho c + 2\rho_f c_f (Nu_D{}^{-1} + Nu_D{}^{-2})$ represents a composite volumetric heat capacity of the system. Assuming again that the physical properties of the system are constant, we can write a modified version of Eq. (13.7):

$$\frac{\partial T}{\partial t} = \frac{2}{\pi R(\rho c)^*}[\eta I - \pi h(T - T_f)], \tag{13.9}$$

with $T(t = 0) = T_f$. From this equation we can estimate the maximum temperature that can be reached by the LCE under continuous illumination in the considered fluid as

$$T_{max} = T_f + \frac{\eta I}{\pi h}, \tag{13.10}$$

or, alternatively, the light intensity required to achieve a specific temperature, e.g. the LCE transition temperature:

$$I^* = \frac{\pi h(T_{NI} - T_f)}{\eta}. \tag{13.11}$$

Equation (13.9) allows us to calculate the system's cooling time constant

$$\tau_c = \frac{(\rho c)^* R}{2h}, \tag{13.12}$$

which depends on the size and thermal properties of the microrobot, as well as on its heat transfer coefficient in the given fluid. The time constant τ_c provides information on the maximum operation frequency of the LCE in the fluid, which can be estimated by

$$f_{max} \sim \tau_c^{-1} = \frac{h}{(\rho c)^* R}. \tag{13.13}$$

This represents a very rough estimation of the maximum operational frequency, but it is still quite informative. In Section 13.3.2 we will get a better understanding of the system's dynamics.

Table 13.1 reports estimates of the light intensity I^* required to heat an LCE cylindrical rod of radius $R = 100\,\mu m$ from 20 to 100 °C in different fluids, as well as of the maximum achievable frequencies of photothermal operation f_{max}. The values of I^* and f_{max} are calculated assuming a mass density $\rho \approx 1200\,kg\,m^{-3}$, a specific heat capacity $c \approx 1500\,J\,(kg\,K)^{-1}$, and an absorption efficiency $\eta = 1$. The heat transfer coefficient is calculated as $h = Nu_D\ k_f/D$, where k_f is the thermal conductivity of the fluid [18]. The light intensity I is related to the laser output power P_{laser} as $I = \eta_{opt}\ P_{laser}/A_{ws}$, where η_{opt} is the overall efficiency of the optical system and A_{ws} is its workspace area. The required laser power P^*_{laser} is therefore calculated from the required light intensity I^* considering the workspace area and efficiency of the optical apparatus presented in [1] ($A_{ws} \sim 1.75\,mm^2$ and $\eta_{opt} \sim 30\%$, respectively).

In Eqs. (13.7), (13.9), and (13.10)–(13.13) we assumed that the physical properties of the system are constant. In reality, however, they depend on the

Table 13.1 Required light intensity I^* (and corresponding laser power P^*_{laser}) and maximum operational frequency f for a LCE of radius $R = 100\,\mu\text{m}$ in different fluids.

Fluid	I^* (W mm^{-2})a)	P^*_{laser} (W)a)b)	f_{max} (Hz)
Air	0.019	0.11	0.4
Water	1.24	7.22	5
Glycerol + water	0.428	2.50	1
Silicone oil	0.168	0.98	0.8

a) Calculated for $T = 100\,°\text{C}$.
b) Considering the workspace area and overall efficiency of the optical apparatus developed in [1].

temperature T, which in turn depends on time t, leading to

$$\frac{\partial}{\partial t}[V\ (\rho c)^*(T)(T - T_{\text{f}})] = 2R(T)L(T)[\eta I - \pi h(T)(T - T_{\text{f}})]. \qquad (13.14)$$

It is important to notice, however, that even if the length L and the radius R of the LCE rod substantially change with the temperature, the volume V can actually be considered constant, such that $V = \pi R(T)^2 L(T) \equiv V_{\text{N}} = \pi R_{\text{N}}^2 L_{\text{N}}$. Indeed, $L(T) = L_N\ \Lambda\ (T)$ and $R(T) = R_N \Lambda(T)^{-\frac{1}{2}}$ (see Section 13.3.2).

Figure 13.2 shows the trends of the photothermal heating and mechanical response of a cylindrical LCE of radius $R = 100\,\mu\text{m}$ in different fluids, according to Eq. (13.14) and considering the light intensities I^* from Table 13.1.

13.3 Light-Controlled Soft Microrobots

Many organisms at different scales adopt locomotion mechanisms based on the common principle of *traveling waves*. These include, for instance, the swimming of a group of tiny, single-celled organisms known as *ciliates*. These microorganisms owe their name to the thousands of cilia that cover their body, and whose coordinated beating results in body-scale traveling-wave propulsion [19].

Traveling-wave propulsion is based on periodic deformations that are actively generated all along an appendage or the body of the organism, i.e. they are based on *distributed actuation* [20]. Achieving distributed, coordinated actuation within the tiny artificial body of a microrobot is particularly difficult, as it requires that the microrobot contains many independent actuators. This poses great challenges from both a fabrication and a control point of view. Whereas it is possible to fabricate micro-actuators, integrating and controlling several of them within a mobile, untethered microrobot is extremely challenging. Moreover, if the mobile microrobot is to be wirelessly powered and controlled by external fields, such fields must be able to selectively address each individual actuator, as well as to coordinate the action of several actuators.

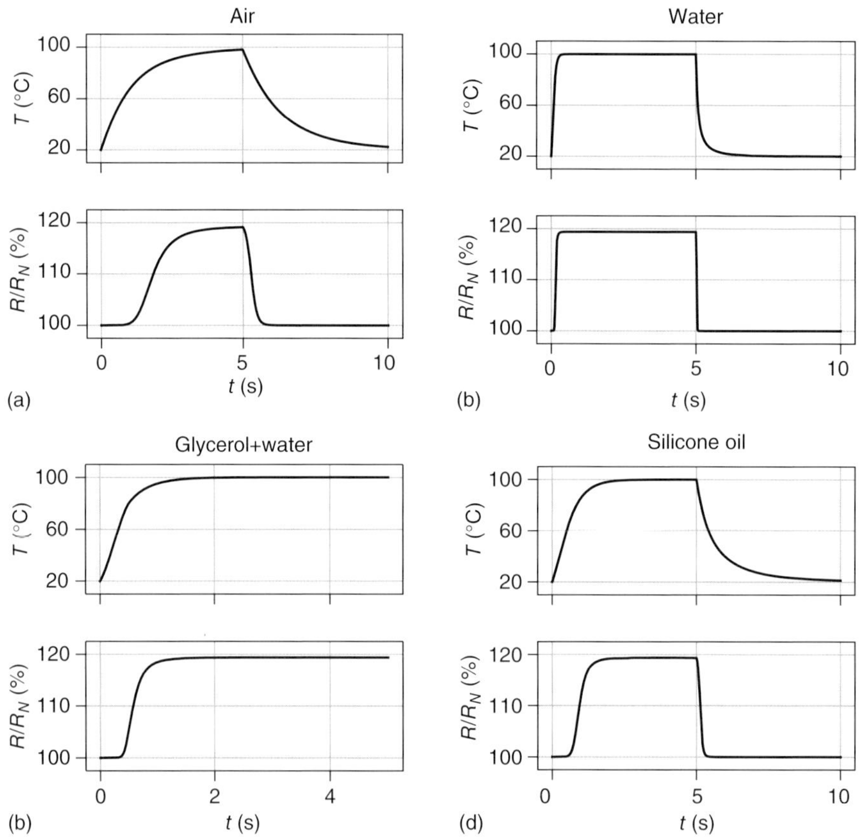

Figure 13.2 Simulated photothermal response of a cylindrical LCE rod ($R = 100\,\mu m$) in different fluids according to Eq. (13.14). The light is turned on at $t = 0$ seconds with intensity $I = I^*$ (different for each fluid – see Table 13.1) and turned off at $t = 5$ seconds. Whereas the time to reach the target temperature (here 100 °C) can be decreased by increasing the light intensity, the cooling time is a property of the system and influences the maximum attainable operational frequency f_{max}. Fluids: (a) air; (b) water; (c) glycerol/water solution; (d) silicone oil.

In this section, I summarize the results of our work on soft microrobots controlled by structured light fields [1]. The microrobots consist in monolithic, microscopic LCE samples with simple geometries (either cylindrical rods or disks), which makes their fabrication relatively easy. The deformation of their LCE body is selectively controlled by structured light fields projected on the microrobot's workspace. The continuous microrobot's body is regarded as an array of many tiny actuators, each being independently addressed and controlled by the local light intensity. In this way, a number of different movements can be achieved, including a variety of traveling-wave deformations. The advantage of this approach is the achievement of complex movements in easy-to-fabricate microrobots, by transferring the burden of power, synchronization, and control from the microrobot into the externally generated light field. Therefore, rather than preprogramming the microrobot's behavior once at the fabrication stage, its action can be dynamically reconfigured during operation via the control software with great flexibility.

Section 13.3.1 introduces the generation of structured light for microrobot's control. Section 13.3.2 reports simulations and experimental results of the controlled actuation of photoresponsive LCEs by structured light. Section 13.3.3 presents the results of the swimming of the soft microrobots based on photoresponsive LCEs and controlled by structured light fields.

13.3.1 Structured Light

The structured light fields are generated by an optical apparatus based on a computer-controlled digital micromirror device (DMD). The DMD modulates the intensity of the incoming laser beam in both time and space, generating a dynamic light field that is projected on the microrobot's workspace through a microscope objective. The DMD module (V-7000, ViaLUX) has 1024 × 768 micromirrors and is addressed by custom software (written in MATLAB, Mathworks) to dynamically modulate the intensity of a 532 nm laser beam (Verdi G10, Coherent). The chosen light wavelength falls in the absorption peak region of the azo-benzene dye in the LCE samples. The beam is expanded upstream of the DMD, to fully illuminate the DMD surface. In the latest upgrade of the apparatus (e.g. in [21]), the laser beam is also homogenized upstream of the DMD by a top-hat beam shaper (πShaper, AdlOptica). The optical system has an overall efficiency η (from laser output to the workspace projection) of about 30%. The beam modulated by the DMD is projected onto the working area containing the microrobots through a 4× microscope objective (Nikon). The workspace is observed by a CMOS camera (resolution 1280 × 1024 – Thorlabs) through the same objective. The images are recorded offline, and, in the case of closed-loop control, they are also processed online to extract the position of the microrobot.

The dynamic light field of intensity I is defined as a two-dimensional rectangular wave in the workspace frame (x_w, y_w) traveling in time t:

$$I(x_w, y_w, t) = \frac{I_0}{2}\{1 + \mathrm{sgn}[\cos(kx_w \cos\varphi + ky_w \sin\varphi - \omega t) - \cos(\pi\delta)]\}, \tag{13.15}$$

where sgn is the sign function, $k = 2\pi/\lambda$, $\omega = 2\pi f$, and $I_0 = \eta_{opt}P_{laser}/A_{ws}$. The parameters λ, δ, f, and φ are the wavelength, duty cycle, frequency, and direction of the two-dimensional rectangular wave, respectively (see Figure 13.3). From a control point of view, the light intensity field is a function of these parameters, that is $I = I(\lambda, \delta, f, \varphi)$. The laser power P_{laser} can be considered as an additional control parameter. The light field parameters are used as control parameters for the microrobots, as the microrobot's deformation depends directly on the projected light field, as described below.

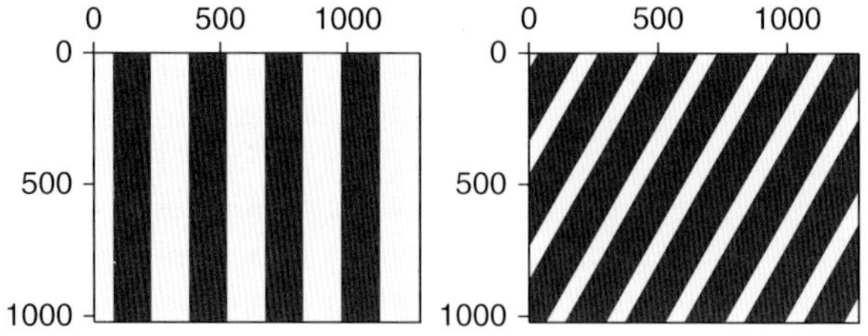

Figure 13.3 Structured light generation. Frames from two different light intensity fields: bright regions are illuminated while dark regions are not. Structured light parameters: (a) $\varphi = 0$ rad, $\delta = 0.5$, $\lambda = 300$ px; (b) $\varphi = \pi/6$ rad, $\delta = 0.3$, $\lambda = 200$ px.

13.3.2 Controlled Actuation

Under the illumination of the structured light field, the photoresponsive LCE deforms according to the following strain tensor:

$$\mathbf{E}_{(R,\Theta,Z)} \equiv \mathbf{E}_{(X,Y,Z)} = \begin{bmatrix} \frac{1-\Lambda}{2\Lambda} & 0 & 0 \\ 0 & \frac{1-\Lambda}{2\Lambda} & 0 \\ 0 & 0 & \frac{\Lambda^2-1}{2} \end{bmatrix}, \tag{13.16}$$

where $\mathbf{E}_{(i,j,k)}$ is the Green–Lagrange strain tensor defined in the LCE Lagrangian cylindrical or cartesian coordinate frame ($Z \parallel \mathbf{n}$), and Λ is the active LCE strain seen in Eq. (13.3).

In the case of horizontal cylindrical LCE microrobots, we can consider that the plane (R, Z) lies in the (x_w, y_w) workspace plane. The microrobot orientation is expressed by the angle ϕ between the Z-axis and the x_w-axis. Given the symmetry of the structure, the light field I is usually generated such that the angle φ is either equal to ϕ or $\phi + \pi$. Ideally, the cylindrical microrobots will only move in one dimension, i.e. along its Z-axis. Here we assume for simplicity that $Z \parallel x_w$. This implies that light field are generated with φ being either 0 rad or π rad. We can thus rewrite Eq. (13.15) as:

$$I(x_w, t) = \frac{I_0}{2}\{1 + \mathrm{sgn}[\cos(kx_w - \omega t) - \cos(\pi\delta)]\}. \tag{13.17}$$

The local deformation of the LCE microrobot body then depends on the local illumination, such as $\Lambda = \Lambda(I) = \Lambda(x_w, t)$. In the ideal case of binary actuation, $\Lambda = 0$ for parts of the microrobot that lie in the dark regions and $\Lambda = \Lambda_{NI}$ for parts in the bright regions (see Figure 13.4). Importantly, as the light field is turned on and the LCE deforms, even in the absence of locomotion or displacement of the whole microrobot, parts of it move from the dark region to the bright region and vice versa. Figure 13.4 shows the simulated deformation of a microrobot, modeled as a linear array of disk elements of radius R and length δL, in a light field, assuming binary actuation and the microrobot's center of mass fixed in (0, 0).

13.3.2.1 Role of Control Parameters

To a first approximation, a microrobot generates a periodic motion pattern that matches the projected structured light field, such that such motion is described

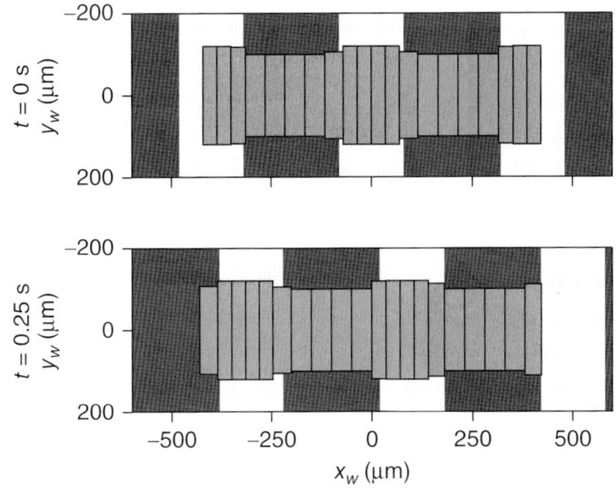

Figure 13.4 Simulated deformation of a microrobot in a light field. Deformation of a cylindrical microrobot ($R_N = 100\,\mu m$, $L_N = 1000\,\mu m$, and $Q_N = 0.396$) in a light field with $\lambda = 400\,\mu m$, $\delta = 0.4$, and $f = 1$ Hz. The dark and bright regions of the background represent the dark and illuminated regions of the light field. The cylindrical microrobot is modeled as a linear array of twenty disk-shaped elements. The deformation of each segment is assumed to be proportional to the fraction of the segment illuminated by the light field.

by the wave parameters that control the light field. The actual motion of the microrobot, however, depends on the complex photothermal and mechanical behavior of their constituent LCE.

The light-controlled actuation in turn influences the locomotion of the microrobots. In the case of a microrobot *swimming* by traveling-wave motions in the bulk of a liquid [1], the speed depends on the frequency f, the wavelength λ, and on the amplitude of the periodic radial deformation $b = \Delta R/2$, according to:

$$V \propto \frac{b^2 f}{\lambda}. \tag{13.18}$$

The amplitude of the radial deformation, however, have a complex dependence on the control parameters, such as $b = b(I_0, f, \delta, \lambda)$, and on the environment. Therefore, the role of the different control parameters must be understood to achieve optimal locomotion performance.

Figure 13.5 shows the effect of the light-control parameters f and δ (see Eq. (13.17)) on the periodic actuation of an individual section of a microrobot at different levels of laser power P_{laser} in silicone oil. For instance, increasing the structured light wave frequency f should in principle increase the motion frequency of the robot (and thus its velocity). In reality, however, an increased frequency can result in a reduced motion amplitude, either by *underexcitation* (see, for instance, the case for $f = 4$ Hz and $\delta = 0.2$ – bottom left graph – with $P_{\text{laser}} = 1$ or 3 W) or *overexcitation* (see same graph, $P_{\text{laser}} = 5$ W). Because of the short period of actuation, in the former case the LCE is not heated up to a temperature that allows for maximum deformation (here ~100 °C), whereas in the latter case the maximum deformation is reached but the short cooling time does not allow for the full shape recovery. The wave duty cycle δ and the light intensity I_0

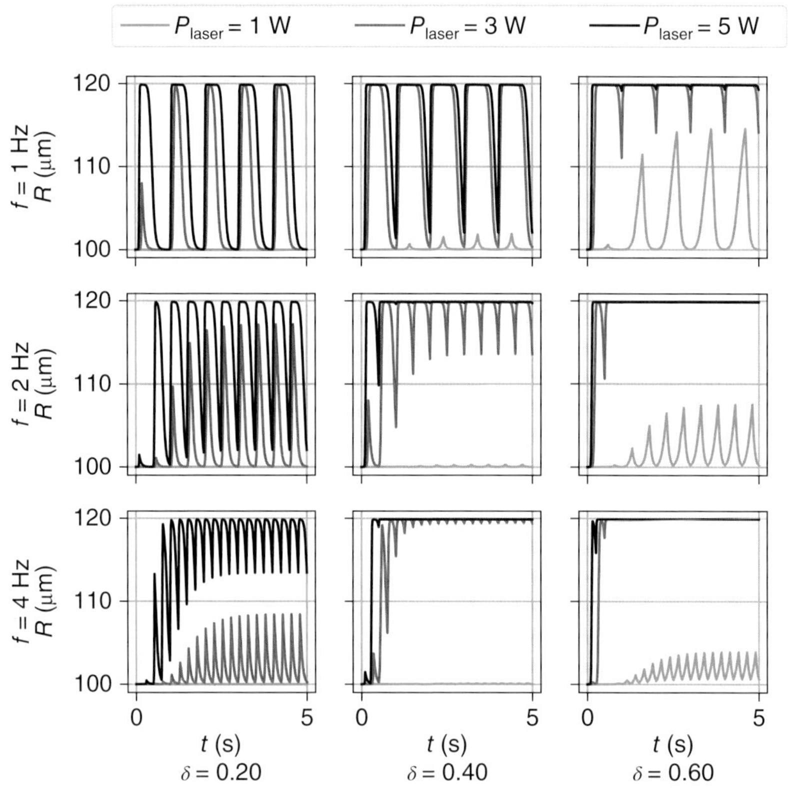

Figure 13.5 Role of the structured light control parameters on the periodic actuation of a microrobot based on photothermal LCEs. The graphs show the simulated periodic actuation of a section of a LCE-based microrobot of initial radius $R_N = 100\ \mu m$ in silicone oil for different values of the structured light frequency f and duty cycle δ at different levels of laser power P_{laser}. It can be noticed that whereas some parameter combinations result in the maximum periodic actuation of the LCE, other combinations lead to either overexcitation or underexcitation of the material and, consequently, in no or limited actuation amplitude.

(related to P_{laser}) also need to be in some specific ranges (that are environment- and frequency-dependent) to result in the maximum amplitude of actuation of the LCE. For instance, we experimentally observed that structured light fields consisting of rectangular waves with relatively low values of δ (usually in the range 0.2–0.3) result in better actuation than square waves ($\delta = 0.5$). This can be observed in the simulations results reported in Figure 13.5, where it is shown that duty cycles of 0.4 and especially 0.6 often result in overexcitation of the material. In these cases, lowering the laser power can reduce the problem of overexcitation, as seen for instance in the top right graph. To summarize, there is a complex dependence of the LCE actuation on the control parameters, which also depends on the environment.

The simulations above do not take into account the role of the light pattern wavelength λ. This was investigated both experimentally and by finite-element simulations in our previous works [1, 22]. Here I summarize the results we obtained in [22], where the deformation of a cylindrical LCE was simulated by considering a predefined sinusoidal spatial distribution of the order parameter Q:

$$Q(z) = Q_{\mathrm{N}} \left[1 - \frac{1 + \cos(kz)}{2}\right], \tag{13.19}$$

where the z-axis corresponds to the axis of the cylinder and is defined in the Eulerian (or spatial) reference frame (i.e. the z-axis is fixed with respect to the global reference frame and the z-position of different section of the LCE changes as the material deforms). The deformation of the LCE due to the order parameter distribution in Eq. (13.19) can be estimated as

$$R(z) \approx R_{\mathrm{avg}} + \frac{\Delta R}{2} \cos(kz), \tag{13.20}$$

where $R_{\mathrm{avg}} = (R_{\mathrm{min}} + R_{\mathrm{max}})/2$ and $\Delta R = R_{\mathrm{max}} - R_{\mathrm{min}} = 2b$. It is worth noticing that here only the mechanical response of the LCE is considered, while its photothermal behavior is neglected. In case of full actuation $\Delta R \equiv \Delta R_{\mathrm{max}} = (\Lambda_{\mathrm{NI}}^{-1/2} - 1)R_{\mathrm{N}}$ and $R_{\mathrm{avg}} = R_{\mathrm{N}} + \Delta R/2$. However, we observed that the continuous LCE material cannot *resolve* patterns with very narrow features, that is light patterns with short wavelengths do not result in corresponding periodic deformations, rather in somewhat homogeneous deformations [22]. This behavior, which is due to the continuity and elasticity of the LCE material, can be described by the following empirical relation:

$$\Delta R \approx \Delta R_{\mathrm{max}} \left[1 - \mathrm{e}^{-\left(\frac{\lambda}{\lambda_{\mathrm{c}}}\right)^{\alpha}}\right], \tag{13.21}$$

where λ_{c} and α are free parameters ($\lambda_{\mathrm{c}} = \lambda_{1/2} \log(2)^{-1/\alpha}$ is the critical wavelength, and $\lambda_{1/2}$ is the wavelength at which the deformation amplitude is half the maximum).

Figure 13.6 reports the trends of the radial response $\Delta R/\Delta R_{\mathrm{max}}$ as a function of λ for LCE microrobots of different radii. The data points are the results of finite-elements mechanical simulations performed in [22] on the basis of Eqs. (13.19) (with $Q_{\mathrm{N}} = 0.25$) and Eq. (13.16). Fitting Eq. (13.21) to the three sets of data points, the following values are obtained for the free parameters: $\lambda_{\mathrm{c}} = 1.29\, R_{\mathrm{N}}$ and $\alpha = 1.26$ (resulting in $\lambda_{1/2} \approx R_{\mathrm{N}}$). The fitted curves are shown in Figure 13.6. It can be noticed that light patterns with short wavelengths ($\lambda \lesssim \lambda_{\mathrm{c}}$) result in attenuated deformations, whereas patterns with long wavelengths ($\lambda > \lambda_{\mathrm{c}}$) generate periodic deformations with full amplitude. In addition, the short-wavelength components in the light pattern will be attenuated by the material response, meaning that sharp changes in the light intensity will be smoothened (i.e. binary light patterns will result in continuous deformations) [22].

13.3.3 *Swimming* Microrobots

This section summarizes the development of the light-controlled *swimming* microrobots based on photothermal LCEs presented in [1]. Each microrobot consists of a cylindrical rod of the photothermal nematic LCEs described in Section 13.2, which have the nematic director **n** aligned along the cylindrical

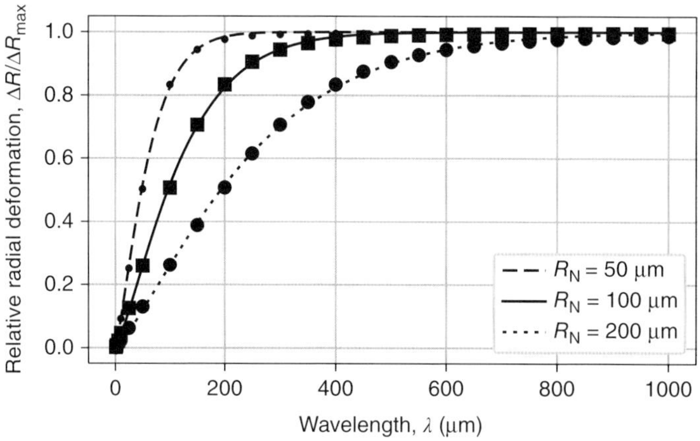

Figure 13.6 Relative radial deformation as a function of wavelength. Relative radial deformation $\Delta R/\Delta R_{max}$ in response to sinusoidal distributions with wavelength λ of the order parameter Q (according to Eq. (13.19)). Data points: results from finite-elements simulations for different values of the radius R_N and $Q_N = 0.25$. Lines: curves obtained from fitting Eq. (13.21) to the data points. Source: Palagi et al. 2016 [22]. Data points from IEEE.

axis. The microrobots are true artificial microswimmers, as they can self-propel in liquids by periodic body shape changes generated by the LCE body. Differently from other microrobots, the light-controlled LCE-based microrobots move without any force or torque applied from outside. The LCE is the *on-board actuator* that generates the propulsive periodic motions of the microrobot. Moreover, the monolithic LCE body is a *continuum actuator* that is selectively controlled by structured light fields, as discussed in Sections 13.3.1 and 13.3.2.

The approach proposed in [1] and based on the combination of photothermal LCEs and structured light enables the generation of complex movement patterns, such as the peristaltic traveling waves shown in Figure 13.4. Such complex movements mimic the locomotion strategy adopted by *ciliates*. These are microorganisms that swim by *metachronal waves*. These consist in the coordinated, wave-like beating of the thousands of cilia that cover the microorganisms' body [19]. Moreover, the dynamic deformation of the LCE microrobot body is not preprogrammed in the material and can thus be controlled by the light field, as discussed in Section 13.3.2. This makes the microrobots extremely versatile, as their movement (or *gait*) can be flexibly tuned to make them adapt to different environments (e.g. [21]).

In [1] a LCE-based microrobot was immersed in a liquid and stabilized in the bulk by a mass density gradient (the liquid consisted of a glycerol–water solution with a concentration gradient). Periodic traveling-wave deformations were induced in the microrobot, controlled by the external light field. These resulted in the force-free and torque-free locomotion of the microrobot in the bulk of the liquid. The direction of movement was controlled by the traveling direction of the light pattern: usually, when the pattern moved to the right, the microrobot swam to the left, and vice versa. That is, the swimming velocity V was *anti-parallel* to the traveling-wave velocity (in this case the swimming velocity was considered positive). However, we observed that in the case of long wavelengths λ the swimming velocity was instead *parallel* to the traveling-wave velocity (negative swimming velocity – see Figure 13.7). This is due to the intrinsic mechanical response of the LCE: the periodic deformation in the radial direction has a maximum amplitude $b \approx \Delta R_{max}/2 = (\Lambda^{-1/2} - 1)R_N/2$, whereas the maximum amplitude of the axial periodic deformation is proportional to the wavelength of the light pattern, $d \propto \lambda$. This means that for short wavelengths the radial component of the deformation is dominant, whereas at long wavelengths the axial one becomes more important. It is important to notice that transversal (radial) traveling-wave deformations are associated with anti-parallel (positive) swimming velocity, whereas longitudinal (axial) deformations with parallel (negative) velocity. A fluid-dynamics model, originally developed for the swimming of ciliates and adapted to our light-controlled LCE soft microrobots [1], predicts the trend of the swimming velocity and also the occurrence of negative velocity at long wavelengths (see Figure 13.7). The difference in the zero-crossing wavelength between the experiments and the model predictions is due to the model considering the microrobot as an *infinitely long* cylinder, whereas it is actually a finite-length cylindrical rod (with length comparable to the longest wavelength among those we tested).

As the movement of the LCE-based soft microrobots is directly controlled by the structured light field, the approach enables the possibility of tuning the swimming gait of the microrobots and their dynamic adaptation to different environments. In [22], a set of tests was performed to study the dependence of the swimming speed on the wave parameters in the case of a microrobot moving on the bottom of a container filled with silicone oil, obtaining different trends compared to the case of the microrobot swimming in the bulk of a glycerol–water solution. A similar investigation was performed in [23], where, by tuning the light field parameters, some locomotion was achieved even within a solid agarose hydrogel. In [21] we demonstrated the adoption of *machine learning* techniques for automatically finding the optimal values of the control parameters in an unknown environment. These studies demonstrate the

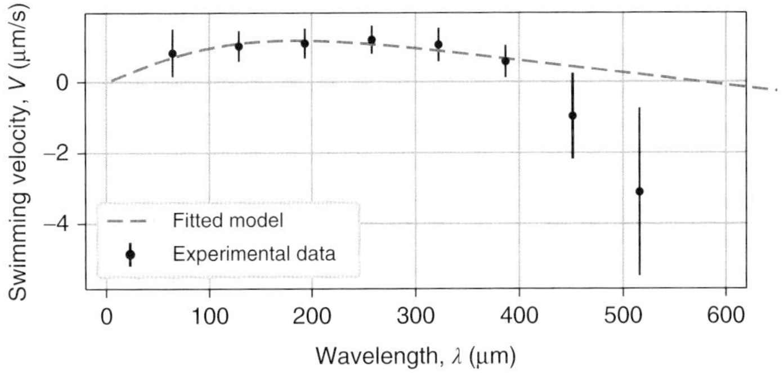

Figure 13.7 Light-controlled LCE-based microrobot swimming in the bulk of a liquid. The graph shows experimental results (dots: average; error bars: standard deviation) on the swimming velocity of a LCE-based microrobot controlled by light patterns of different wavelengths. Source: Palagi et al. 2016 [1]. Data and model from Springer Nature.

flexibility and versatility of using photoresponsive polymers and materials in combination with structured light to develop novel mobile soft microrobots.

13.4 Outlook

This chapter discussed the development of microrobots based on photoresponsive materials, and, in particular, of microrobots based on photothermal LCEs and controlled by structured light fields. The photoresponsive LCEs act as embedded yet wirelessly controlled and powered actuators on board of the microrobots, whereas the structured light fields represent a convenient mean for their control. These light-controlled *soft microrobots* have a behavioral richness that stems from the complex mechanical response of LCEs to the light fields and from their interaction with the environment. This richness can be exploited to achieve complex movements and robotic functionalities, as discussed in this chapter. For instance, it has been shown how simple *pieces* of a LCE can be turned into advanced microrobots that perform sophisticated movements and locomotion reminiscent of that of swimming biological microorganisms.

This chapter provided a basic discussion and modeling of the photothermal response of LCEs intended for microrobotics applications, and highlighted the importance of such understanding on the design and control of the microrobots. As the field of microrobotics is progressing, and with the emergence of *soft microrobots*, soft and stimuli-responsive materials are being pervasively adopted as integrated sensors and actuators on board the microrobots [10]. This will generate an increasing exchange and coevolution between the science of mechanically responsive materials and the engineering of advanced microrobots, eventually leading to novel microrobotic technologies and stimulating the development of innovative smart materials.

References

1 Palagi, S., Mark, A.G., Reigh, S.Y. et al. (2016). Structured light enables biomimetic swimming and versatile locomotion of photoresponsive soft microrobots. *Nat. Mater.* 15 (6): 647–653.
2 Abbott, J.J., Nagy, Z., Beyeler, F., and Nelson, B.J. (2007). Robotics in the small, Part I: microbotics. *IEEE Robot. Autom. Mag.* 14 (2): 92–103.
3 Nelson, B.J., Kaliakatsos, I.K., and Abbott, J.J. (2010). Microrobots for minimally invasive medicine. *Annu. Rev. Biomed. Eng.* 12 (1): 55–85.
4 Sánchez, S., Soler, L., and Katuri, J. (2015). Chemically powered micro- and nanomotors. *Angew. Chem. Int. Ed.* 54 (5): 1414–1444.
5 Ghosh, A. and Fischer, P. (2009). Controlled propulsion of artificial magnetic nanostructured propellers. *Nano Lett.* 9 (6): 2243–2245.
6 Zhang, L., Abbott, J.J., Dong, L. et al. (2009). Artificial bacterial flagella: fabrication and magnetic control. *Appl. Phys. Lett.* 94 (6): 64103–64107.
7 Hu, C., Pané, S., and Nelson, B.J. (2018). Soft micro- and nanorobotics. *Annu. Rev. Control Robot. Auton. Syst.* (February): 1–23.

8 Palagi, S., Pensabene, V., Beccai, L. et al. (2011). Design and development of a soft magnetically-propelled swimming microrobot. In: *2011 IEEE International Conference on Robotics and Automation*, 5109–5114. IEEE.
9 Medina-Sánchez, M., Magdanz, V., Guix, M. et al. (2018). Swimming microrobots: soft, reconfigurable, and smart. *Adv. Funct. Mater.* 28 (25): 1707228.
10 Palagi, S. and Fischer, P. (2018). Bioinspired microrobots. *Nat. Rev. Mater.* 3 (6): 113–124.
11 Breger, J.C., Yoon, C., Xiao, R. et al. (2015). Self-folding thermo-magnetically responsive soft microgrippers. *ACS Appl. Mater. Interfaces* 7 (5): 3398–3405.
12 Fusco, S., Huang, H.W., Peyer, K.E. et al. (2015). Shape-switching microrobots for medical applications: the influence of shape in drug delivery and locomotion. *ACS Appl. Mater. Interfaces* 7 (12): 6803–6811.
13 Magdanz, V., Stoychev, G., Ionov, L. et al. (2014). Stimuli-responsive microjets with reconfigurable shape. *Angew. Chem. Int. Ed.* 53 (10): 2673–2677.
14 Huang, H.W., Sakar, M.S., Riederer, K. et al. (2016). Magnetic microrobots with addressable shape control. In: *2016 IEEE International Conference on Robotics and Automation (ICRA)*, 1719–1724. IEEE.
15 Zeng, H., Wasylczyk, P., Parmeggiani, C. et al. (2015). Light-fueled microscopic walkers. *Adv. Mater.* 27 (26): 3883–3887.
16 Warner, M. and Terentjev, E.M. (2007). *Liquid Crystal Elastomers*, vol. 120. Oxford University Press.
17 Zeng, H., Martella, D., Wasylczyk, P. et al. (2014). High-resolution 3D direct laser writing for liquid-crystalline elastomer microstructures. *Adv. Mater.* 26 (15): 2319–2322.
18 Lienhard, J. IV, and Lienhard, J. V, (2018). *A Heat Transfer Textbook*, version 2.12, 4e. Cambridge, MA: Phlogiston Press.
19 Palagi, S., Jager, E.W., Mazzolai, B., and Beccai, L. (2013). Propulsion of swimming microrobots inspired by metachronal waves in ciliates: from biology to material specifications. *Bioinspir. Biomim.* 8 (4): 046004.
20 Lauga, E. and Powers, T.R. (2009). The hydrodynamics of swimming microorganisms. *Rep. Prog. Phys.* 72 (9): 96601.
21 von Rohr, A., Trimpe, S., Marco, A. et al. (2018). Gait learning for soft microrobots controlled by light fields.https://arxiv.org/abs/1809.03225 In: *2018 IEEE/RSJ International Conference on Intelligent Robots and Systems (IROS)*, 6199–6206. IEEE.
22 Palagi, S., Mark, A.G., Melde, K. et al. (2016). Soft continuous microrobots with multiple intrinsic degrees of freedom. In: *Proceedings of the International Conference on Manipulation, Automation and Robotics at Small Scales*, MARSS, Paris, France, 18–22 July 2016.
23 Palagi, S., Mark, A.G., Melde, K. et al. (2017). Locomotion of light-driven soft microrobots through a hydrogel via local melting. In: *Proceedings of the International Conference on Manipulation, Automation and Robotics at Small Scales*, MARSS, Montreal, QC, Canada, 17–21 July 2017.

14

4D Printing: An Enabling Technology for Soft Robotics

Carlos Sánchez-Somolinos [1,2]

[1] *Instituto de Ciencia de Materiales de Aragón (ICMA), CSIC-Universidad de Zaragoza Departamento de Física de la Materia Condensada Facultad de Ciencias, Calle de Pedro Cerbuna, 12, 50009 Zaragoza, Spain*
[2] *CIBER in Bioengineering, Biomaterials and Nanomedicine (CIBER-BBN), Calle Mariano Esquillor s/n, Zaragoza, 50018, Spain*

14.1 Introduction

Three-dimensional (3D) printing has penetrated in many different research and application domains, including healthcare, automotive, and aerospace, among others, as a key enabling technology allowing the rapid digital fabrication of complex 3D objects, difficult to create or simply not accessible some years ago. 3D printers are present nowadays in many engineering studios or research and development laboratories as a versatile tool that helps in the engineering and in the rapid transformation of ideas and designs into real, tangible, and useful 3D objects [1, 2]. The advent of additive manufacturing (AM) technologies has changed the paradigm of fabrication frequently substituting drills, lathes, or expensive molds by a 3D printer. Conventional 3D printers can be used to create parts made on common thermoplastics with a high degree of precision starting from a graphic file. Besides thermoplastics, other materials including metals, ceramics, or composites, as well as functional materials can be structured to generate complex 3D objects through AM techniques [1–6].

As many other engineering domains, robotics has notably benefited from the advent of AM technologies and nowadays roboticists make use of 3D printing of thermoplastics for the prototyping of parts for robotic devices in a quick and reliable fashion. Engineers can design and, immediately just after, print the conceived components to implement a new part of a robot in few hours, instead of ordering complex pieces fabricated by conventional molding techniques that would be remarkably much more expensive and logistically not practical and time consuming. In this sense, the use of AM technologies empowers engineers to introduce adjustments from the very initial stage of the robot design process, accelerating the creation of the final product.

Materials used in conventional 3D printing, as thermoplastics mentioned above, usually lead to inanimate 3D objects. Four-dimensional (4D) printing goes one step beyond introducing time as a fourth dimension by generating

Mechanically Responsive Materials for Soft Robotics, First Edition. Edited by Hideko Koshima.

Figure 14.1 Sequence of photographs showing the self-folding of a 4D printed multi-material one-dimensional strand into a three-dimensional structure. Source: Reproduced with permission of Tibbits 2014 [8]. Copyright 2014, John Wiley & Sons.

objects that respond to a certain stimulus, for example by changing their shape. The general concept of 4D printing, introduced by Skylar Tibbits in a TED conference, consists in the generation of 3D objects that can change their shape or self-assemble over time [7]. This active capability of 4D printed objects enables them to change their shape in a predictable and programmed fashion, usually relies on the precise positioning of materials with well-defined morphologies in the 3D space (Figure 14.1) [8]. Despite it is not a must, 4D printing usually makes use of smart materials to embed the shape-changing capability in the printed object upon triggering. While smart materials, such as shape memory materials, hydrogels, or liquid crystalline elastomers (LCEs), have been studied for many years for the preparation of shape-changing objects, their integration with AM technologies is quite recent [8]. This capability of smart materials to generate objects that move or shape change under an external stimuli has been embraced by robotics as an enabling tool to generate new functions that eventually could lead to complete robotic systems [9].

The traditional image of a robot, usually made of articulated rigid parts provided with electrical motors or hydraulic or pneumatic actuators able to perform complicated tasks with a high degree of precision and repeatability, is substituted in the area of soft robotics by a system provided with soft materials actuators [10–12]. The ability of 4D printing to program soft stimuli-responsive materials has been identified as an excellent opportunity to further develop this concept of soft robotics toward more sophisticated embodiments. After introducing several concepts and some of the most relevant advanced manufacturing technologies in this field, the concept of 4D printing will be further developed in this chapter with special focus in the materials being employed. The implementation of 4D printed soft responsive structures able to perform targeted and relevant robotic functions will also be briefly presented.

14.2 3D Printing Techniques

AM techniques, first introduced in the late 1980s and commonly known as 3D printing technologies, cover a whole family of systems that essentially can create a complex 3D object starting from a digital design and adding material layer by layer at specific points. Compared with traditional fabrication methods such as casting, requiring molds, or lithography, needing complicated and expensive masks, AM technologies digitally convert a computer-aided design (CAD) into a real 3D complex object. Despite AM techniques cannot compete nowadays in speed production with conventional processing techniques based on molding, 3D printing allows the rapid, flexible, and cost-effective creation of objects on demand, with customized size and shape, multiple materials, and minimal or no post-processing. Besides, objects created by 3D printing can have hollow functional inner structures, impossible to create with conventional manufacturing methods based on molding or subtractive manufacturing methods such as milling. With such attractive characteristics, AM technologies have been widely adopted across academia and also industry, and in different application areas including healthcare, biomedical, aerospace, and robotics [1, 13, 14].

Despite the wide variety of developed techniques to print in three dimensions, using diverse materials and different principles to solidify the applied material, all of these techniques share common principles and features: Usually a design is transformed into a virtual object through CAD. The CAD model is digitally sliced into layers and the coordinates of the object are transferred to computer-controlled translation stages that, in most of the cases, move the device that fixes the material at the desired position.

With a large variety of processes being advanced and companies developing innovative equipment, it is important to set a well-defined terminology related to the fabrication processes. In this sense the American Society for Testing and Materials (ASTM) initiated a standardization process defining seven different technologies under the AM umbrella [15]. According to the process used, this classification distinguishes between: (i) binder jetting that uses a liquid bonding agent that is selectively applied on top of powder materials to locally join them together. (ii) Directed energy deposition, using focused thermal energy, for example coming from a light or electron beam, or a plasma arc. The energy in this case is used to locally melt and fuse materials as they are deposited. (iii) Material jetting, in which droplets of photopolymer or wax, for example are selectively applied. (iv) Powder bed fusion applies selectively heat to fuse a material powder bed. (v) Sheet lamination, in which the object is built by gluing sheets of material together to create a 3D object. (vi) Material extrusion, selectively applying material deposited through a nozzle. (vii) Vat photopolymerization, in which a photopolymer resin in a vat is selectively photocured.

Due to their extensive use, especially in the emerging field of 4D printing, a more detailed description on the processes themselves and the employed materials will be done for the two last categories.

14.2.1 Material Extrusion-Based Techniques

Among this family of techniques, we find the popular and widely extended fused deposition modeling (FDM) machines that make use a solid thermoplastic filaments feeding a hot extrusion printhead that enables the selective deposition of molten polymer (Figure 14.2). After polymer melt application, this immediately solidifies upon cooling below its glass transition temperature becoming fixed at the target position. Among the most common thermoplastics employed we find acrylonitrile butadiene styrene (ABS), polylactic acid (PLA), high-impact

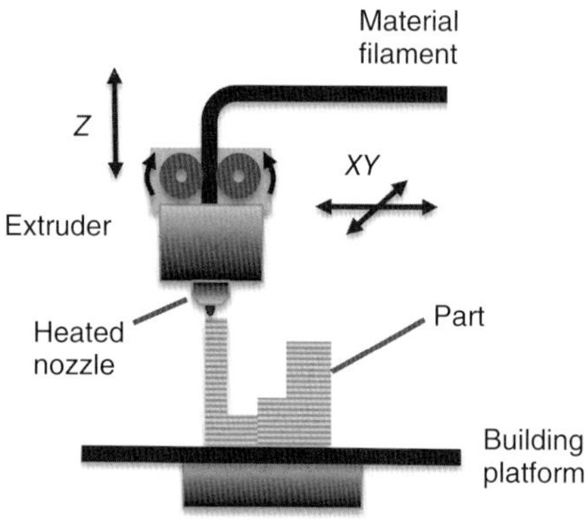

Figure 14.2 Fused deposition modeling of a thermoplastic filament.

polystyrene (HIPS), or polycarbonate (PC), sometimes filled with additives (e.g. conductive) to lead to objects with added functionalities [4].

Beyond thermoplastics, materials showing viscoelastic properties at room temperature (RT) can also be extruded and deposited using this technology. Special considerations need to be taken into account for this type of materials. Besides filament formation characteristics, the printed lines of material should have adequate consistency after deposition to enable feature definition and structural integrity of the 3D printed structures. To achieve these dimensionally accurate and self-supporting structures after printing of many layers of material, ink materials with quite specific properties need to be considered. Inks need to encompass a suitable extrudability that enables precise deposition together with fast gelation upon deposition, allowing the formation of well-defined 3D structures. The conjunction of these two special characteristics has been achieved for example by using yield-stress inks. In these systems, the application of pressure above certain threshold induces flow on the ink; however, once the material leaves the printhead, the material quickly solidifies leading to distortion-free filaments with self-supporting characteristics and capacity to lead to stable 3D structures [6]. This type of yield-stress inks have been obtained by the inclusion of rheology modifiers such as nanoclays or nanosilicates [16, 17]. Interaction of these nanoparticles with charged and also neutral polymers or even with themselves can lead to physically cross-linked 3D networks that present the required shear thinning properties. Molecular additives in polymer formulations or mixtures of polymers that supramolecularly self-assemble leading to physically cross-link and gelation have also been proposed as suitable systems for 3D printing [18]. Besides the inclusion of rheology modifiers, gelation can also be quickly produced after printing by using photopolymerizable inks. Immediately after the ink is printed, actinic light is applied to the deposited materials fixing the printed filament [19].

14.2.2 Vat Photopolymerization Techniques

In this process a liquid photopolymer in a vat is selectively cured by activating polymerization with light. As an example, in a conventional stereolithographic (SLA) system, a focused laser beam is steered to locally photopolymerize the

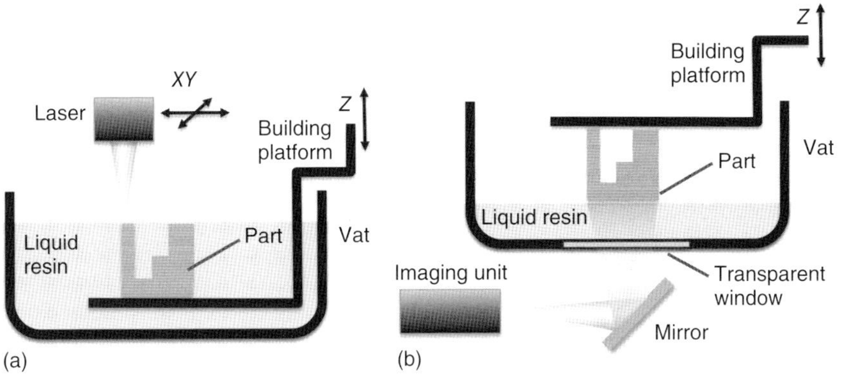

Figure 14.3 (a) Stereolithography and (b) digital light processing of photopolymer resins.

resist and generate a solid layer of an object (Figure 14.3a) [20]. After this layer is generated, the substrate and the created layer is lowered into the resin vat up to a predefined distance and a new layer of liquid photopolymer is applied on top. Later, the exposure process is repeated again layer by layer to generate 3D objects. The use in SLA of laser-focused beams scanning the surface of the liquid photopolymer to locally transform it into a solid cross-linked material provides with high spatial resolution, in the order of few micrometers for the *XY* plane, while resolution in the *Z* direction typically ranges from tens to hundreds of micrometers, being this determined by the light penetration in the photoresist. Actually, this *Z* resolution can be optimized by using suitable additives in the resist that absorb light in a controlled fashion. Two-photon photopolymerization (2PP) can also be used to dramatically improve the resolution of light-based fabrication technologies. Submicron resolutions can be achieved as the probability of two-photon absorption is proportional to the square of light intensity, leading to very small volume pixel (voxel) in the order of tens of nanometers, well below the diffraction limit of light [21, 22].

Instead of using focused laser beams for photopolymer curing, digital light processing (DLP) systems project a full image at once, rather than point by point, by using digital micromirror devices (DMDs) (Figure 14.3b). This technology, typically used to project presentation slides on a screen, enables to produce one complete layer of the object in one single light exposure step significantly reducing printing time with respect to laser-focused SLA technology. Usually illumination is carried out from the bottom irradiating the resin through an optically clear vat bottom provided with an anti-stick layer. This prevents the adhesion of the created object that would provoke deformation or breaking of the generated object. A major issue in this type of light-based technologies is the significant oxygen inhibition that can take place in free-radical polymerization of photocurable resists when in contact with atmosphere. This is critical in SLA systems with illumination from the top in which the cured layer is in direct contact with the atmospheric air, while irradiation from the bottom eliminates this issue. In any of the cases, after UV exposure of a complete layer, object movement and renewal of the photoresists needs to be carried out before UV exposure of the subsequent layers significantly increasing the overall object processing time. Continuous liquid interface production (CLIP) introduced by the group of De Simone introduces significant improvements to simplify processing and reduce processing time [23]. CLIP makes use of a simple DLP bottom illumination configuration (as shown in Figure 14.3b), however, innovatively introduces an oxygen permeable amorphous fluoropolymer (Teflon AF) transparent window. This facilitates the creation of a "dead zone" of few tens of microns in which free-radical polymerization is inhibited. In this way the 3D printed polymeric object does not adhere to the bottom glass of the vat. This enables a continuous creation of the 3D printed object, as the recoating step after curing of each layer is not needed anymore, significantly increasing in this way the production speed of printed parts.

Overall, the different presented printing techniques involve various deposition and fixation processes, which jointly define the feature size, object production speed, and, very importantly, the type of material that can be used in the object preparation. Despite the large advances done in this booming field of AM, still many challenges have to be overcome in order for these technologies to truly enter the ground of industrial applications. More specifically, improvements in production speed, resolution in large printed volumes, and the development of suitable performance printable materials are urgently needed for the further deployment of 3D printing in application fields spanning from engineering, aerospace, automotive, energy, consumer products, and even the food industry.

14.3 4D Printing of Responsive Materials

In 4D printing, time is the fourth dimension for 3D printed structures. These are able to change their shape in a preprogrammed fashion when exposed to the right stimuli, such as heat, humidity, light, and electric or magnetic fields. Conventional 3D printing makes use of materials that after deposition and fixation result in immutable objects, however the use of responsive materials in combination with AM technologies enables the unprecedented introduction of smart character to complex built structures that shift their shape in a prescribed manner in response to a trigger. As in the case of conventional 3D printing technologies, one of the bottlenecks found in the further development of 4D printing is the lack of suitable materials, presenting appropriate mechanical properties and response. Moreover, there is a lack of robust methodologies to provide the structures with a controlled material morphology and therefore structure and function upon exposure to the trigger. Due to their large deformability, smart responsive polymers are being extensively investigated as the most prominent candidate materials for 4D printing, mainly shape memory polymers (SMPs), hydrogels, and LCEs.

14.3.1 Shape Memory Polymers

SMPs are a class of stimuli-responsive materials for which their shape-shift can be programmed [24]. This is generally done by deforming a material structure at a temperature above certain transition temperature of the material (typically the melting point, for semicrystalline polymers or the glass transition temperature for amorphous systems) and then cool down again below this temperature, while keeping the deformation. Once cooled down, the deforming load is eliminated and the structure deformation remains. Once this structure programming is completed, shape-shifting can be triggered by simply heating the system above the transition temperature of the polymer. In this respect SMPs typically present one-way actuation as the original shape recovered after actuation remains after the stimulus disappears.

The direct printing of these SMPs through AM technologies has been investigated seeking the preparation of active structures with controlled shape-shifting. Actually commercially available conventional photocurable resins for SLA typically present this shape memory feature despite their behavior is far from optimum. In this respect, Zarek et al. introduced methacrylated polycaprolactone (PCL) semicrystalline macromers molten in a heated vat to create complex structures through SLA printing [25]. Once printed the 3D created structures are rigid at RT, while they are rubbery at temperatures above 55 °C (the melting point of PCL). Deformations applied to the structures in the high temperature elastomeric state can be fixed by cooling down the object below the melting point of PCL. Once unloaded, heating of the structure above this transition temperature induces shape-shift to the original 3D printed shape (Figure 14.4). Combination of these structures with inkjet printed conductive tracks enabled the preparation of thermally triggered electrical devices.

Tailored photopolymers have been developed to improve the thermomechanical performance of the generated structures. Ge et al. prepared a family of photocurable methacrylate that were structured using projection microstereolithography to create high-resolution SMP complex objects [26]. The adequate formulation of the resin, comprising benzyl methacrylate (BMA) as chain extender and cross-linking agents such as poly(ethylene glycol)dimethacrylate (PEGDMA), bisphenol A ethoxylate dimethacrylate (BPA), and di(ethylene glycol)dimethacrylate (DEGDMA), led to cross-linked networks with tailored properties and improved performance with respect to conventional highly cross-linked photoresists. These materials showed for example rubbery modulus from ~1 MPa to ~100 MPa, glass transition temperatures ranging from −50 to −180 °C, and failure strains with values above 300%, significantly larger than those of commercially available conventional printable materials. This

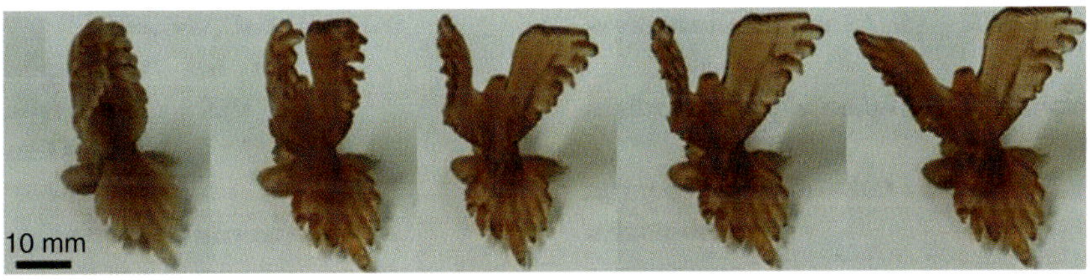

Figure 14.4 Polycaprolactone-based bird prepared by SLA showing shape memory behavior. The bird shape-shifts from the deformed state at RT to the original shape at 70 °C. Source: Zarek et al. 2016 [25]. Copyright 2016, Reproduced with permission of John Wiley & Sons.

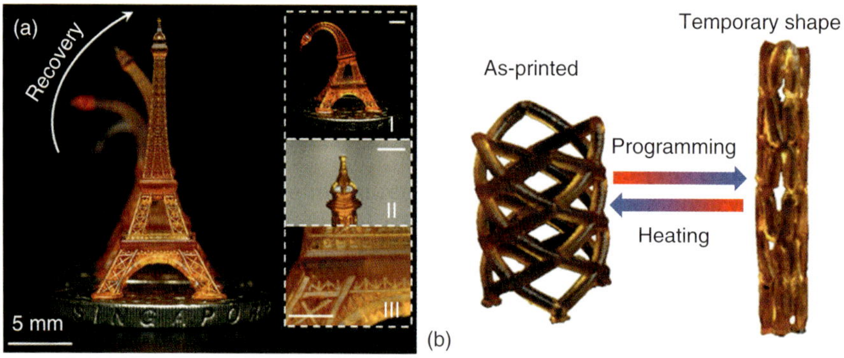

Figure 14.5 (a) An Eiffel tower model and (b) a stent created using projection microstereolithography, exhibiting one-way shape-shifting behavior on heating. Source: Ge et al. 2016 [26]. Copyright 2016, Reproduced with permission of Springer Nature. https://creativecommons.org/licenses/by/4.0/.

tailorability enabled to generate single component 3D structures exhibiting large deformation and controlled shape memory behavior (Figure 14.5).

Besides light-based structuring techniques, extrusion-based printing has also been explored for the preparation of shape-shifting structures of SMPs. Yang et al. have recently employed for this purpose FDM printing in combination with photoresponsive shape memory composites [27]. These essentially comprise SMPs and carbon black as an efficient photothermal transducer. Once the structures are 3D printed and programmed into the temporary shape, external illumination (e.g. sunlight) act as a trigger to recover the original shape (Figure 14.6), enabling remote triggering of the transformation [27].

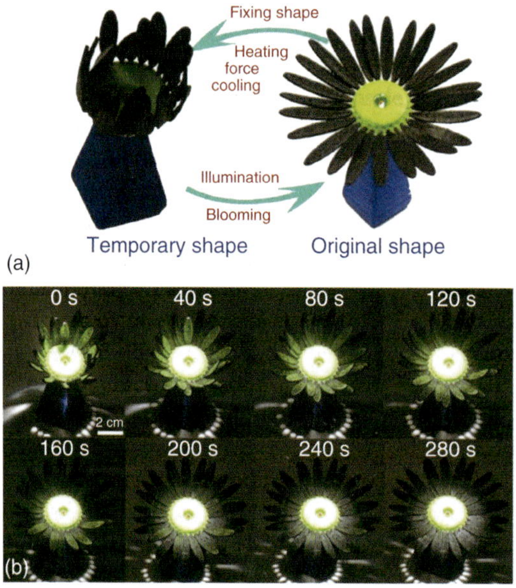

Figure 14.6 Shape memory behavior of 3D printed sunflower triggered by light. Source: Yang et al. 2017 [27]. Copyright 2016, Reproduced with permission of John Wiley & Sons.

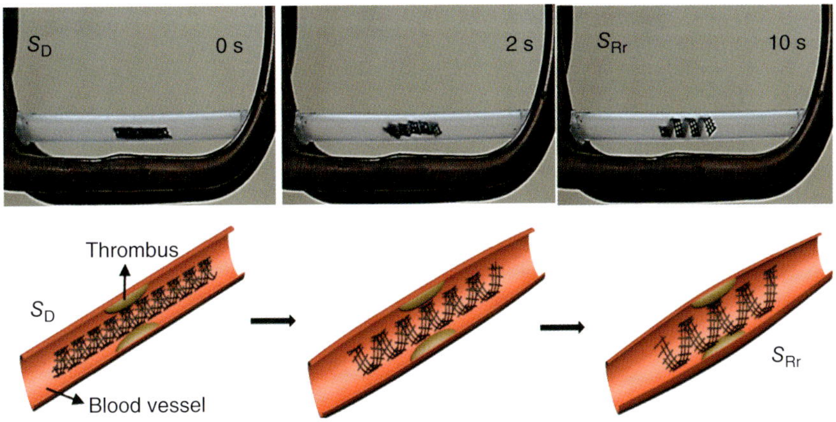

Figure 14.7 4D printed stent based on PLA-Fe_3O_4 composite ink. (a) Shape-shifting triggered by an alternating magnetic field and (b) potential application as an intravascular stent. Source: Wei et al. 2017 [28]. Copyright 2017, Reproduced with permission of American Chemical Society.

Magnetic fields can also be used as remote trigger for shape-shifting. Wei et al. demonstrated the extrusion printing of photocross-linkable PLA-based inks with shape memory characteristics. The inclusion of iron oxide (Fe_3O_4) nanoparticles that can be remotely heated by the use of alternating magnetic fields enabled the remote heating of the 3D printed structure and therefore the recovery of its permanent shape (Figure 14.7) [28].

As mentioned above, programming of SMPs is typically needed to induce shape-shifting in these systems. Direct 4D printing employing SMPs can also be carried out without the need of thermomechanical training by integrating the programming step into the 3D printing process. To do this, a bilayer is printed on top of a substrate with two materials: a SMP that is glassy and stiff at RT and an elastomer that contains compressive stresses that are controlled through the printing process. Heating of the bilayer, after this is released from the supporting substrate, leads to deformation into the final, permanent shape by heating [29].

14.3.2 Hydrogels

Hydrogel actuation and shape deformation rely on local isotropic expansion due to water swelling. Advantageously, with respect SMPs, hydrogels can offer two-way actuation; however, mechanical performance is limited and their response is typically very slow. These highly hydrophilic materials are at the origins of 4D printing, with the first printed shape-shifting elements created by Tibbits containing water-swellable materials. Tibbits first introduced the 4D printing concept by creating, using a multi-material printer, one-dimensional strands that upon exposure to water transformed into 3D objects. First, a rigid material was used to generate a structure provided with angle limiters. The inclusion through 3D printing of a hydrophilic polymer, which expands 150% when exposed to water, incorporated in the structure enables the straight strand to transform into another 3D shape. By properly designing the rigid elements and adequately embedding the active material, the expansion of the hydrogel in the presence of water forces the rigid materials to bend until the rigid parts reach the limit position of the final shape configuration. Figure 14.1 shows one of the first and most famous examples of 4D printed structures with a one-dimensional straight strand that transforms upon water exposure into a 3D cube [8]. Beyond strands, photopatterning of thin polymeric films, regulating the cross-linking density and therefore the local swelling, has also demonstrated to be a powerful tool to program shape changes in hydrogels [30, 31].

Besides local isotropic expansion on swelling, anisotropic characteristics can also be incorporated in hydrogels via the extrusion of hydrogel inks containing anisotropic nano-objects. Gladman et al. demonstrated the extrusion printing of photopolymerizable ink incorporating nanofibrillated cellulose [17]. During extrusion, anisotropic significant alignment of the nanofibrils is achieved, being this fixed during the photopolymerization process just after deposition. The resultant filaments present anisotropic elastic behavior and also anisotropic swelling characteristics that can be controlled and optimized through the adequate selection of the printing parameters (namely nozzle diameter and printing speed). This anisotropic morphology conferred to the structure through the printing process introduces new opportunities in hydrogel structure actuation. A full collection of examples of biomimetic deformations found in plants were demonstrated in response to humidity by these authors [17].

14.3.3 Liquid Crystalline Elastomers

LCE materials have been thoroughly studied as stimuli-responsive materials as they present large anisotropic mechanical response when exposed to a suitable trigger [32–35]. Appropriate design of these materials enables fast and reversible response to different stimuli including heat, light, moisture, pH, or electrical fields. Actuation in these systems relies on the anisotropic stresses generated in the material as a result of a change of the molecular order. For example, an increase of temperature results in a decrease of liquid crystal (LC) order leading to a contraction along the preferential direction of the LC molecules, the director **n**, and an expansion along the perpendicular direction. As a result, precise control of the director enables engineering of the mechanical response in these LCE systems [33, 36, 37].

By using this type of LCE materials, two-way shape-shifting structures has been demonstrated. Typically, thin films of these materials with a precise control of the director have mainly been prepared to implement a wide variety of responsive elements and devices. These films were typically prepared between two glass plates provided by alignment layers in which complex patterns can be inscribed [33–36]. LCs in the cell orient following the alignment layer orientation and are fixed to lead to a cross-linked liquid crystalline system. After releasing the polymeric film from the glass plates programmed actuation, dictated by the director profile, can take place by exposing the film to the appropriate stimulus. Despite the great potential and versatility of these materials, the classical toolbox to process these materials and the thin film character of the so fabricated elements markedly limits

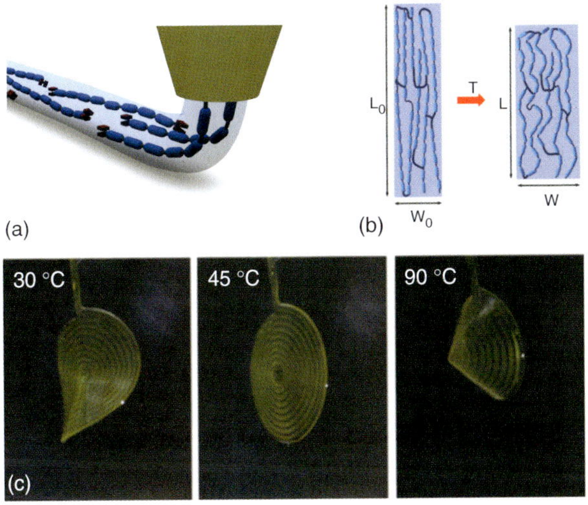

Figure 14.8 (a) Imposed polymer main chain alignment along the printing direction. (b) Thermomechanical response of uniaxially aligned LCEs. (c) Spiral-like director texture transforming from flat into a cone. Source: López-Valdeolivas et al. 2018 [39]. Copyright 2018, Reproduced with permission of John Wiley & Sons.

the energy available to actuate and produce mechanical work precluding their use in real-life applications.

Recently 4D printing of LCEs have been demonstrated at three different laboratories [38–40]. Essentially a main chain liquid crystalline macromer functionalized with acrylate end groups have been employed as an ink. Extrusion printing is employed to deposit filaments that are later fixed through photopolymerization. By appropriate selection of the printing conditions (namely, material flux rate, deposition speed, and nozzle diameter) alignment of the director is reached after printing within the deposited filament (Figure 14.8a). This alignment is fixed through photopolymerization enabling a precise definition of the director architecture in the so created structure. This morphology definition enables digital programming of stresses and deformation upon stimulation with temperature (Figure 14.8b). Although the groups introducing 4D printing in these materials have used temperature as trigger, other stimuli can be envisioned by appropriate design and synthesis of new printable materials with suitable printing and response characteristics.

Engineering of the director enables, as mentioned, precise programming of stresses in the materials leading to controlled deformation. As an example, Figure 14.8c shows a round disc, printed following a spiral printing path and therefore having this same director profile. The disc is nearly flat at RT; however, an increase of the temperature produces stresses within the film. This tends to contract in the azimuthal direction and to expand in the radial one. To accommodate these stresses, the system escapes out of the plane, into the third dimension, forming a cone whose angle becomes stepper as temperature increases [39].

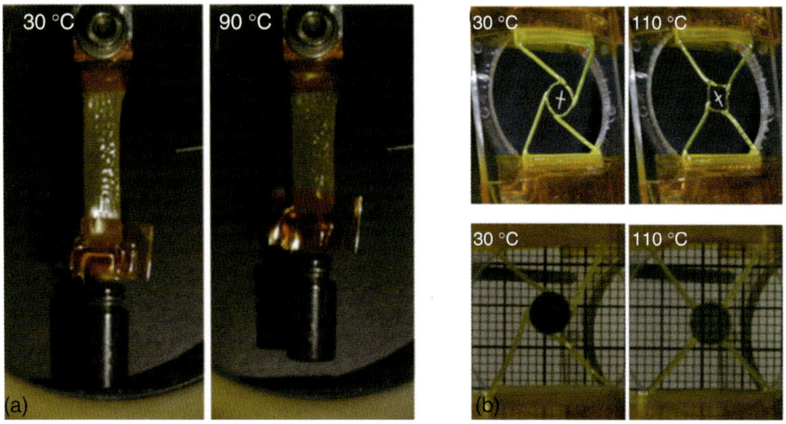

Figure 14.9 (a) Printed LCE artificial muscle (25 × 6 mm). A weight of 5 g attached to the stripe is lifted up on heating. (b) Chiral structure comprising a central ring (4.4 mm diameter) held by four bands (above) and the same chiral structure moving a round piece of linear polarizer on heating. Source: López-Valdeolivas et al. 2018 [39]. Copyright 2018, Reproduced with permission of John Wiley & Sons.

14.4 4D Printing Toward Soft Robotics

The existing toolbox of soft materials and processing methodologies has demonstrated extraordinary potential to generate programmed deformations useful in robotics. Beyond the generation of complex dynamically changing biomimetic shapes, precise programming of soft matter in response to a stimulus can be used to design and create complex purposeful robotic functions. 4D printed linear actuators able to perform mechanical work have been demonstrated using for example LCEs. Figure 14.9a presents a stripe made out of parallel filaments of LCE showing uniaxial orientation along the long direction of the stripe. This orientation of the LC molecules is imposed through the printing process as described in Section 14.3.3. Heating of the stripe, from 30 to 90 °C, results in contraction along the long side of the stripe, that is able to lift an object 30 times heavier than itself over a length of 5 mm [39]. Rotation of objects has also been demonstrated by using 4D printed chiral structures made of LCE as shown in Figure 14.9b. A central ring held by four bands rotates on heating. The ring can be coupled to an external passive element (e.g. a polarizer) so this can be precisely rotated on demand [39]. 4D printing has also been used to create active soft grippers able to grab objects on demand. Figure 14.10 shows a multi-material gripper based on SMPs grabbing an object [26].

Due to their soft nature, this type of robotic actuators has great potential in the field of surgical technologies, as the interaction of these soft elements with living tissues is gentler and safer than that of their rigid counterparts. As an example in this field of application, self-expandable stent structures based on SMPs have been created by 4D printing messes programmed to expand inside a tube widening its diameter as shown already in this chapter (Figures 14.5 and 14.7) [25, 26, 28].

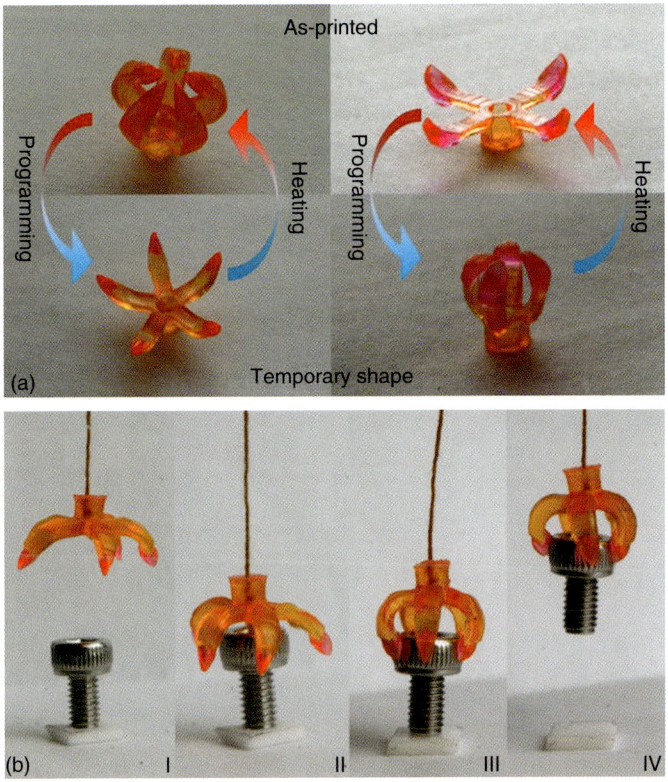

Figure 14.10 3D printed multi-material grippers showing (a) shape-shifting between printed and temporary shapes and (b) gripping ability. Source: Ge et al. 2016 [26]. Copyright 2016, Reproduced with permission of Springer Nature. https://creativecommons.org/licenses/by/4.0/.

14.5 Conclusions

4D printing, still in its infancy, holds the promise to become a key enabling technology for soft robotics, offering endless possibilities for fabrication of active elements and devices. The successful combination of 3D printing and stimuli-responsive polymers offers unparalleled opportunities for the preparation of soft robotic actuators, which can revolutionize the way soft robots are conceived and implemented. The AM character of the methodology allows to prepare multi-material components facilitating, for example the implementation of hybrid soft responsive-rigid and multi-responsive robots. One of the bottlenecks for the further development of 4D printing, as in the case of 3D printing, is the lack of high-performance materials that can be processed using this manufacturing technologies. As a result, intense research needs to be done toward the development of new printable materials responsive to different triggers, with adequate mechanical response and compatible with AM technologies. Developments in this area will allow the preparation of complex elements of different size and geometry, with great freedom of design and with a precise control of the polymer morphology enabling precise programming of the mechanical response upon exposure to stimuli. Overall 4D printing of soft materials open new avenues for the development of new responsive complex soft robotic elements and devices with application in different sectors including medicine, microfabrication, consumer electronics, or haptics.

Acknowledgments

C. Sánchez-Somolinos thanks the Spanish MINECO projects BIO2017-84246-C2-1-R, H2020 FET-OPEN Project 829010 (PRIME), Gobierno de Aragón, and FEDER (EU).

References

1 Tofail, S.A.M., Koumoulos, E.P., Bandyopadhyay, A. et al. (2018). Additive manufacturing: scientific and technological challenges, market uptake and opportunities. *Mater. Today* 21 (1): 22–37.
2 Ligon, S.C., Liska, R., Stampfl, J. et al. (2017). Polymers for 3D printing and customized additive manufacturing. *Chem. Rev.* 117 (15): 10212–10290.
3 Hofmann, M. (2014). 3D printing gets a boost and opportunities with polymer materials. *ACS Macro Lett.* 3 (4): 382–386.
4 Kalsoom, U., Nesterenko, P.N., and Paull, B. (2016). Recent developments in 3D printable composite materials. *RSC Adv.* 6 (65): 60355–60371.
5 Leigh, S.J., Bradley, R.J., Purssell, C.P. et al. (2012). A simple, low-cost conductive composite material for 3D printing of electronic sensors. *PLoS One* 7 (11): e49365.
6 Compton, B.G. and Lewis, J.A. (2014). 3D-printing of lightweight cellular composites. *Adv. Mater.* 26: 5930–5935.
7 Tibbits, S. (2013). The emergence of “4D printing”. https://www.ted.com/talks/skylar_tibbits_the_emergence_of_4d_printing (accessed 17 June 2019).
8 Tibbits, S. (2014). 4D printing: multi–material shape change. *Archit. Des.* 84: 116.
9 de Marco, C., Pané, S., and Nelson, B.J. (2018). 4D printing and robotics. *Sci. Robot.* 3 (18): eaau0449.
10 Kim, S., Laschi, C., and Trimmer, B. (2013). Soft robotics: a bioinspired evolution in robotics. *Trends Biotechnol.* 31: 287–294.
11 Rus, D. and Tolley, M.T. (2015). Design, fabrication and control of soft robots. *Nature* 521: 467–475.
12 Laschi, C., Mazzolai, B., and Cianchetti, M. (2016). Soft robotics: technologies and systems pushing the boundaries of robot abilities. *Sci. Robot.* 1: eaah3690.
13 Truby, R.L. and Lewis, J.A. (2016). Printing soft matter in three dimensions. *Nature* 540 (7633): 371–378.
14 MacDonald, E. and Wicker, R. (2016). Multiprocess 3D printing for increasing component functionality. *Science* 353 (6307): aaf2093.

15 ASTM (2012). ASTM F2792–12a). *Standard Terminology for Additive Manufacturing Technologies*. West Conshohocken, PA: ASTM International.

16 Xavier, J.R., Thakur, T., Desai, P. et al. (2015). Bioactive nanoengineered hydrogels for bone tissue engineering: a growth-factor-free approach. *ACS Nano* 9 (3): 3109–3118.

17 Gladman, S.A., Matsumoto, E.A., Nuzzo, R.G. et al. (2016). Biomimetic 4D printing. *Nat. Mater.* 15 (4): 413–418.

18 Highley, C.B., Rodell, C.B., and Burdick, J.A. (2015). Direct 3D printing of shear-thinning hydrogels into self-healing hydrogels. *Adv. Mater.* 27 (34): 5075–5079.

19 Ouyang, L., Highley, C.B., Rodell, C.B. et al. (2016). 3D printing of shear-thinning hyaluronic acid hydrogels with secondary cross-linking. *ACS Biomater Sci. Eng.* 2 (10): 1743–1751.

20 Hull, C.W. (1986). Apparatus for production of three-dimensional objects by stereolithography. US Patent 4,575,330, filed 08 August 1984.

21 Stampfl, J., Liska, R., and Ovsianikov, A. (eds.) (2017). Multiphoton lithography. In: *Techniques, Materials and Applications*. Wiley-VCH.

22 Barner-Kowollik, C., Bastmeyer, M., Blasco, E. et al. (2017). 3D laser micro- and nanoprinting: challenges for chemistry. *Angew. Chem. Int. Ed.* 56 (50): 15828–15845.

23 Tumbleston, J.R., Shirvanyants, D., Ermoshkin, N. et al. (2015). Continuous liquid interface production of 3D objects. *Science* 347 (6228): 1349–1352.

24 Zhao, Q., Qi, H.J., and Xie, T. (2015). Recent progress in shape memory polymer: new behavior, enabling materials, and mechanistic understanding. *Prog. Polym. Sci.* 49–50: 49–50. 79–120.

25 Zarek, M., Layani, M., Cooperstein, I. et al. (2016). 3D printing of shape memory polymers for flexible electronic devices. *Adv. Mater.* 28 (22): 4449–4454.

26 Ge, Q., Sakhaei, A.H., Lee, H. et al. (2016). Multimaterial 4D printing with tailorable shape memory polymers. *Sci. Rep.* 6: 161.

27 Yang, H., Wan, R.L., Wang, T. et al. (2017). 3D printed photoresponsive devices based on shape memory composites. *Adv. Mater.* 29 (33): 1701627.

28 Wei, H., Zhang, Q., Yao, Y. et al. (2017). Direct-write fabrication of 4D active shape-changing structures based on a shape memory polymer and its nanocomposite. *ACS Appl. Mater. Interfaces* 9 (1): 876–883.

29 Ding, Z., Yuan, C., Peng, X. et al. (2017). Direct 4D printing via active composite materials. *Sci. Adv.* 3 (4): e1602890.

30 Kim, J., Hanna, J.A., Byun, M. et al. (2012). Designing responsive buckled surfaces by halftone gel lithography. *Science* 335 (6073): 1201–1205.

31 Huang, L., Jiang, R., Wu, J. et al. (2017). Ultrafast digital printing toward 4D shape changing materials. *Adv. Mater.* 29: 1605390.

32 White, T.J. and Broer, D.J. (2015). *Nat. Mater.* 14: 1087.

33 de Haan, L.T., Sánchez-Somolinos, C., Bastiaansen, C.M.W. et al. (2012). Engineering of complex order and the macroscopic deformation of liquid crystal polymer networks. *Angew. Chem. Int. Ed.* 51 (50): 12469–12472.

34 de Haan, L.T., Gimenez-Pinto, V., Konya, A. et al. (2013). Accordion-like actuators of multiple 3D patterned liquid crystal polymer films. *Adv. Funct. Mater.* 24 (9): 1251–1258.

35 Gelebart, A.H., Mulder, D.J., Varga, M. et al. (2017). Making waves in a photoactive polymer film. *Nat. Publ. Group* 546 (7660): 632–636.

36 Ware, T.H., McConney, M.E., Wie, J.J. et al. (2015). Voxelated liquid crystal elastomers. *Science* 347 (6225): 982–984.

37 Modes, C.D., Warner, M., Sánchez-Somolinos, C. et al. (2013). Angular deficits in flat space: remotely controllable apertures in nematic solid sheets. *Proc. R. Soc. A* 469 (2153): 20120631.

38 Ambulo, C.P., Burroughs, J.J., Boothby, J.M. et al. (2017). Four-dimensional printing of liquid crystal elastomers. *ACS Appl. Mater. Interfaces* 9 (42): 37332–37339.

39 López-Valdeolivas, M., Liu, D., Broer, D.J., and Sánchez-Somolinos, C. (2018). 4D printed actuators with soft-robotic functions. *Macromol. Rapid Commun.* 39 (5): 1700710.

40 Kotikian, A., Truby, R.L., Boley, J.W. et al. (2018). 3D printing of liquid crystal elastomeric actuators with spatially programed nematic order. *Adv. Mater.* 30: 1706164.

15

Self-growing Adaptable Soft Robots

Barbara Mazzolai, Alessio Mondini, Emanuela Del Dottore, and Ali Sadeghi

Istituto Italiano di Tecnologia, Center for Micro-BioRobotics, Viale Rinaldo Piaggio 34, 56015 Pontedera, PI, Italy

15.1 Introduction

Robotics has achieved impressive results in the last decades, finding vast applications especially within the industrial context, where precision, velocity, and power are primary requirements [1]. As robots have started to move outside of the well-defined frameworks of the factories, researchers had to begin to address new challenges and performance requirements in order to meet the abilities necessary to move in dynamically changing environments, and to safely interact with humans [2]. In this view, the research for a new class of robotic systems recently has opened to the approach of *soft robotics* [3–6]. In soft robotics, components and systems are designed to be more compliant, and consequently more reliable when facing unpredictable situations as well as intrinsically safer during interactions [7–9]. Soft, mechanically responsive, adaptable, and variable stiffness materials, which are unconventional in respect to the traditional "rigid" robotics, are key elements of soft robotics mechanisms. Their structural features and biomechanics facilitate intrinsically the dynamic adaptability of the whole system and are the enabler of new abilities, such as squeezing, stretching, self-healing, morphing, and even growing, peculiar of soft-bodied robots [10].

Since its beginning, several technologies for soft robotics have been designed by looking at animals in order to extract design principles useful to improve the performance of the system's movement [2, 4]. Recently, roboticists have started to look also at the plant kingdom, as a new source of inspiration for effective strategies of dynamic adaptability and resistance within external, challenging, environments [11–15]. Plants have in fact unique abilities to move and adapt their body to various ecosystems, using the mechanism of growth and a wide range of structural materials. The primary growth in plants is a localized process, typically based on the production of new material (the cells) in apical areas of roots and shoots ("meristematic zones") [16]. There is a huge variety of materials showcased by plants, characterized by nonlinear mechanical behavior and often acting as soft matter [17]. The growth process expands for the entire lifetime of a plant, with a biologically "indeterminate" mechanism influenced by both genetics

Mechanically Responsive Materials for Soft Robotics, First Edition. Edited by Hideko Koshima.

and environment. Through growth, plants are able to explore and colonize their surroundings, overtaking obstacles and moving in hard medium like roots in soil, or traveling relatively large distances and negotiating voids like climbing plants do. The changes in the body shape and size depends on food availability, metabolic needs, and population density [18]. Physical and structural diversity and complexity are therefore displayed in a variety of ecosystems. All the forms of trees, climbers, epiphytes, and so on are the producers of three-dimensional (3D) structures that support and host other plants and animals. Plants are stressed by wind-induced tilting and acclimation among different habitats [19], and are able to manage complex interactions between mechanical stress distribution, sensing (e.g. strain, gravity), and integrative and differential developmental growth responses [20]. This strong adaptation to environmental conditions is called *plasticity*, and it is a peculiar characteristic for the plant kingdom [21], which allows plants to strongly adapt their bodies, across both overground and underground organs, to survive and react to environmental changes.

The structural and functional properties of the materials of plants bodies, and their strategy to move across unknown and challenging environments through material deposition and adaptive growth, make the plants a unique and novel model for roboticists, opening the doors to the emerging family of growing robots, based on a new paradigm for machines movement, which follows the "growing from the tip" strategy for a more efficient adaptability in unknown "ecosystems."

The so-called *growing robots* can be defined [22] as robotic entities able to modify their bodies' structure by an incremental addition of material. By "material" can be intended any components, modules, or matter that can be supplied (e.g. from a storage) to the robot, or that the robot can find directly in the environment. Such materials can be connected (e.g. interlocked, fused, incorporated) by the robot in its own structure by using an embedded self-building process, which enables the incremental construction of the body, in real-time, without the need for a predefined design dictating the complete morphology, or architecture, of the body.

The functional characteristics and form of the material become therefore of paramount importance. A set of materials endowed with multi-functionalities to sense and interact with the environment can be used not only to increase the structural mass of the body and to vary its morphological features, but also for integrating new capacities in the system. For instance, material multi-functionality has been successfully exploited in bioinspired soft robotic demonstrators (e.g. artificial *Mimosa pudica*) [23]. In this example, a conductive polymer was exploited for its capability to simultaneously combine active/passive actuation and multimodal sensing within a single composite, and also acts as structural material. Other scientific literature reports numerous examples of soft materials able to respond disparate stimuli, including humidity, temperature, and mechanical stimuli, changing their fundamental properties, such as volume, electrical conductance, stiffness, and so on [24, 25].

The research for the integration of multifunctional materials in robotics is evolving together with the study of novel methodologies for their fabrication. Particularly relevant are additive manufacturing and related technologies, which allow a direct fabrication of complex 3D structures. These include photolithography techniques [26], fused deposition modeling (FDM) [27], powder-bed technology, and several newly emerging direct-write techniques [28, 29]. The so-called 4D printing, in which the fourth dimension is the time, combines the 3D printing with the use of functional materials, so the 3D printed object changes over the time to perform programmed functions, simply based on material properties [30, 31]. The use of multifunctional materials in combination with conductive materials and embedded fabrication techniques open the route for the possibility to self-build structures with complex architecture, routing electric paths and creating structures able to react to environmental stimuli on a well-defined way.

By looking at the importance of responsive adaptability for soft robots, their materials, and the concept of 3D printing, this chapter presents an overview of technological advancements achieved in recent years for self-growing adaptable soft robots. First, the authors' perspective about how "growth" has been historically translated into machines will be presented, passing then to the main hints provided by plants, which have led the research toward growing robots. The evolution of plant-inspired growing technologies is then reported, starting from the first works in robotics using *plants as model*, dating back to 2008, up to the last implementation of plant-inspired growing robot. The ability of plants to adapt reflects into morphological and physiological changes, both ways of adaptation have been translated into artificial solutions and summarized here. In the end, authors present possible applications and the perspectives of this research aimed at improving the adaptability of artificial systems through plant-inspired technologies exploiting soft and functional materials, as well as new hints for distributed intelligence.

15.2 Evolution of Growing Robots

The idea of growing machines can probably be dated back to the mid-twentieth century, when John Von Neumann was discussing about *self-reproductive automata* and the "concept of complication." The "concept of complication" derives from the idea that natural organisms always evolve in more complex systems with respect to their parents, never decreasing in complexity. Von Neumann speculated if and how such a concept could have been reproduced into automata. After Turing's theory on computing automata and Ulam's cellular model [32], Von Neumann formalized his idea on how an automaton can generate another identical automaton starting from a reservoir of floating elements [33].

While Von Neumann's concepts of reproduction and growth remained at theoretical or mostly at computational level, Fukuda et al., was the first to propose a dynamically reconfigurable robotic system (DRRS) implementing CEBOT, a cell-like structured robot [34]. With this work, he introduced the concept of a distributed robotic system composed of separate autonomous and heterogeneous units, which he called "cells." These robotic cells can communicate with each other, approach, connect, and disconnect automatically. His idea was to realize a system able to adapt its structure according to the environment and to perform a task just by being able to differently combine its cells. CEBOT was an important milestone opening up the area of *cellular robotics*, posing a step forward toward Von Neumann's idea of self-reproductive automata. The challenge in this case was to obtain a system not limited by a predefined design, but with the possibility to extend itself via construction methods, using the combination or connection of cells. This dynamic reconfiguration allows the robot to adapt its structure and functions specifically for the demanded task. Now, self-assembly, self-reconfigurability, and distributed control have become features of many works carried out by the robotics community. Among these works, those actually incorporating properties of growth have been discussed in [22].

Successively, Lipson and Pollack in [35] proposed an alternative approach toward self-generating machines, based on the concept of a continuously self-designing reconfigurable robot. They proposed to combine the power of evolutionary computation to co-develop design of the body and control, using additive manufacturing techniques to fabricate the robot. Their idea was to embed all these functionalities directly on the robot and to use recycled material and components that could be fused and readapted for new morphologies. Although the system was able to autonomously conceive a design according to the task to be accomplished, human intervention was still needed for body fabrication, assembling, and coding. The main limitation for them seemed to be the available technology. Additive manufacturing equipment was impractical as part of a robot, limiting its autonomy. On the other hand, in Lipson's vision, additive manufacturing could enable unpredictable structures to be created, and to explore designs that would otherwise be potentially difficult to manufacture with classical approaches.

The complete autonomous evolution of a body was practically possible only for virtual systems in simulated environments [36–39]. These investigations on the evolution and growth of virtual "creatures" converged into the field of *evolutionary robotics*, this approach was used also for biological studies in order to understand morphogenetic properties and evolutionary transitions of real creatures. The tools provided by this field seemed to fit for the investigation of morphogenesis in soft-bodied creatures [40] and consequently to suit for an investigation on the material properties and their effects on the behavior of soft-bodied robots [39]. All these works, and many others, demonstrate the feasibility of obtaining an intellectually growing and adaptable robot. Yet, the step from simulation to the physical realization of a growing autonomous working robot is still ongoing. Again, the main challenge seemed to lie in the physical transduction of evolutionary processes for the morphological evolution and growth of a real robot.

A consistent help for stepping further toward the physical implementation of robots able to grow came from the investigation of *growth in plants*. The observation of their high adaptability to harsh environments and their capability to colonize soil was particularly interesting to take inspiration for the design and development of new artificial systems for adaptive soil penetration and exploration [41]. By investigating the strategies used by plants, a series of energetically efficient technological solutions for movement in soil have been proposed

[11, 12, 42], paving the way for a new generation of robots inspired by plants and demonstrating the physical implementation of the concept of *self-building and growing* in robotics.

15.3 Mechanisms for Adaptive Growth in Plants

It has been estimated that about 350 000 species of plants exist [43], with a huge variety of properties and characteristics. Among these, two of the most interesting features are the capability of roots to penetrate hard soil and the capability of climbing plants to move fast by using existing structures in their environment. At the base of these exploration and motion abilities there is the growth: the addition of new structural materials, which are synthesized and allocated at the inner side of existing cell wall [44]. This happens by means of two processes: cell division and cell elongation. They are located respectively in the meristematic and elongation regions (Figure 15.1). Newly generated cells shift with time from the meristematic to the elongation region, where they axially elongate by water absorption. When this cell elongation is symmetric along the circumference of the root or shoot, the growth is straight, while asymmetric cells elongation is responsible for a bending (differential growth/elongation) [45].

When moving in soil, roots have to cope with a variety of chemical, physical, and mechanical stimuli [46–48]. Particularly, to overcome soil resistance, they use a series of strategies allowing them to find the low-resistant path, adapting their morphology and developmental organ to soil texture and exploiting cracks [49–51].

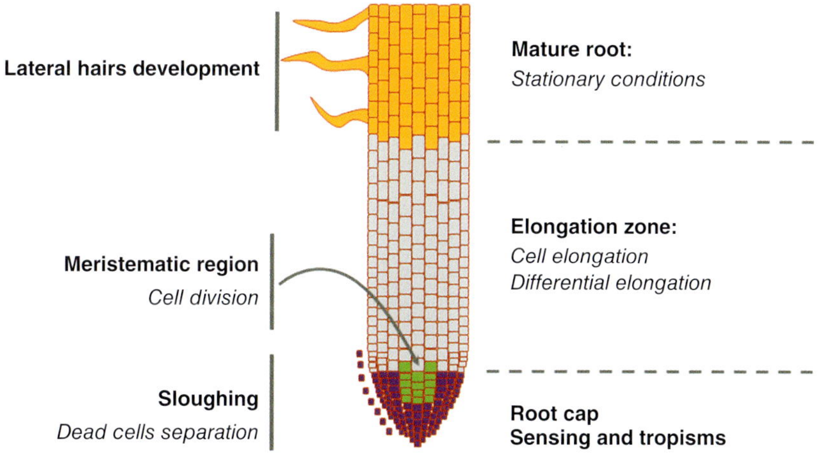

Figure 15.1 Schematic view of meristematic and elongation zones in a plant root. The *root cap* is responsible for sensing and behavior, characterized by the releasing of death cells that creates an interface between soil and root to reduce soil friction. The *meristematic region* is where new cells divide (it is the area of growth). The *elongation zone* is where the cells elongate by water absorption based on the root cap commands. The *mature zone* is where the root is stationary (it does not move with respect to the soil) and it anchors and absorbs nutrients by means of lateral hairs.

The growth provides them the pressure (up to 1 MPa [52–54]) required for the forward advancement. Cap cells are continuously produced in the meristematic region, while the dead cells are released to the environment. These cells slough off from the root cap while producing a lubricant called mucus [55]. The combination of dead cells and mucus creates an interface between the soil and the root apex, which protects the new production of cells in the meristematic region, while promoting root penetration by reducing friction [55–58]. The mature region of the root is strongly anchored to the soil [59], thus allowing the apex to move forward. This anchorage is achieved by root hairs, secondary roots, and with complex root architecture [59–61]. Stronger root–soil adhesion may enable the root to penetrate harder soils [62]. The mechanical impedance of the soil strongly influences the root growth rate and shapes the apex morphology and the overall root architecture [63, 64]. Generally, when a root encounters a higher mechanically impeding soil, the elongation rate decreases and the apex diameter increases [65], and the path of the root follows a decreasing gradient of impedance manifested by the root tortuosity [66]. Additionally, it has been speculated that circumnutation movements, which are circular movements induced by an unequal growing rate at the opposite sides of the plant [67], might be another mechanism adopted by plant roots to facilitate soil penetration [68]. Experiments with *Oryza sativa* L. associated a good establishment of seeds with the rotation at high frequencies of the root tips in paddy fields, which was defining a spiral growth of the root [49].

All these mechanisms allow the root to penetrate in soil only with a small part of its structure, the apex or tip, while the rest of the structure (the mature part) remains stationary with the soil. Differently, in climbing plants, the meristematic and elongation regions are much wider, and movements actuated in shoots and roots have different purposes and dynamics, supported by different anchoring strategies. For instance, tropical climbers show a sensational range of attachment mechanisms and adhesive structures that vary from hooks, spines, tendrils, and twining stems. Diverse mechanisms correspond to different kinds of stem biomechanics, e.g. a quick attachment mechanism, like in stem twining, is usually linked with high flexible stems, whereas species that use adhesive structures like hooks usually develop only an highly flexible stem after an initial growth phase with an initially stiff "searcher" [69]. Many climbing plant species use exaggerated circumnutation movements of the stem branches and specialized organs like twining and winding attachment devices [70, 71]. This periodic oscillatory movement of the shoot around its main axis allows them to come into contact with and climb over neighboring objects. Without an external support, they eventually flop and may creep, continuing growth along the ground. Compared to most self-supporting plant stems, climbing plants have highly heterogeneous structures. Lianas, for instance, are capable of surviving extreme stresses and strains having evolved structural organization and material properties allowing them to survive extreme bending, shear, and torsion [72]. The natural functional biodiversity of climbing mechanisms thus offers a rich source of highly adaptive mechanisms that can be selected and tested for technological applications. Stiffness, flexibility, and toughness are key characteristics of plant materials, which develop in response to different mechanical perturbations.

As another mechanism to allow growth and guided movement, shoots and roots have tropic behavior, a mechanism that allows them to steer their growth toward attractive stimuli and away from repulsive stimuli. For instance, roots are the organs delegate for foraging following increasing moisture gradients [73, 74] or chemicals [75]. Responses to a stimulus may differ between roots and shoots. For example gravity is attractive in roots and repulsive in shoots whereas light is attractive for shoots and repulsive in roots [76]. Tropisms and circumnutations interact each other and are integrated to lead a unified directional response [76] that shapes the root morphology [77] and guide the plant growth.

These features of plants, and particularly the movements enabled by the growth combined with plant structural properties, perception, and decision-making strategies, represent the key sources of inspiration for the next generation of growing robots.

15.4 Plant-Inspired Growing Mechanisms for Robotics

15.4.1 Challenges in Underground Exploration

Soil contains the greatest part of resources useful for humanity and other living creatures of the globe: it is a rich source of water, nutrients, and minerals, and its monitoring is therefore of fundamental importance. Exploring the soil with autonomous technologies is a challenge, but their use could open new horizons for a more precise and eco-sustainable environmental monitoring, agriculture activity, space exploration, mining, or for rescue tasks after accidents or natural disasters. Commonly, soil monitoring is performed by sensorized probes pushed from the top [78–80], with integrated drilling systems [81, 82]. This approach requires big infrastructures and energy sources; and it could require the use of lubricants or fluids to remove soil particles, dirt, and to reduce heating at the tip of the probe induced by drilling. These drawbacks limit the monitoring activity and increase the accumulation of waste (e.g. in agriculture or in space missions). Moreover, these tools normally penetrate straight or with very large curvature radius, making difficult to follow gradient in soil (e.g. for source tracking). The requirement for a higher dexterity and smaller-scale solutions in underground exploration has led to the study of biologically inspired solutions [83–86]. The inspiration in these works comes from burrowing and drilling animals, which exploit their muscular movements to penetrate substrates. Among the various bioinspired models, one of the most efficient in moving under a variety of soils and conditions is the plant root. Plant roots have conquered almost any area on our planet and have the unique feature of being able to reduce penetration resistance, as already discussed in Section 15.3, by growing from their apical region [87, 88]. Because of this, they are able to shape their form during penetration and to adapt to the external framework, and are excellent new models for the development of soil-explorative autonomous robotic solutions.

15.4.2 The "Evolution" of Plantoids

First pioneering studies on the realization of bioinspired artificial roots were reported in 2008. In these works, the focus was in the implementation of an innovative actuator based on electro-osmotic principle for the realization of the elongation and bending regions (still the concept of growing was missing) [41, 89, 90]. By osmosis, it was possible to generate high pressure resulting in high elongation and bending forces, with low power consumption and in the plant timescale.

Later, in 2013, inspired by root-sloughing-cells process and elongation from the tip concept, a new root design was presented [11]. The proposed device translates at the same time the cells elongation and outward flow of sloughing cells in plant root by means of skin eversion at its artificial tip, enabling a more efficient penetration into the soil with respect to the same artificial probe pushed from the top. The main drawback of the proposed solution was the tissue flexibility, used as interface between internal probe and soil, which does not create a solid structure permitting the soil pressure to act on the inner shaft during penetration (problem that becomes more important with increasing depth).

To overcome this limit, a new artificial solution based on the root-growing strategy was realized [12]. Similar to natural roots, the proposed device was able to penetrate the medium imitating the growth of new cells by adding layers of artificial material at its tip level. This growth process provides to the tip of the root the capability to overcome the soil pressure while penetrating the soil. The device that performs the material deposition process (e.g. the growing mechanism) creates a tubular structure while moving down pushed by the new added layer. The developed structure, which forms the body of the robot, does not move with respect to the soil and only the tip performs the penetration. The limit of this solution was the lack of a continuum structure to improve the maximum growth length and the ability of the robot to bend.

In parallel, another approach was proposed to study the bending in an artificial root [13]. A soft bending mechanism based on three springs assembled at 120° was realized. Here the springs could simulate the cells elongation by generating and transmitting linear motions to a sensorized tip. Thanks to the flexibility of the springs and their arrangement and differential elongation, it was possible to obtain the bending of the root and emulate the real root tropic responses. This prototype allowed to deepen the study of root-adaptive behaviors, even though it could not enable long growth, in fact the springs could allow just few centimeters of extension.

A solution to combine growing and bending capabilities was proposed recently in [14]. This plant root-like robot creates its body structure through additive manufacturing techniques. Specifically, the robotic root is composed of a growing body, a growing mechanism, and a tip with sensors for environmental perception and an embedded control to imitate plant-like behaviors. The growing mechanism is a customized 3D printer-like system able to build tubular hollow structures by depositing circular layers of fused thermoplastic material at the tip level. By applying a differential deposition of the material, it is possible to create an asymmetry in the structure that results in a bending of the body. With this implementation it is possible to achieve both straight and curvilinear penetration. The developed structure is stiff and anchors to the soil acting as support for further penetration of the tip. In theory, this system can grow until material is supplied.

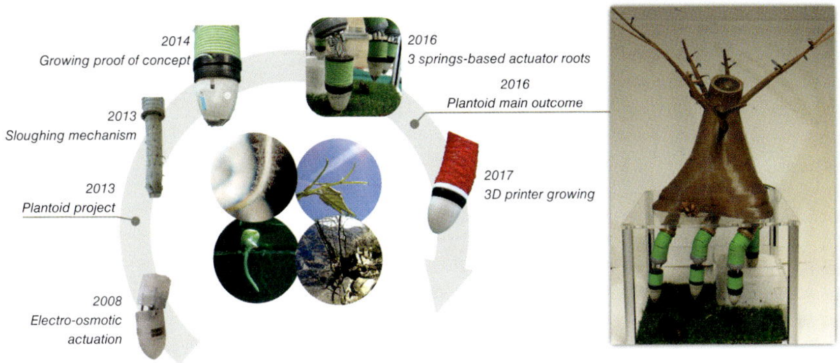

Figure 15.2 Plantoid evolution. The first studies on plant-inspired technologies were reported in 2008. The investigations lead to an evolution of several prototypes whose main milestones are represented in the figure. The Plantoid project (EU-FP7-FETOpen grant no. 29343) started in 2013 and concluded in 2016 presenting a prototype integrating several technologies developed during the project (right side of the picture) [13].

The timely evolution and the characteristics of each prototype are schematically summarized in Figure 15.2, whereas in the following paragraphs each element is better described.

15.4.3 Sloughing Mechanism

This mechanism is inspired by the natural concept of root cap sloughing cells and elongation from the tip. In natural plants, death cells are released by the root cap and create an easier path for the root to penetrate. At the same time, the root grows and elongates only from the apical level, keeping its body anchored to the soil. A similar mechanism has been proposed in [11], in which a soft skin works as a cylindrical track. The skin is released in the soil by a mechanism composed mainly by a textile skin and a cylindrical hollow shaft that pushes the skin in the soil. Initially the skin is located inside the shaft then, during the penetration induced by the forward movement of the shaft, it is released from the hole at the tip toward the external surface of the shaft (Figure 15.3). This outward movement creates a channel anchored to the soil and exploited by the shaft itself to further penetrate. The skin is also endowed with lateral hairs that helps in anchoring and remove soil particles in front of the tip.

The resulting movement is only an outward releasing of the skin at the tip level while the old part is stationary with respect to the soil. The result is a lower friction path that helps in soil penetration reducing the total energy needs [11].

15.4.4 First Growing Mechanism

The growing mechanism in [12] can be considered as a miniature burrowing system that creates the tunnel structure while it moves forward by growing. This system has a cylindrical body that consists of a rotating deposition head that moves new material from the top to the growing zone and shapes a tubular structure out

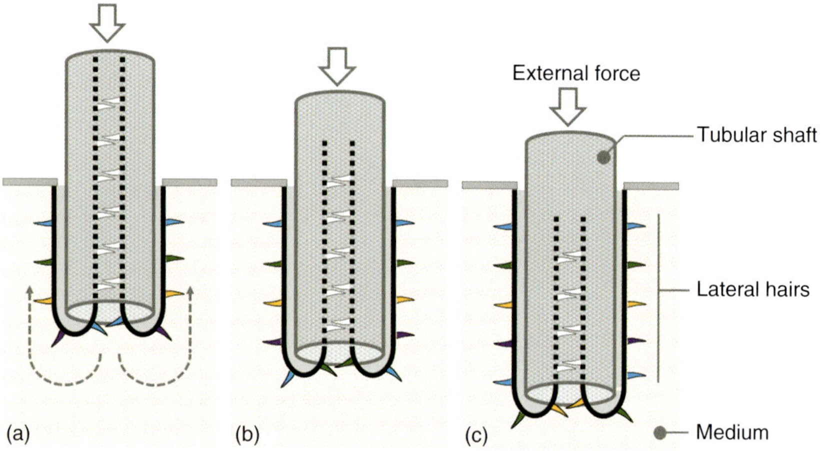

Figure 15.3 Graphical representation of the sloughing mechanism. The sequence (a–c) shows successive steps of penetration in soil. The tubular internal shaft is pushed into soil; during this pushing the skin is everted and anchored to the soil by means of lateral hairs. The skin creates an interface between the soil and internal shaft, which reduces lateral friction.

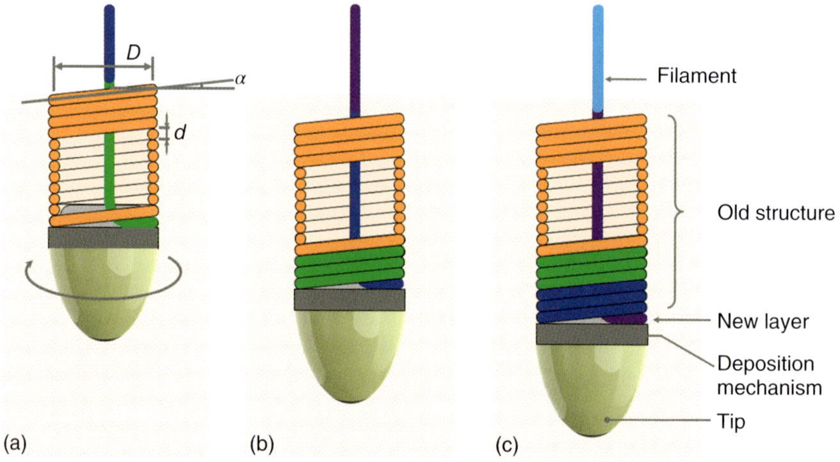

Figure 15.4 Schematic representation of the growing mechanism. The sequence (a–c) shows successive addition of new layers of material at the apical level. Deposition of a new layer creates the force required for the penetration of the tip into the soil.

of its body. The body, fixed with the soil, supports the further construction of the body and the axial movement of the deposition head into the soil (Figure 15.4).

The rotary motion of the deposition head, converted into a linear motion at the tip, provides the force for penetrating soil. The force generated by this screw-like mechanism produces an ideal mechanical advantage:

$$MA_{\text{ideal}} = \frac{\pi D}{d} = \tan^{-1}\alpha \tag{15.1}$$

Assuming that the diameter of the filament (d) remains constant during the growth process (i.e. one deposition cycle of the tip penetrates by a distance equal to the filament diameter), D is the mean diameter of the tubular structure and α the lead angle.

Due to the internal friction, in particular the one between filament and nozzle of the deposition head during extrusion, the torque (T) that should be applied to overcome the soil pressure can be calculated as:

$$T = \frac{\mathrm{FD}}{2} \tan(\alpha + \gamma), \tag{15.2}$$

where γ is the angle of friction and collects all the frictions resulting from the deposition mechanism.

Thanks to the addition of material at the tip level, this mechanism facilitates the soil penetration by omitting peripheral friction and therefore reducing the energy consumption (up to 70%), in comparison with a penetration obtained by pushing into the soil the whole root from the base.

15.4.5 Artificial Roots with Soft Spring-Based Actuators

Natural roots bend and follow stimuli by means of differential elongations of the cells at the apical level, a mechanism that improves the penetration capabilities because it decreases the lateral friction.

To achieve the same functionalities, a novel soft spring-based actuator (SSBA) system has been proposed in [13], making use of helical springs to transmit the motor power in a compliant manner. This novel pan-tilt mechanism is able to bend in each direction, by modulating the length of the springs that connect two planes, and at the same time is flexible and compliant thanks to the use of soft springs.

The mechanism is based on three helical springs (Figure 15.5), located at 120° to each other, used as screws in a nut–screw mechanism that translates the rotational motion in a linear displacement. This soft linear actuator can transmit the motion even after buckling (bending of the spring) and thanks to its compliance, a differential elongation between the three springs results in a bending of the system. Each SSBA can elongate and retract by changing the rotation of the motor connected to it. The robotic root elongates straightly when all three

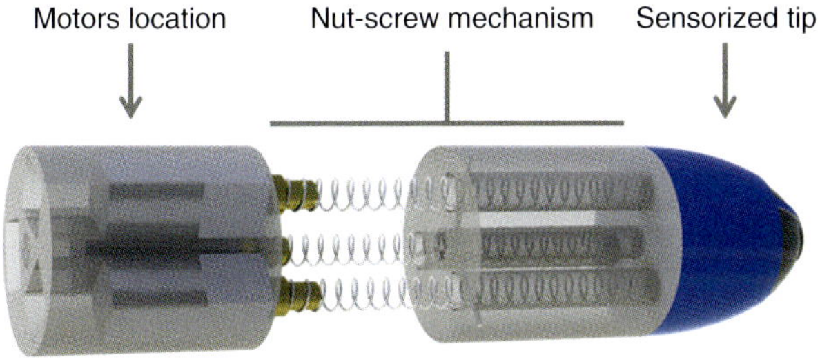

Figure 15.5 Schematic representation of the three SSBAs. The robotic root is composed of a module with the three motors driving the springs, a three-shaft module where the springs slide in and a sensorized tip for controlling behavior.

SSBAs elongate at the same speed, and performs the bending when the elongation velocity is different (differential elongation). Each complete rotation of the spring causes an elongation equal to the pitch (P) of the helix in the spring and, in particular, the SSBA length can be calculated as:

$$l = l_{\mathrm{n}} + \Delta l = l_{\mathrm{n}} + \frac{\beta}{2\pi}P \tag{15.3}$$

where β is the angle of rotation of the spring, l_{n} and Δl are the initial length and elongation of each SSBA respectively, and l the resulting whole length.

As in Section 15.4.2, SSBA can be considered a lead screw and thus its angle can be calculated as:

$$\alpha = \tan^{-1}\frac{2\pi r}{P} \tag{15.4}$$

where r is the radius of the spring. The linear force (F) resulting from the rotation is:

$$F = \frac{T}{r\dfrac{\mu \tan\alpha + 1}{\tan\alpha - \mu}} = \frac{T}{r\dfrac{\mu 2\pi r + P}{2\pi r - \mu P}} \tag{15.5}$$

where T is the applied torque and μ the friction coefficient of the spring with the nut.

This describes an approximation of the real spring behavior, since for the spring the pitch is not constant: under compression the pitch decreases and under pulling increases. This variation affects the force that the actuator can contrast: it increases when the spring pushes (lower pitch) and it increases in the pulling case. However, Eq. (15.5) is applicable if no buckling occurs.

The root bending can be described in terms of arc parameters, namely curvature (k), plane (φ), and arc length (l) by [91]:

$$l = \frac{l_1 + l_2 + l_3}{3} \tag{15.6}$$

$$\varphi = \tan^{-1}\left(\frac{\sqrt{3}(l_2 + l_3 - 2l_1)}{3(l_2 - l_3)}\right) \tag{15.7}$$

$$k = \frac{\sqrt{l_1^2 + l_2^2 + l_3^2 - l_1 l_2 - l_1 l_3 - l_2 l_3}}{d(l_1 + l_2 + l_3)} \tag{15.8}$$

where l_i are the lengths of SSBAs and d is the distance of each actuator from the body center.

This model works if neither compression nor elongation occurs in each spring. This is nearly true in air, where only the weight of the tip causes a small deviation between the theoretical position and the real position. This can permit controlling the bending mechanism in a 3D space for manipulation tasks. The same is not true in soil, even in a relative loose medium. However, in order to

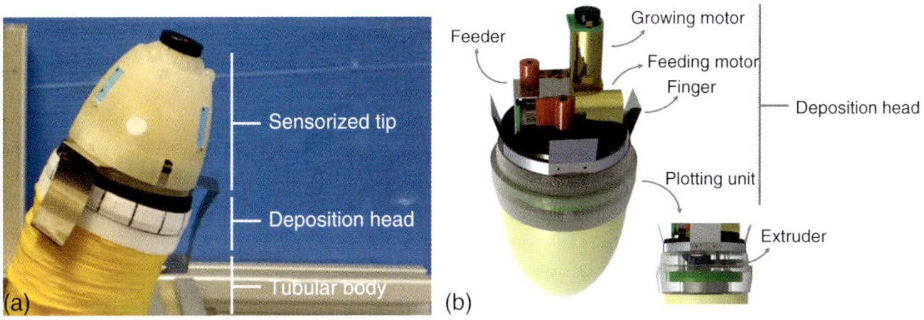

Figure 15.6 (a) Closed view of the growing robot, with main components and (b) inside view of the deposition head reproducing the 3D printer-like mechanism.

control a robotic root in a subsoil application where environment is unknown, the implementation of a plant root-inspired behavior looks more suitable (see Section 15.5). Such robotic mechanism allows the implementation and testing of plant-inspired behaviors.

15.4.6 Growing Robots via Embedded 3D Printing

Taking inspiration from the ability of plants to adapt their shape and structure while growing, in [14] an onboard 3D printing mechanism for robotic roots has been investigated. The system is designed by exploiting the classic approach of FDM and integrates a customized 3D printer inside the artificial root to extrude thermoplastic material on purpose.

The growing system is composed by a tubular body, a growing head, and a sensorized tip that controls the robot behaviors (Figure 15.6a). The principle of the growing head is similar to the one described in Section 15.4.4, but in this case the 3D printer-like system is composed of an extruder and a plotting unit (Figure 15.6b). The extruder unit includes a feeder mechanism that pulls a filament from outside, and then pushes it through a guiding tube toward a heater, which in turn externally fuses the filament that can be extruded from a nozzle. The plotting unit provides a rotational motion to the growing head generating a circular deposition of the extruded material. The growing head is then interfaced with the root tubular body through four flexible clamping fingers installed on its circumferential sides to prevent the rotation of the growing mechanism inside the printed tubular body. Melting of the material combined with the pressure generated by the soil on the deposited layers guarantee the realization of a solid structure. The whole system is controlled by three parameters: feeding speed, plotting speed, and heater temperature. By regulating the two speeds it is possible to generate straight growing and bending in each direction with different curvatures. By tuning the heater temperature, an optimal structure strength is obtained in both air and soil.

The material plays an important role for an effective penetration of the soil. It needs to be externally sticky enough to permit the layers to attach, and at the same time needs to be semisolid in the internal part, in order to be strong enough to overcome soil pressure and push the tip ahead.

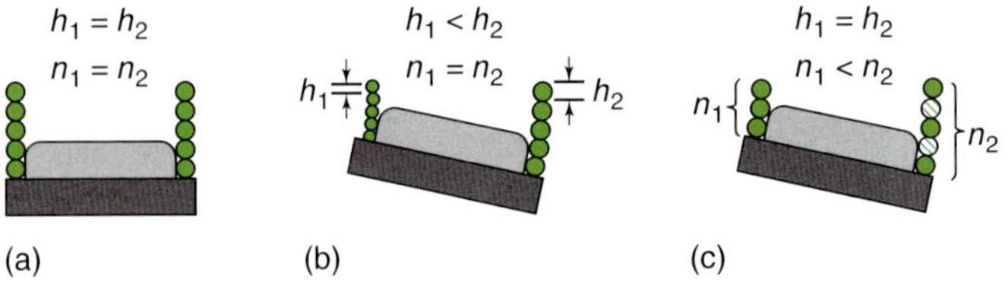

Figure 15.7 Possible strategies to grow straight or bend: (a) a symmetric deposition of material results in a straight growth, continuous deposition; (b) a bending obtained by the deposition of the same amount of layers with different thickness; and (c) a bending obtained by the deposition of a different amount of layers with the same height, reversing deposition.

15.4.6.1 Deposition Strategies

When the material is deposited uniformly along the plotting circumference, the resulting structure is straight (Figure 15.7a). Adding a different amount of material on opposite sides of the robotic root structure creates an asymmetry that results in a bending in the lower deposition direction. In particular, the bending can be obtained by two methods:

(A) *Continuous deposition*: A deposition of the same number of layers k with different thicknesses h_1 and h_2 (Figure 15.7b)
(B) *Reversing deposition*: A deposition of different number of layers n and m with the same thickness h (Figure 15.7c).

Continuous Deposition (Method A) With this deposition strategy, the direction of the rotation is never inverted and the bending is obtained by tuning the layer height along the deposition circumference. A curvature (Figure 15.8) can be obtained by setting properly the layer thickness h at a generic point P_α along the circumference (with α angle of P with x-axis):

$$P_\alpha = \left(\frac{d}{2}\cos\alpha, \frac{d}{2}\sin\alpha\right) \tag{15.9}$$

Considering θ the growing direction with respect to the x-axis, in that position the deposition should be minimum (height h_1) and in the opposite position $(\theta + \pi)$

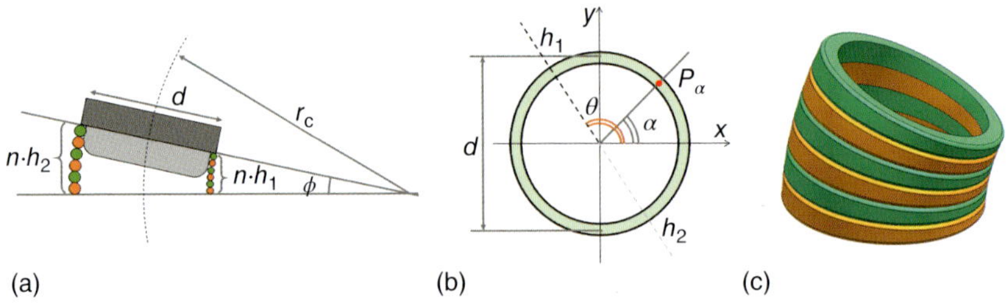

Figure 15.8 Overview of the bending parameters for method A: (a) side view with typical curvature achieved by a deposition of an equal number of layers (n) with different thickness (h_1, h_2); (b) top view of the plotting area depicting relevant parameters for the model of the deposition; and (c) typical resulting structure.

the deposition should be maximum (height h_2). The height of the layer can be calculated in a generic point P_α as a function of the angle α:

$$h_\alpha = \frac{h_1 - h_2}{d}\left(\frac{d}{2}\cos(\alpha - \theta) - \frac{d}{2}\right) + h_1 \tag{15.10}$$

To obtain this height variation, a control on the velocity of plotting and feeding of the deposition mechanism is needed. However, the change in the feeding speed is challenging to be achieved due to the need of reaching the optimal parameters for melting the thermoplastic filament. In fact, the extruder temperature needs time to change and consequently is not able to manage a fast variation of the speed for maintaining the correct extrusion temperature. At the end, only the plotting speed is modulated keeping the input volumetric flow rate ($mm^3\ s^{-1}$) of material (Q) constant.

The plotting speed ($rad\ s^{-1}$) at each α can be expressed as:

$$v_\alpha = \frac{Q}{h_\alpha \cdot w_\alpha}\frac{2}{d} \tag{15.11}$$

where w_α is the width (mm) of the deposited filament in P_α, and can be approximated [92] with:

$$w = 2.5\,h - 0.25 \tag{15.12}$$

The growing step will result in:

$$\text{Grow} = n\frac{h_2 + h_1}{2} \tag{15.13}$$

with a curvature radius of:

$$r_c = \frac{d}{2}\frac{h_2 + h_1}{h_2 - h_1} = \frac{d}{2}\frac{\frac{h_2}{h_1} + 1}{\frac{h_2}{h_1} - 1} \tag{15.14}$$

and a bending angle of:

$$\phi = \frac{\text{Grow}}{r_c} = n\frac{h_2 - h_1}{d} \tag{15.15}$$

Reversing Deposition (Method B) With this strategy, the reversion of the plotting direction is used to deposit a different number of layers on the two sides of the bending plane. This process is characterized by a sequence of steps needed to achieve the required amount of the material along the circumference in order to obtain the bending in the desired growing direction (θ).

Reversion of motion requires (i) to assure that there is an empty gap in front of the extrusion nozzle before inverting the motion to avoid deposition internally to the structure instead of versus it and (ii) to cool down the old material to preserve its shape and position. To address these points, after finishing the deposition in the previous direction, it is necessary to keep the plotting mechanism rotating for a few degrees without feeding. And then to stop for a few seconds also the

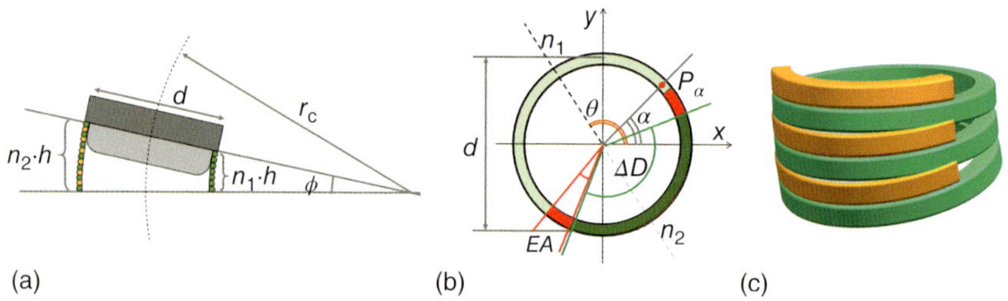

Figure 15.9 Overview of the bending parameters for method B: (a) side view with typical curvature achieved by a deposition of a different amount of layers (n_1, n_2) with equal thickness (h); (b) top view of the plotting area depicting relevant angles and sectors for the deposition process; and (c) typical resulting structure.

plotting to guarantee the cooling and fixing of the previous layer, before restarting deposition in the reverse direction.

A combination of complete layers with half deposited layers generates an asymmetry that results in bending of the structure.

The relevant parameters for the growing task are (Figure 15.9) the bending direction (θ), which is the central angle of the sector without material; the amplitude of the sector with material named the differential deposition sector (ΔD) that has the center in the direction opposite to the bending angle ($\theta + 180°$); the angle before the inversion process between the stop of the feeding and the stop of the plotting, named the extra angle (EA); and the deposition ratio (K), the number of full deposition circle between a single differential deposition.

The differential deposition task can be summarized in three steps:

(1) Plotting and feeding motor are turned on (Figure 15.10a). The thermoplastic material is extruded and deposited on the old structure, building the body and pushing the tip ahead;
(2) When the position of the extruder goes out from the differential deposition sector, the feeding motor is stopped (Figure 15.10b) while the plotting rotation continues for an extra EA angle before the plotting motor is stopped (Figure 15.10c);
(3) After a few seconds (three to five seconds), necessary for cooling the old layer, the plotting direction is reversed and restarted together with feeding (Figure 15.10d). The filament rises above the old layer (Figure 15.10e) and K complete layers are deposited until the next inversion (Figure 15.10f).

The growing step will result:

$$\text{Grow} = h\frac{n_1 + n_2}{2} \tag{15.16}$$

with a curvature radius of:

$$r_c = \frac{d}{2} \cdot \frac{n_1 + n_2}{n_2 - n_1} = \frac{d}{2} \cdot \frac{\frac{n_2}{n_1} + 1}{\frac{n_2}{n_1} - 1} \tag{15.17}$$

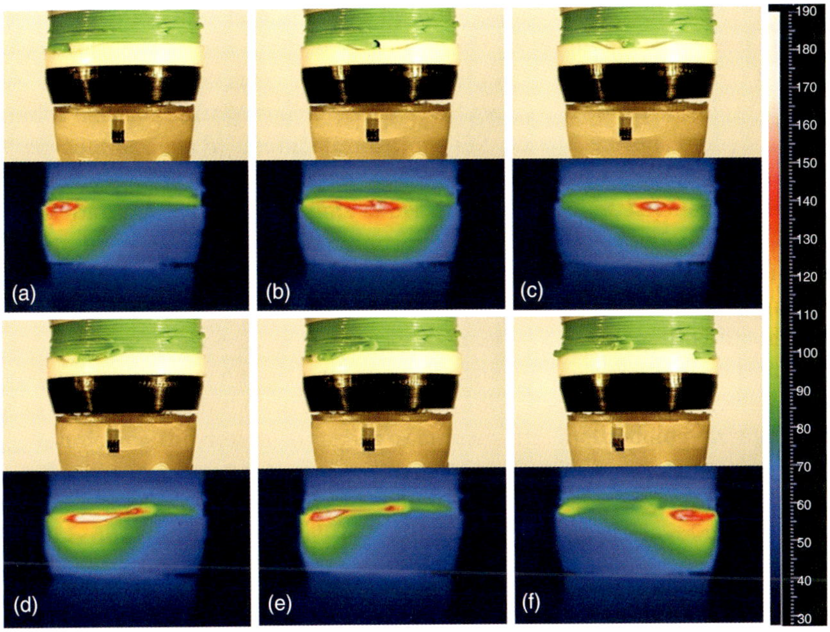

Figure 15.10 Deposition phases of method B with relative thermographic camera images: (a) the filament is deposited toward the right; (b) the feeding is stopped while plotting continues for an EA angle to remove the material in front of the nozzle; (c) the plotting is stopped for three to five seconds to cool down the old layer; and (d–f) the process restarts in the opposite direction.

and a bending angle of:

$$\phi = \frac{\text{Grow}}{r_c} = h\frac{n_2 - n_1}{d} \tag{15.18}$$

15.5 Adaptive Strategies in Plant for Robot Behavior

The previous reported growing robot mechanism (Section 15.4.6) is a non-holonomic system, having five degree of freedom in configuration space (head and pitch angles, and position in 3D space), but only three degrees of maneuverability (two degrees of steerability, which are the angles defining growth orientation, and one degree of mobility, that is the growth velocity). Therefore, it can be compared with mobile robots moving in 3D space.

The control system and behavioral architecture of growing robots depend on the specific task to be accomplished, and define their adaptation capability. The robot can be teleoperated using an external or embedded vision, or it can be controlled using a plant-inspired autonomous behavior, based on tropisms or circumnutation movements, or can use some other control systems to identify the target to reach, on the base of information of the surrounding coming from its own perception (e.g. vision, tactile, proximity sensors). In any case, a good knowledge of the kinematics of the robot is fundamental to understand and control its movement.

15.5.1 A Plant-Inspired Kinematics Model

An introduction of the forward kinematics model, inspired by plant growth, describing the motion obtained by growing from the tip in 3D space through homogeneous transformation matrices is reported in [93]. This model can be generalized and adopted for describing the motion of any growing robot, since the kinematics of the growing from the tip in plants is confined to the description of the motion of the tip in space. If i is a moving coordinate frame, solid with the robot's tip, and j the inertial frame (Figure 15.11), the origin of coordinate frame i relative to coordinate frame j can be denoted by:

$$ {}^{j}p_i = \begin{pmatrix} {}^{j}x_i \\ {}^{j}y_i \\ {}^{j}z_i \end{pmatrix} \tag{15.19} $$

A generic point ${}^{i}r \in \mathbb{R}^3$ in frame i can be expressed in frame j as ${}^{j}r \in \mathbb{R}^3$ knowing the transformation matrix ${}^{j}R_i \in \mathrm{SO}(3)$, with the equation:

$$ \begin{pmatrix} {}^{j}r \\ 1 \end{pmatrix} = \begin{pmatrix} {}^{j}R_i & {}^{j}p_i \\ 0_{1\times 3} & 1 \end{pmatrix} \begin{pmatrix} {}^{i}r \\ 1 \end{pmatrix}, \tag{15.20} $$

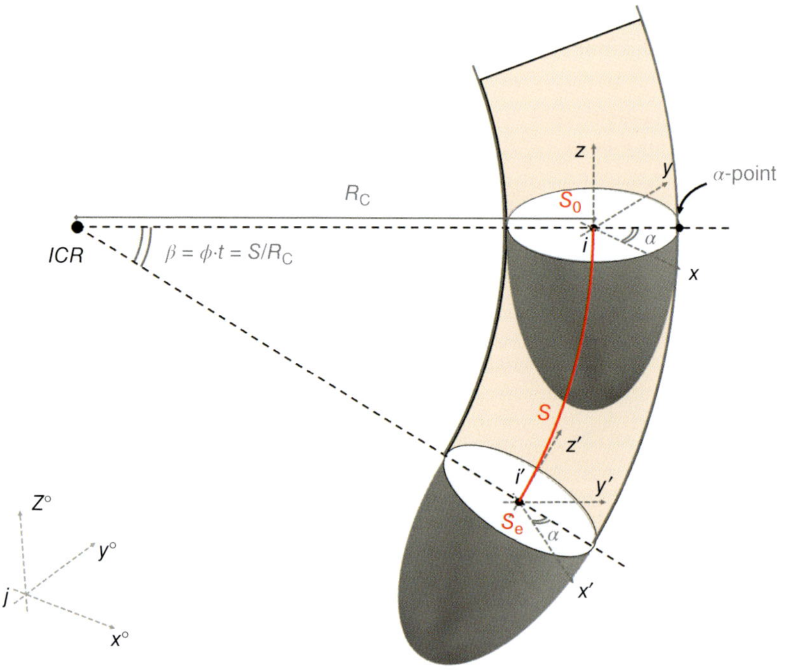

Figure 15.11 Schematic of growing robot moving in space. The tip is represented in an initial position s_0 and moved in a final position s_e for a displacement S. Local frame i (solid with the tip) and the inertial frame j are represented, as well as the key parameters for a turn.

where the first factor of the right hand is the homogenous transformation matrix ${}^{j}T_{i} \in SO(4)$. The forward kinematics of a growing robot can be described by following the tip position in space.

From a frame i, the next frame is obtained as a function of the growth velocity (g), position for the actuation of greatest material deposition (expressed as angle α w.r.t. the x-axis in frame i) and intensity of bending in a unit of time (ϕ). With these actuation parameters, frame i can be obtained with the sequence of post-multiplied transformations:

$$
{}^{i}T = T_{z,\alpha} T_{\mathrm{tr},v} T_{y,\phi} T_{\mathrm{tr},-v} T_{z,-\alpha} =
$$

$$
\begin{pmatrix}
C_\alpha{}^2 C_\phi + S_\alpha{}^2 & C_\alpha S_\alpha C_\delta - C_\alpha S_\alpha & C_\alpha S_\phi & \dfrac{gC_\alpha(1 - C_\phi)}{\phi} \\
C_\alpha C_\phi S_\alpha - C_\alpha S_\alpha & C_\alpha{}^2 + C_\phi S_\alpha{}^2 & S_\alpha S_\phi & \dfrac{gC_\alpha(1 - S_\phi)}{\phi} \\
-C_\alpha S_\phi & -S_\alpha S_\phi & C_\phi & \dfrac{gS_\phi}{\phi} \\
0 & 0 & 0 & 1
\end{pmatrix}, \tag{15.21}
$$

where the first subscript of T indicates if T is a translation (tr) or a rotation matrix (by indicating around which axis), and the second subscript gives the angle of rotation or direction of translation (v); and by convention: $C_\alpha = \cos\alpha$ and $S_\alpha = \sin\alpha$. In Eq. (15.21), the greatest deposition is applied at the α-point along the circumference of the robot's tip, with respect to its x–y plane (Figure 15.11); the rotation $T_{z,\alpha}$ is used to localize this point and $T_{z,-\alpha}$ is used to rotate back the tip after the transformations $T_{\mathrm{tr},v} T_{y,\phi} T_{\mathrm{tr},-v}$ that are used to identify the center of rotation in space (v is the vector $[-R_c\ 0\ 0\ 1]^{\mathrm{T}}$) and then apply the rotation of ϕ about the y-axis for obtaining the bending (Figure 15.11). Since R_c can be expressed as a relation between the intensity of bending and growth velocity ($R_c = g/\phi$), the matrix in Eq. (15.21) is obtained by substitution. When the root goes straight $\phi = 0$, then $\alpha = 0$ and the last column defines a straight growth of g along the z-axis of the root. The kinematic chain, which describes the movement of the tip from an initial configuration s_0 to a final configuration s_e, is obtained by consecutive multiplication of the above homogeneous transformation matrices:

$$
{}^{j}T_{i}(s_e) = \prod_{s_0}^{s_e} {}^{i}T(s_t). \tag{15.22}
$$

As for mobile robots, also for a growing robot it is possible to identify a feasible path from s_0 to s_e composed of a sequence of turns and straight lines [94, 95]. For each segment there is the corresponding transformation matrix ${}^{i}T(s_t)$, and the sequence of those matrices describes the kinematic chain for the growing robot from s_0 to s_e.

The pose of the robot at the curvature point s_0 can be represented by its position (x_s, y_s, z_s), pitch γ_s, and heading θ_s angles. The starting pose is then identified with $X_s = \langle x_s, y_s, z_s, \theta_s, \gamma_s \rangle$. Analogously for final pose $X_e = \langle x_e, y_e, z_e, \theta_e, \gamma_e \rangle$. The problem of finding a feasible path in 3D between X_s and X_e can be reduced to the problem of finding two optimal paths in 2D [96]. Two planes (one where X_s lies and the other where X_e lies) should be identified and must intersect each other. An intermediate target pose X_t can then be localized on the intersecting line of the two planes, thus, when the robot reaches this intermediate target pose, it can easily pass from the first to the second plane just by changing the position of material deposition. The first optimal problem is then to find the shortest path between X_s and X_t, while the second problem is to find the optimal path between X_t and X_e.

It should also be considered that the maneuverability space in growing robots can be limited by mechanical constraints, which provides for a minimum curvature radius R_{min} that the robot can perform. Such R_{min} should be taken into account when looking for the shortest path. Resolution of shortest path problems with curvature constraints are well explored in literature [94], and the proposed resolution approach allowes finding a suboptimal path for growing robots in 3D. An example of path identified between two poses in space is reported in Figure 15.12, together with the final robot configuration.

15.5.2 Plant-Inspired Behavioral Control

As reported in Section 15.3, plants show different behaviors that help them to explore efficiently the environment for finding nutrients and water, for anchoring and in general for improving the wellness of the whole plant. In particular, they show the ability to follow or escape from external stimuli (tropisms) by sensing gradients through the receptors located in the root cap.

The apex is considered the center of commands for tropic responses and there are several integrated receptors for many different stimuli (e.g. humidity, temperature, touch, light, gravity, and chemicals for nutrients). Each stimulus has its own priority written in the plant DNA. In a simplified hypothesis where these priorities are constant and independent from the history of the plant, it is possible to define a stimulus-oriented algorithm able to follow the resultant of the applied stimuli [13]. The proposed algorithm has been implemented and tested in a robotic apex that embeds three sensors, positioned at 120° along the circumference, for each implemented stimulus (i.e. humidity, temperature, and tactile), plus a sensor for gravity. Thanks to this sensor configuration it is possible to measure a gradient or a stimulus in the 3D space [13].

Based on this concept, a robotic root can be configured with its starting DNA encoding information about the optimal environmental conditions as target values to reach. The interplay among tropisms is implemented through priorities associated with each stimulus, defined by this DNA. The phenotype, as for real plant (phenotypic plasticity [77]), is then defined by the interaction with the environment.

The bending direction is obtained by combining each tropic result (obtained as a resultant of the acquired sensors data along the apex) with the appropriate weight given by the priority. This procedure results in the following equation:

$$\overrightarrow{R(t)} = \sum \overrightarrow{S_i(t)}, \tag{15.23}$$

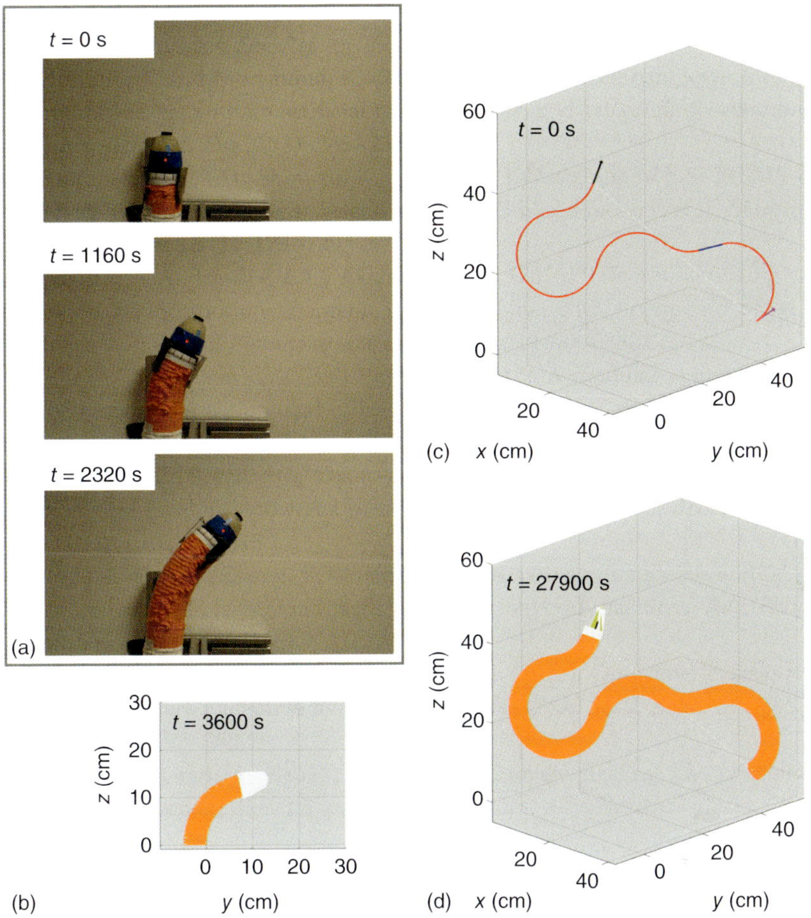

Figure 15.12 Example of (a) a time sequence for the curvature (of radius 12 cm) performed in air by the plant-inspired growing robot with (b) the corresponding final configuration obtained in simulation with the kinematics model. Example of (c) Dubin's path found over two intersecting planes from a starting pose (magenta arrow) to a final pose (black arrow). Curvilinear segments are in red while straight segments are in blue. (d) The final configuration achieved by the robot in simulation.

where $\overrightarrow{S_i(t)}$ is the resulting vector of each of the stimuli (e.g. humidity, temperature, gravity, or touch) at time t, obtained by the intensity of the stimulus weighted with its priority toward the perceiving direction. The root will move with intensity $\|\overrightarrow{R(t)}\|$ toward the corresponding direction obtained as angle from the reference axis on the tip.

The output of the algorithm can be used to drive the deposition of the material in the 3D printing-like mechanism, or in driving spring motor power in the SSBA-based robotic roots.

With this stimulus-oriented algorithm, each individual root only considers its instantaneous perception. However, the process of growth in plants is commanded also by a historical evolution of internal processes (e.g. nutrients uptake, photosynthesis, resources redistribution), which notably affects the direction of root's growth [77]. To take into account this historical evolution of plant behavior, a dynamicity of stimuli priorities should be introduced.

This can be done by taking into account eight basic properties extracted by the biological model and integrating them in an algorithm (uptake–kinetic control algorithm [97]) that adapts stimuli priorities, specifically nutrients, according to the root knowledge about the level of each nutrient.

The formalized properties are:

Property 1: Plant growth is principally driven by the need for balancing the internal nutrient concentrations rather than by collecting the closest and most available nutrient in soil;
Property 2: A feedback control modifies the uptake rate of a nutrient according to its internal state;
Property 3: Roots show directional responses toward or away from a stimulus;
Property 4: The directional response is induced by the perception of a gradient;
Property 5: Tropic responses are combined to obtain a single directional response;
Property 6: There is a fast and direct highway where nutrients are transported immediately from roots to shoot;
Property 7: Nutrients are distributed among organs according to the strength of their requests;
Property 8: Root tissues and cell vacuoles are local memories storing information on nutrient status.

While the previously developed stimulus-oriented algorithm was based only on properties 3–5, the uptake–kinetic control considers all the eight properties.

The basic steps of the control are:

1) Update of internal state and uptake–kinetic parameters;
2) Perception of the environment;
3) Uptake of nutrients;
4) Nutrients sent to shoot;
5) Evaluation of growing direction;
6) Growth.

Each root is an autonomous agent, taking decisions independently from the others, only on the base of its internal status and local environmental perception. The stronger or weaker attraction/repulsion to different nutrients is obtained by a continuous update of nutrients priority. The mechanism of priority update has been inspired by the uptake-kinetic mechanism in plants, which is described by the Michaelis–Menten equation, and defines the uptake rate, or absorption velocity, as:

$$I = \frac{I_{\mathrm{max}} \cdot C}{K_{\mathrm{m}} + C}, \tag{15.24}$$

with I_{max} identifying the capacity factor (maximal rate of absorption), which is approached asymptotically when the ion concentration in the medium increases; K_{m} represents the concentration with half of the maximal rate of absorption; and

C is the concentration perceived in soil. It has been shown that the parameters of the uptake kinetics (I_{max} and K_m) are strongly influenced by the internal concentration status of the plant. For instance, in both *Zea mays* (corn) and soybean, it has been shown that with an increasing concentration of phosphorus in plants, both parameters decrease linearly (I_{max} more rapidly than K_m) [98]. Thus, I_{max} is a value that dynamically changes for each nutrient and it can be an indicator of the priority for the nutrient:

$$P = \frac{1.1 \cdot I_{max}^2}{I_{max}^{+2}} - 0.1, \tag{15.25}$$

where the I_{max}^+ defines the maximum value that I_{max} can assume. Once resources are acquired by the root agent, they are redistributed among all the agents according to their available memory (imitating osmosis–transmission principle, Property 7).

Exploiting the proposed control on three robotic root agents, each root is able to regulate the growth direction, toward the nutrients that are most needed, and to adjust nutrient uptake, by decreasing the absorption rate of the most plentiful one. Results from computer simulations and implementation of the proposed control on the Plantoid robotic platform demonstrate an emergent swarming behavior aimed at optimizing the internal equilibrium among nutrients through the self-organization of the roots.

15.5.3 Circumnutation Movements in Natural and Artificial Roots

As already stated in Section 15.3, circumnutation is a circular movement in plant performed by both peripheral aerial parts and root tips during their growth [99]. This movement in roots has still an unclear role, but it seems that it has significant implications in the optimization of soil penetration [49, 68, 100]. Starting from this hypothesis, experimental studies have been carried out using artificial roots, based on SSBA actuators (Section 15.4.3), implementing circumnutation behavior in order to better understand the capability of such movement to facilitate soil exploration [42]. The study compares the axial forces necessary for pushing the robotic root from its top into an artificial soil in the case of straight penetration and circumnutation movements. Circumnutation behavior was integrated in the previously described stimulus-oriented behavior by adding an oscillatory stimulus ($\overrightarrow{S_{cn}(t)}$) with the possibility to modulate period and amplitude.

Experiments show up to 80 times less force when circumnutations have been used, with respect to a straight penetration. This demonstrates that such movements are convenient in terms of forces, inducing less stress on the penetration device and reaching higher depths compared to a straight probe movement with the same amount of externally applied force. However, in this work, the power needed to generate the probe tip's rotary motion was neglected.

To study the energy efficiency of root penetration while performing circumnutations, in [101] comparative experiments using a setup purposively developed were performed. The circumnutation has been simplified in a rotational movement around a linear one (the penetration direction). The axial forces applied at the probe tip and the torque were isolated to evaluate the total energy consumed by the system in different conditions (i.e. compactness level of the soil, circumnutation periods, and amplitudes). In the proposed setup the energy consumed by the artificial root during penetration can be calculated as:

$$W_{\text{Tot}} = W_{\text{V}} + W_{\text{R}} = \int (F_{\text{V}(t)} \cdot v + \tau_{(t)} \cdot \omega) \cdot \mathrm{d}t \tag{15.26}$$

where $F_{\text{V}(t)}$ and $\tau_{(t)}$ are the measured axial force and torque, and v and ω are the penetration and rotation speed. Changing the period of circumnutation (and consequently ω) it is possible to find optimal parameters for energy consumption with the soil under test. More precisely the key parameter is the lead angle of the resulting helix:

$$\varepsilon = \frac{P}{\pi d} = \frac{vT}{\pi d} \tag{15.27}$$

where P and d are the pitch and diameter of helix, and T is the circumnutation period. The main parameters for this movement are depicted in Figure 15.13a. As demonstrated by results in [101], decreasing much ε (many rotation for unit of penetration), the axial force F_{V} tends to became zero, because the soil in front of the tip is removed by the rotational motion, but the component of energy due to rotation increases with a consequent negative effect on the total energy consumed by the system. When instead ε increases a lot (almost not rotating), it is practically a straight penetration with no benefit obtained from circumnutation. In the middle of these boundary conditions there is a point of minimum for energy (Figure 15.13b). Results confirmed that combinations of period and amplitude of circumnutation exist giving an advantage in terms of energy needed to penetrate. The tests demonstrated a reduction up to 33% of work done by using circumnutations with respect to straight penetration, while the axial force at the tip level results were 78% less.

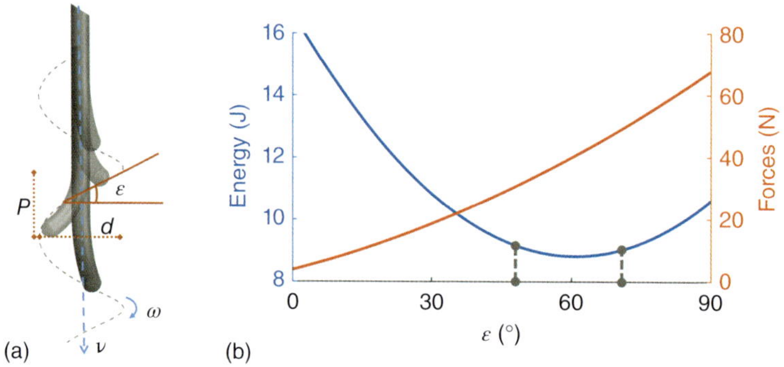

Figure 15.13 (a) Schematic representation of circumnutation movements with characteristic parameters: P represents pitch of the helix, d diameter of the helical path, ε lead angle, v penetration velocity, and ω angular velocity. (b) Graph of total energy and vertical forces obtained on average from the experiments in [101]. The total energy considers the work done by the system at steady state (when the critical depth where stabilization of forces is reached) from vertical and rotational movements. The forces plotted include only the averaged vertical force perceived at the tip level in the same condition of steady state. The graph shows the advantages of using circumnutation, especially in the range of ε between 46° and 65° (range between gray lines), where the energy shows a valley, while the vertical forces show an increasing behavior.

Experiments also show that this point of minimum lie in the range of ε between 46° and 65°. From previous biological experiments [42], circumnutations period in *Z. mays* primary roots were estimated between 60 and 80 minutes and amplitudes with a radius of 0.2 mm on average. Knowing that circumnutations occur in the central elongation zone [102] placed in maize at about 4 mm from the tip [87], a radius of 0.2 mm corresponds to ~3° of nutation amplitude. Considering a growing rate for *Z. mays* of 1.59 mm h^{-1} [42], its ε is ~52°, a value perfectly in the range of optimal ε as previously found.

The obtained results lead to the assumption that circumnutations may represent a key mechanism, for both plants and artificial probes, to generate cavity expansion and crack propagation in the soil plastic zone, i.e. the cylindrical zone surrounding the penetration cavity where stresses are sufficient to cause failure [103]. This investigation provides an explanation for circumnutations in plant roots and proposes the applicability of such movement for artificial soil-explorative probes.

15.6 Applications and Perspective

The sessile nature of plants leads us to think that they do not move, but instead they move greatly, on a different timescale compared to animals, purposively, effectively, and efficiently. Growing mechanism in plants has evolved as powerful strategy for their movement, both underground and above-ground. Plants continuously adapt to the external environment, evolving their shapes following diverse directions on the base of a series of stimuli and conditions. The structural functionality of the material of their bodies, the plasticity and perception, and the capability to move and resist also in high-perturbed ecosystems make them a perfect model of inspiration in the design, structural materials, and control strategies for the new class of *growing robots*.

There are several existing classes of robots able to move in different environments, like wheeled and legged robots for terrestrial applications and drones for air. However, each of them has some limitations: wheeled robots need normally quite structured terrain to move; legged robots can move in a wider range of terrains but the obstacles that they can negotiate depends on their body size; aerial robotic vehicles are the one that can negotiate voids, however, their abilities are limited by the available energy [104]. Instead, growing robots have ideally the possibility to reach and maintain every position in a 3D space while moving and adapting their bodies to the environment since they always maintain a contact with the base and they may exploit the built structure for energy and data transmission.

Systems based on plant behaviors can create their own structures by growing, with no predefined design. This ability makes them suitable to move in unpredictable, cluttered, and dangerous environments (e.g. searching after disasters, such as earthquakes, or localizing objects that are difficult to reach), to wrap around and create a support to existing structures (e.g. precarious buildings), to explore and detect resources, to monitor and explore various terrains, to investigate possible contamination or mineral deposits, or to assess the general status of the soil, on Earth or other planets. Agriculture, earth science, and space exploration are direct areas of application in this sense. The approach of using plant-inspired growing robots could represent another way to create pipes or to colonize areas for searching specific targets. Moving within tunnels, across debris, or across unstable constructions requires extreme sensibility and capability of adaptation in order to avoid additional risks of collapses or explosions. Underground tunneling, caves exploration, and mining represent also other sectors of application, in which small-scale robotic solution can improve the precision and energy-efficient methodology of operations. By using new biocompatible materials and deposition techniques, the medical area can also benefit from soft-growing investigation and operation tools that use their compliant and adaptable bodies to reach different parts of human body in a noninvasive manner.

In all these cases, the importance of the architectural characteristics and the functionality of the used materials play a fundamental role for an effective adaptive behavior and deployment of continuously growing structures. Structures made of soft matter, progressively added, and with multi-perceptive and -mechanical properties enable the soft robots to undergo elastic deformation at any point in their structures for producing structural movements and to adapt the shape on the base of the perception and task needs. In this case, studying biomechanical and functional properties in plant tissues (such as wood, bamboo, textile fibers, fruits, and vegetables), in relation with the plant ecological development, will help developing new materials and mechanical designs.

We still need to improve our knowledge on the functional principles and physical strategies used by plants to grow, explore the environment using their movements, anchor to external supports, change shape, and physically adapt to the surrounding environment. A deeper research is also needed to better understand their decision-making behaviors for the translation into design guidelines for engineers and roboticists. A cross-disciplinary effort is needed as methodological approach to address these challenges, bringing together science and engineering, increasing the scientific knowledge and its translation into effective and sustainable engineering solutions to materialize the huge impact promised by the field of plant-inspired robotics.

Acknowledgments

Authors are thankful to their colleague Dr. Laura Margheri for her support and suggestions during manuscript drafting.

References

1 Siciliano, B. (2008). *Springer Handbook of Robotics*. Springer.
2 Rus, D. and Tolley, M.T. (2015). Design, fabrication and control of soft robots. *Nature* 521 (7553): 467–475.
3 Trivedi, D., Rahn, C.D., Kier, W.M., and Walker, I.D. (2008). Soft robotics: biological inspiration, state of the art, and future research. *Appl. Bionics Biomech.* 5 (3): 99–117.

4 Kim, S., Laschi, C., and Trimmer, B. (2013). Soft robotics: a bioinspired evolution in robotics. *Trends Biotechnol.* 31 (5): 287–294.
5 Majidi, C. (2014). Soft robotics: a perspective – current trends and prospects for the future. *Soft Robotics* 1 (1): 5–11.
6 Laschi, C. and Cianchetti, M. (2014). Soft robotics: new perspectives for robot bodyware and control. *Front. Bioeng. Biotechnol.* 2 https://doi.org/10.3389/fbioe.2014.00003.
7 Pfeifer, R., Bongard, J., and Grand, S. (2007). *How the Body Shapes the Way We Think: A New View of Intelligence.* Cambridge, MA: MIT Press.
8 Paul, C. (2006). Morphological computation. *Robot. Auton. Syst.* 54 (8): 619–630.
9 Laschi, C. and e Mazzolai, B. (2016). Lessons from animals and plants: the symbiosis of morphological computation and soft robotics. *IEEE Robot. Autom. Mag.* 23 (3): 107–114.
10 Laschi, C., Mazzolai, B., and e Cianchetti, M. (2016). Soft robotics: technologies and systems pushing the boundaries of robot abilities. *Sci. Robot.* 1 (1): eaah3690.
11 Sadeghi, A., Tonazzini, A., Popova, L., Mazzolai, B. (2013) Robotic mechanism for soil penetration inspired by plant root. 2013 IEEE International Conference on Robotics and Automation, 6–10 May 2013, Karlsruhe, Germany, pp. 3457–3462.
12 Sadeghi, A., Tonazzini, A., Popova, L., and Mazzolai, B. (2014). A novel growing device inspired by plant root soil penetration behaviors. *PLoS One* 9 (2): e90139.
13 Sadeghi, A., Mondini, A., Dottore, E.D. et al. (2017). A plant-inspired robot with soft differential bending capabilities. *Bioinspir. Biomim.* 12 (1): 015001.
14 Sadeghi, A., Mondini, A., and Mazzolai, B. (2017). Toward self-growing soft robots inspired by plant roots and based on additive manufacturing technologies. *Soft Robotics* 4 (3): 211–223.
15 Wooten, M. and Walker, I. (2018). Vine-inspired continuum tendril robots and circumnutations. *Robotics* 7 (3): 58.
16 Soni, N.K. (2010). *Fundamentals of Botany.* Tata McGraw-Hill Education.
17 Moulia, B. (2013). Plant biomechanics and mechanobiology are convergent paths to flourishing interdisciplinary research. *J. Exp. Bot.* 64 (15): 4617–4633.
18 Sebens, K. (1987). The ecology of indeterminate growth in animals. *Annu. Rev. Ecol. Syst.* 18 (1): 371–407.
19 Martone, P.T., Boller, M., Burgert, I. et al. (2010). Mechanics without muscle: biomechanical inspiration from the plant world. *Integr. Comp. Biol.* 50 (5): 888–907.
20 Moulia, B. and Fournier, M. (2009). The power and control of gravitropic movements in plants: a biomechanical and systems biology view. *J. Exp. Bot.* 60 (2): 461–486.
21 Sultan, S.E. (2000). Phenotypic plasticity for plant development, function and life history. *Trends Plant Sci.* 5 (12): 537–542.

22 Del Dottore, E., Sadeghi, A., Mondini, A. et al. (2018). Toward growing robots: a historical evolution from cellular to plant-inspired robotics. *Front. Robot. AI*: 5.

23 Taccola, S., Greco, F., Sinibaldi, E. et al. (2015). Toward a new generation of electrically controllable hygromorphic soft actuators. *Adv. Mater.* 27 (10): 1668–1675.

24 Minko, S. (2006). *Responsive Polymer Materials: Design and Applications.* Wiley-Blackwell.

25 Stuart, M.A.C., Huck, W.T., Genzer, J. et al. (2010). Emerging applications of stimuli-responsive polymer materials. *Nat. Mater.* 9 (2): 101.

26 Feldmann, M., Waldschik, A., and Büttgenbach, S. (2007). Electromagnetic micro-actuators, micro-motors, and micro-robots. *Proc. SPIE* 6798 https://doi.org/10.1117/12.759482.

27 Yamada, A., Niikura, F., and Ikuta, K. (2008). A three-dimensional micro-fabrication system for biodegradable polymers with high resolution and biocompatibility. *J. Micromech. Microeng.* 18 (2): 025035.

28 Lebel, L.L., Aissa, B., Khakani, M.A.E., and Therriault, D. (2010). Ultraviolet-assisted direct-write fabrication of carbon nanotube/polymer nanocomposite microcoils. *Adv. Mater.* 22 (5): 592–596.

29 Guo, S., Gosselin, F., Guerin, N. et al. (2013). Solvent-cast three-dimensional printing of multifunctional microsystems. *Small* 9 (24): 4118–4122.

30 Gladman, A.S., Matsumoto, E.A., Nuzzo, R.G. et al. (2016). Biomimetic 4D printing. *Nat. Mater.* 15 (4): 413.

31 Ge, Q., Qi, H.J., and Dunn, M.L. (2013). Active materials by four-dimension printing. *Appl. Phys. Lett.* 103 (13): 131901.

32 Beyer, W.A., Sellers, P.H., and Waterman, M.S. (1985). Stanislaw M. Ulam's contributions to theoretical theory. *Lett. Math. Phys.* 10 (2–3): 231–242.

33 von Neumann, J. (1951). The general and logical theory of automata. In: *Cerebral Mechanisms in Behavior; the Hixon Symposium*, 1–41. Oxford: Wiley.

34 Fukuda, T., Nakagawa, S., Kawauchi, Y., and Buss, M. (1988). "Self organizing robots based on cell structures-CKBOT," in 1988 IEEE International Workshop on Intelligent Robots (Tokyo: IEEE), 145–150.

35 Lipson, H. and Pollack, J.B. (2000). Towards continuously reconfigurable self-designing robotics. In: *Proceedings 2000 ICRA. Millennium Conference. IEEE International Conference on Robotics and Automation. Symposia Proceedings (Cat. No. 00CH37065), 24–28 April 2000*, vol. 2, 1761–1766. San Francisco, USA: IEEE.

36 Sims, K. (1994). Evolving 3D morphology and behavior by competition. *Artif. Life* 1 (4): 353–372.

37 Marbach, D. and Ijspeert, A.J. (2004). Co-evolution of configuration and control for homogenous modular robots. In: *Proceedings of the Eighth Conference on Intelligent Autonomous Systems (IAS8)*, 712–719. IOS Press.

38 von Haller, B., Ijspeert, A., and Floreano, D. (2005). Co-evolution of structures and controllers for Neubot underwater modular robots. In: *Advances in Artificial Life. ECAL 2005. Lecture Notes in Computer Science*, vol. 3630

(eds. M.S. Capcarrère, A.A. Freitas, P.J. Bentley, et al.), 189–199. Berlin, Heidelberg: Springer.

39 Bongard, J. (2011). Morphological change in machines accelerates the evolution of robust behavior. *Proc. Natl. Acad. Sci.* 108 (4): 1234–1239.

40 Joachimczak, M. and Wróbel, B. (2012). Co-evolution of morphology and control of soft-bodied multicellular animats. In: *Proceedings of the 14th Annual Conference on Genetic and Evolutionary Computation* (ed. T. Soule), 561–568. New York, USA: ACM.

41 Mazzolai, B., Mondini, A., Corradi, P. et al. (2011). A miniaturized mechatronic system inspired by plant roots for soil exploration. *IEEE/ASME Trans. Mechat.* 16 (2): 201–212.

42 Del Dottore, E., Mondini, A., Sadeghi, A. et al. (2016). Circumnutations as a penetration strategy in a plant-root-inspired robot. *IEEE Xplore*: 4722–4728. https://doi.org/10.1109/ICRA.2016.7487673.

43 Joppa, L.N., Roberts, D.L., Myers, N., and Pimm, S.L. (2011). Biodiversity hotspots house most undiscovered plant species. *Proc. Natl. Acad. Sci.* 108 (32): 13171–13176.

44 Dumais, J. (2013). Modes of deformation of walled cells. *J. Exp. Bot.* 64 (15): 4681–4695.

45 Verbelen, J.-P., De Cnodder, T., Le, J. et al. (2006). The root apex of arabidopsis thaliana consists of four distinct zones of growth activities: meristematic zone, transition zone, fast elongation zone and growth terminating zone. *Plant Signal. Behav.* 1 (6): 296–304.

46 Gilroy, S. and Masson, P. (2008). *Plant Tropisms*. Wiley.

47 Arnaud, C., Bonnot, C., Desnos, T., and Nussaume, L. (2010). The root cap at the forefront. *C. R. Biol.* 333 (4): 335–343.

48 Baluška, F., Mancuso, S., Volkmann, D., and Barlow, P.W. (2010). Root apex transition zone: a signalling–response nexus in the root. *Trends Plant Sci.* 15 (7): 402–408.

49 Inoue, N., Arase, T., Hagiwara, M. et al. (1999). Ecological significance of root tip rotation for seedling establishment of *Oryza sativa* L. *Ecol. Res.* 14 (1): 31–38.

50 Brown, A.H. (1993). Circumnutations: from Darwin to space flights. *Plant Physiol.* 101 (2): 345–348.

51 Vollsnes, A.V., Futsaether, C.M., and Bengough, A.G. (2010). Quantifying rhizosphere particle movement around mutant maize roots using time-lapse imaging and particle image velocimetry. *Eur. J. Soil Sci.* 61 (6): 926–939.

52 Clark, L.J., Whalley, W.R., and Barraclough, P.B. (2003). How do roots penetrate strong soil? *Plant Soil* 255 (1): 93–104.

53 Greacen, E.L. and Oh, J.S. (1972). Physics of root growth. *Nat. New Biol.* 235 (53): 24–25.

54 Barley, K.P. and Greacen, E.L. (1967). Mechanical resistance as a soil factor influencing the growth of roots and underground shoots. In: *Advances in Agronomy*, vol. 19, 1–43. Elsevier.

55 Hawes, M.C., Bengough, G., Cassab, G., and Ponce, G. (2002). Root caps and rhizosphere. *J. Plant Growth Regul.* 21 (4): 352–367.

56 Bengough, A.G. and McKenzie, B.M. (1997). Sloughing of root cap cells decreases the frictional resistance to maize (*Zea mays* L.) root growth. *J. Exp. Bot.* 48 (4): 885–893.
57 Iijima, M., Higuchi, T., Barlow, P.W., and Bengough, A.G. (2003). Root cap removal increases root penetration resistance in maize (*Zea mays* L.). *J. Exp. Bot.* 54 (390): 2105–2109.
58 Bengough, A.G. and Kirby, J.M. (1999). Tribology of the root cap in maize (*Zea mays*) and peas (*Pisum sativum*). *New Phytol.* 142 (3): 421–425.
59 Goodman, A. (1999). The effects of soil bulk density on the morphology and anchorage mechanics of the root systems of sunflower and maize. *Ann. Bot.* 83 (3): 293–302.
60 Bengough, A.G., McKenzie, B.M., Hallett, P.D., and Valentine, T.A. (2011). Root elongation, water stress, and mechanical impedance: a review of limiting stresses and beneficial root tip traits. *J. Exp. Bot.* 62 (1): 59–68.
61 Czarnes, S., Hiller, S., Dexter, A. et al. (1999). Root: soil adhesion in the maize rhizosphere: the rheological approach. *Plant Soil* 211 (1): 69–86.
62 Bingham, I. and Bengough, A. (2003). Morphological plasticity of wheat and barley roots in response to spatial variation in soil strength. *Plant Soil* 250 (2): 273–282.
63 Iijima, M., Barlow, P.W., and Bengough, A.G. (2003). Root cap structure and cell production rates of maize (*Zea mays*) roots in compacted sand. *New Phytol.* 160 (1): 127–134.
64 Bengough, A., Croser, C., and Pritchard, J. (1997). A biophysical analysis of root growth under mechanical stress. In: *Plant Roots-From Cells to Systems*, 107–116. Springer.
65 Materechera, S., Alston, A., Kirby, J., and Dexter, A. (1992). Influence of root diameter on the penetration of seminal roots into a compacted subsoil. *Plant Soil* 144 (2): 297–303.
66 Popova, L., van Dusschoten, D., Nagel, K.A. et al. (2016). Plant root tortuosity: an indicator of root path formation in soil with different composition and density. *Ann. Bot.* 118 (4): 685–698.
67 Darwin, C. (1897). *The Power of Movement in Plants*. Appleton.
68 Fisher, J.E. (1964). Evidence of circumnutational growth movements of rhizomes of Poa pratensis L. that aid in soil penetration. *Can. J. Bot.* 42 (3): 293–299.
69 Rowe, N.P., Isnard, S., Gallenmüller, F., and Speck, T. (2006). Diversity of mechanical architectures in climbing plants: an ecological perspective. In: *Ecology and Biomechanics: A Mechanical Approach to the Ecology of Animals and Plants* (eds. A. Herrel, T. Speck and N.P. Rowe), 35–59. Florida, USA: Taylor & Francis, Boca Raton.
70 Darwin, C. (1865). On the movements and habits of climbing plants. *J. Linn. Soc. Lond. Bot.* 9 (33–34): 1–118.
71 Darwin, C. (1881). Movements of plants. *Nature* 23 (592): 409.
72 Fisher, J. and Ewers, F. (1991). Structural responses to stem injury in vines. In: *The Biology of Vines* (eds. F.E. Putz and H.A. Mooney), 99–124. Cambridge: Cambridge University Press.

73 Takano, M., Takahashi, H., Hirasawa, T., and Suge, H. (1995). Hydrotropism in roots: sensing of a gradient in water potential by the root cap. *Planta* 197 (2).
74 Eapen, D., Barroso, M.L., Ponce, G. et al. (2005). Hydrotropism: root growth responses to water. *Trends Plant Sci.* 10 (1): 44–50.
75 Newcombe, F.C. and Rhodes, A.L. (1904). Chemotropism of roots. *Bot. Gaz.* 37 (1): 22–34.
76 Hart, J.W. (1990). *Plant Tropisms: and Other Growth Movements*. Springer Science & Business Media.
77 Hodge, A. (2004). The plastic plant: root responses to heterogeneous supplies of nutrients. *New Phytol.* 162 (1): 9–24.
78 Maurer, W.C. (1980). *Advanced Drilling Techniques*. Tulsa, OK: Petroleum Pub. Co.
79 Bar-Cohen, Y. and Zacny, K. (eds.) (2009). *Drilling in Extreme Environments: Penetration and Sampling on Earth and Other Planets*. Weinheim: Wiley-VCH.
80 Cooper, G.A. (1994). Directional Drilling. *Sci. Am.* 270 (5): 82–87.
81 Zacny, K., Paulsen, G., Chu, P. et al. (2012). Mars drill for the Mars sample return mission with a Brushing and Abrading bit, regolith and powder bit, core PreView Bit and a coring bit. In: *2012 IEEE Aerospace Conference*, 1–8. IEEE.
82 Kömle, N.I., Kaufmann, E., Kargl, G. et al. (2008). Development of thermal sensors and drilling systems for lunar and planetary regoliths. *Adv. Space Res.* 42 (2): 363–368.
83 Bar-Cohen, Y. (2006). Biomimetics—using nature to inspire human innovation. *Bioinspir. Biomim.* 1 (1): P1–P12.
84 Gouache, T.P., Gao, Y., Coste, P., and Gourinat, Y. (2011). First experimental investigation of dual-reciprocating drilling in planetary regoliths: Proposition of penetration mechanics. *Planet. Space Sci.* 59 (13): 1529–1541.
85 Gao, Y., Ellery, A., Sweeting, M., and Vincent, J. (2007). Bioinspired drill for planetary sampling: literature survey, conceptual design and feasibility study. *J. Spacecr. Rocket.* 44 (3): 703–709.
86 Winter, A.G. and Hosoi, A.E. (2011). Identification and evaluation of the Atlantic Razor Clam (Ensis directus) for biologically inspired subsea burrowing systems. *Integr. Comp. Biol.* 51 (1): 151–157.
87 Ishikawa, H. and Evans, M.L. (1995). Specialized zones of development in roots. *Plant Physiol.* 109 (3): 725–727.
88 Barlow, P.W. (2002). The root cap: cell dynamics, cell differentiation and cap function. *J. Plant Growth Regul.* 21 (4): 261–286.
89 Mazzolai, B., Corradi, P., Mondini, A., Mattoli, V., Laschi, C., Mugnai, S., Dario, P. (2008) Proceedings of Biological Approaches for Engineering, University of Southampton, pp. 50–3, Presented at the Biological Approaches for Engineering Conference, 17–19 March 2008, University of Southampton, Southampton, UK.
90 Mondini, A., Mazzolai, B., Corradi, P. et al. (2008). A preliminary study of a robotic probe for soil exploration inspired by plant root apexes. In: *2008*

2nd IEEE RAS EMBS International Conference on Biomedical Robotics and Biomechatronics, 19–22 Oct. 2008, 115–120. Scottsdale, USA: IEEE.
91 Webster, R.J. III, and Jones, B.A. (2010). Design and kinematic modeling of constant curvature continuum robots: a review. *Int. J. Robot. Res.* 29 (13): 1661–1683.
92 Del Dottore, E., Sadeghi, A., Mondini, A., Mazzolai, B. (2018) Continuous growth in plant-inspired robots through 3D additive manufacturing 1–7. https://doi.org/10.1109/ICRA.2018.8460616
93 Del Dottore, E., Mondini, A., Sadeghi, A., and Mazzolai, B. (2018). A plant-inspired kinematic model for growing robots. In: *2018 IEEE International Conference on Soft Robotics (RoboSoft), 24–28 April 2018*, 20–24. Livorno, Italy: IEEE.
94 Dubins, L.E. (1957). On curves of minimal length with a constraint on average curvature, and with prescribed initial and terminal positions and tangents. *Am. J. Math.* 79 (3): 497.
95 Siegwart, R., Nourbakhsh, I.R., and Scaramuzza, D. (2011). *Introduction to Autonomous Mobile Robots*. Cambridge, MA: MIT Press.
96 Babaei, A. and Mortazavi, M. (2010). Three-dimensional curvature-constrained trajectory planning based on in-flight waypoints. *J. Aircr.* 47 (4): 1391–1398.
97 Del Dottore, E., Mondini, A., Sadeghi, A., and Mazzolai, B. (2018). Swarming behavior emerging from the uptake–kinetics feedback control in a plant-root-inspired robot. *Appl. Sci.* 8 (1): 47.
98 Jungk, A., Asher, C., Edwards, D., and Meyer, D. (1990). Influence of phosphate status on phosphate uptake kinetics of maize (*Zea mays*) and soybean (*Glycine max*). In: *Plant Nutrition – Physiology and Applications*, 135–142. Springer.
99 Migliaccio, F., Tassone, P., and Fortunati, A. (2013). Circumnutation as an autonomous root movement in plants. *Am. J. Bot.* 100 (1): 4–13.
100 Kislev, M., Korach, E., and Negbi, M. (1979). Mechanisms of root penetration of seeds germinating on the soil surface. *Ann. Bot.* 43 (1): 87–92.
101 Del Dottore, E., Mondini, A., Sadeghi, A. et al. (2017). An efficient soil penetration strategy for explorative robots inspired by plant root circumnutation movements. *Bioinspir. Biomim.* 13 (1): 015003.
102 Okada, K. and Shimura, Y. (1990). Reversible root tip rotation in Arabidopsis seedlings induced by obstacle-touching stimulus. *Science* 250 (4978): 274–276.
103 Salgado, R., Mitchell, J., and Jamiolkowski, M. (1997). Cavity expansion and penetration resistance in sand. *J. Geotech. Geoenviron.* 123 (4): 344–354.
104 Walker, I.D. (2015). Biologically inspired vine-like and tendril-like robots. In: *2015 Science and Information Conference (SAI)*, 714–720. IEEE Xplore.

16

Biohybrid Robot Powered by Muscle Tissues

Yuya Morimoto[1,2] *and Shoji Takeuchi*[1,2]

[1] *The University of Tokyo, Department of Mechano-Informatics, Graduate School of Information Science and Technology, Hongo 7-3-1, Bunkyo-ku, Tokyo, 113-8656, Japan*
[2] *The University of Tokyo, Center for International Research on Integrative Biomedical Systems, Institute of Industrial Science, Komaba 4-6-1, Meguro-ku, Tokyo, 153-8505, Japan*

16.1 Introduction

Biohybrid robotics has been studied as a way to integrate biological components with synthetic components to engineer and use biological systems like traditional mechanical systems. As mechanically responsive materials to drive biohybrid robots, muscles have attracted attention in recent years [1–3]. One of the advantages of using muscles in biohybrid robotics is high driving performance and wide controllability [4]. Muscle contractions have good efficiency (10%–40% during cyclic motion) to conserve energy [5]. In the case of skeletal muscle, the power-to-weight ratio of muscle contractions is $\sim$1 kW kg^{-1}, superior to traditional robotic actuators requiring an energy source [2]. The contractions can be controlled in various ways, such as neuronal transmission from motor neurons via neuromuscular junctions [6] and electrical stimulation inducing activation of voltage-dependent ion channels [7, 8]. By producing recombinant light-sensitive muscle, the contractions can be controlled also by optical stimulation [9]. In addition, muscles make complex actuations possible by characteristic muscular structures seen in such organisms as octopuses and earthworms [10]. Incorporating design strategies of the muscular structures into biohybrid robotics makes biomimetic movements possible. Furthermore, uses of muscles as driving elements allow construction of biohybrid robots with various sizes. At the micrometer scale (10^{-5} to 10^{-4} m), a single muscle cell works as a microsized linear actuator to drive microstructures. Muscle tissue, an aggregate of muscle cells, can be used as a drive element for a robot having a size of millimeter-to-centimeter scale (10^{-3} to 10^{-1} m).

Muscles such as cardiac muscles and skeletal muscles are mainly used in biohybrid robots. Cardiac muscles are composed of cardiomyocytes (heart muscle cells). Cardiomyocytes are easy-to-use driving elements for biohybrid robots because they can contract spontaneously without external stimulation, including

Mechanically Responsive Materials for Soft Robotics, First Edition. Edited by Hideko Koshima.

electrical stimulation [11]. Because of the characteristics of contractions, a single cardiomyocyte can be used as a biological actuator. Furthermore, contractions of cardiomyocytes are synchronized when cardiomyocytes contact each other, making large contractile force and long contractile length possible in cardiac muscle (an aggregate of cardiomyocytes). Therefore, cardiomyocytes and cardiac muscle are useful to spontaneously actuate biohybrid robots of various sizes ranging from micrometer size (using a single cardiomyocyte) to centimeter size (using cardiac muscle). On the other hand, skeletal muscles are appropriate for actuations with complete on–off controllability by external stimulations, because they do not contract spontaneously [2]. They are bundles of contractile skeletal muscle fibers formed from the fusion of skeletal myoblasts. Because skeletal myoblasts do not have contractility, differentiation from myoblasts to muscle fibers is necessary to realize contractions. Therefore, although mounting skeletal muscle on robots is not easy, it is a promising bioactuator because of its high controllability and high contractile force. Moreover, insect cells and tissues are attractive bioactuators because of their temperature durability between 5 and 40 °C [12], in contrast to mammalian cells and tissues durable around 37 °C only.

Biohybrid robots powered by muscles are classified into three types according to the morphologies of muscle: biohybrid robots with a single muscle cell, a muscle cell monolayer, and a muscle tissue. In the biohybrid robots with a single muscle cell, both ends of the muscle cell are connected to the substrate of the robot, allowing deformation of the substrate by contractions of the muscle cell. Its contractile force and contraction length are limited, corresponding to the inherent contractility of the muscle cell, and the configuration is not suitable for driving a large structure. A muscle cell monolayer is an appropriate driving element for biohybrid soft robotics at millimeter-to-centimeter scales. By arranging a muscle cell monolayer on a flexible substrate, contractions of the muscle cell monolayer induce wide deformation of the substrate. On the other hand, a muscle tissue can be used as a driving element for various types of biohybrid robots from soft robots to hard robots with joints at millimeter-to-centimeter scales by connecting both ends of the muscle tissue to the flexible substrate or the robot skeleton.

In this chapter, the recent developments of biohybrid robots are summarized. First, a brief overview of preparation methods and characteristics of muscles (i.e. cardiac, skeletal, and insect ones) usable in biohybrid robots is provided, especially of the introduction of engineered muscle tissues in vitro. Next, the functionality of biohybrid robots powered by muscles is introduced according to the classification based on muscle morphology. Finally, future directions of biohybrid robots powered by muscle are discussed.

16.2 Muscle Usable in Biohybrid Robots

Muscles have different characteristics, such as contractility, controllability, lifetime, and appropriate condition for actuation, depending on their types, cell sources, and morphology. To achieve the driving of biohybrid robots with muscle, selection of the muscle that is suitable for configurations of biohybrid robots is necessary. In this section, a brief overview of the characteristics of each type of muscle and the in vitro fabrication method of muscle tissues from muscle cells is provided.

16.2.1 Cardiomyocyte and Cardiac Muscle Tissue

In biohybrid robotics, primary cardiomyocytes extracted from rat or mouse hearts are mainly used due to their high contractility and procurability. The primary cardiomyocytes can be adhered to substrates [13, 14] or pillars [15, 16] usable as bodies of biohybrid robots by a standard culture method. Although the primary cardiomyocytes have good capabilities as driving elements, experimental animals are necessary, leading to ethical issues. To escape the problem, pluripotent stem cells, such as embryonic stem cells and induced pluripotent stem cells (iPSCs), have been available as alternatives because they can differentiate into cardiomyocytes [17]. However, the preparation of cardiomyocytes derived from pluripotent stem cells requires a complicated differentiation procedure and brings cost-related issues. This is difficult for engineers who are inexperienced in cell culture work to prepare them.

Seeding cardiomyocyte suspensions onto robot bodies is a general method for the arrangement of single cardiomyocytes and cardiomyocyte monolayers in biohybrid robots. Because of the very low proliferation potency of cardiomyocytes [18], other than fetal cardiomyocytes, the concentration of cardiomyocytes in the suspensions makes the control of a single cardiomyocyte or cardiomyocyte monolayer possible – a single cardiomyocyte when the concentration is low and a cardiomyocyte monolayer when the concentration is high. In another method, a cardiomyocyte monolayer is constructible using temperature-responsive polymers, poly(*N*-isopropylacrylamide) (PIPAAm) [19]. Since PIPAAm coated on a culture dish is dissolved at lower than 32 °C, a cardiomyocyte monolayer cultured on the dish is released and is free-floating.

Cardiac muscle tissues are constructible by culturing cardiomyocytes in extracellular matrices (ECMs), such as type I collagen [20], a mixture of type I collagen and Matrigel [21–24], and a mixture of fibrin gel and Matrigel [25, 26]. Molding cardiomyocyte-laden ECMs enables the formed cardiac muscle tissues to have various shapes, such as a rod [17, 23, 24] (Figure 16.1a), mesh [25, 26] (Figure 16.1b), and ring [21, 22] (Figure 16.1c). The orientation of cardiomyocytes in the cardiac muscle tissue is important to determine the direction of its contractile movement, and cardiomyocytes can be aligned in one direction by embedding synthetic fibers in the cardiomyocyte-laden ECMs [20] or making the cardiomyocyte-laden ECMs into striped shapes with narrow width [24]. Mounting the tissues on the biohybrid robot is possible by hooking the ring-shaped tissues [21, 22] (Figure 16.1c) or fixing both ends of the rod-shaped and the mesh-shaped tissues [17, 23, 24, 26].

Although cardiomyocytes and cardiac muscle tissues contract spontaneously, the paces of their contractions are controllable with electrical stimulations [20, 22, 24, 25]. In addition, by optogenetic modification to express a light-sensitive ion channel, channelrhodopsin-2 (ChR2), the contraction paces of the recombinant cardiomyocytes can be controlled by optical stimulations [27]. However,

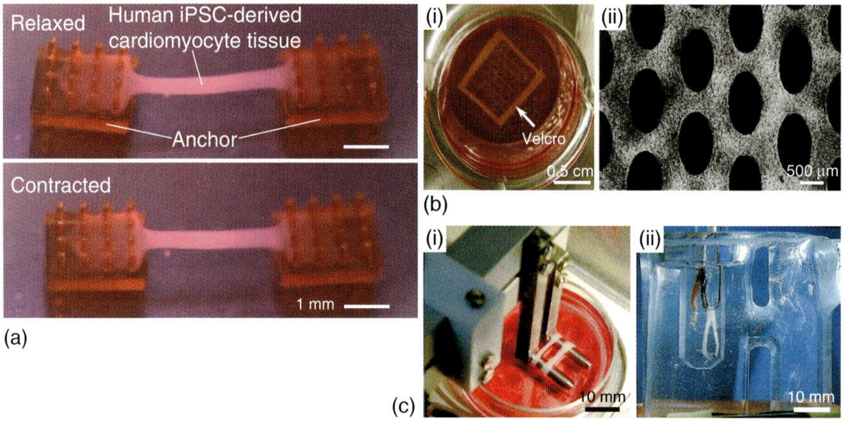

Figure 16.1 Cardiac muscle tissues formed by culturing cardiomyocytes in ECMs: (a) rod-shaped tissue of human iPSC-derived cardiomyocytes before and after its contraction, (b) mesh-shaped human cardiac tissue anchored within a Velcro frame, and (c) ring-shaped cardiac tissues mounted on (c, i) double mechanical arms and (c, ii) a mechanical arm and ring. Source: (a) Reprinted with permission of [24], 2016 licensed under Creative Commons Attribution-NonCommercial 3.0 Unported License. (b) Reprinted with permission of [26]. Copyright 2013, Elsevier. (c) Reprinted with permission of [21]. Copyright 2001, Wolters Kluwer Health, Inc.

only the contraction pace is controllable with both electrical stimulation and optical stimulation, and switching contractions on or off is not possible.

16.2.2 Skeletal Muscle Fiber and Skeletal Muscle Tissue

In order to mount skeletal muscle fibers to biohybrid robots, the formation of skeletal muscle fibers by fusion of a plurality of skeletal myoblasts is necessary. In the biohybrid robotics field, primary myoblasts extracted from rat limbs are a primary choice, because skeletal muscle fibers made from them have high contractility. However, preparation of primary myoblasts requires experimental animals, involving ethical issues similar to those for primary cardiomyocytes. As another option that does not involve experimental animals, C2C12 cells, cell lines of mouse skeletal myoblast, are available. Although the contractility of skeletal muscle fibers made from C2C12 cells is lower than that of those made from primary myoblasts, in terms of usability, we think that C2C12 cells will mainly be used in biohybrid robotics in the future.

In the process of mounting skeletal muscle fibers in biohybrid robots, myoblasts can adhere to substrates and proliferate until the myoblasts contact each other. When differentiation is induced in the presence of myoblast contacts, a monolayer of skeletal muscle fibers is formed on the substrates. Although reducing the number of formed skeletal muscle fibers is possible by linearly patterning the culture area of myoblasts, it is difficult to construct only a single skeletal muscle fiber. In addition, because skeletal muscle fibers spontaneously shrink by their intrinsic traction force, a certain level of substrate stiffness is necessary to prevent the shrinkage for maintenance of the morphology and contractility of the constructed monolayer of skeletal muscle fibers.

In the construction of skeletal muscle tissues, there are mainly two methods: self-assembly of skeletal muscle fibers and culturing myoblasts in ECM similar to the formation of cardiac muscle tissues. In the self-assembly method, after construction of a muscle fiber monolayer on a culture dish, the progressed culture causes spontaneous peeling of the monolayer from the dish by its intrinsic traction force; thereby, skeletal muscle tissue is constructed by catching both ends at pins arranged on the dish [28–30] (Figure 16.2a). This method has been widely used in the biological field because of easy muscle tissue formation. However, since the tissue is fixed to the pins on the dish, it is difficult to mount the tissue on a biohybrid robot. On the other hand, skeletal muscle tissues can be constructed without culture dishes by culturing myoblasts in ECMs, such as type I collagen [31], Matrigel [32], a mixture of type I collagen and Matrigel [33, 34], a mixture of fibrin gel and Matrigel [35], and a mixture of type I collagen, fibrin gel, and Matrigel [36, 37] under fixation of their ends to prevent shrinkage. According to the shapes of molded myoblast-laden ECMs, the shapes of skeletal muscle tissues can be changed to a rod shape [31, 32, 34–36] (Figure 16.2b) and a mesh shape [37] (Figure 16.2c). Skeletal muscle fibers in the skeletal muscle tissue are generally aligned between fixed ends by the unidirectional traction force of the muscle fibers. Moreover, alignment of the skeletal muscle fibers become higher when formation of skeletal muscle fibers with narrow striped myoblast-laden ECMs [32]. In biohybrid robotics, since skeletal muscle tissues shrink spontaneously, it is difficult to manipulate skeletal muscle tissues constructed at other places for installation in robots. Forming myoblast-laden ECMs on biohybrid robots or installing the ECMs in the robots is mainly used to mount skeletal muscle tissues on the robots.

Contractions of the muscle fiber monolayer and skeletal muscle tissue can be controlled by electrical stimulation [28–30, 32, 34–36] and optical stimulation with recombination for ChR2 expression [33]. In the case of electrical stimulation, contractile forces increase according to the frequency of electrical pulses and the magnitude of the electrical field; as the frequency and/or magnitude increases, the contractile force increases [28]. Furthermore, in contrast to cardiomyocytes and cardiac muscle tissues, applying electrical pulses with higher frequency induces a continuous contraction by superimposing twitches.

16.2.3 Cell and Tissue Other Than Mammals

Isolated muscles from organisms other than mammals are also usable in biohybrid robots. Especially, isolated frog muscles have been studied for a long time in the field of biology [38] and can be applied to biohybrid robotics based on established fundamental knowledge. However, the use of isolated muscles causes limitations in the dimensions and architectures of biohybrid robots, because customization of the robots is necessary depending on the size and contractile force of the available isolated muscle.

Furthermore, insect dorsal vessel tissues (DVTs) are available as robust driving elements in biohybrid robots. The DVTs can be used between approximately

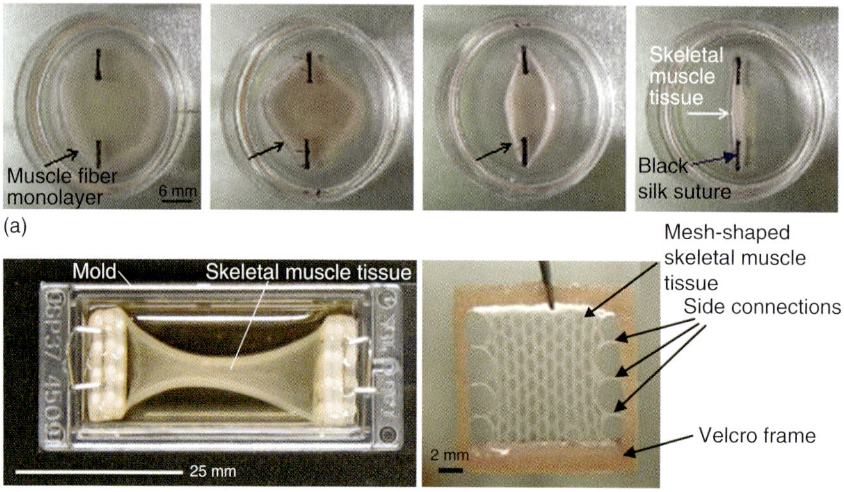

Figure 16.2 Skeletal muscle tissues formed by (a) self-assembly of skeletal muscle fibers and (b, c) culturing myoblasts in ECMs: (a) sequential images for formation of skeletal muscle tissue whose edges were anchored to pins (black silk sutures), (b) rod-shaped skeletal muscle tissue formed by culturing myoblasts in a rectangular mold, and (c) mesh-shaped skeletal muscle tissues anchored on two opposite horizontal sides and via three-side connections on perpendicular sides to a Velcro frame. Source: (a) Reprinted with permission of [30]. Copyright 2009, Elsevier. (b) Reprinted with permission of [31]. Copyright 2012, John Wiley and Sons. (c) Reprinted with permission of [37]. Copyright 2009, Elsevier.

5 and 40 °C over a few months without medium exchange, and they contract autonomously, similar to mammalian cardiac muscle. Electrical stimulation can induce contractions of DVTs, and their contractile force increases with increasing voltage of the stimulations [12]. The main drawback of using DVTs is their low controllability resulting from autonomous contractions [2]. In addition, muscle fibers derived from insect eggs are promising candidate for environmentally robust driving elements [39]. Culturing the insect muscle cells in molds enabled insect muscle tissues to be formed by self-assembly of the cells without supporting materials such as ECMs, and the tissues survived for a month without changes in media [40].

16.3 Actuation of Biohybrid Robots Powered by Muscle

Many researchers have demonstrated actuations of biohybrid robots with muscle, aiming to reproduce movements of living organisms. By changing the dimensions and architectures of the biohybrid robots and muscles, the biohybrid robots achieved various biomimetic motions such as pumping, grasping, walking, and swimming, by using the basic design strategies of hearts, claws, inchworms, and aquatic organisms. In this section, the actuation characteristics of biohybrid robots with a single muscle cell, a muscle cell monolayer, or muscle tissues are introduced.

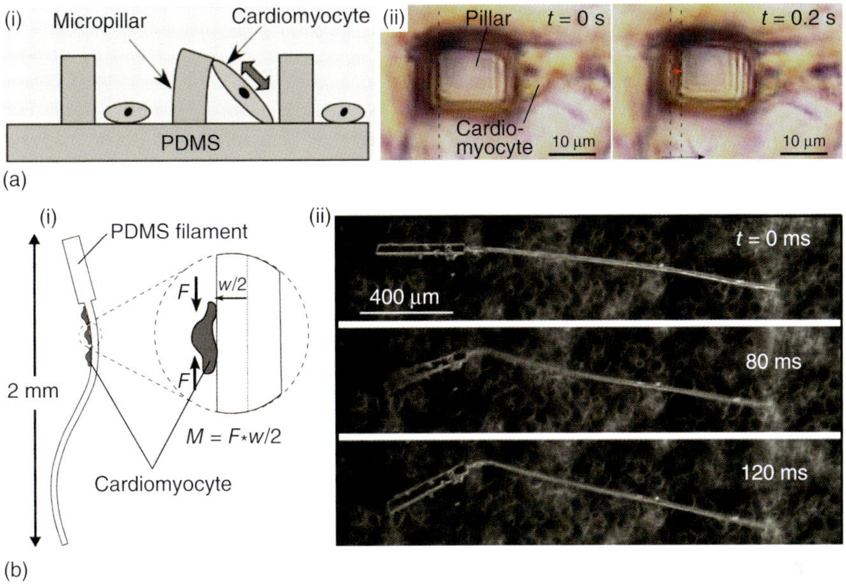

Figure 16.3 Biohybrid robots with a single cardiomyocyte: (a) deformation of micropillars by contractions of a single cardiomyocyte – (a, i) schematic illustration and (a, ii) sequential images of the deformation and (b) biohybrid robot powered by cardiomyocytes to mimic flagella swimming – (b, i) schematic illustration for configurations of the biohybrid robot and (b, ii) sequential images of swimming of the biohybrid robot with a mean speed of 9.7 $\mu m\ s^{-1}$ by producing flagellar propulsion. Source: (a) Reprinted with permission of [42]. Copyright 2006, Royal Society of Chemistry. (b) Reprinted with permission of [43]. Copyright 2014, Springer Nature.

16.3.1 Biohybrid Robot with a Single Muscle Cell

In biohybrid robots with a single muscle cell, cardiomyocytes are promising driving elements because they contract spontaneously even in a single-cell state. Combining cultured cardiomyocytes with mechanical flexible microstructures makes possible the spontaneous conversion of chemical energy to mechanical force, thereby enabling the structure to move.

The biohybrid robots with a single cardiomyocyte mainly have a driving strategy for attaching the cardiomyocyte to a flexible pillar. Attaching the ends of a single cardiomyocyte to a sidewall and a root of a micropillar enabled contractions of cardiomyocytes to induce the deformation of the micropillar [41, 42] (Figure 16.3a). Displacements of the micropillar tip in the deformation were a few micrometers in the cases in which acrylamide pillars of 50 μm× 50 μm (cross section) × 200 μm (height) [41] and polydimethylsiloxane (PDMS) pillars of 10 μm× 15 μm (cross section) × 25 μm (height) [42] were used. As described above, although the deformation of the micropillars by cardiomyocyte contractions is not large, the driving system enables the microstructure to actuate periodically.

As an example of the application of the driving system to biohybrid robotics, a long PDMS filament with cardiomyocytes adhering on its slender tail was proposed to produce microscale flagellar propulsion [43] (Figure 16.3b). By contractions of cardiomyocytes, a filament of 20–30 × 7–10 μm (cross section) was bent periodically, resulting in swimming of the biohybrid robot at 5–10 μm s^{-1}. In addition, the robot with two tails swam at 81 μm s^{-1}. This research indicates that cardiomyocytes are useful driving elements in microsized robotics by mimicking flagella behavior. Another biohybrid robot that mimicked flagella behavior was developed by culturing skeletal muscle fibers on a long PDMS filament [44]. Displacement of the filament tip was more than 17 μm, seeming to be smaller than that with cardiomyocytes. Moreover, a gear-shaped microstructure with cardiomyocytes was proposed as a swimming biohybrid robot [45], but the authors did not show actuations of the robot. As mentioned previously, muscle cells, especially cardiomyocytes, are usable driving elements to actuate microsized robots.

16.3.2 Biohybrid Robot with Monolayer of Muscle Cells

Because a monolayer of muscle cells generates a larger contractile force than a single muscle cell, biohybrid robots with a monolayer can perform large actuations. The monolayer is cultured on a flexible substrate, so that its contractions induce deformation of the substrate. When a monolayer of cardiomyocytes was on a PDMS cantilever (150 μm × 750 μm) (Figure 16.4a), the cantilever was bent by its contractions, and its displacement was ~35 μm [46], larger actuations than that of the driving system with a single cardiomyocyte (~3 μm). In the actuations, the orientation of cardiomyocytes in a monolayer is important, because the monolayer contracts in the direction of cell orientation, resulting in the generation of a large contractile force. When microgrooves were made on the cantilever, cardiomyocytes aligned in the direction of the microgrooves [46, 47]. Similarly, line-patterned proteins, such as fibronectin or laminin, also promote cellular alignment in the line direction on the cantilever. In the case of using a thin PDMS cantilever with line-patterned proteins, the tip displacement of the thin cantilever achieved ~1 mm by contraction of a monolayer of aligned cardiomyocytes [48] (Figure 16.4b). Using the same method, a monolayer of aligned skeletal muscle fibers on a thin PDMS cantilever can be prepared. Although contractions of the monolayer enabled the thin cantilever to bend [49] (Figure 16.4c), the deformation amount was smaller than that induced by the cardiomyocyte monolayer. Comparing the deformation of the skeletal muscle fiber monolayer and the cardiomyocyte monolayer, cardiomyocyte monolayer is found to be preferable for bending a flexible substrate, because the traction force of skeletal muscle fibers also bends the substrate without the contractions. In the case of using skeletal muscle fibers, although bending of a silicon cantilever was achieved by bridging muscle fibers between the cantilever tip and a base, its displacement induced by muscle contractions was also small [50]. In addition, the cardiomyocyte monolayer culture on a tubular substrate with pillars and microgrooves was usable as a driving system for linear actuations [51]. The performance of the driving system showed that spontaneous contractions of the cardiomyocyte monolayer achieved ~2.2% average linear contractions of the system.

As examples of using a driving system in which a cardiomyocyte monolayer bends a flexible substrate, biohybrid robots based on a driving system achieved

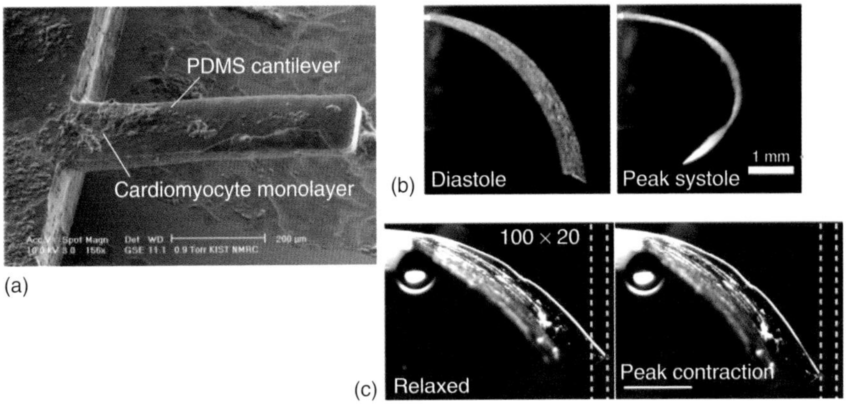

Figure 16.4 Deformation of substrates by contractions of muscle monolayers: (a) PDMS cantilever with cardiomyocyte monolayer, (b) deformation of a thin PDMS cantilever by contractions of a monolayer of aligned cardiomyocytes attached to the cantilever, and (c) deformation of a thin PDMS cantilever by contractions of a monolayer of aligned skeletal muscle fibers attached to the cantilever. Source: (a) Reprinted with permission of [46]. Copyright 2005, American Chemical Society. (b) Reprinted with permission of [48]. Copyright 2010, Elsevier. (c) Reprinted with permission of [49]. Copyright 2013, Elsevier.

various movements such as pumping, walking, gripping, and swimming. For pumping, a diaphragm pump was developed. The diaphragm pump had a push bar covered by a cardiomyocyte monolayer formed using a PIPAAm coated dish (Figure 16.5a). Pushing the bar by contractions of the monolayer-induced deformation of a diaphragm, similar to the way in traditional diaphragm pumps, results in generation of microfluidic flows ($0.1\,\mu l\,min^{-1}$) [52]. In addition, as another type, a diaphragm pump in which the cardiomyocyte monolayer is attached directly to a diaphragm has been proposed. This pump also achieved making a flow of $0.2\,\mu l\,min^{-1}$ by deformation of the diaphragm with the contractions [53]. Moreover, a spherical pump with a cardiomyocyte monolayer wrapping around a PDMS hollow sphere (diameter 5 mm, thickness 250 μm) was fabricated to mimic the pumping behavior of a heart (Figure 16.5b). The biohybrid pump made it possible to feed media at $0.01–0.1\,\mu l\,min^{-1}$ [54]. Although the flow rates of the above biohybrid pumps are much smaller than those of traditional microfluidic pumps, the biohybrid pumps have a great advantage of driving spontaneously using only chemical energy.

Several walking biohybrid robots have been proposed using bending of flexible substrates by contractions of the cardiomyocyte monolayer. At a microscale, a biohybrid robot composed of a silicon-based scaffold, a thin gold layer working as two legs, and a cardiomyocyte monolayer on the gold layer was developed to produce movements like inchworms on a culture dish [55] (Figure 16.6a). The walking speed depended on the contraction frequency of the cardiomyocyte monolayer and the step size, and the resulting maximum speed was $38\,\mu m\,s^{-1}$. At a millimeter scale, a PDMS structure with three short front legs and three long rear legs was used for a flexible substrate of a walking biohybrid robot [56] (Figure 16.6b). By contractions of a cardiomyocyte monolayer on the PDMS structure, the biohybrid robot walked in the same manner as inchworms at a

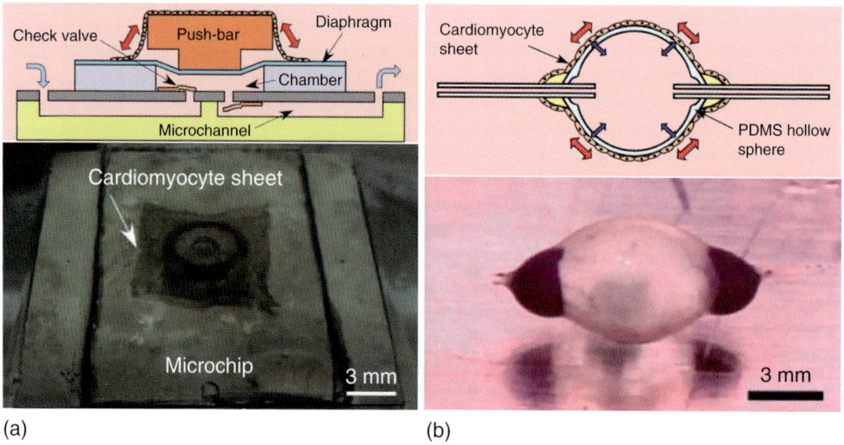

Figure 16.5 Pumping biohybrid robot with a cardiomyocyte monolayer: (a) illustration and image for a diaphragm pump driven by contractions of a cardiomyocyte monolayer and (b) illustration and image for a spherical pump driven by contractions of a cardiomyocyte monolayer. Source: (a) Reprinted with permission of [52]. Copyright 2006, Royal Society of Chemistry. (b) Reprinted with permission of [54]. Copyright 2007, Royal Society of Chemistry.

speed of 86.8 $\mu m\, s^{-1}$. The estimated walking distance for seven days was more than 50 m. Similarly, a triangle PDMS thin film with a cardiomyocyte monolayer folded manually into a 3D shape could walk on a culture dish at a speed of ~8 $mm\, min^{-1}$ at 1 Hz contractions paced by electrical stimulations [57]. Because the triangle-shaped biohybrid robots were formed with an isotropic cardiomyocyte monolayer rather than an anisotropic cardiomyocyte monolayer formed with line patterns of fibronectin, the performance indicated that alignment of cardiomyocytes was not required for the robot to walk. Furthermore, polyethylene glycol diacrylate (PEGDA)-based hydrogel was proposed as a material for a walking robot substrate to mimic the elasticity of the heart muscle (22–50 kPa) [58] (Figure 16.6c). The PEGDA-based biohybrid robot achieved a maximum walking speed of ~236 $\mu m\, s^{-1}$. As mentioned previously, a combination of the cardiomyocyte monolayer with various types of substrate achieved locomotion of biohybrid robots.

Gripping was achieved by a long oval PDMS thin film with a monolayer of aligned cardiomyocytes [57]. The biohybrid robot demonstrated motions of open/close (contact/release of film ends) and could control the open/close time by the electrical frequency pacing contractions of the cardiomyocyte monolayer. In the case of a triangle PDMS thin film with a monolayer of cardiomyocytes, the biohybrid robot swam with a maximum speed of ~24 $mm\, min^{-1}$ at 1 Hz pacing [57]. A swimming biohybrid robot was also constructible using a PDMS cantilever with a cardiomyocyte monolayer extending from the robot body as an autonomous driving fin (Figure 16.7a), and its maximum swimming speed was 142 $\mu m\, s^{-1}$ [59]. As a swimming biohybrid robot that mimics aquatic organisms, a biohybrid jellyfish composed of a jellyfish-shaped PDMS substrate and cardiomyocyte monolayer patterned to mimic the muscle configuration of a jellyfish has been proposed [60] (Figure 16.7b). As a result, the swimming of the

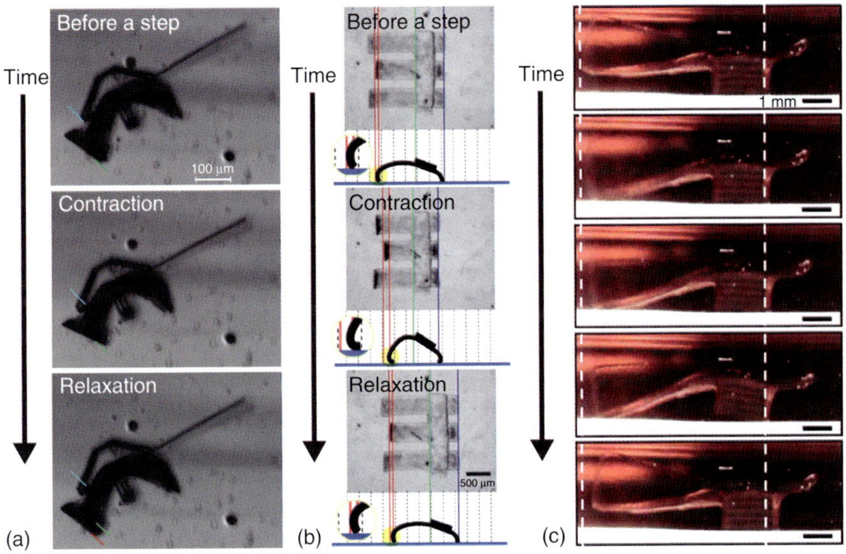

Figure 16.6 Walking biohybrid robot with a cardiomyocyte monolayer: (a–c) sequential images for a single step of (a) a biohybrid robot with two legs, (b) a biohybrid robot with three front legs and three rear legs, and (c) a biohybrid robot with two hydrogel legs. Source: (a) Reprinted with permission of [55]. Copyright 2005, Springer Nature. (b) Reprinted with permission of [56]. Copyright 2007, Royal Society of Chemistry. (c) Reprinted with permission of [58]. 2012 licensed under Creative Commons Attribution-NonCommercial-No Derivative Works 3.0 Unported License.

biohybrid jellyfish generated similar stroke motions and flow fields to those of living jellyfish. As another example of mimicking aquatic organisms, a biohybrid ray was developed with a ray-shaped PDMS-based substrate and a monolayer mimicking the muscle configuration of a ray [27]. Because cardiomyocytes were optogenetically modified in the biohybrid ray, the speed and direction of its swimming were controlled by changing the frequency of excitation light and placements (left or right of the substrate) exposed to light. This biohybrid ray was one-tenth of the size of a living ray, so its scale-up remained an issue, but mimicking of mollusk movements is being realized, as described previously.

As a biohybrid robot based on a monolayer of skeletal muscle fibers, a crawling robot was developed with multiwalled carbon nanotube sheets coated with poly(3,4-ethylenedioxythiophene) [61]; the carbon nanotubes promoted alignment of skeletal muscle fiber to the longitudinal direction of the rectangular sheet. By dividing the cell culture area and the nonculture area on the sheet using a mold, the biohybrid robot generated hornworm-like actuations by contractions of skeletal muscle fibers (Figure 16.8a). As a gripping robot, a microtweezer was fabricated by combining a skeleton made from dog hairs and skeletal muscle fibers [62]. Although muscle contractions bent the skeleton, displacement of the tips was less than 1 µm, which is not enough for operation of a gripper. In walking robots powered by a monolayer of skeletal muscle fibers, a silicon microdevice and a collagen sheet were used as flexible substrates [63, 64] (Figure 16.8b). Deformation of the substrates by muscle contractions enabled the robots to walk,

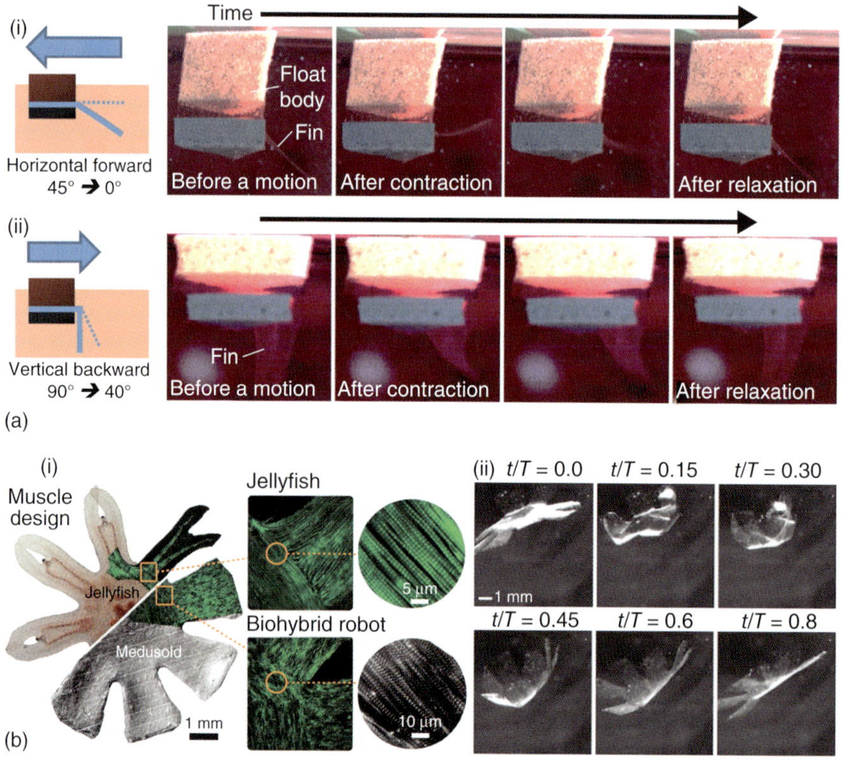

Figure 16.7 Swimming biohybrid robot powered by a cardiomyocyte monolayer: (a) schematic illustration and image for swimming of a biohybrid robot with (a, i) a horizontal fin to move forward and (a, ii) a vertical fin to move backward and (b) jellyfish-shaped biohybrid robot – (b, i) comparison of muscle design of a living jellyfish and the biohybrid robot and (b, ii) sequential images for swimming of the jellyfish-shaped biohybrid robot (t, time passed since start of stroke cycle; T, duration of stroke cycle). Source: (a) Reprinted with permission of [59]. Copyright 2016, Royal Society of Chemistry. (b) Reprinted with permission of [60]. Copyright 2012, Springer Nature.

in a way similar to that of robots with cardiomyocyte monolayers. In addition, biohybrid robots consisting of a collagen sheet show that skeletal muscle fibers are appropriate driving elements for their walking, because the average walking speed of skeletal muscle fibers powering the robots was $\sim 80\,\mu m\,min^{-1}$ and that of a cardiomyocyte powering the robot was $\sim 10\,\mu m\,min^{-1}$. However, the walking speed of the robot powered by skeletal muscle fibers was slow compared with that of other robots powered by cardiomyocytes [55–58]. Thus, it is difficult to achieve large actuations of biohybrid robots based on bending substrates with skeletal muscle fibers, because substrates need a certain stiffness against the traction force of muscle fibers.

16.3.3 Biohybrid Robot with Muscle Tissues

Skeletal muscle tissues have been mainly used for biohybrid robots using engineered muscle tissues. In the biohybrid robots with skeletal muscle tissues,

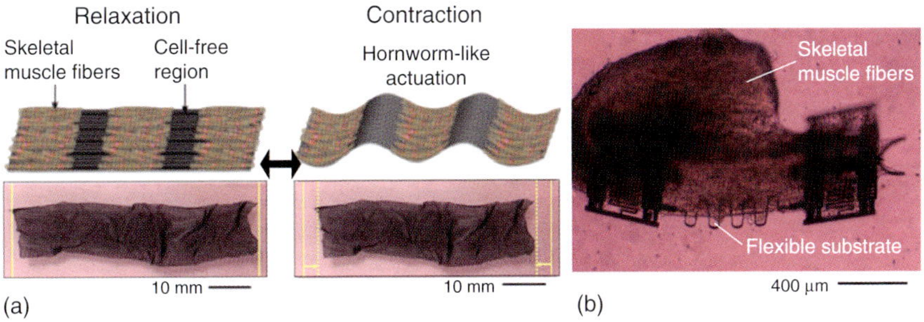

Figure 16.8 Biohybrid robots powered by a monolayer of skeletal muscle fibers: (a) crawling biohybrid robot to actuate hornworm-like motions (the culture area of skeletal muscle fibers and the cell-free region were alternately patterned on the robot) and (b) walking biohybrid robot with a monolayer of skeletal muscle fibers on a flexible substrate. Source: (a) Reprinted with permission of [61]. 2016 licensed under Creative Commons Attribution 4.0 International License. (b) Reprinted with permission of [63]. Copyright 2010, Springer Nature.

walking is a typical actuation. Bashir and coworkers have developed several types of the walking biohybrid robot. In the walking robots, a flexible PEGDA-based hydrogel substrate with two hard pillars was used as a skeleton [65]. Because the formed skeletal muscle tissues were bridged between the pillars, their contractions induced deformation of the flexible substrate, resulting in locomotion of the robot at a maximum speed of ~156 μm s^{-1} (more than 1.5 body length min^{-1}) [66] (Figure 16.9). In addition, the locomotion was predictable by computational simulations for geometry and performance of the skeletal muscle tissue and the scaffold, indicating that a forward engineering approach is usable in developments of biohybrid robots powered by skeletal muscle tissues [67]. Furthermore, when two optogenetically modified skeletal muscle tissues were mounted on two pairs of two hard pillars, the walking robot could deform its flexible substrate by light exposure and achieved directional locomotion at a speed of ~310 μm s^{-1} (1.3 body length min^{-1}) [68]. Due to the high spatial resolution of the light stimulation, only half of the skeletal muscle tissue could be excited by light exposure, and the excitation induced rotational locomotion at a speed of ~2 deg s^{-1}. A robot powered by optogenetically modified skeletal muscle tissue was used as a remodeling and healing model for mechanical damage induced via laceration by microscissors [69], indicating that biohybrid robots with skeletal muscle tissues are applicable to the study of muscle physiological reactions. The other actuation made possible by a biohybrid robot with a skeletal muscle tissue is gripping. Flexible PDMS microtweezers with a skeletal muscle tissue achieved movements of its tips by contractions of the tissue [70]. In the above biohybrid robots, skeletal muscle tissues had a large intrinsic traction force inducing spontaneous muscle shrinkage and malfunction of muscle contractility, so flexible robot skeletons should have appropriate stiffness to counter the force and to be deformed by muscle contractions.

To escape the requirement of optimizing the skeleton stiffness, an antagonistic pair of skeletal muscle tissues has been proposed as a driving system in which it is possible to balance the traction forces between them. When the antagonistic pair

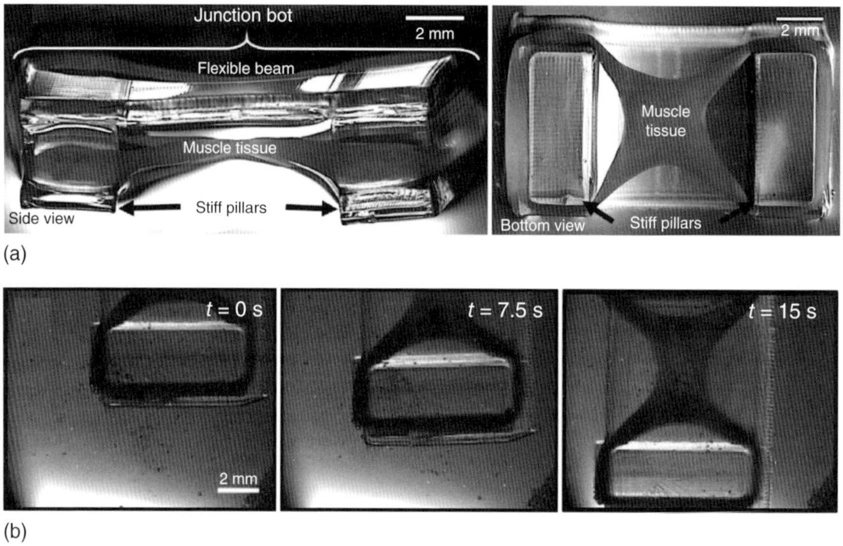

Figure 16.9 Walking biohybrid robots with a skeletal muscle tissue: (a) side and bottom views of the robot and (b) sequential bottom views for the walking of the robots by contractions of the skeletal muscle tissue. Source: Reprinted with permission of [67]. Copyright 2018, John Willey and Sons.

of skeletal muscle tissues was mounted on a skeleton with a joint similar to mammalian musculoskeletal systems, the biohybrid robot achieved ~90° of the joint rotation by selective contractions and extensions of each skeletal muscle tissue [71] (Figure 16.10a). Moreover, the biohybrid robot demonstrated pick-and-place manipulation of a ring (Figure 16.10b), indicating that the biohybrid robot has the ability to replicate various finger-like or arm-like movements. In the case of an antagonistic pair of skeletal muscle tissues mounted on a flexible substrate, actuations of the biohybrid robot were predictable using computational simulations with the stiffness of the substrate and performance of the skeletal muscle tissues [72]. This indicates that a forward engineering approach can also be used in biohybrid robots powered by an antagonistic pair of skeletal muscle tissues.

Isolated whole muscles are also usable as driving elements for biohybrid robots. A biohybrid robot composed of a skeleton with a joint and an antagonistic pair of two isolated frog muscles succeeded in joint rotation by contractions of the muscles [73] (Figure 16.11a). As a result, the robot performed simple swimming maneuvers and directional swimming at a maximum speed of more than 1/3 body length s^{-1}. As another type of a biohybrid robot with an explanted whole muscle, a diaphragm pump using an earthworm muscle as a driving element was also reported [74] (Figure 16.11b). As walking robots, biohybrid robots composed of flexible skeletons and muscles of the California sea hare were developed, and contractions of the muscles resulted in locomotion of the robots at a speed of ~0.5 cm min^{-1} [75, 76]. Although the above biohybrid robots with isolated whole muscle can be developed without cell culture techniques, the size and performance of the muscles limit the shapes and dimensions of biohybrid robots.

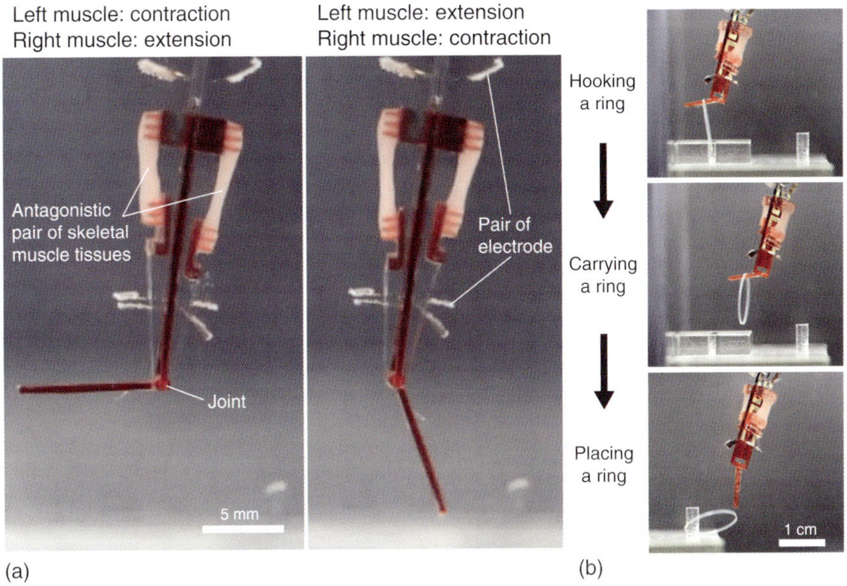

Figure 16.10 Biohybrid robot powered by an antagonistic pair of skeletal muscle tissues: (a) joint rotation of the robot by selective contractions of each skeletal muscle tissue and (b) pick-and-place manipulation of a ring by controlling the joint rotation. Source: Reprinted with permission of [71]. Copyright 2018, The American Association for the Advancement of Science.

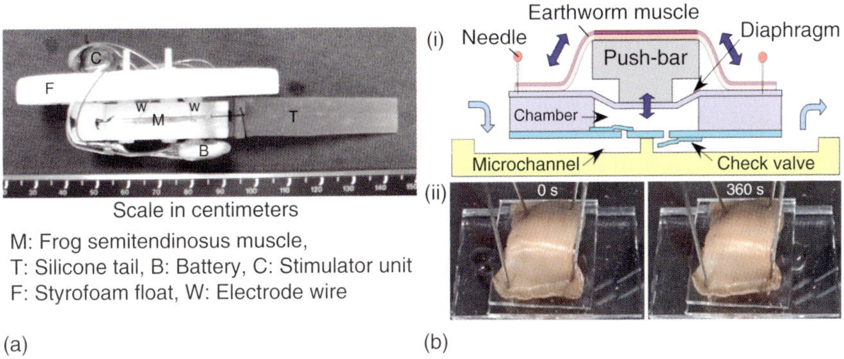

Figure 16.11 Biohybrid robots with isolated whole muscles: (a) swimming biohybrid robot with an antagonistic pair of frog muscles and (b) diaphragm pump powered by an earthworm muscle – (b, i) schematic illustration of the pump and (b, ii) images of the pump before and after delivering water from the left side to the right side. Source: (a) Reprinted with permission of [73]. 2004 licensed under Creative Commons Attribution 4.0 International License. (b) Reprinted with permission of [74]. Copyright 2017, Elsevier.

Insect DVTs are usable as temperature-adaptable driving elements for biohybrid robots, because the contractions of DVTs make possible the deformation of a flexible substrate even at 25 °C when connecting DVTs on it [77]. Biohybrid robots with DVTs succeeded in room temperature operations such as inchworm locomotion caused by deformation of a rectangular PDMS substrate [78]

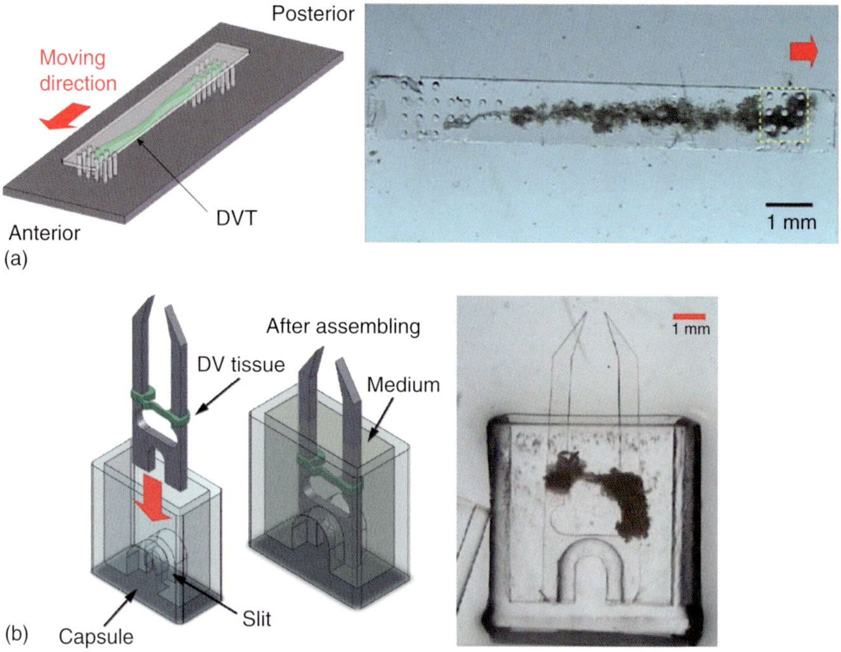

Figure 16.12 Biohybrid robots with insect DVTs: (a) inchworm-like walking biohybrid robot composed of a flexible substrate and an insect DVT and (b) gripping biohybrid robot composed of PDMS microtweezers and an insect DVT. Source: (a) Reprinted with permission of [78]. 2012 licensed under Creative Commons Attribution 2.0 Generic License. (b) Reprinted with permission of [79]. Copyright 2013, Royal Society of Chemistry.

and gripping by the tip movements of PDMS microtweezers [79] (Figure 16.12). Because DVTs are dissected from insects, the flexibility of the robot shapes and dimensions is not high compared to that of engineered skeletal muscle tissues. Using engineered insect muscle tissues derived from insect eggs is one of the solutions; however, biohybrid robots with engineered insect muscle tissues have not been proposed yet.

16.4 Summary and Future Directions

Muscles have been widely used as a driving element of biohybrid robots, from a single muscle cell for microsized robots to muscle tissue for macrosized robots. To show a proof of concept of robotic motions powered by muscle, many researchers have demonstrated various actuations such as walking and swimming. Although the performance of biohybrid robots has been insufficient compared with that of animals, recent experiments have shown their potential to reproduce lifelike motions. As examples expressing lifelike motions, swimming robots with a cardiomyocyte monolayer reproduced motions of aquatic animals such as jellyfish and ray, and robots with an antagonistic pair of skeletal muscle tissues achieved joint rotation similar to a finger or an arm. Actuations of these biohybrid robots were controlled by not only electrical stimulation, but also optical stimulation, in contrast to actuations of traditional mechanical machines mainly requiring electrical supply. In addition, because cocultured tissues of motor neurons and skeletal muscle tissues have already been proposed to induce their contractions by neural stimulations [32, 80, 81], various biohybrid robots driven by neural stimulations will be constructed in the future.

To mimic lifelike movements more, construction of large-sized biohybrid robots is necessary using large muscles with high contraction force, because animals have from millimeter- to meter-sized muscles, unlike current biohybrid robots with a maximum centimeter size. For engineering large muscles on robots in the future, it is necessary to clarify the methods of integrating blood vessels or flow paths with muscles allowing transport of oxygen and nutrients. Furthermore, although the current biohybrid robots have provided certain results by mimicking the design strategies of animals, the methodology of muscle and synthetic component integration has not been established. When procedures for the integration of muscles and synthetic components are optimized, muscles will become available as well as mechanical actuators. Therefore, by solving these remaining challenges, biohybrid robots will develop into a new robotics field to realize lifelike movements.

References

1 Feinberg, A.W. (2015). Biological soft robotics. *Annu. Rev. Biomed. Eng.* 17: 243–265.

2 Ricotti, L., Trimmer, B., Feinberg, A.W. et al. (2017). Biohybrid actuators for robotics: a review of devices actuated by living cells. *Science Robotics* 2 (12): eaaq0495.

3 Chan, V., Asada, H.H., and Bashir, R. (2014). Utilization and control of bioactuators across multiple length scales. *Lab Chip* 14 (4): 653–670.

4 Van Ham, R., Sugar, T.G., Vanderborght, B. et al. (2009). Compliant actuator designs review of actuators with passive adjustable compliance/controllable stiffness for robotic applications. *IEEE Rob. Autom Mag.* 16 (3): 81–94.

5 Ricotti, L. and Menciassi, A. (2012). Bio-hybrid muscle cell-based actuators. *Biomed. Microdevices* 14 (6): 987–998.

6 Witzemann, V. (2006). Development of the neuromuscular junction. *Cell Tissue Res.* 326 (2): 263–271.

7 Endo, M. (2011). Reiji Natori, Setsuro Ebashi, and excitation-contraction coupling. *Prog. Biophys. Mol. Biol.* 105 (3): 129–133.

8 Fujita, H., Nedachi, T., and Kanzaki, M. (2007). Accelerated de novo sarcomere assembly by electric pulse stimulation in C2C12 myotubes. *Exp. Cell. Res.* 313 (9): 1853–1865.

9 Asano, T., Ishizua, T., and Yawo, H. (2012). Optically controlled contraction of photosensitive skeletal muscle cells. *Biotechnol. Bioeng.* 109 (1): 199–204.

10 Kim, S., Laschi, C., and Trimmer, B. (2013). Soft robotics: a bioinspired evolution in robotics. *Trends Biotechnol.* 31 (5): 23–30.

11 Tanaka, Y., Sato, K., Shimizu, T. et al. (2007). Biological cells on microchips: new technologies and applications. *Biosens. Bioelectron.* 23 (4): 449–458.
12 Uesugi, K., Shimizu, K., Akiyama, Y. et al. (2016). Contractile performance and controllability of insect muscle-powered bioactuator with different stimulation strategies for soft robotics. *Soft Rob.* 3 (1): 13–22.
13 Elvassore, N., Cimetta, E., Pizzato, S. et al. (2009). Production of arrays of cardiac and skeletal muscle myofibers by micropatterning techniques on a soft substrate. *Biomed. Microdevices* 11 (2): 389–400.
14 Boudreau-Beland, J., Duverger, J.E., Petitjean, E. et al. (2015). Spatiotemporal stability of neonatal rat cardiomyocyte monolayers spontaneous activity is dependent on the culture substrate. *PLoS One* 10 (6): 0127977.
15 Zhao, Y. and Zhang, X. (2006). Cellular mechanics study in cardiac myocytes using PDMS pillars array. *Sens. Actuators A Phys.* 125 (2): 398–404.
16 Oyunbaatar, N.E., Lee, D.H., Patil, S.J. et al. (2016). Biomechanical characterization of cardiomyocyte using PDMS pillar with microgrooves. *Sensors* 16 (8): 1258.
17 Ralphe, J.C. and de Lange, W.J. (2013). 3D engineered cardiac tissue models of human heart disease: learning more from our mice. *Trends Cardiovasc. Med.* 23 (2): 27–32.
18 Zacchigna, S., Martinelli, V., Moimas, S. et al. (2018). Paracrine effect of regulatory T cells promotes cardiomyocyte proliferation during pregnancy and after myocardial infarction. *Nat. Commun.* 9: 2432.
19 Shimizu, T., Sekine, H., Isoi, Y. et al. (2006). Long-term survival and growth of pulsatile myocardial tissue grafts engineered by the layering of cardiomyocyte sheets. *Tissue Eng.* 12 (3): 499–507.
20 Nunes, S.S., Miklas, J.W., Liu, J. et al. (2013). Biowire: a platform for maturation of human pluripotent stem cell-derived cardiomyocytes. *Nat. Methods* 10 (8): 781–787.
21 Zimmermann, W.H., Schneiderbanger, K., Schubert, P. et al. (2002). Tissue engineering of a differentiated cardiac muscle construct. *Circ. Res.* 90 (2): 223–230.
22 Horiguchi, H., Imagawa, K., Hoshino, T. et al. (2009). Fabrication and evaluation of reconstructed cardiac tissue and its application to bio-actuated microdevices. *IEEE Trans. Nanobiosci.* 8 (4): 349–355.
23 de Lange, W.J., Hegge, L.F., Grimes, A.C. et al. (2011). Neonatal mouse-derived engineered cardiac tissue a novel model system for studying genetic heart disease. *Circ. Res.* 109 (1): 8–19.
24 Morimoto, Y., Mori, S., Sakai, F., and Takeuchi, S. (2016). Human induced pluripotent stem cell-derived fiber-shaped cardiac tissue on a chip. *Lab Chip* 16 (12): 2295–2301.
25 Liau, B., Christoforou, N., Leong, K.W., and Bursac, N. (2011). Pluripotent stem cell-derived cardiac tissue patch with advanced structure and function. *Biomaterials* 32 (35): 9180–9187.
26 Zhang, D.H., Shadrin, I.Y., Lam, J. et al. (2013). Tissue-engineered cardiac patch for advanced functional maturation of human ESC-derived cardiomyocytes. *Biomaterials* 34 (23): 5813–5820.

27 Park, S.J., Gazzola, M., Park, K.S. et al. (2016). Phototactic guidance of a tissue-engineered soft-robotic ray. *Science* 353 (6295): 158–162.
28 Dennis, R.G. and Kosnik, P.E. (2000). Excitability and isometric contractile properties of mammalian skeletal muscle constructs engineered in vitro. *In Vitro Cell. Dev. Biol. Anim.* 36 (5): 327–335.
29 Huang, Y.C., Dennis, R.G., Larkin, L., and Baar, K. (2005). Rapid formation of functional muscle in vitro using fibrin gels. *J. Appl. Physiol.* 98 (2): 706–713.
30 Lam, M.T., Huang, Y.C., Birla, R.K., and Takayama, S. (2009). Microfeature guided skeletal muscle tissue engineering for highly organized 3-dimensional free-standing constructs. *Biomaterials* 30 (6): 1150–1155.
31 Smith, A.S.T., Passey, S., Greensmith, L. et al. (2012). Characterization and optimization of a simple, repeatable system for the long term in vitro culture of aligned myotubes in 3D. *J. Cell. Biochem.* 113 (3): 1044–1053.
32 Morimoto, Y., Kato-Negishi, M., Onoe, H., and Takeuchi, S. (2013). Three-dimensional neuron-muscle constructs with neuromuscular junctions. *Biomaterials* 34 (37): 9413–9419.
33 Sakar, M.S., Neal, D., Boudou, T. et al. (2012). Formation and optogenetic control of engineered 3D skeletal muscle bioactuators. *Lab Chip* 12 (23): 4976–4985.
34 Shima, A., Morimoto, Y., Sweeney, H.L., and Takeuchi, S. (2018). Three-dimensional contractile muscle tissue consisting of human skeletal myocyte cell line. *Exp. Cell. Res.* 370 (1): 168–173.
35 Juhas, M., Engelmayr, G.C., Fontanella, A.N. et al. (2014). Biomimetic engineered muscle with capacity for vascular integration and functional maturation in vivo. *Proc. Natl. Acad. Sci. U.S.A.* 111 (15): 5508–5513.
36 Hinds, S., Bian, W.N., Dennis, R.G., and Bursac, N. (2011). The role of extracellular matrix composition in structure and function of bioengineered skeletal muscle. *Biomaterials* 32 (14): 3575–3583.
37 Bian, W.N. and Bursac, N. (2009). Engineered skeletal muscle tissue networks with controllable architecture. *Biomaterials* 30 (7): 1401–1412.
38 Ramsey, R.W. and Street, S.F. (1940). The isometric length-tension diagram of isolated skeletal muscle fibers of the frog. *J. Cell. Comp. Physiol.* 15 (1): 11–34.
39 Baryshyan, A.L., Woods, W., Trimmer, B.A., and Kaplan, D.L. (2012). Isolation and maintenance-free culture of contractile myotubes from *Manduca sexta* embryos. *PLoS One* 7 (2): e31598.
40 Baryshyan, A.L., Domigan, L.J., Hunt, B. et al. (2014). Self-assembled insect muscle bioactuators with long term function under a range of environmental conditions. *RSC Adv.* 4 (75): 39962–39968.
41 Morishima, K., Tanaka, Y., Ebara, M. et al. (2006). Demonstration of a bio-microactuator powered by cultured cardiomyocytes coupled to hydrogel micropillars. *Sensors and Actuators B-Chemical* 119 (1): 345–350.
42 Tanaka, Y., Morishima, K., Shimizu, T. et al. (2006). Demonstration of a PDMS-based bio-microactuator using cultured cardiomyocytes to drive polymer micropillars. *Lab Chip* 6 (2): 230–235.

43 Williams, B.J., Anand, S.V., Rajagopalan, J., and Saif, M.T.A. (2014). A self-propelled biohybrid swimmer at low Reynolds number. *Nat. Commun.* 5: 3081.
44 Anand, S.V., Ali, M.Y., and Saif, M.T.A. (2015). Cell culture on microfabricated one-dimensional polymeric structures for bio-actuator and bio-bot applications. *Lab Chip* 15 (8): 1879–1888.
45 Kim, D.H., Park, J., Suh, K.Y. et al. (2006). Fabrication of patterned micromuscles with high activity for powering biohybrid microdevices. *Sens. Actuators B Chem.* 117 (2): 391–400.
46 Park, J., Ryu, J., Choi, S.K. et al. (2005). Real-time measurement of the contractile forces of self-organized cardiomyocytes on hybrid biopolymer microcantilevers. *Anal. Chem.* 77 (20): 6571–6580.
47 Lind, J.U., Busbee, T.A., Valentine, A.D. et al. (2017). Instrumented cardiac microphysiological devices via multimaterial three-dimensional printing. *Nat. Mater.* 16 (3): 303–309.
48 Alford, P.W., Feinberg, A.W., Sheehy, S.P., and Parker, K.K. (2010). Biohybrid thin films for measuring contractility in engineered cardiovascular muscle. *Biomaterials* 31 (13): 3613–3621.
49 Sun, Y., Duffy, R., Lee, A., and Feinberg, A.W. (2013). Optimizing the structure and contractility of engineered skeletal muscle thin films. *Acta Biomater.* 9 (8): 7885–7894.
50 Shimizu, K., Sasaki, H., Hida, H. et al. (2010). Assembly of skeletal muscle cells on a Si-MEMS device and their generative force measurement. *Biomed. Microdevices* 12 (2): 247–252.
51 Vannozzi, L., Ricotti, L., Cianchetti, M. et al. (2015). Self-assembly of polydimethylsiloxane structures from 2D to 3D for bio-hybrid actuation. *Bioinspiration Biomimetics* 10 (5): 056001.
52 Tanaka, Y., Morishima, K., Shimizu, T. et al. (2006). An actuated pump on-chip powered by cultured cardiomyocytes. *Lab Chip* 6 (3): 362–368.
53 Park, J.Y., Kim, I.C., Baek, J.G. et al. (2007). Micro pumping with cardiomyocyte-polymer hybrid. *Lab Chip* 7 (10): 1367–1370.
54 Tanaka, Y., Sato, K., Shimizu, T. et al. (2007). A micro-spherical heart pump powered by cultured cardiomyocytes. *Lab Chip* 7 (2): 207–212.
55 Xi, J.Z., Schmidt, J.J., and Montemagno, C.D. (2005). Self-assembled microdevices driven by muscle. *Nat. Mater.* 4 (2): 180–184.
56 Kim, J., Park, J., Yang, S. et al. (2007). Establishment of a fabrication method for a long-term actuated hybrid cell robot. *Lab Chip* 7 (11): 1504–1508.
57 Feinberg, A.W., Feigel, A., Shevkoplyas, S.S. et al. (2007). Muscular thin films for building actuators and powering devices. *Science* 317 (5843): 1366–1370.
58 Chan, V., Park, K., Collens, M.B. et al. (2012). Development of miniaturized walking biological machines. *Sci. Rep.* 2: 857.
59 Holley, M.T., Nagarajan, N., Danielson, C. et al. (2016). Development and characterization of muscle-based actuators for self-stabilizing swimming biorobots. *Lab Chip* 16 (18): 3473–3484.
60 Nawroth, J.C., Lee, H., Feinberg, A.W. et al. (2012). A tissue-engineered jellyfish with biomimetic propulsion. *Nat. Biotechnol.* 30 (8): 792–797.

61 Kim, T.H., Kwon, C.H., Lee, C. et al. (2016). Bio-inspired hybrid carbon nanotube muscles. *Sci. Rep.* 6: 26687.
62 Hoshino, T. and Morishima, K. (2010). Muscle-powered cantilever for microtweezers with an artificial micro skeleton and rat primary myotubes. *J. Biomech. Sci. Eng.* 5 (3): 245–251.
63 Fujita, H., Van, T.D., Shimizu, K. et al. (2011). Designing of a Si-MEMS device with an integrated skeletal muscle cell-based bio-actuator. *Biomed. Microdevices* 13 (1): 123–129.
64 Webster, V.A., Hawley, E.L., Akkus, O. et al. (2016). Effect of actuating cell source on locomotion of organic living machines with electrocompacted collagen skeleton. *Bioinspiration Biomimetics* 11 (3): 036012.
65 Raman, R., Cvetkovic, C., and Bashir, R. (2017). A modular approach to the design, fabrication, and characterization of muscle-powered biological machines. *Nat. Protoc.* 12 (3): 519–533.
66 Cvetkovic, C., Raman, R., Chan, V. et al. (2014). Three-dimensionally printed biological machines powered by skeletal muscle. *Proc. Natl. Acad. Sci. U.S.A.* 111 (28): 10125–10130.
67 Pagan-Diaz, G.J., Zhang, X.T., Grant, L. et al. (2018). Simulation and fabrication of stronger, larger, and faster walking biohybrid machines. *Adv. Funct. Mater.* 28 (23): 1801145.
68 Raman, R., Cvetkovic, C., Uzel, S.G.M. et al. (2016). Optogenetic skeletal muscle-powered adaptive biological machines. *Proc. Natl. Acad. Sci. U.S.A.* 113 (13): 3497–3502.
69 Raman, R., Grant, L., Seo, Y. et al. (2017). Damage, healing, and remodeling in optogenetic skeletal muscle bioactuators. *Adv. Healthc. Mater.* 6 (12): 1700030.
70 Kabumoto, K., Hoshino, T., Akiyama, Y., and Morishima, K. (2013). Voluntary movement controlled by the surface EMG signal for tissue-engineered skeletal muscle on a gripping tool. *Tissue Eng. Part A* 19 (15–16): 1695–1703.
71 Morimoto, Y., Onoe, H., and Takeuchi, S. (2018). Biohybrid robot powered by an antagonistic pair of skeletal muscle tissues. *Sci. Robot.* 3 (18): eaat4440.
72 Morimoto, Y., Onoe, H., and Takeuchi, S. (2019). Biohybrid device with antagonistic skeletal muscle tissues for measurement of contractile force. *Adv. Robot.* 33 (5): 208–218.
73 Herr, H. and Dennis, R.G. (2004). A swimming robot actuated by living muscle tissue. *J. NeuroEng. Rehabil.* 1 (1): 6.
74 Tanaka, Y., Noguchi, Y., Yalikun, Y., and Kamamichi, N. (2017). Earthworm muscle driven bio-micropump. *Sensors and Actuators B-Chemical* 242: 1186–1192.
75 Webster, V.A., Chapin, K.J., Hawley, E.L. et al. (2016). *Aplysia californica* as a novel source of material for biohybrid robots and organic machines. *Biomim. Biohybrid Syst. Living Mach.* 2016 (9793): 365–374.
76 Webster, V.A., Young, F.R., Patel, J.M. et al. (2017). 3D-printed biohybrid robots powered by neuromuscular tissue circuits from *Aplysia californica. Biomim. Biohybrid Syst. Living Mach.* 10384: 475–486.

77 Akiyama, Y., Iwabuchi, K., Furukawa, Y., and Morishima, K. (2009). Long-term and room temperature operable bioactuator powered by insect dorsal vessel tissue. *Lab Chip* 9 (1): 140–144.
78 Akiyama, Y., Hoshino, T., Iwabuchi, K., and Morishima, K. (2012). Room temperature operable autonomously moving bio-microrobot powered by insect dorsal vessel tissue. *PLoS One* 7 (7): e38274.
79 Akiyama, Y., Sakuma, T., Funakoshi, K. et al. (2013). Atmospheric-operable bioactuator powered by insect muscle packaged with medium. *Lab Chip* 13 (24): 4870–4880.
80 Uzel, S.G.M., Platt, R.J., Subramanian, V. et al. (2016). Microfluidic device for the formation of optically excitable, three-dimensional, compartmentalized motor units. *Sci. Adv.* 2 (8): e1501429.
81 Cvetkovic, C., Rich, M.H., Raman, R. et al. (2017). A 3D-printed platform for modular neuromuscular motor units. *Microsys. Nanoeng.* 3: 17015.

Index

Mechanically Responsive Materials for Soft Robotics, First Edition. Edited by Hideko Koshima.

r

s